Chemical Transmission in the Mammalian Central Nervous System

Chemical Transmission in the Mammalian Central Nervous System

Edited by
Charles H. Hockman, Ph.D.,
and
Detlef Bieger, Dr.Med.

Department of Physiology and Biophysics
and
School of Basic Medical Sciences
University of Illinois, Urbana-Champaign

with the organizing assistance of:
George M. Ling

Department of Pharmacology
Faculty of Medicine
University of Ottawa
and
W.H.O., Geneva, Switzerland

University Park Press
Baltimore · London · Tokyo

UNIVERSITY PARK PRESS
International Publishers in Science and Medicine
Chamber of Commerce Building
Baltimore, Maryland 21202

Copyright © 1976 by University Park Press

Typeset by The Composing Room of Michigan, Inc.

Manufactured in the United States of America by Universal Lithographers, Inc., and The Maple Press Co.

All rights, including that of translation into other languages, reserved. Photomechanical reproduction (photocopy, microcopy) of this book or parts thereof without special permission of the publisher is prohibited.

Library of Congress Cataloging in Publication Data
Main entry under title:

Chemical transmission in the mammalian central nervous system.

Includes index.
1. Neural transmission. 2. Central nervous systems.
3. Mammals—Physiology. I. Hockman, Charles H.
II. Bieger, Detlef. III. Ling, George M. [DNLM:
1. Neural transmission. 2. Synapses—Physiology.
3. Psychopharmacology. 4. Mammals—Physiology.
WL102 H685c]
QP363.C46   599'.01'88   75-46565
ISBN 0-8391-0863-X

Contents

Contributors   vii
Preface   ix

Chapter One
The Synapse: Some Reflections on the Relationships between Structure and Function in the Central Nervous System
Hugh McLennan   1

Chapter Two
Amino Acid Inhibitory Transmitters in the Central Nervous System
Graham A. R. Johnston   31

Chapter Three
Excitatory Amino Acids
Peter N. R. Usherwood   83

Chapter Four
Acetylcholine and Synaptic Transmission in the Central Nervous System
John W. Phillis   159

Chapter Five
On the Physiology and Pharmacology of Cerebral Dopamine Neurons
Detlef Bieger and Charles H. Hockman   215

Chapter Six
Norepinephrine and Central Neurons
Barry J. Hoffer and Floyd E. Bloom   327

Chapter Seven
Serotonin and the Central Nervous System
Thaddeus J. Marczynski   349

Index   431

Contributors

Detlef Bieger, Department of Physiology and Biophysics and School of Basic Medical Sciences, University of Illinois, Urbana, Illinois

Floyd E. Bloom, Arthur Vining Davis Center for Behavioral Neurobiology, The Salk Institute, San Diego, California

Charles H. Hockman, Department of Physiology and Biophysics and School of Basic Medical Sciences, University of Illinois, Urbana, Illinois

Barry J. Hoffer, Laboratory of Neuropharmacology, Division of Special Mental Health Research, NIMH, St. Elizabeth's Hospital, Washington, D.C.

Graham A. R. Johnston, Department of Physiology, The John Curtin School of Medical Research, Australian National University, Canberra, Australia

George M. Ling, Department of Pharmacology, University of Ottawa, Ottawa, Canada, and World Health Organization, Geneva, Switzerland

Thaddeus J. Marczynski, Department of Pharmacology, University of Illinois College of Medicine, Chicago, Illinois

Hugh McLennan, Department of Physiology, University of British Columbia, Vancouver, British Columbia, Canada

John W. Phillis, Department of Physiology, University of Saskatchewan, Saskatoon, Saskatchewan, Canada

Peter N. R. Usherwood, Department of Zoology, University of Nottingham, England

Preface

Frequent queries from colleagues and students about a systematic account of the various substances which qualify as transmitters at mammalian central nervous synapses made us conscious of the absence of a comprehensive yet manageable volume that would encompass biochemical, pharmacological, physiological, and behavioral aspects of neural transmission. While reviews and treatises on this topic are pouring forth in prolific numbers, a swelling tide of "neuroscientific" publications is leaving the reader adrift in a flotsam of experimental data and hypotheses. With *Chemical Transmission in the Mammalian Central Nervous System,* we have striven to bridge the gap between compendium and handbook with the assistance of distinguished workers in the field.

The first edition of this volume could hardly appear at a more auspicious time as knowledge about neurohumoral transmission is approaching a new steady state following its logarithmic growth in the past two decades. During this period, the exploitation of new methodologies has opened up fresh territories for the experimental study of neural function, and the exploration of these uncharted regions is challenging traditional concepts of neuroanatomy and neurophysiology. And neurochemistry, emerging from its infancy, has joined other disciplines at the frontiers of neuroscience.

While technological advances have enabled us to dissect neural function at a molecular level, these are but infinitesimal steps towards a goal made elusive by the bewildering intricacies of neuronal organization. Nevertheless, some neuroscientists will not despair in their audacious attempts to unravel the mysteries of the brain and, for better or for worse, without such lofty goals their natural curiosity might soon be dissipated. Yet pragmatic interests continue to loom large. Neurological and mental disorders exact a terrible toll on our society and, notwithstanding the facility with which the function of certain brain structures can be explicated in pathophysiological terms, we are still groping for an understanding of the most elementary aspects of normal brain function. We are confident that methodological progress holds more than a glimmer of hope in providing rational bases upon which causal treatments for mental and neurological diseases can be developed.

In the pages that follow, we have an opportunity to take stock of the significant events of the past two decades as we assess current research into synaptic transmission within the central nervous system. The reader will note that each contributor has, in his own way, attempted to provide a functional interpretation of the information with which he deals. As editors, we believe that some success has been achieved in this endeavor, and we hope that, as readers, you will share in this belief.

We are grateful to Mrs. Geraldine Swift and the staff of the Word Processing Center of the School of Basic Medical Sciences for their most efficient assistance in typing manuscripts.

Chemical Transmission in the Mammalian Central Nervous System

CHAPTER ONE

The Synapse: Some Reflections on the Relationships between Structure and Function in the Central Nervous System

Hugh McLennan

It is both interesting and instructive to recall that less than 70 years ago, the anatomical nature of the communications of cell with cell within the central nervous system (CNS) was still under active debate, although admittedly at that time the proponents of the so-called reticular theory were fighting a last-ditch action. At the Nobel Symposium in 1906, the two positions were clearly delineated:

"Les corbeilles péricellulaires et les plexus grimpants et d'autres dispositions morphologiques, dont la forme varie selon les centres nerveux que l'on étudie, attestent que les éléments nerveux possèdent des relations réciproques de *contiguité* et non de *continuité*, et que ces rapports de contact plus ou moins intime s'établissent toujours, non entre les arborisations nerveuses seules, mais entre ces ramifications d'une part et le corps et les prolongements protoplasmiques d'autre part." (78)

"... je ne puis m'empêcher de déclarer que, lorsque la théorie du neurone fit triomphalement et presque par consentement unanime son entrée dans la science, je me suis trouvé dans l'impossibilité de suivre le courant, parce que devant moi se dressait un fait anatomique bien concret: c'était l'existence de cette formation à laquelle j'ai donné le nom de réseau nerveux diffus. A ce réseau, que je n'ai pas hésité à considerer comme un organe nerveux, je devais attacher une importance d'autant plus grande, que sa signification m'avait été clairement révélée par la façon même dont il est constitué. En effet, à sa formation contribuent à la fois, bien qu'avec des modalités diverses et en différente mesure, tous les éléments nerveux du système nerveux central." (*44*)

Such are the ways of science that two skilled and imaginative scholars could interpret the microscopic appearance of the same specimen (Figure 1) as lending support to their two highly divergent views.

Of course the "neuronal theory" expounded by Cajal and others triumphed, and it was recognized that the unions between nerve cells, while close, are not absolute:

> "[An axon's] mode of termination as well as that of the collaterals to which it may give rise is in the form of an arborescent tuft, which is applied to the body or dendrites of some other cell. So far as our present knowledge goes we are led to think that the tip of a twig of the arborescence is not continuous with but merely in contact with the substance of

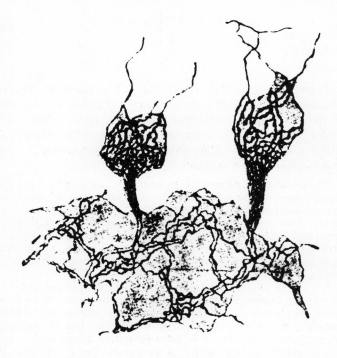

Figure 1. Pericellular baskets surrounding Purkinje cells of cerebellum, and network of axons in subjacent granular layer. (From Ramón y Cajal, S.: *Neuron Theory or Reticular Theory?* Madrid: Consejo Superior de Investigaciones Cientificas, 1954; see also Golgi, C.: *La doctrine du neurone. Theorie et faits*. Les prix Nobel en 1906, Stockholm, 1906.)

the dendrite or cell-body on which it impinges. Such a special connection of one nerve-cell with another might be called a *synapsis*." (*83*)

Much became known of the function and modes of action of Sherrington's "synapses" in the years that followed. Our present knowledge indicates that the majority of synapses within and without the CNS, operates through the elaboration within and release of specific chemical substances from the presynaptic axonal terminals, and it is with the actions of drugs on these chemical processes that the subject of neuropharmacology is largely concerned. It is the intent of this chapter to outline the morphological and electrical characteristics of synapses.

Studies of the detailed morphology of synapses have been possible only since the advent of electron microscopy, and in 1954 brief statements by de Robertis and Bennett (*26*) and by Palay (*74*) appeared which have remained largely unchallenged. They may be summarized briefly in Palay's words:

> "All synapses possess the same fundamental fine structure as follows:
> 1. Close apposition of the limiting membranes of presynaptic and postsynaptic cells without any protoplasmic continuity across the synapse. The two apposed membranes are separated by a cleft about 200 Å wide, and display localized regions of thickening and increased density.
> 2. The presynaptic expansion of the axon, the endfoot or bouton terminal, contains a collection of mitochondria and clusters of small vesicles about 200–650 Å in diameter."

He went on to state:

> "The small synaptic vesicles provide the morphological representation of the prejunctional, subcellular units of neurohumoral discharge at the synapse demanded by physiological evidence."

Modifications in the detail of this original picture have of course been made. These include: 1) the categorization of synapses into two types on the basis of the electron density and apparent thickness of the pre- and postsynaptic membranes (*46*), the functional significance of which remains in some doubt; 2) the recognition that in some presynaptic terminals the majority of vesicles do not possess the circular profiles originally described, but rather appear elliptical or "flattened" (*9, 86*); 3) the description of other larger and more electron-dense granules interspersed among the "typical" synaptic vesicles in certain endings; and 4) the discovery that there exist both a physical or chemical bond between the pre- and postsynaptic membranes and some material in the synaptic cleft which is not discernible elsewhere in the perineuronal space (*89*). Since certain of these matters have been subjects of considerable speculation and discussion in terms of function, a statement of the present position seems in order.

In his 1959 paper, Gray (46) distinguished two morphological synaptic types in the cerebral cortex of the rat, which were classified as Type 1 and Type 2. In the former the contact region exhibited a rather uniform electron density throughout its extent but the thickness of the dense zone appeared greater on the postsynaptic side owing to the association of material on the cytoplasmic surface, and in the latter the thickened portions covered only a small part of the total synaptic area, and where these occurred there was no widening of the cleft or interposition of extracellular material (Figure 2). Later the differentiation in the morphology of the synaptic vesicles was described, the division being between the spherical (S-type) and flattened (F-type) structures, and the correlation of S-type vesicles with Type 1 synapses and F-type with Type 2 was subsequently made (1). Furthermore, it was claimed that the first category could

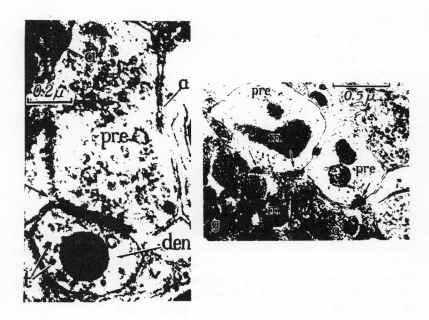

Figure 2. Synapses from cerebral cortex of rat. Left: Type 1 synapse on dendrite (*den*.). Arrow indicates thickened membrane at contact region, and asymmetry of electron density in region of contact is obvious. Right: axosomatic synapses of Type 2. Arrows indicate membrane thickenings. *Legend:* m, mitochondria; g, granules of endoplasmic reticulum; a, nonthickened regions of membrane. (From Gray, E. G.: *J. Anat.* 93:420–433, 1959.)

be associated with an excitatory function and the second with postsynaptic inhibition.

More recent studies, using different techniques for the preparation of tissue samples for electron microscopy (EM), show that the appearance of flattened vesicles is an artifact of fixation with aldehydes, since this type of differentiation was not usually observed when osmium fixation was used (26, 41), and flattened vesicles are not seen in freeze-etched material in which, presumably, the vesicles are preserved in their "nearest-to-native state" (1). Nevertheless there are underlying differences in the vesicle populations. Five main classes of synaptic endings have been distinguished in primate spinal cords, where the differences are in part determined by the extent to which flattening of the vesicles occurs after treatment with aldehydes (10), and similar reports have been made for feline material (14).

The osmotic pressure of the fixative used can also affect the proportion of flattened vesicles found in an ending (88). In the freeze-etched material in which all vesicles appear spherical, whether aldehyde-fixed or not, investigators have shown that the mean diameter of the vesicles in endings apparently of Gray's Type 1 are about 480 Å, whereas those presumed to be in Type 2 category average about 390 Å (1) (Figure 3.). Thus the possibility of a correlation between morphology and function remains, although all of the questions cannot be regarded as being completely answered at the present time.

There are of course other organelles visible in presynaptic terminals. Many endings contain small numbers of vesicles possessing an electron-dense core, and it is reported that the number of them increases during regeneration (75). There is also convincing evidence to indicate that endings containing marked quantities of granular vesicles function through the mediation of one of the catecholamines or of 5-hydroxytryptamine (5-HT) as the chemical transmitters of synaptic action (8, 49), and that these unusual subcellular structures represent the intracellular storage sites for the monoamines (24, 52).

In the light of all of this inferential knowledge, questions relating to the formation of the vesicles assume some importance. There is a considerable volume of evidence, derived in large part from studies of known adrenergic systems (53), demonstrating a somatofugal flow of transmitter along axons, which in turn implies that the synthesis of transmitter—and presumably also its incorporation into storage vesicles—is accomplished within the cell body. Conversely there is also evidence (again most convincingly demonstrated for adrenergic systems) for the reuptake of the transmitter following its release from an activated terminal, with subse-

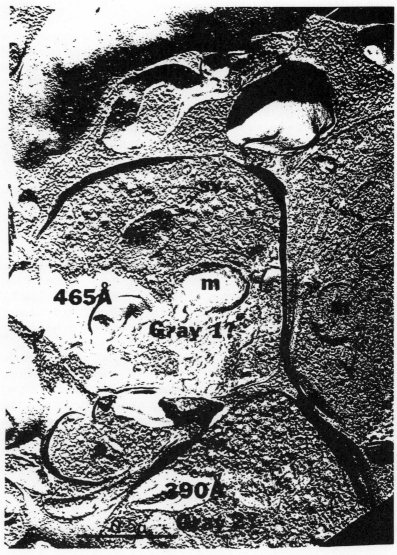

Figure 3. Freeze-etched terminals from monkey spinal cord. Lower terminal contain spherical vesicles of mean diameter 390 A and is possibly Type 2; upper one contains vesicles of 465 A mean diameter and may be Type 1. Glutaraldehyde fixation. *Legend:* sv. synaptic vesicle; m, mitochondria. (From Akert, K., Pfenninger, K., Sandri, C., et al.: *In* Pappas, G. D., and Purpura, D. P. (Eds.), *Structure and Function of Synapses*. New York: Raven Press, 1972, pp. 67–86.)

quent reuse of the chemical substance, and this finding in turn suggests that a "peripheral" formation of vesicles must be entertained.

In this latter connection, investigators have described another type of organelle, contained within a variety of synaptic endings, which has been called a complex vesicle (*48*). These bodies appear to be composed of ordinary vesicles enclosed within a shell formed by hexagonal and pentagonal lattices, and it is guardedly suggested that the complex vesicles may be precursors of the more common smooth-surfaced variety (*92*).

Gray and Willis (*48*) have demonstrated that both empty shells and shell fragments can be detected within synaptic endings (Figure 4) and have postulated that:

> "Synaptic vesicles probably form from invagination and pinching off of the surface membrane and at this stage acquire a shell. The shell is thought to consist initially of hexagons lying on the membrane which become rearranged when the membrane invaginates to form a sphere of hexagons and pentagons. The formed complex vesicle then moves into the cytoplasm."

Following this, it is suggested that the shell is largely or wholly lost to yield mature vesicles (Figure 5). The possibility that an empty shell may generate a vesicle ("?" in Figure 5) was also considered but was felt to be unlikely.

It thus appears that two modes of formation of vesicles are operative, one occurring in the cell body with a subsequent movement by axoplasmic flow to the terminals, and a second within the nerve endings themselves. That the nerve terminals may be involved with the uptake and subsequent reuse of liberated transmitter must also be viewed as an additional mechanism.

It is now universally accepted that the operation of the synapses the morphology of which has been briefly considered thus far is mediated by the release of specific chemical substances from presynaptic terminals which are invaded by action potentials, and that the liberated transmitter diffuses across the cleft separating the pre- and postsynaptic membranes and reacts with the latter in such a way as to modify its resting permeability to various ions. The identification of the chemical transmitters, their metabolism, mode of reaction with the postsynaptic receptors, and their inactivation have been and continue to be areas of study attracting the attention of many investigators.

Although the details are beyond the scope of this introductory chapter, there are nevertheless certain fundamental and general aspects of the process that require mention at this point. Since the discovery by Fatt and Katz (*37*) of a "synaptic noise" at the vertebrate neuromuscular junction,

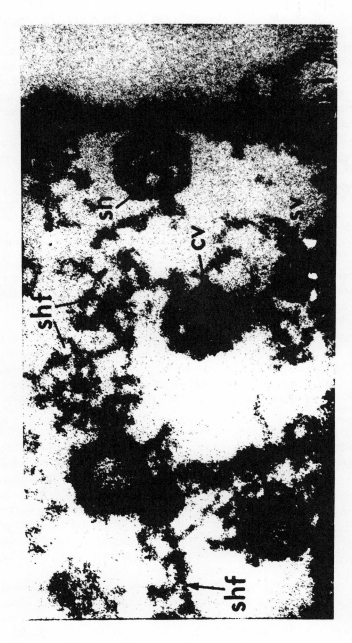

Figure 4. Portion of presynaptic ending from rat cerebral cortex, fixed with osmium smooth-surfaced "typical" synaptic vesicle; sh, empty shell; and shf, shell fragments. (From Gray, E. G., and Willis, R. A.: *Brain Res.* 24:149–168, 1970.)

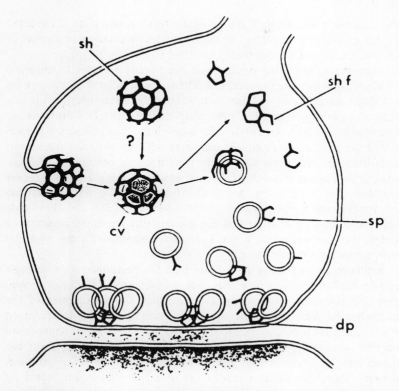

Figure 5. Postulated mode of formation of synaptic vesicles within terminal. *Legend:* sp, spinous material adhering to synaptic vesicle and thought to be shell fragment; dp, dense projection lying on presynaptic membrane. Other abbreviations as in Figure 4. (From Gray, E. G., and Willis, R. A.: *Brain Res.* 24:149–168, 1970.)

it is recognized that there is a resting release of transmitter into the synaptic cleft that is not dependent on activation of the presynaptic terminal, and there is suggestive evidence that this property is likely to be common to all synapses (*55, 62*). Statistical analysis of the "noise" indicated that it was attributable to the release of multimolecular units of transmitter of relatively fixed size (quanta), and that the much greater release elicited by activation of the presynaptic terminal results from an enhanced probability of release of any individual quantum (*54*).

The question that arises therefore is this: Is there an identifiable morphological counterpart of the quantum of transmitter? It was in this context that Palay (*74*) suggested that the synaptic vesicles could represent the stored quanta of transmitter, although the direct evidence for making this correlation was minimal and, as Birks, Huxley, and Katz (*7*) pointed out, the morphological picture in *itself* provides no evidence in support of the concept, although it can readily be fitted to it.

Attempts to settle the matter have been made by using biochemical techniques in which the synaptic vesicles have been isolated by differential centrifugation and analyzed for their content of transmitter, in the majority of experiments specifically for acetylcholine (*50, 94*). Unfortunately these studies too have been essentially inconclusive, for although a proportion of the total acetylcholine is associated with the vesicular fraction of tissue homogenates, there also exists a cytoplasmic component that appears to be present physiologically and which has a more rapid turnover than does the vesicular fraction (*68*). The synthesizing enzyme choline acetyltransferase is also located in the cytoplasm (*38*), and these data could be interpreted to infer that the nonvesicular fraction represents the pool of transmitter available for immediate release, with the vesicular portion forming a reserve store.

Evidence derived from the extensive studies performed on adrenergic systems implies that this view may indeed be correct. At adrenergic synapses the situation seems quite clear in that the major part of the catecholamine content of the appropriate nerve endings is contained within organelles which, as mentioned earlier, have been called dense-cored granules because of their appearance in EM pictures. From them the amines can be liberated by pharmacological agents such as reserpine. However, there also exists a small store of amine that is unaffected by reserpine treatment but which may be liberated by electrical stimulation of the nerve terminals (*82*), and this pool is normally replenished from the larger reserpine-sensitive store.

The morphological site of this readily available transmitter is uncertain. The catecholamine cannot be "free" in the cytoplasm of the nerve endings, for there it would be subject to catabolism by the monamine oxidase (MAO) associated with the mitochondria, and there is also experimental evidence that nerve impulses are incapable of causing the release of cytoplasmic amine even when the oxidase has been inactivated. Nevertheless some compartmentation of the amine in sites presumably different from the dense-cored granules seems to be indicated, and although it is true that adrenergic endings in EM section contain the usual clear vesicles in addition to the granules (Figure 6), there is no indication that these latter represent the store of available transmitter either.

One is therefore forced back to the question of the likely equation of vesicle with quantum of transmitter, without any certain answer being in evidence at the present time. Ginsborg (*43*), in discussing the situation for

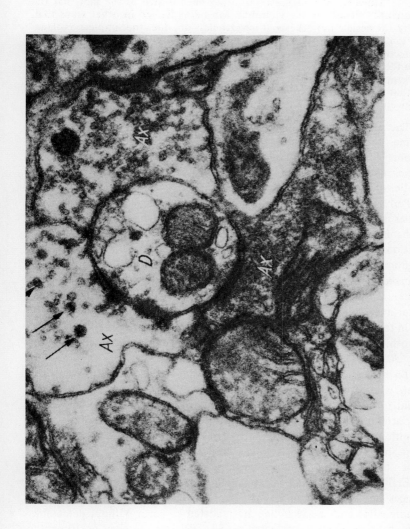

Figure 6. Three axonal terminals (Ax) in close apposition to dendritic process (D) from substantia nigra of mouse. Two terminals certain only clear vesicles; third contains in addition dense-cored vesicles (arrows). (From Bak, I. J.: *Exp. Brain Res. 3*:40–57, 1967.)

acetylcholine to which much attention has been devoted in studies of this matter, summarized the current state of knowledge in these terms:

> "However although attention has been drawn to the fact that the quantal release of acetylcholine can be explained in other ways than by the vesicle hypothesis, it nevertheless provides the simplest explanation for the existence of quanta, especially when one takes into account the constancy of the amount of acetylcholine per quantum under widely different experimental conditions . . . . In fact the vesicle hypothesis is so attractive that the main justification for suggesting that quantitative evidence is wanting is that some credit should be given to those who in the future provide more compelling data."

The electrical consequences of synaptic activation have, in general outline, been well understood since the mid-1950s, and it does not seem reasonable to undertake here more than a brief summary of the facts. For further detailed information and references the reader is referred to a number of reviews including (*27, 28, 54,* and *70*).

Perhaps one of the more striking conclusions from the numerous studies which have been executed was the general applicability of the mechanisms which were originally described for the frog neuromuscular junction and cat spinal motoneuron, i.e., that throughout the animal kingdom only two basic mechanisms appeared to be involved. In one, an increased permeability of the postsynaptic membrane to sodium and potassium ions is induced by the action of the liberated synaptic transmitter and leads to depolarization and excitation of the postsynaptic cell; and in the second, an enhanced permeability to potassium or chloride ions, or both, commonly, although not invariably, produces postsynaptic inhibition (*30*).

To these two fundamental processes it may now be necessary to add others whose time courses of action are considerably slower, although the extent to which these may be of general importance is not yet known. In this latter type of action, the changes in the characteristics of the postsynaptic membrane lead not to increases but to decreases in its resting permeability to small ions. Thus, a diminished conductance of the membrane to potassium ions is believed to underlie the excitation of cerebral cortical neurons by acetylcholine (*61*), whereas the slow potential changes occurring in sympathetic ganglion cells under the influence of noradrenaline and acetylcholine are reported to occur without detectable changes in membrane conductance (*58, 59*), like the slow postsynaptic potentials that can be evoked in these cells.

It is important, however, to consider briefly the significance of the electrical parameters that can be measured in neurons during synaptic activation. All of the methods involve the penetration of a postsynaptic cell by one or more microelectrodes of a size sufficiently small that the signals thereafter recorded show little or no decay attributable to neuronal damage, and the total number of hours that neurophysiologists the world over have devoted to the achievement of this evanescent goal must be enormous.

The simplest observations to make when a neuron has been successfully impaled by an electrode are the changes in the transmembrane potential. It is by this means that the depolarizing excitatory postsynaptic potentials (EPSPs) (Figure 7, upper panel) (*18*) and end-plate potentials (*37*) were discovered, together with the spontaneous quantal potentials which are the basic constituents of the evoked response (*11, 37*). The hyperpolarizing inhibitory postsynaptic potentials (IPSPs) of the spinal cord were similarly examined (*19*) (Figure 7, lower panel), and many of the properties of these fundamental synaptic responses—time courses, delays, summation, and so forth—were described early (Figure 7).

With suitable sophistication of technique, other valuable data can be acquired from cells penetrated by single electrodes. Records of the voltage changes produced in response to brief current pulses yield measurements of the conductance of the cell membrane, and such measurements indicate that the resting conductance is increased during both excitatory and inhibitory synaptic events. More prolonged applications of current delivered through an intracellular pipette filled with an appropriate salt solution produce modifications of the ionic composition of the internal milieu through the electrophoretic ejection of an ion from the electrode tip. Largely by these means the ascription of the conductance changes during EPSPs to sodium and potassium ions and during IPSPs to chloride and potassium has been possible (*29*). Finally, the observation of synaptic potentials evoked during the passage of continuous currents, the effect of which is artificially to alter the "resting potential" across the membrane, has permitted estimates to be made of the equilibrium potentials for the excitatory and inhibitory PSPs (*17, 18*), which may thereafter be compared with the corresponding potentials for the individual ions calculated from the Nernst equations. These estimates of synaptic equilibrium potentials have assumed even greater importance when attempts have been made to identify the chemical natures of the synaptic transmitters active at individual cells (*3, 91*).

It was only to be expected that with additional experimentation certain difficulties of interpretation would come to light. Thus, e.g., in experiments wherein the activity evoked by a single Ia afferent fiber has

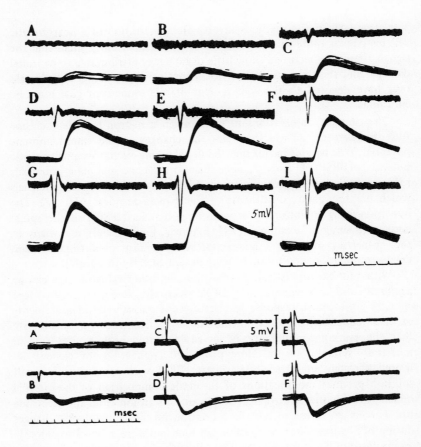

Figure 7. Postsynaptic potentials in feline spinal motoneurons. Upper panel: EPSPs generated in medial gastrocnemius motoneuron by afferent volley in medial gastrocnemius nerve of gradually increasing intensity, as indicated by potentials obtained from dorsal root shown in upper of each pair of records. (From Eccles, J. C.: The Physiology of Synapses. Berlin: Springer, 1964.) Lower panel: IPSPs from biceps–semitendinosus motoneuron recorded in similar fashion with increasing volleys in quadriceps nerve. (From Coombs, J. S., Eccles, J. C., and Fatt, P.: *J. Physiol. (Lond.)* 130:396–413, 1955.)

been recorded in a motoneuron rather than the response to stimulation of a whole dorsal rootlet or peripheral nerve, the EPSPs appear much less constant in amplitude and time course (*12, 13, 64*), and this phenomenon is attributable at least in part to the distance of the active synapses from the site of the recording electrode. It is now recognized that the same consideration may apply to the determination of the equilibria for synaptic potentials, specifically in the sense that a current applied through an electrode impaling the soma of a neuron (which is the most probable site of penetration)—with which one attempts to reverse a synaptic potential generated far out on the dendrites—may be considerably greater than if the active synapses are less distally situated. In the former case the equilibrium potential for the synaptic event will appear to be farther removed from the resting potential of the cell than is in fact the case. Such considerations also underlie the discrepancy reported (20) between the reversal potential for motoneuronal EPSPs and that for the excitatory action of L-glutamic acid, which has been proposed as the chemical mediator of certain central excitatory synapses.

In these experiments the EPSP always appeared to have a reversal potential more positive than that for glutamate electrophoretically applied to the region of the soma. The discrepancy could be fully explained if the excitatory synapses were largely or wholly axodendritic, and for this there is certain experimental evidence. The importance of correlation between structure and function thus once more becomes apparent. One may summarize the situation by saying that it is no longer sufficient to know that the terminals of one neuron make contact with another cell; rather one must seek additional information not only on the type of presynaptic element involved but also on the location of the synapses formed on the postsynaptic neuron.

Despite the problems of interpretation, the general principles already stated remain true. For postsynaptic inhibition, in which many synapses appear to be axosomatic rather than axodendritic—and thus the distortion imposed by long paths of electrotonic conduction is not a factor—the equilibrium potential for IPSPs lies near or is slightly more negative than the equilibrium potential for chloride ions, and thus it is inferred that potassium ions are likely to be involved as well. The analogous situation for EPSPs is less clear, but the equilibrium potential for the process is near zero, suggesting that the permeability of the membrane to both sodium and potassium ions (at least) is increased. A final point regarding the location of synapses may be summed up very simply. Those synapses placed distantly from the site of action potential generation—which in the majority of mature neurons at least is near the soma-axonal junction (76)—are less effective in determining the overall excitability of the postsynaptic cell than are those more proximally situated.

The characteristic feature of the synaptic events just discussed and illustrated by Figure 7 is rapidity of action, with which is to be contrasted the very much slower change in neuronal excitability that may be observed under suitable circumstances and which has been called "remote," "presynaptic," or more descriptively "prolonged" inhibition.

Because of the continuing controversy over the underlying structural and functional mechanisms, I have left for the last a more detailed consideration of this inhibitory process. It was first reported for spinal motoneurons (*39, 40*) but has since been described at many other sites (*81*).

Briefly, one physiological observation is that stimulation of muscle or cutaneous afferent fibers elicits a very prolonged inhibition of motoneurons (Figure 8), an inhibition which, unlike the much briefer postsynaptic type, is unaffected by strychnine. It is however antagonized by the alkaloids picrotoxin (*34*) and bicuculline (*22*). The inhibition is manifested as a reduced amplitude of evoked EPSPs, which cannot be attributed to a smaller size of each quantum of transmitter (*63*), and it occurs without a detectable change in the conductance of the motoneuronal membrane (*35*). Associated with the inhibition is a depolarization of the incoming primary afferent terminals (*31, 33*).

Among the possibilities, not necessarily mutually exclusive, suggested to explain these physiological and pharmacological findings, one is that the inhibition is of a standard postsynaptic type, but where the synapses responsible are situated far out on the dendrites and therefore beyond the electrotonic range at which a conductance change would be detected by an electrode impaling the neuronal soma (*39*). Since the inhibition is antagonized by picrotoxin and bicuculline, this possibility envisages that the inhibitory transmitter involved is γ-aminobutyric acid (GABA), a suggestion supported by the finding that in the hypoglossal nucleus GABA is more effective as an inhibitory agent than is glycine in the dendritic regions of the nucleus, whereas the reverse is true at the level of the cell somata (*2*). This proposal of a remote postsynaptic inhibition does not of course directly explain the observed depolarization of the afferent terminals which accompanies the prolonged inhibition.

A second possibility, proposed by Curtis et al. (*21*), is in effect an extension of the first and seeks to account for the terminal depolarization in terms of an increased intercellular potassium ion level:

> "Elevation of the extracellular potassium concentration, as a result of the efflux of this ion at inhibitory synapses on neurones and terminals, may modify the excitability of afferent terminals, particularly if transmitter action is prolonged in duration and the disposition of neuronal and glial elements in the environment of neighboring excitatory and inhibitory endings restricts diffusion.... It is also possible that localized changes in extracellular potassium levels occur

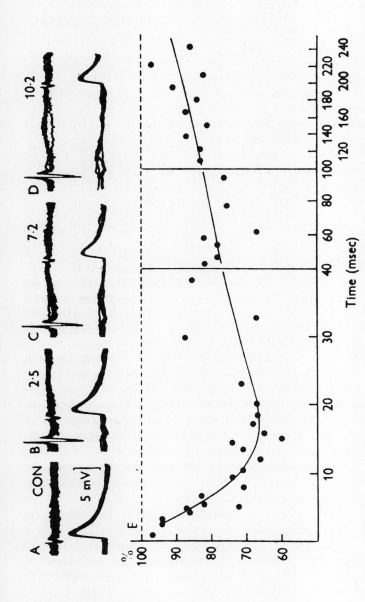

Figure 8. Time course (E) of prolonged inhibitions of monosynaptic EPSPs in gastrocnemius motoneuron, elicited by volley in biceps–semitendinosus nerve. *A* shows control EPSP; *B*, *C*, and *D*, responses where inhibitory stimulus was delivered 2.5, 7.2, and 10.2 msec, respectively, before test shock. Upper traces in *A–D* recorded from dorsal root. (From Eccles, J. C., Eccles, R. M., and Magni, F.: *J. Physiol. (Lond.) 159*:147–166, 1961.)

as a result of conductance increases at excitatory synapses, and thus synaptic terminals may be depolarized because of their proximity to either excitatory or inhibitory synapses without the involvement of a genuine axo-axonic synaptic mechanism."

The third proposed mechanism is that of Eccles and his colleagues, and is based on the observation originally made by Liley (*66*) that the amount of transmitter released from an ending is determined by the amplitude of the presynaptic action potential which invades it, i.e., on the "resting" polarization of that ending. The suggestion was that the primary afferent terminals impinging on motoneurons were themselves subject to a synaptic depolarization, which is turn implies that there should exist synaptic endings making contact with other terminals, a proposal for which Eccles, Kostyuk, and Schmidt (*32*) admitted that there was then no experimental proof. Because the inhibition did not involve a change in excitability of the motoneurons but was believed to be exerted only at the level of the primary afferent terminals, it was named "presynaptic," and since picrotoxin antagonized the action, it was further suggested that GABA might be the depolarizing transmitter active at the postulated axoaxonic synapses.

What is the evidence to be adduced in favor of these hypotheses? The fact that both picrotoxin and bicuculline are able to antagonize prolonged inhibition speaks powerfully for the involvement of GABA at some stage of the mechanism; however, its site of action is not to be simply fixed. At other sites in the CNS, GABA fulfills the role of a transmitter of postsynaptic inhibition, i.e., it enhances the permeability of the postsynaptic membrane to chloride ions, and it could of course exert the same type of effect at remote dendritic sites, in line with the first of the proposals already made. The importance of this mechanism has been argued particularly by Granit (*45*) and by Kellerth (*56*) who favor the concept that all prolonged inhibition of motoneurons is of the remote postsynaptic type. Although it is now admitted that stimulation of higher threshold afferents can indeed evoke a remote dendritic mechanism, it seems that presynaptic events must continue to receive consideration to explain fully all of the features of prolonged inhibition (*16, 81*).

There are certain difficulties with the proposition that GABA is the transmitter released at an axoaxonic synapse to yield a depolarization of an afferent terminal, as was originally suggested by Eccles et al. (*34*). As previously noted, the conductance changes and other reactions elicited by this amino acid are those characteristic of postsynaptic inhibition (*22, 60*); and although a depolarizing action on certain neurons has been shown (*25*), its effect when applied directly to afferent terminals has been to hyperpolarize rather than to depolarize them (*23, 42*). Depolarizations have been noted when the substance was topically applied to the spinal cord (25), but several explanations other than that GABA has had a direct effect on the terminals are possible in these cases. One such is the second hypothesis already put forward (21), namely, that terminal depolarization is secondary to an increased extracellular potassium concentration. At present there is essentially no experimental evidence available from research on vertebrates to support this concept, although it is entirely plausible theoretically, and indeed changes in the extracellular environment of neurons as a consequence of synaptic activity nearby may influence subsequent events much more generally than is presently recognized.

Finally, what of the morphological evidence? As was mentioned, when Eccles et al. (33) proposed axoaxonic synapses to explain afferent depolarization and thus "presynaptic" inhibition, no such structures had been described. In 1962 Gray published a brief report of an axonal terminal apparently in synaptic contact with another, which suggested that this arrangement could form the morphological basis for presynaptic inhibition. Since that date numerous reports have been published of similar observations both in the spinal cord (Figure 9) and in other regions of the CNS where prolonged strychnine-insensitive inhibitions were known (90).

Two major questions arise in attempting to assess the morphological data in relation to prolonged inhibition: 1) What information can be obtained by examining the shapes of the vesicles in the complex synaptic structures? 2) To what extent can one rely on the ascription of both elements in an "axoaxonic" synapse as axonal terminals? The situation for both of these propositions is complicated.

If one recalls the earlier discussion of the likely functional significance of spherical vs flattened vesicles in terms of excitation and inhibition respectively, one might expect that if, on the one hand, the presynaptic element of an axoaxonic synapse is to have a depolarizing action it should contain spherical vesicles. On the other hand, if GABA—which elsewhere acts as an inhibitory transmitter—is involved the vesicle should appear flattened. Both situations are known—thus Khattab (57), Szentágothai (84), and Conradi (15) have all published photographs obtained from spinal cord material, showing that both elements of synapses described as axoaxonic and located in the vicinity of spinal motoneurons contained round vesicles; conversely Bodian (10) and Réthelyi (80) report that flattened vesicles occur in the presynaptic elements, and to compound the problems McLaughlin (69) has failed to discern axoaxonic synapses in regions where they have been described by other authors.

The many recent demonstrations of dendrites which are presynaptic to other dendrites, i.e., which form dendrodendritic junctions, have rendered

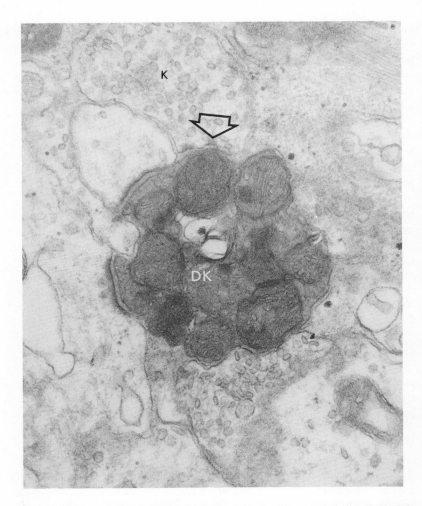

Figure 9. Axonal terminal (K) in synaptic contact with second terminal (DK) in feline spinal cord. Second terminal is degenerating as consequence of dorsal root section five days previously. Direction of transmission indicated by arrow. (By courtesy of Henry J. Ralston.)

unacceptable as the sole definition for an axonal terminal the demonstration of vesicles and differentiated membranes, and it is largely on this basis alone that the description of axoaxonic synapses in the CNS rests. Thus, e.g., despite the vesicles which it contains, the structure labeled "?" in Figure 10 is possibly a dendrite which appears to make synaptic contact with other dendritic profiles.

However, axoaxonic contacts have been demonstrated, although rarely; thus in the convincing micrographs of Ralston (77) and Westrum and Black (93), the intermediate elements in "axoaxodendritic" complexes are beyond doubt axonal terminals; in the first case, from the spinal cord, the presynaptic terminal contains round and flattened vesicles (Fig-

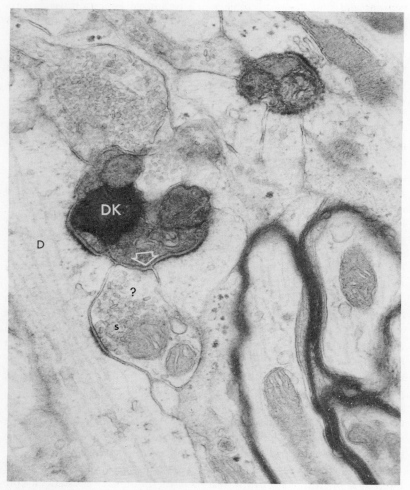

Figure 10. An example of a possible axo-dendro-dendritic synapse in the feline spinal cord. A degenerating terminal (DK) of dorsal root origin forms contacts with the dendritic profile (D) adjacent to it. The structure marked "?" which contains synaptic vesicles (s) may be a dendrite which makes contact with one of the same dendritic profiles. Serial sections of the profile "?" reveal it to contain ribosomes which are associated with dendrites, not axons. (By courtesy of Henry J. Ralston.)

ure 9); in the second, from the spinal trigeminal nucleus, they appear flattened. This statement, of course, should not imply that all other descriptions of apparent axoaxonic contacts are in error but simply that the case for them has not been adequately proved.

This then is a summary of the present situation: 1) Spinal motoneurons and other cells are subject to a prolonged inhibition that occurs without a detectable change in the conductance of the cell membrane. 2) The inhibition is accompanied by a depolarization of the afferent terminals which impinge on the cell. 3) The inhibition is diminished by agents that antagonize the effects of GABA. 4) Axoaxonic junctions in which the presynaptic element may contain either round or flattened vesicles exist.

Any attempt to put together these apparently conflicting pieces of evidence leads to the conclusion that more than one process is likely involved. First, because GABA must be implicated, and in line with the other known modes of action of this amino acid, it seems reasonable to accept the operation of a remote postsynaptic inhibition. Second, the existence of axoaxonic synapses in which the presynaptic element contains round vesicles bespeaks the operation of an excitatory transmitter, and among the amino acids only glutamate has a direct depolarizing action on afferent terminals (23). Axoaxonic synapses with flattened vesicles presynaptically logically should be the morphological counterpart of afferent hyperpolarization, and this phenomenon has been detected in group Ia terminals in the spinal cord (51).

There remains the possibility that GABA may indeed act also as a depolarizing transmitter, although in my view this is unlikely. To the argument that picrotoxin (34) or bicuculline (25, 65) is reported to reduce certain signs of terminal depolarization, one may reply that this has been an inconstant observation (85), and indeed it is reported that bicuculline increases terminal depolarization elicited from supraspinal sources (6). The pharmacological analysis is therefore complicated. All of the changes in terminal polarization could be secondary to potassium accumulations, as Curtis et al. (21) have suggested, occurring in part through interference with the operation of nearby inhibitory synapses involving GABA. The elucidation of these possibilities awaits further experimentation.

From the neuropharmacological viewpoint, the principal focus of interest is on the postsynaptic reactions of transmitter substances, confirmed or suspected, with the receptor sites situated on the membranes of the target cells. Much of what is discussed in the subsequent chapters is concerned with these reactions, and the volume of research on the postsynaptic actions of acetylcholine, the catecholamines (noradrenaline and dopamine), 5-HT, the inhibitory amino acids (glycine and GABA), and the excitants (glutamic and aspartic acid) is too great to review here in a few words.

It would, however, be improper in the present discussion not to consider briefly the other sites of potential pharmacological assault on synaptic mechanisms. The integrity of these mechanisms requires the proper functioning of a number of steps: 1) transmitters must be synthesized within the presynaptic cells; 2) the newly formed materials must be stored in such fashion that they can be released in quantal amounts in response to a suitable stimulus; and 3) transmitters, liberated and having fulfilled their postsynaptic roles, must be inactivated so that the synapses are capable of further transmission. For some of the potential transmitters previously mentioned the compendium of pharmacological knowledge is fairly complete (71); for others it is far less so, but for all synapses operating through the release of a chemical agent the steps involved are common.

Finally it should be noted that there may exist in the mammalian nervous system synapses that are relatively immune to the actions of pharmacological agents. At the time when the classic experiments of Loewi (67) indicated that mediation of synaptic transmission by a chemical means was possible, the concept was not widely accepted by neurophysiologists, and an electrotonic mode of coupling between neurons continued to be favored. In the years that followed, the pendulum swung completely to the side of chemical mediation, and it is only recently that electrically mediated junctions have been unequivocally demonstrated in invertebrate and certain lower vertebrate species. The question of the extent to which similar processes may occur in mammals is unknown and may indeed by minimal. However, the demonstration of electrotonic coupling between cells in the mesencephalic root of the fifth sensory nucleus in the rat (5) indicates that this type of transmission is no longer merely a theoretical possibility; and an earlier report (72) indicates that electrical coupling can also occur between feline motoneurons, even though the interaction may not be "synaptic" in the usual sense of the term. Since transmission in these cases is accomplished electrically, occurring across low resistance bridges between the coupled cells, the complex gamut of chemical events is unnecessary, and it is obvious that junctions which operate in this way will be insensitive to drugs that interfere with one or more of the steps in the chemical transmission process already mentioned.

In the following chapters many of the matters mentioned briefly and in general terms here will be discussed in much greater detail. What I have attempted to emphasize is the importance of the correlations between morphology and function; I have also tried to draw attention to some of the difficulties that have been faced in the past, and which continue to perplex those seeking to understand such correlations. The neurosciences have advanced immeasurably since the great debate of 1906; nonetheless the problems requiring solution continue to appear and, like those which face the climbers of previously unscaled peaks, seem ever to lie above and ahead of those who seek to overcome them.

LITERATURE CITED

1. Akert, K., Pfenninger, K., Sandri, C., et al.: Freeze etching and cytochemistry of vesicles and membrane complexes in synapses of the central nervous system. *In* Pappas, G. D. and Purpura, D. P. (Eds.): *Structure and Function of Synapses.* New York: Raven Press, 1972, pp. 67–86.
2. Altmann, H., Bruggencate, G. ten, and Sonnhof, U.: Differential strength of action of glycine and GABA in hypoglossus nucleus. *Pflügers Arch. 331*:90–94, 1972.
3. Aprison, M. H., and Werman, R.: A combined neurochemical and neurophysiological approach to identification of central nervous system transmitters. *Neurosciences Res. 1*:143–174, 1968.
4. Bak, I. J.: The ultrastructure of the substantia nigra and caudate nucleus of the mouse and the cellular localization of catecholamines. *Exp. Brain Res. 3*:40–57, 1967.
5. Baker, R., and Llinas, R.: Electrotonic coupling between neurones in the rat mesencephalic nucleus. *J. Physiol.* (Lond.) *150*:134–144, 1971.
6. Benoist, J. M., Besson, J. M., Conseiller, C., et al.: Action of bicuculline on presynaptic inhibition of various origins in the cat's spinal cord. *Brain Res. 43*:672–676, 1972.
7. Birks, R., Huxley, H. E., and Katz, B.: The fine structure of the neuromuscular junction of the frog. *J. Physiol* (Lond.) *150*:134–144, 1960.
8. Bloom, F. E., and Aghajanian, G. K.: An electron microscopic analysis of large granular synaptic vesicles of the brain in relation to monoamine content. *J. Pharmacol. Exp. Ther.159*:261–273, 1968.
9. Bodian, D.: Synaptic types on spinal motoneurons: An electron microscopic study. *Bull. Johns Hopkins Hosp. 119*:16–45, 1966.
10. Bodian, D.: An electron microscopic characterization of classes of synaptic vesicles by means of controlled aldehyde fixation. *J. Cell. Biol. 44*:155–124, 1970.
11. Boyd, I. A., and Martin, A. R.: The end-plate potential in mammalian muscle. *J. Physiol.* (Lond.) *132*:74–91, 1956.
12. Burke, R. E.: Composite nature of the monosynaptic excitatory postsynaptic potential. *J. Neurophysiol. 30*:1114–1137, 1967.

13. Burke, R. E.: Group Ia synaptic input to fast and slow twitch motor units of cat triceps surae. *J. Physiol.* (Lond.) *196*:605–630, 1968.
14. Conradi, S.: Ultrastructure and distribution of neuronal and glial elements on the motoneuron surface in the lumbosacral spinal cord of the adult cat. *Acta Physiol. Scand.* (Suppl. 332):5–48, 1969.
15. Conradi, S.: Ultrastructure of dorsal horn boutons on lumbosacral motoneurons of the adult cat, as revealed by dorsal root section. *Acta Physiol. Scand.* (Suppl. 332):85–115, 1969.
16. Cook, W. A., Jr., and Cangiano, A.: Presynaptic inhibition of spinal motoneurons. *J. Neurophysiol. 35*:389–403, 1972.
17. Coombs, J. S., Eccles, J. C., and Fatt, P.: The specific ionic conductances and the ionic movements across the motoneuronal membrane that produce the inhibitory postsynaptic potential. *J. Physiol.* (Lond.) *130*:326–373, 1955.
18. Coombs, J. S., Eccles, J. C., and Fatt, P.: Excitatory synaptic action in motoneurones. *J. Physiol.* (Lond.) *130*:374–395, 1955.
19. Coombs, J. S., Eccles, J. C., and Fatt, P.: The inhibitory suppression of reflex discharges from motoneurones. *J. Physiol.* (Lond.) *130*: 396–413, 1955.
20. Curtis, D. R.: The actions of amino acids upon mammalian neurones. In Curtis, D. R. and McIntyre, A. K. (Eds.): *Studies in Physiology.* New York: Springer, 1965, pp. 34–42.
21. Curtis, D. R., Duggan, A. W., Felix, D., et al.: Bicuculline, an antagonist of GABA and synaptic inhibition in the spinal cord of the cat. *Brain Res. 32*:69–96, 1971.
22. Curtis, D. R., Duggan, A. W., Felix, D., et al.: Antagonism between bicuculline and GABA in the cat brain. *Brain Res. 33*:57–73, 1971.
23. Curtis, D. R. and Ryall, R. W.: Pharmacological studies upon spinal presynaptic fibres. *Exp. Brain Res. 1*:195–204, 1966.
24. Dahlström, A.: Observations on the accumulation of noradrenaline in the proximal and distal parts of peripheral adrenergic nerves after compression. *J. Anat. 99*:677–689, 1965.
25. Davidoff, R. A.: The effects of bicuculline on the isolated spinal cord of the frog. *Exp. Neurol. 35*:179–193, 1972.
26. de Robertis, E. and Bennett, H. S.: Submicroscopic vesicular component in the synapse. *Fed. Proc. 13*:35, 1954.
27. Eccles, J. C.: *The Physiology of Nerve Cells.* Baltimore: Johns Hopkins Press, 1957.
28. Eccles, J. C.: *The Physiology of Synapses.* Berlin: Springer, 1964.
29. Eccles, J. C.: The ionic mechanisms of excitatory and inhibitory synaptic action. *Ann. N. Y. Acad. Sci. 137*:473–494, 1966.
30. Eccles, J. C.: Postsynaptic inhibition in the central nervous system. In Quarton, G. C., Melnechuk, T., and Schmitt, F. O. (Eds.): *The Neurosciences.* New York: Rockefeller University Press, 1967, pp. 408–427.
31. Eccles, J. C., Eccles, R. M., and Magni, F.: Central inhibitory action attributable to presynaptic depolarization produced by muscle afferent volleys. *J. Physiol.* (Lond.) *159*:147–166, 1961.

32. Eccles, J. C., Kostyuk, P. G., and Schmidt, R. F.: Central pathways responsible for depolarization of primary afferent fibres. *J. Physiol.* (Lond.) *161*:237–257, 1962.
33. Eccles, J. C., Magni, F., and Willis, W. D.: Depolarization of central terminals of group I afferent fibres from muscle. *J. Physiol.* (Lond.) *160*:62–93, 1962.
34. Eccles, J. C., Schmidt, R., and Willis, W. D.: Pharmacological studies on presynaptic inhibition. *J. Physiol.* (Lond.) *168*:500–530, 1963.
35. Eide, E., Jurna, I., and Lundberg, A.: Conductance measurements from motoneurones during presynaptic inhibition. In von Euler, C., Skoglund, S., and Soderberg, U. (Eds.): *Structure and Function of Inhibitory Neuronal Mechanisms.* Oxford: Pergamon Press, 1968, pp. 215–219.
36. Fatt, P. and Katz, B.: Some observations on biological noise. *Nature 166*:597–598, 1950.
37. Fatt, P. and Katz, B.: An analysis of the end-plate potential recorded with an intracellular electrode. *J. Physiol.* (Lond.) *115*:320–370, 1951.
38. Fonnum, F.: The "compartmentation" of choline acetyltransferase within the synaptosome. *Biochem. J., 103*:262–270, 1967.
39. Frank, K.: Basic mechanisms of synaptic transmission in the central nervous system. *I.R.E. Tr. Med. Electron., ME-6*:85–88, 1959.
40. Frank, K. and Fuortes, M. G. F.: Presynaptic and postsynaptic inhibition of monosynaptic reflexes. *Fed. Proc. 16*:39, 1957.
41. Fukami, Y.: Two types of synaptic bulb in snake and frog spinal cord: The effect of fixation. *Brain Res. 14*:134–145, 1969.
42. Galindo, A.: GABA-picrotoxin interaction in the mammalian central nervous system. *Brain Res. 14*:763–767, 1969.
43. Ginsborg, B. L.: The vesicle hypothesis for the release of acetylcholine. In Andersen, P. and Jansen, J. K. S. (Eds.): *Excitatory Synaptic Mechanisms.* Oslo: Universitetsforlaget, 1970. pp. 77–82.
44. Golgi, C.: *La doctrine du neurone. Theorie et faits.* Les prix Nobel en 1906, Stockholm, 1906.
45. Granit, R.: The case for presynaptic inhibition by synapses on the terminals of motoneurons. In von Euler, C., Skoglund, S., and Soderberg, U. (Eds.): *Structure and Function of Inhibitory Neuronal Mechanisms.* Oxford, Pergamon Press, 1968, pp. 183–185.
46. Gray, E. G.: Axo-somatic and axo-dendritic synapses of the cerebral cortex: An electron microscope study. *J. Anat. 93*:420–433, 1959.
47. Gray, E. G.: A morphological basis for pre-synaptic inhibition? *Nature 193*:82–83, 1962.
48. Gray, E. G. and Willis, R. A.: On synaptic vesicles, complex vesicles and dense projections. *Brain Res. 24*:149–168, 1970.
49. Hand, A. R.: Adrenergic and cholinergic nerve terminals in the rat parotid gland. Electron microscopic observation on permanganate-fixed glands. *Anat. Rec. 173*:131–140, 1972.
50. Hebb, C. O. and Whittaker, V. P.: Intracellular distributions of acetylcholine and choline acetylase. *J. Physiol.* (Lond.) *142*:187–196, 1958.

51. Hongo, T., Jankowska, E., and Lundberg, A.: The rubrospinal tract. III. Effects on primary afferent terminals. *Exp. Brain Res. 15*:39–53, 1972.
52. Kapeller, K. and Mayor, D.: The accumulation of noradrenaline in constricted sympathetic nerves as studied by fluorescence and electron microscopy. *Proc. Roy. Soc. B:167*:282–292, 1967.
53. Kapeller, K. and Mayor, D.: An electron microscopic study of the early changes proximal to a constriction in sympathetic nerves. *Proc. Roy. Soc. B:172*:39–51, 1969.
54. Katz, B.: *Nerve, Muscle and Synapse.* New York: McGraw-Hill, 1966.
55. Katz, B. and Miledi, R.: A study of spontaneous miniature potentials in spinal motoneurones. *J. Physiol.* (Lond.) *168*:389–422, 1963.
56. Kellerth, J. O.: Aspects on the relative significance of pre- and postsynaptic inhibition in the spinal cord. In von Euler, C., Skoglund, S., and Soderberg, U. (Eds.): *Structure and Function of Inhibitory Neuronal Mechanisms.* Oxford: Pergamon Press, 1968. pp. 197–212.
57. Khattab, F. I.: A complex synaptic apparatus in spinal cords of cats. *Experientia 24*:690–691, 1968.
58. Kobayashi, H. and Libet, B.: Generation of slow postsynaptic potentials without increases in ionic conductance. *Proc. Natn. Acad. Sci. U.S.A.* 60:1304–1311, 1968.
59. Kobayashi, H. and Libet, B.: Actions of noradrenaline and acetylcholine on sympathetic ganglion cells. *J. Physiol.* (Lond.) *208*:353–372, 1970.
60. Krnjević, K.: Glutamate and γ-aminobutyric acid in brain. *Nature 228*:119–124, 1970.
61. Krnjević, K., Pumain, R., and Renaud, L.: The mechanism of excitation by acetylcholine in the cerebral cortex. *J. Physiol.* (Lond.) *215*:247–268, 1971.
62. Kuno, M.: Quantal components of excitatory synaptic potentials in spinal motoneurones. *J. Physiol.* (Lond.) *175*:81–99, 1964.
63. Kuno, M.: Mechanism of facilitation and depression of the excitatory synaptic potential in spinal motoneurones. *J. Physiol.* (Lond.) *175*:100–112, 1964b.
64. Kuno, M. and Miyahara, J. T.: Non-linear summation of unit synaptic potentials in spinal motoneurones of the cat. *J. Physiol.* (Lond.) *201*:465–477, 1969.
65. Levy, R. A. and Anderson, E. G.: The effect of the GABA antagonists bicuculline and picrotoxin on primary afferent terminal excitability. *Brain Res. 43*:171–180, 1972.
66. Liley, A. W.: The effects of presynaptic polarization on the spontaneous activity at the mammalian neuromuscular junction. *J. Physiol.* (Lond.) *134*:427–443, 1956.
67. Loewi, O.: Über humorale Übertragbarkeit der Herznervenwirkung. I. Mitteilung. *Pflügers Arch. 189*:239–242, 1921.
68. Marchbanks, R. M.: Exchangeability of radioactive acetylcholine with the bound acetylcholine of synaptosomes and synaptic vesicles. *Biochem. J. 106*:87–95, 1968.

69. McLaughlin, B. J.: The fine structure of neurones and synapses in the motor nuclei of the cat spinal cord. *J. Comp. Neurol. 144*:429–460, 1972.
70. McLennan, H.: *Synaptic Transmission* (2d ed.). Philadelphia: Saunders, 1970.
71. McLennan, H.: Studies on synaptic transmission using drugs as tools. *Triangle 10*:85–92, 1971.
72. Nelson, P. G.: Interaction between spinal motoneurones of the cat. *J. Neurophysiol. 29*:275–287, 1966.
73. Palay, S. L.: Electron microscope study of the cytoplasm of neurons. *Anat. Rec. 118*:336, 1954.
74. Palay, S. L.: Synapses in the central nervous system. *J. Biophys. Biochem. Cytol.* 2 Suppl. 193–201, 1956.
75. Pellegrino de Iraldi, A. and de Robertis, E.: The neurotubular system of the axon and the origin of granulated and non-granulated vesicles in regenerating nerves. *Z. Zellforsch. 87*:330–344, 1968.
76. Purpura, D. P.: Comparative physiology of dendrites. In Quarton, G. C., Melnechuk, T., and Schmitt, F. O. (Eds.): *The Neurosciences.* New York: Rockefeller University Press, 1967, pp. 372–393.
77. Ralston, H. J., III: Dorsal root projections to dorsal horn neurons in the cat spinal cord. *J. Comp. Neurol. 132*:303–329, 1968.
78. Ramón y Cajal, S.: *Structures et connexions des neurones.* Les prix Nobel en 1906, Stockholm, 1906.
79. Ramón y Cajal, S.: *Neuron Theory or Reticular Theory*? Madrid: Consejo Superior de Investigaciones Cientificas, 1954.
80. Réthelyi, M.: Ultrastructural synaptology of Clarke's column. *Exp. Brain Res. 11*:159–174, 1970.
81. Schmidt, R. F.: Presynaptic inhibition in the vertebrate central nervous system. *Ergebn. Physiol. 63*:20–101, 1971.
82. Sedvall, G. and Thorson, J.: The effect of nerve stimulation on the release of noradrenaline from a reserpine resistant transmitter pool in skeletal muscle. *Biochem. Pharmacol.* 12 Suppl. 65–66, 1963.
83. Sherrington, C. S.: From Foster, M. (Ed.): *A Text Book of Physiology* (7th ed.). Part III. London: Macmillan, 1897.
84. Szentágothai, J.: Synaptic structure and the concept of presynaptic inhibition. In von Euler, C., Skoglund, S., and Soderberg, U. (Eds.): *Structure and Function of Inhibitory Neuronal Mechanisms.* Oxford: Pergamon Press, 1968, pp. 15–31.
85. Tebēcis, A. K. and Phillis, J. W.: The use of convulsants in studying possible functions of amino acids in the toad spinal cord. *Comp. Biochem. Physiol. 28*:1303–1315, 1969.
86. Uchizono, K.: Characteristics of excitatory and inhibitory synapses in the central nervous system of the cat. *Nature 207*:642–643, 1965.
87. Uchizono, K.: Synaptic organization of the Purkinje cells in the cerebellum of the cat. *Exp. Brain Res. 4*:97–113, 1967.
88. Valdivia, O.: Methods of fixation and the morphology of synaptic vesicles. *J. Comp. Neurol. 142*:257–274, 1971.
89. van der Loos, H.: Fine structure of synapses in the cerebral cortex. *Z. Zellforsch. 60*:815–825, 1963.

90. Walberg, F.: Axoaxonic contacts in the cuneate nucleus, probable basis for presynaptic depolarization. *Exp. Neurol. 13*:218–231, 1965.
91. Werman, R.: A review—criteria for identification of a central nervous system transmitter. *Comp. Biochem. Physiol.. 18*:745–766, 1966.
92. Westrum, L. E.: On the origin of synaptic vesicles in the cerebral cortex. *J. Physiol.* (Lond.) *179*:4–6P, 1965.
93. Westrum, L. E. and Black, R. G.: Changes in the synapses of the spinal trigeminal nucleus after ipsilateral rhizotomy. *Brain Res. 11*:706–709, 1968.
94. Whittaker, V. P.: The use of synaptosomes in the study of synaptic and neural membrane function. In Pappas, G. D. and Purpura, D. P. (Eds.): *Structure and Function of Synapses.* New York: Raven Press, 1972, pp. 87–100.

CHAPTER TWO

Amino Acid Inhibitory Transmitters in the Central Nervous System

Graham A. R. Johnston

Synaptic excitation of neurons can be rendered less effective by the activity of inhibitory synapses which alter the membrane properties of the postsynaptic neuron (postsynaptic inhibition) or of the presynaptic excitatory terminals (presynaptic inhibition). (See Ch. 1.) The former generally involves a change in membrane conductance and a membrane hyperpolarization, both of which reduce membrane excitability. Presynaptic inhibition is considered to be effected by depolarization of the presynaptic excitatory terminals, which reduces the amount of excitatory transmitter released. The amino acids, γ-aminobutyric acid (GABA) and glycine, are hyperpolarizing postsynaptic inhibitory transmitters that increase membrane conductance, in contrast to catecholamines which, in some neurons, apparently produce postsynaptic inhibition associated with a decrease in membrane conductance. (See Chs. 5 and 6.) GABA may also be involved in presynaptic inhibition where it would have a depolarizing action on presynaptic terminals.

GABA appears to be the dominant inhibitory transmitter in the rostral neuraxis, whereas glycine probably plays a role no less significant in the spinal cord and brainstem. In addition, diverse substances with characteristic distribution patterns, including noradrenaline, dopamine, 5-hydroxytryptamine, and acetylcholine, likely act as inhibitory transmitters. And finally, other amino acids, such as taurine, may also be inhibitory transmitters.

How is a transmitter identified? In brief, information is sought on the presence, synthesis, release, action, and inactivation of the putative transmitter, as well as on its specific antagonists (*60*). In early investigations of amino acids in the central nervous system (CNS), demonstration of the mechanism of action of the putative inhibitory transmitters loomed as the largest problem. Postsynaptic inhibition was known to be associated with a membrane hyperpolarization, but GABA could not be shown to have a hyperpolarizing action. This obstacle was removed in 1965 when Obata, Ito, and their colleagues (*194, 195*) demonstrated that GABA hyperpolarized Deiters' neurons. This work was soon extended to other central neurons by many groups of workers.

Another significant advance in this field was also made in 1965 when Aprison and Werman (*7*) suggested that glycine may be an inhibitory transmitter in the spinal cord. Following this memorable year, many new observations were made and many old ones reinterpreted, resulting in an impressive body of information regarding amino acids as inhibitory transmitters in the CNS; however, much work remains to be done, because the clinical pharmacology of these substances has yet to be knowingly exploited.

γ-AMINOBUTYRIC ACID (GABA)

The identification of GABA as an inhibitory transmitter in the mammalian CNS is based on experimental observations pertaining to its presence, synthesis, release, postsynaptic action, inactivation, and specific antagonists. To date, numerous reviews concerning aspects of the synaptic function of GABA have been published (*17, 30, 105, 124, 201, 213, 214*).

GABA is a zwitterion at neutral pH with pK values of 4.03 (proton gain) and 10.27 (proton loss) at 35° (*159*). It is a flexible molecule, and structure activity studies on GABA analogues implicate both the intramolecular distance between the zwitterionic centers and the rotational freedom of the molecule as important factors governing the physiological activity of GABA. The fully extended and folded conformations of GABA are shown in Figure 1. X-ray crystallography indicates that GABA exists in a partially folded conformation in the solid state (*239, 255*), whereas molecular orbital calculations (*156*) and proton magnetic resonance studies (*203*) favor extended conformations for GABA in solution.

Neurochemistry

Levels in the CNS In mammals GABA is found in appreciable amounts only in the CNS, with the highest levels being localized in the substantia nigra (SN) and globus pallidus (*85, 198*).

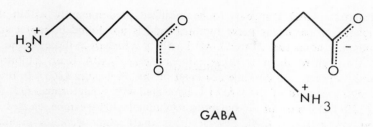

Figure 1. Extended and folded conformations of GABA.

Most investigators have found that GABA does not pass the blood-brain barrier in adult mammals to raise the levels of GABA in the brain, although it is recently reported that subcutaneous injection of large doses of GABA (up to 60 μmol/g) increases the brain levels in mice of all ages (*170*). The anticonvulsant action of hyperosmotic solutions of GABA injected intraperitoneally has been interpreted as resulting from brain dehydration rather than as being a specific effect of GABA (*73*). Brain levels of GABA are altered by a variety of drugs, many of which inhibit GABA-metabolizing enzymes (*17, 124*).

Synthesis GABA appears to be synthesized essentially by only one enzymatic activity, glutamate decarboxylase (GAD, L-glutamate-1-carboxylase, EC 4.1.1.15), which catalyzes the irreversible decarboxylation of L-glutamate to GABA. In addition putrescine (*224*) and 4-hydroxybutyric acid (*185*) are possible precursors of GABA.

GAD exists in at least two forms: GAD I and GAD II. GAD I, the form more widely studied, has been partially purified from brain (*135, 241*). It requires pyridoxal phosphate as a cofactor and is strongly inhibited by anions such as chloride, and by carbonyl trapping reagents such as thiosemicarbazide (*213*). The levels of the enzyme change during hibernation, after adrenalectomy and electroconvulsive shock treatment (*17*), and in Parkinson's disease, in which decreased levels of GAD I are found in the SN and globus pallidus (*174*). GAD I is relatively insensitive to product inhibition by GABA, but the latter may repress the synthesis of GAD I (*102, 244, 246*). The induction of GAD I synthesis by large doses of glutamate has been reported (*162*). Adenosine triphosphate (ATP) inhibits GAD I activity, and this inhibition is antagonized by inorganic phosphate (*259*).

The regional and subcellular distribution of GAD I is consistent with that expected of an enzyme catalyzing the synthesis of an inhibitory transmitter: GAD I appears to be localized predominantly within the axoplasm of particular nerve terminals and is not found in appreciable amounts in nonneural tissues. GAD I activity is highest in those regions of the CNS where there is other evidence that GABA is an inhibitory transmitter, e.g., the Purkinje cell layer of the cerebellum (*165*). In brain homogenates most of the GAD I is associated with synaptosomes (*11, 88, 192, 221*), and most of the synaptosome-bound activity can be released by hypoosmotic shock (*88*). Several drugs that cause decreases in brain GABA, including thiosemicarbazide, allylglycine, and 3-mercaptopropionic acid, are likely to act by inhibiting the activity of GAD I (*17*). No direct causal relation has yet been established, however, between the convulsant effects of these substances and an inhibition of GABA synthesis (*245*).

In contrast to GAD I, GAD II is relatively insensitive to inhibition by chloride, is stimulated rather than inhibited by carbonyl trapping reagents (*101*), and is found in various nonneuronal tissues (*164*). The physiological role of GAD II is at present unknown.

Degradation GABA is catabolized in the CNS by a specific GABA transaminase (GABA-T, 4-aminobutyrate:2-oxoglutarate aminotransferase, EC 2.6.1.19) to succinic semialdehyde, which is in turn oxidized to succinic acid by a specific dehydrogenase (SSA-dehydrogenase). GABA-T can catalyze GABA synthesis by the reverse reaction from succinic semialdehyde, but this is unlikely to be a significant source of GABA in vivo, because the high SSA-dehydrogenase activity effectively removes succinic semialdehyde and ensures that GABA-T functions as a GABA degrading enzyme under normal conditions (*124*).

Multiple forms of GABA-T may exist (*268*). GABA-T has been extensively purified from brain, and its kinetic properties and substrate specificity have been studied (*243, 269*). GABA-T will transaminate some other substances related to GABA, including β-alanine and δ-aminovaleric acid, but is inactive on γ-amino-β-hydroxybutyric acid and taurine (*21, 243, 269*). GABA-T is found in appreciable amounts in nonneural tissues, particularly in liver and kidney (*41*). Both GABA-T and SSA-dehydrogenase are mitochondrial enzymes (*11, 221*). Many inhibitors of GABA-T activity, including aminooxyacetic acid, hydroxylamine, and 3-hydrazinopropionic acid, cause increases in brain GABA (*17, 124*). As with inhibitors of GAD I, cause and effect are difficult to correlate.

Although there seems to be little doubt that the major pathway of GABA catabolism is that mediated by GABA-T and SSA-dehydrogenase, a variety of alternative pathways has been postulated. GABA may be converted to γ-aminobutyrylcholine, γ-butyrobetaine, γ-guanidinobutyric acid, various GABA-peptides, γ-amino-β-hydroxybutyric acid, and γ-hydroxybutyric acid (*17, 124*).

The metabolic pathway from 2-oxoglutarate to succinate by way of L-glutamate, GABA, and succinic semialdehyde constitutes the "GABA-shunt," i.e., an alternative pathway to that through succinyl-coenzyme A in the normal tricarboxylic acid cycle. By pulse-labeling GABA in rat cerebral slices, it is estimated that the GABA-shunt constitutes about 8% of the total carbon flux of the tricarboxylic acid cycle (*12*). This indicates a half-life of brain GABA of approximately ten min. In vivo experiments with rats on the rate of increase in brain GABA following GABA-T inhibition indicate that the rate of GABA synthesis is approximately 5 μmol/g/hr (*264*).

Uptake Mammalian brain slices actively accumulate exogenous GABA by a saturable "high affinity" process (K_m ca. 10^{-5} M), which is structurally specific, dependent on temperature and requires sodium ions in the external medium (*29, 82, 128, 213*).

The structural specificity of GABA uptake ensures that other metabolites are not transported into GABA pools; GABA uptake is not influenced by high concentrations of a variety of amino acids and other putative transmitters (*126, 128*). Some close structural analogues of GABA do inhibit GABA uptake, and the competitive inhibitors of uptake (Figure 2) may be alternative substrates for the transport process. Some of them may thus find pharmacological application as "false transmitters." Various

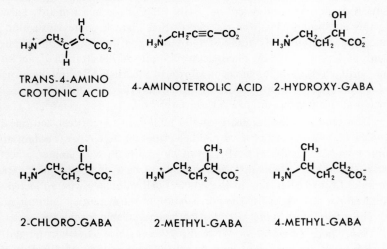

Figure 2. Competitive inhibitors of GABA uptake—possible "false transmitters."

drugs inhibit GABA uptake in a relatively nonspecific manner, including p-chloromercuriphenylsulfonate, chlorpromazine, imipramine, haloperidol, juglone, dibutyryl cyclic AMP, and the protoveratrines (*22, 96, 126*).

The subcellular localization of GABA uptake sites may be in a unique population of nerve terminals. ^{3}H-GABA taken up into rat cerebral slices has a similar subcellular distribution to that of endogenous GABA and GAD I (*192, 248*), with about 40% of the radioactivity being recovered in synaptosomes after homogenization and density-gradient centrifugation. Synaptosomes containing ^{3}H-GABA can be partially separated from those accumulating other substances such as catecholamines (*129*), glycine (*126*), and L-glutamate (*277*). Autoradiographic studies of the ^{3}H-GABA accumulated by slices of rat cerebral cortex indicate that the major site of uptake is a population of nerve terminals that accounts for approximately 30% of all terminals (*125*). Qualitatively similar results are obtained for ^{3}H-GABA uptake into other areas of the rat CNS (*125*), but the major site of uptake in the rat retina appears to be glial cells (*193*).

At present the physiological role of GABA uptake is unclear and is likely to remain so until specific inhibitors of this process are discovered. A probable function is the maintenance of low extracellular concentrations of GABA. On the basis of existing evidence, it seems very likely that GABA is taken up, at least in part, into the nerve terminals that release GABA, thus serving to reconstitute the presynaptic store(s) of the transmitter. The vital question is, does this uptake also terminate the action of GABA on postsynaptic receptors? Some support for an affirmative answer to this question comes from the observation that the depressant action of microelectrophoretically administered GABA, on the firing of spinal interneurons, is potentiated by microelectrophoretic administration of p-chloromercuriphenylsulfonate, an agent known in inhibit GABA uptake in vitro (*13, 52*). Potentiation of GABA-mediated synaptic inhibition by this mercurial agent, however, has yet to be observed.

The duration and intensity of the inhibition of cortical neurons are enhanced, and the response to microelectrophoretic GABA is potentiated following the increase of GABA levels by treatment with aminooxyacetic acid (*97*). These effects may be the result of delayed GABA inactivation produced by an alteration in the rate of GABA uptake owing to increased intracellular GABA levels. The potentiation of the synaptic inhibition may also be attributable to increased GABA being available for release.

Subcellular Localization While the gross regional variations in central GABA levels are consistent with its role as an inhibitory transmitter, the precise intracellular localization of the GABA store(s) has yet to be demonstrated. Under certain conditions, as much as 60% of the total GABA in homogenates of rat cerebral cortex is particle bound (*81, 192*), but conventional subcellular fractionation procedures lead to very low recoveries of GABA in isolated synaptosomes (*175, 220*). It seems likely that GABA is lost from synaptosomes during fractionation (*192*) and that extensive redistribution of GABA takes place (*248*). Synaptosomes actively take up exogenous GABA (*126*), and synaptic vesicles bind GABA by a sodium-dependent process (*213*). Most of the GABA in synaptosomes can be released by hypoosmotic treatment (*175*), but a small amount remains associated with the synaptic vesicles (*166*).

Degeneration studies indicate that GABA is associated with the terminals of Purkinje cell axons on neurons in Deiters' nucleus, and that the concentration of GABA associated with isolated Purkinje cells is much higher than that associated with isolated spinal motoneurons (*202*). Such studies are clearly limited in application, and there is an obvious need for a sensitive and specific histochemical procedure to localize endogenous GABA.

Release The in vivo release of endogenous GABA in cats has been demonstrated 1) into the fourth ventricle following stimulation of Purkinje cells (*196*), and 2) from the surface of the cerebral cortex following surface, thalamic, and brainstem stimulation (*127, 134*). GABA release from the surface of the cortex following surface stimulation is calcium dependent (*127*). The release of endogenous GABA from synaptosomes has also been demonstrated (*72*).

Exogenous GABA can also be released from the cerebral cortex in vivo (*184*) and in vitro (*237*) by electrical stimulation. Thus exogenous GABA appears not only to be selectively taken up into nerve terminals but also to enter a GABA pool from which it is released by suitable stimulation.

Effects of GABA on General Brain Metabolism GABA stimulates protein synthesis in a ribosomal system from immature rat brain (*18*). Intraperitoneal GABA decreases brain levels of noradrenaline and 5-HT and is without influence on DA (*279*). GABA inhibits the pyridoxal kinase activity in extracts of rat brain; this may represent a feedback mechanism regulating GAD activity through the levels of pyridoxal phosphate (*260*).

GABA and Synaptic Plasticity Several investigators have commented on the possibility that the GAD activity in a synaptic terminal may determine whether or not the terminal releases GABA—an inhibitory substance—or glutamate—an excitatory substance—or mixtures of both (*60, 124, 213*). This is particularly intriguing since GAD I activity can be modulated on a short-term basis by the concentration of chloride and ATP (*259*) and on a long-term basis by alterations in the rate of synthesis of this inducible-repressible enzyme.

There are perhaps three observations that indicate that this GABA-glutamate "switching" via GAD I is not a general property of a majority of neurons in the cerebral cortex: First, inhibitors of GAD I activity reduce GABA levels but, in general, do not increase glutamate levels. Second, GAD I does not seem to be a constituent of a majority of synaptosomes in brain homogenates (*221*). Third, GABA uptake appears to be a property of a certain population of nerve terminals that represent approximately one-third of the total nerve terminals (*125*) and which differ from those that take up glutamate by a high affinity process (*277*).

Neuropharmacology

Postsynaptic Action of GABA When administered microelectrophoretically, GABA depresses the firing of neurons throughout the mammalian CNS. The ionic basis of this depression is a reversible membrane hyperpolarization, attributable mainly to an increased permeability to chloride ions and perhaps also to potassium ions.

GABA-induced hyperpolarization was first demonstrated by Obata, Ito, and their colleagues for neurons in Deiters' nucleus (*194, 195*) and subsequently for neurons in the pericruciate cortex (*163*), spinal motoneurons (*59*), spinal interneurons (*37*), and cerebellar Purkinje cells (*231*).

The reversal potentials for both synaptically and GABA-induced hyperpolarization of cerebral, Deiters', and spinal neurons are similar and at a more hyperpolarized level than the resting potential. Thus it is probable that the equilibrium potentials for the associated conductance increases are similar or identical (*38, 59, 163, 197*).

Intracellular injection of a series of anions and cations has essentially the same effects on synaptically and GABA-induced potential changes, indicating that the permeability changes induced by GABA are identical to those occurring during certain synaptic inhibitions. Elevation of the intracellular chloride ion concentration converts synaptically and GABA-induced hyperpolarization into depolarizations.

Antagonists At least three classes of compounds appear to antagonize the postsynaptic effects of GABA. The structures of representatives of the classes—picrotoxinin, bicuculline, and benzyl penicillin—are shown in Figure 3.

Picrotoxin This convulsant, an equimolar mixture of picrotoxinin and picrotin, is isolated from *Anamirta cocculus* and related to poisonous plants native to India. The commercially available picrotoxin can be

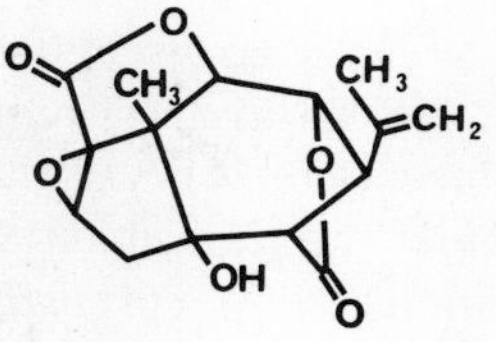

PICROTOXININ

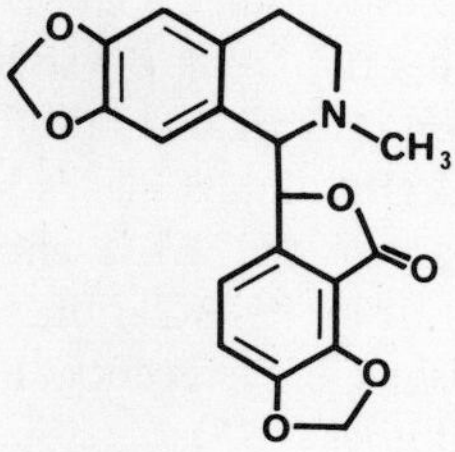

BICUCULLINE

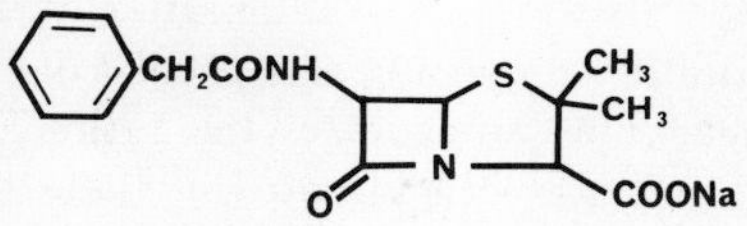

BENZYL PENICILLIN

Figure 3. GABA antagonists.

readily separated into its components by column chromatography on silica gel (*132*). Picrotoxinin is approximately 50 times more active than picrotin as a convulsant, whereas the structurally related bitter principals, tutin and coriamyrtin, are as potent as picrotoxinin (*133*). The chemistry of these substances has recently been reviewed (*210*).

Picrotoxin is a noncompetitive inhibitor of the action of GABA on the crayfish inhibitory neuromuscular junction (*249*) and is known to antagonize a number of inhibitions in the mammalian CNS on systemic administration. The microelectrophoretic administration of picrotoxin is hampered by the lack of readily ionizable groups on the molecules which thus can only be carried out of the micropipettes by electroosmosis. Nevertheless there have been reports of picrotoxin reversibly antagonizing the action of GABA on neurons in many areas of the CNS, including the cuneate nucleus (*91*), Deiters' nucleus (*38*), the hypoglossus nucleus (*39*), the spinal cord (*83*), and the cerebral cortex (*107*). Picrotoxin has also been reported to block consistently the action of glycine on spinal neurons (*64*) and to influence the action of both glycine and GABA on approximately 25% of spinal interneurons, being ineffective on the remainder (*51*).

It is apparent that technical difficulties limit the usefulness of picrotoxin as a GABA antagonist. Perhaps tutin or coriamyrtin would be more useful, or a structural analogue bearing a charged group. The alkaloid dendrobine shows some structural similarity to these substances and is as active as picrotoxin as a convulsant; however, it is a glycine antagonist without influence on GABA (Figure 10).

Bicuculline This convulsant is a phthalide-isoquinoline alkaloid first isolated from *Dicentra cucullaria* and subsequently from a variety of *Corydalis* species (*176, 177*). Its convulsant action in frogs and rabbits was soon recognized (*272*), but its function as a GABA antagonist is a relatively recent discovery (*48*). In a similar manner the gross chemical structure of bicuculline was worked out soon after its isolation (*177*); however, its absolute configuration was established only recently by optical rotatory dispersion and circular dichroism studies (*235*).

Bicuculline antagonizes the action of GABA and a variety of central inhibitions. As with picrotoxin, there are some experimental difficulties associated with its use, particularly its low solubility in water. Bicuculline also antagonizes the action of a number of depressant amino acids structurally related to GABA but is without appreciable influence on the action of glycine and related amino acids (*19, 49, 50*). Of particular interest as bicuculline-sensitive depressants are the conformationally restricted GABA analogues, muscimol, 4-aminotetrolic, and trans-4-aminocrotonic acid (Figures 2 and 7). As these are all analogues of extended conformations of GABA, it appears that GABA interacts with bicuculline-sensitive postsynaptic receptors in an extended conformation.

Structure-activity studies on bicuculline have been initiated (*48, 140*). Corlumine and bicucine methyl ester show similar potency to bicuculline as convulsants and as GABA antagonists (Figure 4). The mirror image of bicuculline, (−)-bicuculline, is relatively inactive as a GABA antagonist, as are adlumine, bicucine, bicucullinediol, capnoidine, hydrastine, and narcotine (Figure 5). Laudanosine and *N*-methylbicuculleine are glycine antagonists (Figure 10). These results are consistent with the original proposal (*48*) regarding the structural similarity between bicuculline and GABA illustrated in Figure 6: this suggests that GABA and bicuculline interact with the same receptors and may provide information about the manner by which GABA acts on these receptors. Other proposals on the interaction of bicuculline with receptors have been made by several investigators (*118, 234, 239*).

Quaternization of the heterocyclic nitrogen atom in bicuculline produces more active compounds. Bicuculline methiodide ("*N*-methylbicuculline" *208*) is a more powerful convulsant than bicuculline hydrochloride when injected intracisternally. Bicuculline methochloride (Figure 4) is approximately 100 times more soluble in water than bicuculline hydro-

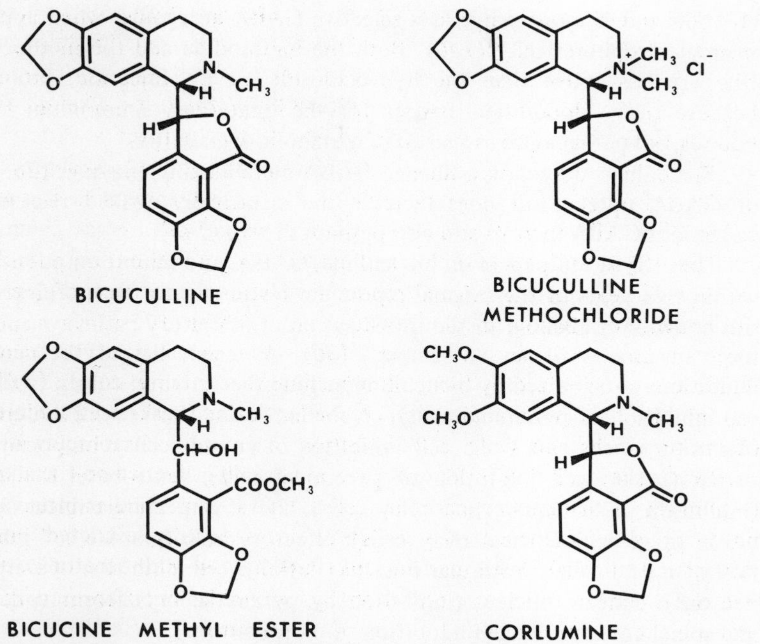

CNS Inhibitory Transmitters   41

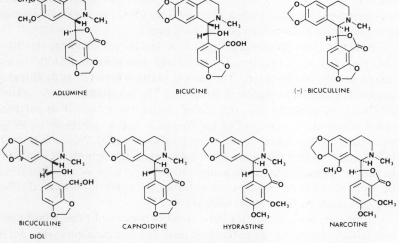

Figure 4. Bicuculline and related compounds that antagonize action of GABA.

Figure 5. Analogues of bicuculline that are inactive as GABA antagonists.

chloride and is more active as a selective GABA antagonist when applied microelectrophoretically (140). Both the methiodide and the methochloride are less active than the hydrochloride on i.v. injection, probably because of a blood-brain barrier for the quaternary ammonium compounds, but perhaps also as a result of metabolic degradation.

Bicuculline does not influence GABA-metabolizing enzymes (20, 68) or GABA uptake but does increase the electrically evoked release of exogenous GABA in vitro and also perhaps in vivo (145).

The fifty-seven papers on bicuculline, GABA, and inhibition published within two years of the original report are testimony to the usefulness of this convulsant alkaloid in the investigation of inhibitory pathways and to the "advantages of an antagonist" (80). A recent list of the central inhibitions antagonized by bicuculline include the: cerebral cortex (stellate cell inhibition of pyramidal cells); cerebellar cortex (basket cell inhibition of Purkinje cells and Golgi cell inhibition of granule cells); hippocampal cortex (basket cell inhibition of pyramidal cells); ventrobasal thalamus (inhibition of thalamocortical relay cells); lateral geniculate nucleus (inhibition of geniculocortical relay cells); olfactory bulb (granule cell inhibition of mitral cells); vestibular nucleus (Purkinje cell inhibition of Deiters' neurons); cuneate nucleus (inhibition by pyramidal or afferent volleys); and spinal cord (prolonged inhibition of motoneurons).

It must be emphasized that *a priori* GABA is not to be equated with the transmitter(s) mediating these inhibitions: much more evidence is needed for positive identification, particularly as other bicuculline-sensitive depressants are known to occur in the CNS, e.g., imidazole-4-acetic acid, γ-amino-β-hydroxybutyric acid, and taurine.

Penicillin The epileptogenic action of benzyl penicillin on the brain following topical, intracerebral, or systemic administration of this compound is well documented. This action seems likely to result from an antagonism of GABA-mediated inhibition. The antagonism of the effects of GABA by penicillin was first noted during recordings of dorsal root potentials in the spinal cord of the frog (63) and subsequently on single neurons in the cat spinal cord and cerebral cortex (56). Penicillin is much less effective than bicuculline as a GABA antagonist in both preparations. In the cat, penicillin has little action on the depression of neuronal firing by glycine; and penicillin does not influence the uptake of GABA, glycine, or L-glutamate by slices of cat spinal cord (13).

Structural similarities have been noted between penicillin, bicuculline, and GABA (Figure 6), suggesting that penicillin and bicuculline may act in similar ways at the same receptors.

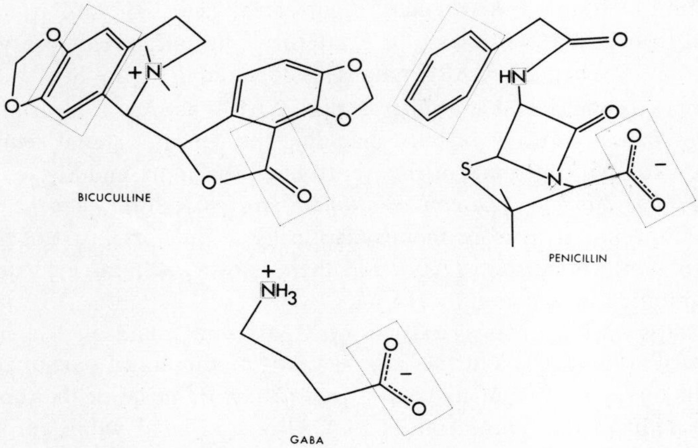

Figure 6. Structural similarities between bicuculline, GABA, and penicillin.

Tetanus Toxin Tetanus toxin reduces the prolonged inhibition of reflexes (*242*) which may involve GABA synapses. It seems possible that the toxin may act presynaptically on GABA release, since it does not antagonize the postsynaptic action of GABA on Renshaw cells (*46*) and does not alter spinal levels of GABA (*87, 143, 225*).

Synapses Using GABA as Transmitter

The most fully documented case for GABA as a transmitter is at the synapses of cerebellar Purkinje cell axons on neurons in Deiters' nucleus. Less but sufficiently adequate evidence is available for GABA as a transmitter in the cerebral, cerebellar, and hippocampal cortices: an anatomical feature common to these regions is the stellate cells, which have relatively short axons spreading tangentially and terminating with "baskets" around the bodies and larger dendrites of pyramidal and Purkinje cells. These morphologically similar basket-type neurons appear to be GABA-releasing inhibitory interneurons (*54*). Discussion of inhibitory influences in the spinal cord serves as an introduction to the problems of "presynaptic inhibition," a poorly understood process likely to involve GABA. (See Ch. 1, "The Synapse.")

This section is thus broadly concerned with the role of GABA in three types of synaptic inhibition, including: 1) long axon neurons mediating postsynaptic inhibition from one brain area to another (cerebellum to brainstem in the case cited); 2) short axon neurons mediating postsynaptic inhibition within a particular brain area; and 3) inhibition that results from a reduced release of excitatory transmitter—true presynaptic inhibition. A schematic GABA synapse is depicted in Figure 8.

There is considerable evidence that GABA acts as a transmitter at many synapses in other regions, including the cuneate, lateral geniculate, oculomotor and trigeminal nuclei, ventrobasal thalamus, and SN.

Deiters' Nucleus (Lateral Vestibular Nucleus) Purkinje cells in the cerebellum inhibit neurons monosynaptically in the dorsal part of Deiters' nucleus in the brainstem (*123*), and there is overwhelming evidence that this inhibition is mediated by GABA.

Relatively high concentrations of GABA are found associated with isolated Purkinje cells and the large neurons of the dorsal part of Deiters' nucleus but not those of its ventral part (*202*). Removal of the cerebellar vermis reduces the concentration of GABA associated with neurons isolated from the dorsal part of the nucleus but not that associated with neurons from its ventral part (*202*). The GAD I content of the dorsal part of Deiters' nucleus is also reduced after cerebellar lesions that cause a degeneration of Purkinje cell terminals. Subcellular distribution studies indicate that GAD I is contained in the synaptic terminals *on* rather than *in* the cell bodies of Deiters' neurons (*89*).

Neurons in Deiters' nucleus are hyperpolarized by electrophoretically administered GABA (*38, 195*) in a manner similar to synaptically induced hyperpolarization. Picrotoxin (*38*) and bicuculline (*50*) antagonize the action of GABA on Deiters' neurons, whereas strychnine is without effect. Strychnine does not modify Purkinje cell inhibition of Deiters' neurons (*195*), whereas bicuculline, administered electrophoretically or intravenously, readily suppresses the inhibition of Deiters' neurons produced by stimulation of the cerebellum (*50*). Relatively high doses of picrotoxin are required to suppress this inhibition (*197*).

Repetitive stimulation of the cerebellar cortex produces an increased release of GABA into the perfused fourth ventricle, most likely as a result of liberation of GABA from the terminals of Purkinje cells in Deiters' and cerebellar nuclei lying close to the ventricle (*196*).

Cerebellum There is evidence that GABA is an inhibitory transmitter in the cerebellum and is released by basket and stellate cells onto Purkinje cells. Golgi cells may also release GABA and, since Purkinje cells release GABA at their terminals in Deiters' nucleus, it seems likely that their terminals on neurons of the intracerebellar nuclei also release GABA (*54*).

Within the cerebellar cortex the highest levels of GABA and GAD I are found in the Purkinje cell layer (*165*). The relatively high GABA levels associated with isolated Purkinje cell bodies may be owing to the adhering terminals of basket cells (*202*). Only a relatively small proportion (12%) of nerve terminals in the cerebellum appears to accumulate exogenous GABA (*110, 125*).

Purkinje cells are hyperpolarized and their membrane conductance is increased by electrophoretically administered GABA (*231*). GABA-induced depression of the firing of Purkinje cells is antagonized by bicuculline but not by strychnine (*50*). Basket cell inhibition of Purkinje cells is antagonized by bicuculline and unaffected by strychnine (*27, 54*). Golgi cell inhibition of granule cells is blocked by bicuculline and picrotoxin and unaffected by strychnine (*27*).

Cerebral Cortex The inhibition of pyramidal tract neurons, in layers V and VI of the cerebral cortex, produced by electrical stimulation of the pyramids or of the surface of the cerebral cortex or by chemical stimulation of intracortical neurons is likely to be mediated by GABA (*54*).

The hyperpolarization of pyramidal tract neurons by GABA results from an increase in membrane conductance similar to that produced by synaptic inhibition of these neurons following surface stimulation (*79, 163*), and GABA release from the cortex is promoted by such surface stimulation (*127*). The action of GABA on, and the synaptic inhibition of pyramidal tract neurons are antagonized by bicuculline and unaffected by strychnine (*50, 54*).

Hippocampus GABA may also be an inhibitory transmitter mediating basket cell inhibition of pyramidal tract neurons in the hippocampus.

GAD I levels in the hippocampus (areas CA1 and CA2) are not influenced by lesions in any of the known input pathways to this area, indicating that GAD I is localized to intrinsic cells of the hippocampus (*240*). Subcellular distribution studies indicate that most of the GAD I is in nerve endings. The regional distribution of GAD I activity is bimodal; i.e., one region of high GAD I activity is found in the cellular layer which contains the pyramidal cell bodies, and a second is in the molecular layer (*240*). GABA-T has a similar regional distribution (*207*). GABA uptake is concentrated in clusters around pyramidal cell bodies (*109*), and about 40% of the total number of synaptosomes accounts for the GABA taken up by hippocampal homogenates (*125*).

The strychnine-insensitive basket cell inhibition of hippocampal pyramidal neurons is antagonized by microelectrophoretic bicuculline in doses that block the depressant action of GABA, but not that of glycine, on the firing of these neurons (*55*). It is not known if GABA hyperpolarizes hippocampal neurons.

Spinal Cord: GABA and "Presynaptic Inhibition" GABA and GAD I are found in higher concentrations in the dorsal areas of spinal gray matter than in the ventral gray matter or in the white matter (*98, 99, 186*). GABA uptake in the spinal cord (*126*) appears to be mediated by a unique population of nerve terminals which constitutes about 25% of the total and is different from the 25% that accumulates glycine (*125*).

GABA hyperpolarizes spinal motoneurons (*59*) and interneurons (*37*) in a manner similar to spinal postsynaptic inhibition, which involves permeability changes to chloride and probably also to potassium ions. Bicuculline and picrotoxin block the action of GABA on spinal interneurons and Renshaw cells (*50, 83*). The action of GABA on spinal neurons is potentiated by *p*-chloromercuriphenylsulfonate (*52*), a mercurial that inhibits the uptake of GABA by slices of cat spinal cord (*13*).

Although many inhibitory effects in the spinal cord can be antagonized by strychnine (*45*), the long latency and prolonged inhibition (often exceeding 100 msec) of extensor motoneurons that follow repetitive stimulation of muscle and cutaneous afferent fibers are reduced by bicuculline and picrotoxin (*49, 120, 153, 171, 222*). This prolonged inhibition is frequently referred to as "presynaptic" inhibition because of 1) the observed lack of changes in the excitability of the postsynaptic membrane and 2) the failure to detect a membrane hyperpolarization associated with this type of effect. Some evidence has been obtained, however, for such postsynaptic effects (*44*), and it is possible that "presynaptic" inhibition in the spinal cord (and in the cuneate nucleus) may involve: 1) conventional postsynaptic hyperpolarization by GABA at axodendritic synapses, 2) hyperpolarization by GABA at axoaxonic synapses on excitatory terminals, and 3) depolarization of such terminals by GABA, and perhaps by potassium ions, leading to a reduced release of excitatory transmitter (*50*). (See Ch. 1.) GABA is known to depolarize a number of neural structures, including autonomic ganglia (*76*) and sensory ganglion cells (*75*), and this action is abolished by bicuculline and picrotoxin.

Depletion of GABA levels in the spinal cord by treatment with semicarbazide is correlated with decreased "presynaptic inhibition" (*23*). It is not known if this treatment alters the spinal levels of any other amino acid. Exogenous GABA is released from slices of rat spinal cord by electrical stimulation (*104, 112*).

Compounds Structurally Related to GABA

Diverse compounds that may occur as natural, if minor, constituents of normal central nervous tissue are structurally related to GABA and may

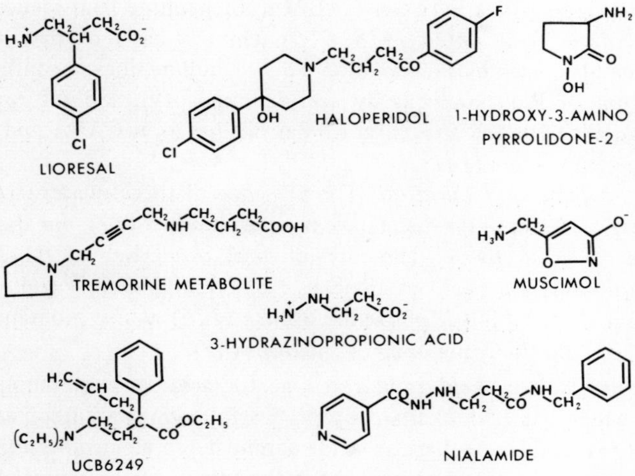

Figure 7. Centrally active drugs with structural similarities to GABA.

play as yet unknown roles in the synaptic functions of GABA. A variety of drugs (Figure 7) that act on the CNS show structural similarity to GABA, and some of them may act on central synapses, using GABA as an inhibitory transmitter.

γ-Amino-β-hydroxybutyric Acid (GABOB) The occurrence of GABOB in the mammalian CNS has been disputed (*267*), although its betaine, carnitine (see below), undoubtedly occurs in the brain. GABOB is not a substrate for GABA-T (*21, 243*) and can be only a relatively weak substrate for the GABA uptake system (*126*). GABOB is a somewhat weaker depressant of the firing of central neurons than is GABA, and its action is antagonized by bicuculline but not by strychnine (*49, 50*).

γ-Aminobutyrylcholine (GABACh) GABACh has been found in the brain and is reported to have depressant effects on the electrical and convulsive activity of the cerebral cortex (*141*). Topically administered GABACh does not appear to have a reproducible effect, however, on the electrical responses of the cortex and, in contrast to the powerful depressant action of GABA, electrophoretically administered GABACh has little or no depressant action on the firing of spinal interneurons or neurons of the cerebral cortex. GABACh influences the firing of Renshaw cells in the spinal cord, and these and other observations suggest that the pharmacological actions of GABACh in the CNS are related to cholinergic systems rather than to those involving GABA. GABACh is a weak inhibitor of GABA uptake (*13, 22*).

Structural similarities between GABACh, bicuculline, and bicuculline methiodide have been noted (*118, 208*). On the basis of this model acetylcholine also resembles bicuculline, yet bicuculline does not influence the excitation of Renshaw cells by acetylcholine (*49*), and the original model (Figure 6) for the structural similarities between GABA and bicuculline is to be preferred (*141*).

γ-Butyrobetaine and Carnitine The presence of these quaternary ammonium derivatives in brain tissue have been established (*17*), but they are not unique to brain tissue. The physiological properties of these and related compounds are best described as "acetylcholine-like" (*86, 113*). Carnitine is a weak inhibitor of GABA uptake (*126*); γ-butyrobetaine has no direct action on the firing of spinal neurons (*62*).

Imidazole-4-acetic Acid Imidazole-4-acetic acid is a metabolite of histamine, which has CNS excitant properties following parenteral administration (*178*). Imidazole-4-acetic acid, administered electrophoretically, is a depressant of the firing of neurons in the spinal cord, cerebral cortex, and medulla and is comparable to GABA in potency, and this action is antagonized by bicuculline (*50, 115, 206*). It is a weak inhibitor of GABA uptake into slices of cat spinal cord (*13*) and of rat brain GAD I (*232*).

L-*2,4-Diaminobutyric Acid* The "neurotoxic" amino acid, L-2,4-diaminobutyric acid ("α-amino-GABA") (*200*) is found in very low concentrations in brain (*190*). It is a weak inhibitor of the firing of spinal neurons (*62*) and a comparatively strong noncompetitive inhibitor of GABA uptake in rat brain slices (*126*).

GABA-Peptides Homocarnosine (GABA-histidine) is found in low concentrations in brain (*266*). It has an effect neither on the spontaneous electrical activity nor on the responsiveness of single neurons in the brainstem to GABA (*258*). Furthermore, homocarnosine does not influence GABA uptake (*126*).

Homoanserine (GABA-1-methylhistidine) has been detected in brain (*189*) together with the enzyme carnosine-N-methyltransferase, which catalyzes the synthesis of homoanserine from GABA and 1-methylhistidine. α-(γ-Aminobutyryl)-lysine has also been detected in brain (*188, 205, 266*).

The coupling of pantoic acid and GABA to form homopantothenic acid is catalyzed by brain extracts (*78*), but this product has yet to be isolated as a natural constituent of brain. The physiological significance of these GABA-peptides is unknown.

γ-Guanidinobutyric Acid This substance is present in the brain as well as in many other mammalian organs. It is a convulsant and has an

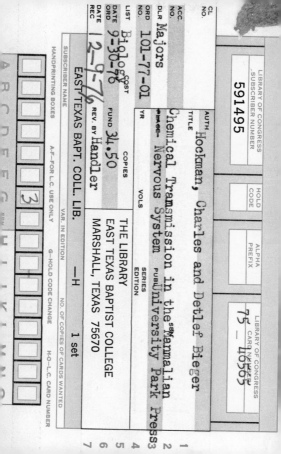

excitatory action when applied topically to the cerebral cortex (136). When applied electrophoretically, it has no direct action either on the firing of spinal neurons (62) or on the action of GABA on these neurons (58).

γ-*Hydroxybutyric Acid* The presence of γ-hydroxybutyric acid in brain has been demonstrated, as has the formation of this substance in small amounts from labeled GABA (74, 216). γ-Hydroxybutyric acid has a strong depressant action on the CNS, which may be attributable to an increase in the activity of the pentose phosphate pathway in the brain (247). Administration of γ-hydroxybutyric acid also leads to an elevation of brain DA levels, perhaps because of an inhibition of DA release (217; see Ch. 5). It does not influence GABA uptake in rat brain slices (126) or the action of GABA on spinal neurons (58).

It is likely that the CNS depressant effects of 1,4-butanediol and of γ-butyrolactone are mediated through metabolism to γ-hydroxybutyric acid (181).

Muscimol This isoxazole is a psychotomimetic compound isolated from the mushroom *Amanita muscaria* (84, 271). It bears remarkable structural similarity to GABA and, like GABA, is a powerful inhibitor of the firing of central neurons when administered electrophoretically (142). This depressant action is antagonized by bicuculline and not by strychnine (49, 50). Unlike GABA, however, muscimol exerts pronounced central effects when administered systemically (0.02 μmol/g i.p.) to healthy adult mammals (223, 253, 271). Muscimol is unlikely to be a substrate for GABA-T (21), for it is a weak noncompetitive inhibitor of GABA uptake and thus unlikely to be a substrate for the GABA uptake system; inefficient removal of muscimol from the environment of GABA postsynaptic receptors may be an important factor contributing to the central effects of this psychotomimetic agent (139).

Haloperidol Haloperidol and other antipsychotic butyrophenones show structural similarities to GABA (131), but their central effects have yet to be related to actions on GABA-mediated synaptic inhibition. Haloperidol is an inhibitor of GABA uptake in brain slices (126), but this action is not specific since haloperidol also inhibits many other transport systems (144).

β-(*p-Chlorophenyl*)-γ-*aminobutyric Acid* This GABA derivative (Lioresal, Ciba-Geigy) is active orally in the treatment of human spasticity (40). It has no effect on GABA uptake (22) or on the firing of medullary reticular neurons (117).

GABA Esters While GABA does not readily pass the blood-brain barrier to produce central effects, certain esters, the octyl ester in particu-

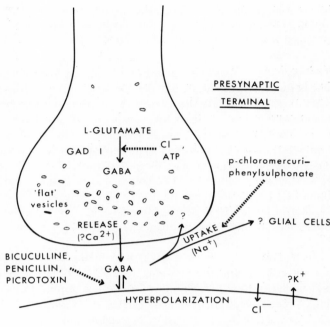

Figure 8. GABA synapse. The operation of a GABA synapse may be summarized as follows: 1) GABA is synthesized in presynaptic terminal by action of GAD I on L-glutamate; this can be inhibited by chloride, ATP, and carbonyl trapping reagents; 2) presynaptic terminal contains large numbers of "flat" vesicles which may bind GABA; 3) following stimulation of presynaptic terminal, GABA is released into synaptic cleft by process that may be calcium dependent; 4) GABA in cleft interacts with postsynaptic neuron, causing membrane hyperpolarization, inward flow of chloride ions, and probably outward flow of potassium ions; this interaction may be antagonized by bicuculline, penicillin, and picrotoxin; 5) extracellular GABA may be taken up into presynaptic terminal and perhaps also into neighboring glial cells by sodium-dependent, structurally specific transport system, which can be inhibited by p-chloromercuriphenylsulfonate; this transport system may maintain low extracellular concentrations of GABA in synaptic cleft and may reconstitute presynaptic stores of transmitter.

lar, suppress motor activity and protect against drug-induced convulsions (26).

1-Hydroxy-3-amino-pyrrolidone-2 This compound has a CNS depressant action, and some of its possible metabolites are structurally related to GABA (32) (See also Chs. 3 and 5.) Its central action is, however, more likely to result from a direct effect on central excitatory transmission, since it antagonizes the excitation of central neurons induced by L-glutamate and L-aspartate (*70*). Some other derivatives of pyrrolidone-2 depress central nystagmus (*95*).

Other Drugs Other centrally active drugs structurally related to GABA include: 1) an active metabolite of tremorine, *N*-(4-pyrrolidino-2-butynyl)-γ-aminobutyric acid (*103*); 2) an antidepressant, ethyl 2-allyl-2-phenyl-4-diethylaminobutyrate (*94*); and 3) a depressant, 3-hydrazinopropionic acid which, in vitro, is the most powerful inhibitor of GABA-T known, and which is likely to act in vivo by raising central GABA levels (*265*). It is a weak inhibitor of other transaminating enzymes (*147, 265*), a weak depressant of the firing of central neurons (*52*), and a noncompetitive inhibitor of GABA uptake (*126*). The monoamine oxidase inhibitor, nialamide, is a diamide derivative of hydrazinopropionic acid (Figure 7).

GLYCINE

The idea that glycine, the simplest amino acid and a constituent of all cells, is an inhibitory synaptic transmitter dates from the observation by Aprison and Werman (*7*) that the ventral gray matter of the spinal cord contains unusually high levels of glycine. Prior to this observation glycine was considered to have merely a GABA-like depressant action on the firing of central neurons. In 1965 Aprison and Werman (*7*) suggested that glycine was worthy of consideration as a synaptic transmitter, and supporting data for their proposal soon followed. Glycine administered electrophoretically hyperpolarized spinal motoneurons in a manner similar to spinal postsynaptic inhibition (*275*), and an association was demonstrated between glycine and spinal interneurons (*65*). Furthermore, the long-known antagonism of certain spinal postsynaptic inhibitions by strychnine was interpreted on the basis of an antagonism of the postsynaptic action of glycine (*57*). The evidence in favor of glycine as a spinal inhibitory transmitter is now substantial, but "release" studies on endogenous glycine are needed. Numerous reviews concerning the inhibitory synaptic function of glycine have appeared in recent years (*5, 8, 30, 60, 105, 273*).

Glycine is a zwitterion at neutral pH with pK values of 2.33 (proton gain) and 9.53 (proton loss) at 35° (*157*). Glycine in the solid state is polymorphic, three forms having been recognized and their X-ray crystal structures determined (*119, 121, 122*). The rotational isomerism of glycine in solution has been studied by proton magnetic resonance spectroscopy (*42*) and has also been the subject of potential energy calculations (*209*).

Neurochemistry

Levels in the CNS In the cat and rat CNS, levels of glycine in excess of 3 µmol/g are encountered only in the medulla and spinal cord (*6, 7, 16, 99, 138, 152, 226, 227*). Within the spinal cord, glycine levels are higher in the lumbar and cervical enlargements and in the sacral cord than in the thoracic cord (*6*). Within the lumbosacral cord, glycine levels are higher in the gray than in the white matter, and higher in the ventral than in the dorsal regions of both gray and white matter (*6, 7, 99, 138*).

Temporary aortic occlusion reduces the number of interneurons in the central gray matter of the spinal cord and diminishes polysynaptic excitatory reflexes (*257*). Following such treatment spinal levels of GABA and glutamine remain unchanged, whereas the levels of glycine, aspartate, and glutamate are reduced, the reduction in glycine and aspartate being correlated with the interneuron count (*65*). These findings suggest an association between glycine, aspartate, and the interneurons destroyed by ischemia.

There is evidence that glycine crosses the blood-brain barrier after systemic administration (13 µmol/g i.p.) to raise the central levels of glycine and serine (*212*). The flux of glycine from the blood into the rat CNS is of the order of magnitude of 0.1 µmol/g/hr (*226*). This flux is 10 to 60 times higher in immature rats (*14*). Glycine has the lowest cerebrospinal fluid (CSF) to plasma ratio of the amino acids found in cat CSF, and this may be accounted for by a mediated transport of glycine from the CSF to the blood (*187*). Peripherally administered glycine relieves experimental hind limb rigidity (*238*), raises the electroshock threshold and potentiates barbiturate anesthesia (*161*), protects against hyperbaric oxygen-induced and drug-induced convulsions (*236, 278*), and enhances ethanol-induced sleeping time (*31*).

At least two types of hyperglycinemia exist. The first is characterized by acidosis and respiratory distress. The second, in which glycine is the only amino acid the level of which is elevated in the blood, is characterized by seizures beginning early in infancy, and, on the basis of decreased urinary oxalic acid excretion, it appears that the conversion of glycine to glyoxylate may be defective (*25*).

Metabolism Glycine can be synthesized in nervous tissue from glucose. The main pathways appear to be through D-3-phosphoglycerate and D-glycerate, and two observations suggest that the "nonphosphorylated" pathway is the more important. First, an extracts of rat cerebral cortex, the enzyme D-glycerate dehydrogenase (D-glycerate:NAD oxidoreductase, EC

1.1.1.29) is inhibited in a noncompetitive manner by glycine, suggestive of end-product inhibition, whereas D-3-phosphoglycerate dehydrogenase is inhibited by glycine only to a very small extent (*261*). Second, the regional distribution of D-glycerate dehydrogenase activity in the cat CNS, but not that of D-3-phosphoglycerate dehydrogenase, correlates closely with the regional distribution of glycine (*262*). This indicates that the rate-controlling step in glycine metabolism may be the conversion of D-glycerate to hydroxypyruvate. The "nonphosphorylated" pathway to glycine from 2-phospho-D-glycerate may involve the steps illustrated in Figure 9.

Although the route from hydroxypyruvate to glycine is not yet established, two major possibilities exist. The more obvious route would be that via serine by the enzymes hydroxypyruvate:α-alanine (glutamate/glutamine) aminotransferase and serine hydroxymethyltransferase (L-serine: tetrahydrofolate 5,10-hydroxymethyltransferase, EC 2.1.2.1). The required aminotransferase activity is, however, very low or absent in brain (*270*). Serine hydroxymethyltransferase, which catalyzes the interconversion of serine and glycine, has been purified from brain (*35*), and aspects of its development (*34*) and regional and subcellular distribution in CNS tissue have been studied (*69*). Glycine and serine are readily interconvertible in vivo in the CNS, isotopically labeled glycine being more rapidly

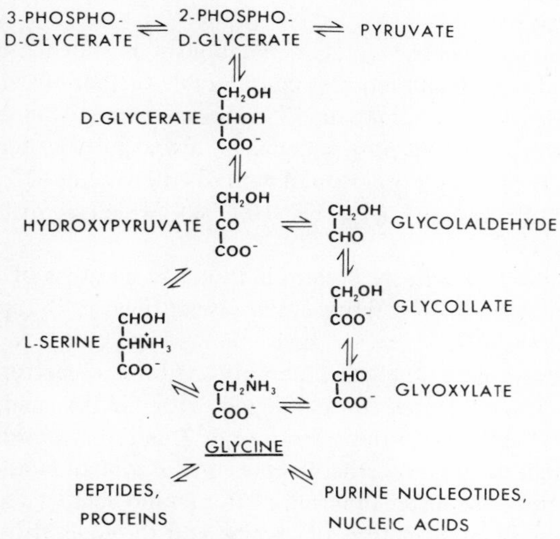

Figure 9. Possible nonphosphorylated pathways of central glycine metabolism.

and to a much greater extent incorporated into serine than the label from serine is incorporated into glycine (*226*). Detailed precursor-product experiments between serine and glycine have yet to be carried out.

It is possible that the route from hydroxypyruvate to glycine may not involve serine. The nonoxidative decarboxylation of hydroxypyruvate to glycolaldehyde by extracts of beef brain has been reported (*106*). Oxidation of glycolaldehyde to glyoxylate via glycollate and subsequent transamination would result in the production of glycine. Glycine: 2-oxoglutarate aminotransferase (EC 2.6.1.4), the required transaminating activity producing glycine from glyoxylate and L-glutamate, has been described in cat and rat CNS tissue, and other amino group donors can also be used (*24, 146, 148*). These transaminase activities are essentially glycine-forming; the reverse reaction from glycine to glyoxylate can be catalyzed by D-amino acid oxidase (*77*), but because of the low activity of this enzyme in the spinal cord, it is considered to be relatively unimportant in central glycine metabolism.

As in other tissues, glycine is incorporated into proteins, various peptides such as glutathione, and purine nucleotides. Several peptide hydrolases capable of metabolizing glycine peptides have been demonstrated in CNS tissue (*179*). Although glycine can be catabolized by oxidative decarboxylation (*280*), it appears that glycine in nervous tissue is not readily converted into intermediates of the tricarboxylic acid cycle or of glycolysis (*5*).

In summary central glycine metabolism is much more complicated than that of GABA and requires considerable further investigation.

Uptake Glycine uptake in CNS tissue is mediated by at least two kinetically distinct transport systems: a "low affinity system" (K_m ca. 10^{-4} M), which is shared by other small neutral amino acids (*28, 29, 233*), and a "high affinity system" (K_m ca. 10^{-5} M), which is specific for glycine (*144, 173, 191*).

While the low affinity system is found in all areas of the CNS and in other tissues, the high affinity system is confined to the spinal cord, pons, and medulla (*144*). These are areas where: 1) glycine is found in relatively high concentrations; 2) glycine, administered microelectrophoretically, is a potent neuronal depressant compared with GABA; and 3) strychnine-sensitive synaptic inhibitions are found. This set of observations suggests that the high affinity specific glycine uptake system is associated in some way with the transmitter function of this amino acid.

The postsynaptic action of glycine and the high affinity transport of glycine are clearly different processes, since strychnine antagonizes the postsynaptic action and does not influence the transport system (*13, 59, 144, 191*). In contrast *p*-chloromercuriphenylsulfonate antagonizes the transport system (*13, 144*) and potentiates the observed postsynaptic action (*52*).

A likely function of the high affinity transport system, in addition to maintaining low extracellular concentrations of glycine, is to recover glycine from the synaptic environment after release and postsynaptic action in order to reconstitute the presynaptic store(s) of the transmitter. Such a recovery system would serve a useful purpose since glycine metabolism appears to be comparatively slow.

There is support for a presynaptic location for the high affinity specific glycine transport system. Density-gradient centrifugation experiments indicate that the particles responsible for the uptake of glycine in homogenates of rat spinal cord include a population of synaptosomes that can be partially separated from those accumulating GABA (*126*) and glutamate (*9*). Autoradiographic studies also indicate that glycine and GABA are taken up into different presynaptic terminals in rat spinal cord (*125*), and that the terminals accumulating glycine may be characterized by containing "flat," as opposed to spherical, vesicles (*180*). Although nerve terminals appear to be the principal site of high affinity glycine uptake in the rat spinal cord, uptake into glial components is also observed (*108*).

As with GABA uptake, the structural specificity of the high affinity glycine uptake system ensures that other metabolites are not transported into what may be described as the "glycine-transmitter pool(s)." High affinity glycine uptake into slices of cat and rat spinal cord is not inhibited by high concentrations of GABA or of any of the other depressant amino acids (α-alanine, cystathionine, and serine) found in appreciable amounts in spinal tissue (*13, 144*).

A variety of centrally acting drugs inhibits the high affinity specific glycine uptake system, including chlorpromazine, imipramine, and haloperidol (*144*), but none of them is particularly specific. *p*-Chloromercuriphenylsulfonate inhibits the uptake of glycine and GABA into slices of cat spinal cord to a similar extent but is much less effective in inhibiting the uptake of L-aspartate and L-glutamate (*13*). This correlates closely with the findings that this mercurial, administered microelectrophoretically, potentiates the depressant actions of glycine and GABA on the firing of cat spinal interneurons much more readily than it potentiates the excitant actions of L-aspartate and L-glutamate on these neurons (*52*). These observations suggest that uptake normally limits the potency of microelectrophoretically applied amino acids. This is not to say that uptake contributes to the inactivation of synaptically released transmitters. Perhaps owing to technical difficulties or low potency, or both, *p*-chloromercuriphenylsulfonate appears not to affect synaptic inhibition (*52*). Until strychnine-sensitive spinal inhibition can be shown to be prolonged by an agent that also modifies the time course of action of applied glycine, there is insufficient evidence for *or* against the proposal that uptake significantly contributes to the inactivation of synaptically released glycine.

Release The release of endogenous glycine in vivo, following stimulation of inhibitory pathways, has yet to be demonstrated. Exogenous glycine is released from the cat spinal cord, in vivo in response to peripheral nerve stimulation, into a perfusate of the central canal (*149*), containing *p*-hydroxymercuribenzoate, an inhibitor of glycine uptake by both the low and high affinity systems. Endogenous acetylcholine was released under similar conditions when eserine was substituted for *p*-hydroxymercuribenzoate.

In vitro the release of exogenous glycine has been demonstrated from slices of rat spinal cord following suitable electrical stimulation or treatment with high potassium concentrations (*104, 112*). There is some evidence that this evoked release is calcium dependent. Evoked release of exogenous glycine in vitro from the hemisected toad spinal cord (*4, 183*) is also reported.

Neuropharmacology

Postsynaptic Action Glycine, administered microelectrophoretically, depresses the firing of neurons throughout the mammalian CNS, but its potency as a depressant relative to that of GABA varies considerably, perhaps reflecting the relative number of synapses at which glycine and GABA may function as inhibitory transmitters. In general, glycine is a more potent depressant than GABA of spinal interneurons and motoneurons, bulbar reticular neurons, and neurons in the red nucleus; glycine and GABA are approximately equipotent on spinal Renshaw cells and cuneate neurons; and glycine is less potent than GABA on neurons in the cerebral cortex, Deiters' nucleus, and medial geniculate nucleus (*38, 58, 59, 71, 92, 116, 154, 252*).

Intracellular studies of the mechanism of the glycine-induced depression of neuronal firing have been carried out on spinal interneurons and motoneurons, on Deiters' neurons, and on neurons in the cerebral cortex

(*37, 38, 59, 154, 276*). Glycine hyperpolarizes these neurons. Onset of hyperpolarization is rapid, readily reversible, and accompanied by an increase in membrane conductance. Glycine also hyperpolarizes spinal neurons grown in tissue culture, causing an increase in membrane conductance (*114*). The ionic basis for the glycine-induced hyperpolarization of spinal neurons and of Deiters' neurons is attributable to an increased permeability to chloride ions (and perhaps also to potassium ions) and cannot be distinguished from synaptically induced hyperpolarization by measurements of reversal potentials and the effects of intracellular injections of a series of anions and cations. It should be emphasized that on the basis of these experiments, it is impossible to distinguish the action of glycine on spinal and Deiters' neurons from that of GABA, β-alanine, or δ-aminovaleric acid. The action of glycine on cerebral neurons, however, appeared to differ somewhat from that of GABA, although the low potency of glycine on these neurons makes interpretation of these results difficult.

Antagonists A variety of compounds (Figure 10) are known to antagonize the depressant action of glycine on spinal neurons and to antagonize certain spinal synaptic inhibitions. Strychnine is the most potent of these compounds and the most widely investigated.

Strychnine Systemic or microelectrophoretic administration of relatively low concentrations of this convulsant indole alkaloid reversibly depresses certain postsynaptic inhibitions in the spinal cord without modifying synaptic excitation (*45*), and reversibly abolishes the postsynaptic action of microelectrophoretically administered glycine on spinal neurons without influencing that of GABA, norepinephrine, or dopamine (*53, 57, 58*).

Strychnine blocks the depressant action of glycine on neurons throughout the CNS, having little or no action on GABA-induced depression. Strychnine also antagonizes the actions of α- and β-alanine, cystathionine, serine, and taurine, which on this basis have been called "glycine-like" amino acids, but does not influence the actions of "GABA-like" amino acids, γ-amino-β-hydroxybutyric acid, 3-aminopropanesulfonic acid, δ-aminovaleric acid, and ϵ-aminocaproic acid (*58*). This remarkably clearcut division of amino acids on the basis of strychnine antagonism is also supported on the basis of bicuculline antagonism, although there are some regions of the CNS where the depressant actions of taurine and β-alanine are antagonized by both bicuculline and strychnine (*49, 50, 61*). The antagonism of glycine by strychnine may be competitive (*53, 137*), but technical difficulties prevent definite conclusions from being made.

Strychnine is reported to inhibit competitively the binding of glycine to a membrane fraction isolated from rat brain (263).

Strychnine antagonizes the action of glycine on spinal motoneurons, both the hyperpolarization and the depolarization after intracellular chloride administration: this and the lack of effect of strychnine on GABA-induced hyperpolarization suggest that strychnine antagonizes the interaction of glycine with its postsynaptic receptors and not the subsequent ionic movements (59).

The central synaptic inhibitions antagonized by strychnine in the cat include: in the spinal cord, motoneurons (direct, polysynaptic, and recurrent inhibition by impulses in segmental afferents; inhibition by impulses from higher centers); and Renshaw cells (inhibition by impulses from hind paw and inhibition by stimulating medullary reticular formation). In the medulla they include inhibition of hypoglossal and masseter motoneurons by impulses in segmental afferents, and inhibition of medullary reticular neurons by recurrent inhibition and by impulses from medial pontine reticular formation. As with GABA and bicuculline-sensitive inhibitions it is to be emphasized that glycine is, *in the absence of other evidence,* not to be equated with the transmitter(s) mediating these inhibitions. Nevertheless of the strychnine-sensitive depressant amino acids found in the spinal cord (138), glycine is: present in the highest concentration, the most powerful depressant, and the only one for which a specific "high affinity" uptake system has been described.

Other Convulsant Indoles A variety of indoles, structurally related to strychnine, all of which block certain spinal postsynaptic inhibitions, are glycine antagonists having little or no action on GABA (58) (Curtis, Duggan, and Johnston, unpublished observations). The relative potencies of these substances, which include brucine, diaboline, the Wieland-Gümlich aldehyde, and some reduction products of strychnine (211), in blocking glycine and the synaptic processes are similar.

Gelsemine is also an antagonist of glycine and of certain spinal inhibitions (49).

Morphine Alkaloids Thebaine is the most powerful convulsant of this group of alkaloids. In comparatively large doses in excess of the usual analgesic doses, morphine and codeine are also convulsants. These three alkaloids and the synthetic compound levorphan are antagonists of various strychnine-sensitive spinal inhibitions and antagonists of the depressant action of glycine, with little or no effect on that of GABA (47, 58) (Curtis, Duggan, and Johnston, unpublished observations).

Tetrahydroisoquinoline Alkaloids The convulsant tetrahydroisoquinoline, laudanosine (182), is a glycine antagonist (49).

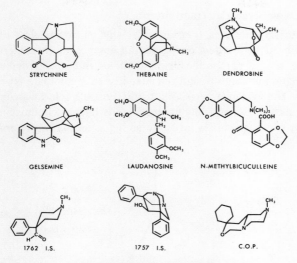

Figure 10. Glycine antagonists.

Dendrobine The convulsant alkaloid from the Chinese medicinal herb Chih-shih-hu, dendrobine *(43)*, shows a remarkable structural similarity to picrotoxinin and a similar potency as a convulsant. It is, however, a glycine antagonist with no action on GABA *(49)*.

Synthetic Strychnine-like Convulsants The following synthetic compounds are all glycine antagonists: 4-phenyl-4-formyl-N-methylpiperidine (1762 I.S.); 5,7-diphenyl-1,3-diaza-adamantan-6-ol (1757 I.S.); hexahydro-2'-methylspiro [cyclohexane-1,8' (6H)-oxazino (3,4A) pyrazine] (C.O.P.); and N-methylbicuculleine *(58, 140)*.

Tetanus Toxin The postsynaptic inhibition of spinal motoneurons and interneurons resulting from impulses in a variety of afferent pathways is blocked by tetanus toxin *(45)*. Since suppression of the synaptic inhibition of Renshaw cells by tetanus toxin is not accompanied by antagonism of the depressant effects of either glycine or GABA on these neurons *(46)*, it seems likely that the toxin acts presynaptically. This effect of tetanus toxin may result from a reduction in the amount of inhibitory transmitter available for release or from an interference with the actual release process. The latter is more likely, since the toxin does not appreciably reduce spinal glycine levels *(87, 143)*.

Glycine injected intraperitoneally into rats can temporarily relieve tetanus spasticity *(87)*.

Synapses Using Glycine as Transmitter

The case for glycine as a transmitter at any particular class of synapse leaves much to be desired owing to the absence of meaningful release studies, the lack of precise localization studies, our poor understanding of central glycine metabolism, and our inability to manipulate glycine levels selectively. The evidence pertaining to the possible role of glycine at any particular class of synapse at present is thus largely circumstantial. The synapses at which glycine is probably acting as an inhibitory transmitter,

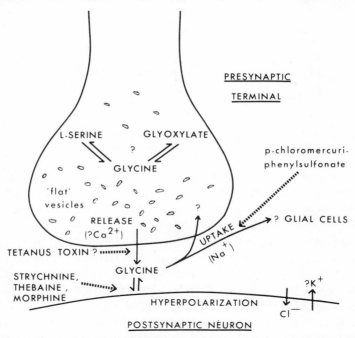

Figure 11. Glycine synapse. Operation of glycine synapse may be summarized as follows: 1) neither source (s) nor site (s) of glycine synthesis is known with certainty but glycine may be synthesized from L-serine or glyoxylate, or both, in presynaptic terminal; 2) presynaptic terminal contains large numbers of "flat" vesicles; 3) following stimulation of presynaptic terminal, glycine is released into synaptic cleft by process that may be calcium dependent and which may be inhibited by tetanus toxin; 4) glycine in cleft interacts with postsynaptic neuron, causing membrane hyperpolarization, inward flow of chloride ions, and probably outward flow of potassium ions; this interaction may be antagonized by strychnine, thebaine and morphine; 5) extracellular glycine may be taken up into presynaptic terminal and perhaps also into neighboring glial cells by sodium-dependent, structurally specific transport system that can be inhibited by p-chloromercuriphenylsulfonate; this transport system may maintain low extracellular concentrations of glycine, inactivate glycine in synaptic cleft, and reconstitute presynaptic stores of transmitter.

and that appear to be among those most amenable to additional experimental study, are those mediating "direct" inhibition of spinal motoneurons following stimulation of group Ia afferents and those mediating "recurrent" inhibition of motoneurons.

Compounds Structurally Related to Glycine

Other α-amino acids in addition to glycine may act as central synaptic transmitters, since they occur in the CNS and have postsynaptic actions when administered microelectrophoretically. In addition, serine and glyoxylate, as close metabolites of glycine, may play roles in the synaptic function of glycine.

L-Cystathionine The neurochemistry of this and related sulfur amino acids has been reviewed (*90*). Although the occurrence of cystathionine in spinal tissue and its depressant action on spinal interneurons led to the proposal that L-cystathionine is a spinal inhibitory transmitter (*274*), the levels of this amino acid were overestimated by factor of 10 (*273*).

The cystathionine levels in the human brain vary widely and the factors governing this variability remain to be elucidated (*230, 250*). The levels of cystathionine in tissue obtained at autopsy are higher than in tissue obtained by biopsy (*205*). High levels of cystathionine have been found in the rat cerebellum (*152, 227*) and in the bovine pineal body (*167*). In the cat spinal cord, cystathionine is found in equal amounts in the gray and white matter in a form not efficiently extracted by aqueous ethanol (*138*).

A selective elevation of cystathionine levels occurs in cystathioninuria, a disease characterized by mental retardation (*228*). The distribution of cystathionine in the tissues of the pyridoxine-deficient rat closely resembles that in the cystathioninuric mental defective patient, suggesting that the genetic defect in cystathioninuria may be a high requirement of the cystathionine-degrading enzyme, cystathioninase, for pyridoxal phosphate (*111, 155*). In homocystinuria there is a deficiency of cystathionine in the brain; this disease is characterized by convulsive seizures and mental retardation (*33*).

L-Cystathionine, the natural isomer (*251*), has a relatively weak, strychnine-sensitive, depressant action on the firing of spinal neurons, weaker than that of a mixture of the four possible isomers (*58*). Intraventricular injection influences the EEG and the onset of sleep (*155*). Intravenously administered L-cystathionine is not transported into the brain, but metabolic degradation products of L-cystathionine do penetrate the blood-brain barrier and are incorporated into brain protein (*36*).

L-α-Alanine L-α-Alanine is present in moderate amounts throughout the mammalian CNS and shows little regional variation (*15, 226, 227, 250*). When it is applied microelectrophoretically, it is a somewhat weaker, strychnine-sensitive, depressant of the firing of spinal neurons than is glycine, but stronger than L-cystathionine and L-serine (*58*); L-α-alanine does not inhibit the high affinity uptake of glycine by slices of cat and rat spinal cord (*13, 144*), and it is itself taken up by a "low affinity" system (K_m ca. 5×10^{-4} M) into slices and homogenates of rat cerebral cortex and spinal cord (*144, 173, 233*).

Systemically administered L-α-alanine (11 μmol/g i.p.) crosses the blood-brain barrier to raise the central levels of this amino acid and also of glycine (*212*), suggesting that it can promote glycine metabolism perhaps by sparing intermediates in glycolysis, such as pyruvate.

Exogenous L-α-alanine is not released from slices of rat spinal cord by electrical stimulation which results in release of exogenous glycine (*112*).

L-Serine In the rat CNS, L-serine is found in lower concentrations than is glycine in all areas other than the telencephalon, where L-serine levels are higher than those of glycine, and the cerebellum, where L-serine and glycine are found in approximately equal concentration (*226, 227*); L-serine and glycine are interconvertible in the CNS and the possible function of L-serine in central glycine metabolism has been discussed earlier. L-Serine is also concerned with L-cystathionine metabolism (*90*).

Systemically administered glycine (13 μmol/g i.p.) raises the central levels of serine (*212*), and systemically administered serine in a dose of 9 μmol/g i.p. enhances ethanol-induced sleeping time (*31*). The rate of entry of serine from the blood into the brain of immature rats is 40 to 120 times that of mature rats (*14*).

L-Serine, applied microelectrophoretically, is a weak, strychnine-sensitive depressant of the firing of spinal neurons (*58*). It does not influence the high affinity uptake of glycine by slices of cat and rat spinal cord (*13, 144*) and is itself taken up by a "low affinity" system (K_m ca. 6×10^{-4} M) into slices and homogenates of rat cord (*144, 173*).

Exogenous L-serine is not released from hemisected toad spinal cord following electrical stimulation which releases exogenous glycine (*183*).

Glyoxylate Glyoxylate may serve as a glycine precursor in the CNS (*146, 147*). The levels of glyoxylate in the brain are normally very low but rise more than 100-fold during thiamine deficiency (*172*); it is not known if glycine levels are altered appreciably under these conditions.

When administered microelectrophoretically, glyoxylate is without direct action on the firing of spinal neurons (*58*), is toxic to mammals, and produces renal failure (*215*). The related glycollate (Figure 10) is less toxic, antagonizes the convulsant action of strychnine, and potentiates the depressant action of γ-hydroxybutyric acid (*168*).

The sources of glyoxylate in mammals deserve further investigation (*219*), particularly in view of the postulated role of glyoxylate in the control of the tricarboxylic acid cycle (*93, 218*).

TAURINE

Taurine, a nonprotein, sulfur-containing, amino acid, may also be an inhibitory synaptic transmitter in the CNS. It depresses the firing of central neurons, its synthesizing enzyme is associated with isolated nerve endings, it is taken up into brain slices by a "high affinity" process, and it is released from the brain in vivo following stimulation.

Aspects of taurine biochemistry, pharmacology, and physiology have been reviewed recently (*90, 130, 150, 256*).

Taurine is a zwitterion at neutral pH with pK values of ca. 1.5 (proton gain) and 8.82 (proton loss) at 35° (*2, 158*). In the solid state, taurine is present in an extended form with the protonated amino and sulfonate groups in the *gauche* conformation (*199*).

Neurochemistry

Levels in CNS Taurine is found in varying concentrations throughout the nervous system but, unlike GABA and glycine, higher levels of taurine are found in tissues other than the CNS, particularly in the heart (*10*). In the cat CNS, relatively high levels are found in the lateral geniculate nucleus (*100*). In the rat brain, higher levels are found in the cerebrum and cerebellum than in the midbrain, pons, and medulla (*1, 152, 227*). In the human brain at autopsy the highest taurine levels are found in the cerebellum (*205*).

Increasing the plasma levels of taurine 100-fold does not significantly elevate the brain levels of this amino acid in adult rats (*169*). Taurine does exchange across the blood-brain barrier, and from pulse-labeling experiments the half-life of peripherally administered taurine in the exchangeable pools in the CNS is of the order of magnitude of 100 min (*169*). The rate of entry of taurine into rat brain from the blood is the same in immature (under two weeks of age) as in mature rats (*14*).

Urinary taurine excretion is low in mongolism (*160*), and a significant positive correlation is found between taurine excretion and two measures of adaptive behavior in mongoloids (*254*).

Metabolism The major route of taurine synthesis in the CNS is probably via decarboxylation of cysteine sulfinic acid to hypotaurine and subsequent oxidation to taurine, although the decarboxylation of cysteic acid to taurine may also be important (*90*).

The enzyme(s) catalyzing the decarboxylation of cysteine sulfinic acid in the brain is associated with synaptosomes in brain homogenates (*1*). Unlike the levels of taurine, the levels of decarboxylase activity increase during postnatal development, and in the adult brain are relatively uniformly distributed (*1*).

Taurine is slowly degraded into isethionic acid in vivo in the rat brain (*204*). It is not a substrate for the GABA-T activity in a crude mitochondrial extract of rat brain and is not transaminated with 2-oxoglutarate as an amino acceptor in such extracts (*21*).

Taurine Uptake Taurine is rapidly taken up by brain slices, and regional variation in uptake has been noted with the highest tissue, medium ratios being obtained in tissue slices from the midbrain and the olfactory bulb (*152, 169*). Taurine uptake is energy dependent, being inhibited by cyanide and ouabain (*15*), and kinetic studies indicate that it is a "high affinity" uptake (K_m ca. 5×10^{-5} M), similar in kinetic properties to those described for glycine and GABA (*71*). The substrate specificity of the system(s) mediating taurine uptake is not known: taurine itself is a relatively weak inhibitor of the high and low affinity glycine uptake systems and the GABA uptake system (*126, 144*), so that its uptake is unlikely to be mediated efficiently by any of these systems.

Release Endogenous taurine is released from the cerebral cortex in vivo as a result of stimulation of the reticular formation, and the rate of release of endogenous taurine is higher during cortical arousal than during sleep (*134*).

Exogenous taurine is released from rat brain slices following electrical stimulation under conditions that result in the release of GABA (*67*).

Neuropharmacology

Postsynaptic Action of Taurine and Taurine Antagonists When tested microelectrophoretically on spinal neurons, taurine has a strychnine-sensitive, bicuculline-insensitive, depressant action, weaker than that of glycine or GABA and comparable in potency to that of L-α-alanine and γ-amino-β-hydroxybutyric acid (*49, 58*). A similar action is observed in the cerebral cortex and the lateral geniculate nucleus, with the important difference that *both* bicuculline and strychnine antagonize this action (*50, 61*). Thus, in the cerebral cortex, taurine would be an excellent candidate for any synaptic inhibitions that could be shown to be antagonized by both bicuculline and strychnine, and it is unlikely to be a transmitter at synapses uniquely sensitive to either of these alkaloids.

Compounds Structurally Related to Taurine

Taurine is 2-aminoethanesulfonic acid. Both the corresponding carboxylic (β-alanine) and phosphonic acids have been found in the CNS in very low concentrations.

β-Alanine Very low levels of β-alanine are found in the CNS (*250, 281*). β-Alanine depresses the firing of central neurons, when administered microelectrophoretically, and is very similar to taurine. In the spinal cord and the lateral vestibular nucleus this action is antagonized by strychnine and unaffected by bicuculline; however, in the thalamus and in the cerebral, cerebellar, and hippocampal cortices this action is antagonized by both alkaloids (*49, 50*).

β-Alanine hyperpolarizes spinal motoneurons in a manner similar to that induced by synaptic inhibition (*59*).

2-Aminoethanephosphonic Acid This compound is found in very low concentrations in the bovine and human brain (*3, 229*); its function in the nervous system is unknown.

CONCLUSION

The evidence has been discussed for the participation of GABA, glycine, and taurine as inhibitory synaptic transmitters in the mammalian CNS. There are cogent reasons for considering GABA to be an established, glycine a highly probable, and taurine a possible central inhibitory transmitter. Although many centrally acting drugs in current use have at least some action on these amino acid transmitter systems, it is not yet apparent whether or not these actions make major contributions to the therapeutic effects of these drugs. Thus, we can look forward to an exciting future concerning the neuropharmacology of these amino acids.

ACKNOWLEDGMENTS

I wish to thank Professor D. R. Curtis, Dr. M. L. Uhr, Dr. A. K. Tebēcis, and Mrs. H. Walsh for their very helpful advice and assistance in the preparation of this chapter.

LITERATURE CITED

1. Agrawal, H. C., Davison, A. N., and Kaczmarek, L. K.: Subcellular distribution of taurine and cysteinesulphinate decarboxylase in developing rat brain. *Biochem. J. 122*:759–763, 1971.

2. Albert, A.: Quantitative studies of the avidity of naturally occurring substances for trace metals. I. Amino-acids having only two ionizing groups. *Biochem. J. 47*:531–538, 1950.
3. Alhadeff, J. A. and Davies, G. D.: Occurrence of acid in human brain. *Biochemistry 9*:4866–4869, 1970.
4. Aprison, M. H.: Evidence of the release of ^{14}C-glycine from hemisectioned toad spinal cord with dorsal root stimulation. *Pharmacologist 12*:222P, 1970.
5. Aprison, M. H., Davidoff, R. A., and Werman, R.: Glycine: Its metabolic and possible transmitter roles in nervous tissue. *Handbook Neurochem. 3*:381–397, 1970.
6. Aprison, M. H., Shank, R. P., and Davidoff, R. A.: A comparison of the concentration of glycine, a transmitter suspect, in different areas of the brain and spinal cord in seven different vertebrates. *Comp. Biochem. Physiol. 28*:1345–1355, 1969.
7. Aprison, M. H. and Werman, R.: The distribution of glycine in cat spinal cord and roots. *Life Sci. 4*:2075–2083, 1965.
8. Aprison, M. H. and Werman, R.: A combined neurochemical and neurophysiological approach to identification of central nervous system transmitters. *Neurosci. Res. 1*:143–174, 1968.
9. Arregui, A., Logan, W. J., and Snyder, S. H.: Specific glycine accumulating synaptosomes in the spinal cord of the rat. *Proc. Nat. Acad. Sci. U. S. A. 69*:3485–3489, 1972.
10. Awapara, J.: The taurine concentration of organs from fed and fasted rats. *J. Biol. Chem. 218*:571–576, 1956.
11. Balazs, R., Dahl, D., and Harwood, J. R.: Subcellular distribution of enzymes of glutamate metabolism in rat brain. *J. Neurochem. 13*:897–905, 1966.
12. Balazs, R., Machiyama, Y., Hammond, B. J., et al.: The operation of the γ-aminobutyrate bypath of the tricarboxylic acid cycle in brain tissue in vitro. *Biochem. J. 116*:445–467, 1970.
13. Balcar, V. J. and Johnston, G. A. R.: High affinity uptake of transmitters: studies on the uptake of L-aspartate, GABA, L-glutamate and glycine in cat spinal cord. *J. Neurochem. 20*:529–539, 1973.
14. Banos, G., Daniel, P. M., Moorehouse, S. R., et al.: The entry of amino acids into the brain of the rat during the postnatal period. *J. Physiol.* (Lond.) *213*:45–46P, 1971.
15. Battistin, L., Grynbaum, A., and Lajtha, A.: Energy dependence of amino acid uptake in brain slices. *Brain Res. 16*:187–197, 1969.
16. Battistin, L., Grynbaum, A., and Lajtha, A.: Distribution and uptake of amino acids in various regions of the cat brain in vitro. *J. Neurochem. 16*:1459–1468, 1969.
17. Baxter, C. F.: The nature of γ-aminobutyric acid. *Handbook Neurochem. 3*:289–353, 1970.
18. Baxter, C. F., Tewari, S., and Raeburn, S.: The possible role of gamma-aminobutyric acid in the synthesis of protein. *Adv. Biochem. Psychopharmacol. 4*:195–216, 1972.

19. Beart, P. M., Curtis, D. R., and Johnston, G. A. R.: 4-Aminotetrolic acid: A new conformationally restricted analogue of γ-aminobutyric acid. *Nature* (New Biol.) *234*:80–81, 1971.
20. Beart, P. M. and Johnston, G. A. R.: Bicuculline and GABA-metabolizing enzymes. *Brain Res. 38*:226–227, 1972.
21. Beart, P. M. and Johnston, G. A. R.: Transamination of analogues of γ-aminobutyric acid by extracts of rat brain mitochondria. *Brain Res. 49*:459–462, 1973.
22. Beart, P. M. and Johnston, G. A. R.: GABA uptake in rat brain slices: Inhibition by GABA analogues. *J. Neurochem. 20*:319–324, 1973.
23. Bell, J. A. and Anderson, E. G.: The influence of semicarbazide-induced depletion of γ-aminobutyric acid on presynaptic inhibition. *Brain Res. 43*:161–169, 1972.
24. Benuck, M., Stern, F., and Lajtha, A.: Regional and subcellular distribution of aminotransferases in rat brain. *J. Neurochem. 19*:949–957, 1972.
25. Berry, H. K.: Hereditary disorders of amino acid metabolism associated with mental deficiency. *Ann. N.Y. Acad. Sci. 166*:66–73, 1969.
26. Bertelli, A., Donati, L., Lami, V., et al.: Gamma-amino acid esters and the CNS. *Int. J. Neuropharmacol. 7*:149–154, 1968.
27. Bisti, S., Iosif, G., Marchesi, G. F., et al. Pharmacological properties of inhibitions in the cerebellar cortex. *Exp. Brain Res. 14*:24–37, 1971.
28. Blasberg, R. G.: Specificity of cerebral amino acid transport: A kinetic analysis. *Prog. Brain Res. 29*:245–256, 1968.
29. Blasberg, R. and Lajtha, A.: Heterogeneity of the mediated transport systems of amino acid uptake in brain. *Brain Res. 1*:86–104, 1966.
30. Bloom, F. E.: Amino acids and polypeptides in neuronal function. *Neurosci. Res. Prog. Bull. 10*:121–251, 1972.
31. Blum, K., Wallace, J. E., and Geller, I.: Synergy of ethanol and putative neurotransmitters: glycine and serine. *Science 176*:292–294, 1972.
32. Bonta, I. L., De Vos, C. J., Grijsen, H., et al.: 1-Hydroxy-3-aminopyrrolidone-2 (HA966): A new GABA-like compound, with potential use in extrapyramidal diseases. *Br. J. Pharmacol. 43*:514–535, 1971.
33. Brenton, D. P., Cusworth, D. C., and Gaull, G. E.: Homocystinuria; metabolic studies on 3 patients. *J. Pediatr. 67*:58–68, 1966.
34. Bridgers, W. F.: Serine transhydroxymethylase in developing mouse brain. *J. Neurochem. 15*:1325–1328, 1968.
35. Broderick, D. S., Candland, K. L., North, J. A., et al.: The isolation of serine transhydroxymethylase from bovine brain. *Arch. Biochem. Biophys. 148*:196–198, 1972.
36. Brown, F. C. and Gordon, P. H.: A study of L-^{14}C-cystathionine metabolism in the brain, kidney and liver of pyridoxine-deficient rats. *Biochim. Biophys. Acta 230*:434–445, 1971.
37. Bruggencate, G. ten and Engberg, I.: Analysis of glycine actions on spinal interneurones by intracellular recording. *Brain Res. 11*:446–450, 1968.

38. Bruggencate, G. ten and Engberg, I.: Iontophoretic studies in Deiters' nucleus of the inhibitory actions of GABA and related amino acids and the interactions of strychnine and picrotoxin. *Brain Res.* 25:431–448, 1971.
39. Bruggencate, G. ten and Sonnhof, U.: Glycine and GABA actions in hypoglossus nucleus and blocking effects of strychnine and picrotoxin. *Experientia* 27:1109, 1971.
40. Burke, D., Andrews, C. J. and Knowles, L.: The action of a GABA derivative in human spasticity. *J. Neurol. Sci.* 14:199–208, 1971.
41. Caciappo, F., Pandolfo, L., and Di Chiara, G.: Transamination reaction between 4-aminobutyric acid and α-ketoglutaric acid in certain rat tissues. *Boll. Soc. Ital. Biol. Sper.* 36:465–467, 1959.
42. Cavanaugh, J. R.: The rotational isomerism of phenylalanine by nuclear magnetic resonance. *J. Am. Chem Soc.* 89:1558–1563, 1967.
43. Chen, K. K. and Chen, A. L.: The pharmacological action of dendrobine. The alkaloid of Chin-shih-hu. *J. Pharmacol. Exp. Ther.* 55:319–325, 1935.
44. Cook, W. A. and Cangiano, A.: Presynaptic and postsynaptic inhibition of spinal motoneurones. *J. Neurophysiol.* 35:389–403, 1972.
45. Curtis, D. R.: The pharmacology of central and peripheral inhibition. *Pharmacol. Rev.* 15:333–364, 1963.
46. Curtis, D. R. and de Groat, W. C.: Tetanus toxin and spinal inhibition. *Brain Res.* 10:208–212, 1968.
47. Curtis, D. R. and Duggan, A. W.: The depression of spinal inhibition of morphine. *Agents Actions* 1:14–19, 1969.
48. Curtis, D. R., Duggan, A. W., Felix, D., et al.: GABA, bicuculline and central inhibition. *Nature* 226:1222–1224, 1970.
49. Curtis, D. R., Duggan, A. W., Felix, D., et al.: Bicuculline, an antagonist of GABA and synaptic inhibition in the spinal cord. *Brain Res.* 32:69–96, 1971.
50. Curtis, D. R., Duggan, A. W., Felix, D., et al.: Antagonism between bicuculline and GABA in the cat brain. *Brain Res.* 33:57–73, 1971.
51. Curtis, D. R., Duggan, A. W., and Johnston, G. A. R.: Glycine, strychnine, picrotoxin and spinal inhibition. *Brain Res.* 14:759–762, 1969.
52. Curtis, D. R., Duggan, A. W., and Johnston, G. A. R.: The inactivation of extracellularly administered amino acids in the feline spinal cord. *Exp. Brain Res.* 10:447–462, 1970.
53. Curtis, D., Duggan, A. W., and Johnston, G. A. R.: The specificity of strychnine as a glycine antagonist in the mammalian spinal cord. *Exp. Brain Res.* 12:547–565, 1971.
54. Curtis, D. R. and Felix, D.: The effect of bicuculline upon synaptic inhibition in the cerebral and cerebellar cortices of the cat. *Brain Res.* 34:301–321, 1971.
55. Curtis, D. R., Felix, D., and McLennan, H.: GABA and hippocampal inhibition. *Br. J. Pharmacol.* 40:881–883, 1970.
56. Curtis, D. R., Game, C. J. A., Johnston, G. A. R., et al.: Convulsive action of penicillin. *Brain Res.* 43:242–245, 1972.

57. Curtis, D. R., Hösli, L., and Johnston, G. A. R.: Inhibition of spinal neurones by glycine. *Nature 215*:1502–1503, 1967.
58. Curtis, D. R., Hösli, L., and Johnston, G. A. R.: A pharmacological study of the depression of spinal neurones by glycine and related amino acids. *Exp. Brain Res. 6*:1–18, 1968.
59. Curtis, D. R., Hösli, L., Johnston, G. A. R., et al.: The hyperpolarization of spinal motoneurones by glycine and related amino acids. *Exp. Brain Res. 5*:235–258, 1968.
60. Curtis, D. R. and Johnston, G. A. R.: Amino acid transmitters. *Handbook Neurochem. 4*:115–134, 1970.
61. Curtis, D. R. and Tebēcis, A. K.: Bicuculline and thalamic inhibition. *Exp. Brain Res. 16*:210–218, 1972.
62. Curtis, D. R. and Watkins, J. C.: The excitation and depression of spinal neurones by structurally related amino acids. *J. Neurochem. 6*:117–141, 1960.
63. Davidoff, R. A.: Penicillin and presynaptic inhibition in the amphibian spinal cord. *Brain Res. 36*:218–222, 1972.
64. Davidoff, R. A. and Aprison, M. H.: Picrotoxin antagonism of the inhibition of interneurones by glycine. *Life Sci. 8*:107–112, 1969.
65. Davidoff, R. A., Graham, L. T., Shank, R. P., et al.: Changes in amino acid concentrations associated with loss of spinal interneurones. *J. Neurochem. 14*:1025–1031, 1967.
66. Davidoff, R. A., Shank, R. P., Graham, L. T., et al.: Association of glycine with spinal interneurones. *Nature 214*:680–681, 1967.
67. Davidson, A. N. and Kaczmarek, L. K.: Taurine—a possible neurotransmitter? *Nature 234*:107–108, 1971.
68. Davies, W. E. and Comis, S. D.: Bicuculline and its effect on γ-aminobutyric acid transaminase in the guinea pig brain stem. *Nature 231*:156–157, 1971.
69. Davies, L. P. and Johnston, G. A. R.: Serine hydroxymethyltransferase in the central nervous system: regional and subcellular distribution studies. *Brain Res. 54*:149–156, 1973.
70. Davies, J. and Watkins, J. C.: Is 1-hydroxy-3-aminopyrrolidone-2 (HA966) a selective excitatory amino-acid antagonist? *Nature* (New Biol.) *238*:61–63, 1972.
71. Davis, R. and Huffman, R. D.: Pharmacology of the brachium conjunctivum red nucleus synaptic system in the baboon. *Fed Proc. 28*:775, 1969.
72. de Belleroche, J. S. and Bradford, H. F.: Metabolism of beds of mammalian cortical synaptosomes: Responses to depolarizing influences. *J. Neurochem. 19*:585–602, 1972.
73. de Feudis, F. V.: Dehydration of the brain by intraperitoneal injections of hyperosmotic solutions of γ-aminobutyric acid and DL-α-alanine. *Experientia 27*:1284–1285, 1971.
74. de Feudis, F. V. and Collier, B.: Amino acids of brain and γ-hydroxybutyrate induced depression. *Arch. Int. Pharmacodyn. 187*:30–36, 1970.
75. de Groat, W. C.: GABA-depolarization of a sensory ganglion: antagonism by picrotoxin and bicuculline. *Brain Res.* 38:429–432, 1972.

76. de Groat, W. C. Lalley, P. M., and Block, M.: The effects of bicuculline and GABA on the superior cervical ganglion of the cat. *Brain Res.* 25:665–668, 1971.
77. de Marchi, W. J. and Johnston, G. A. R.: The oxidation of glycine by D-amino acid oxidase in extracts of mammalian central nervous tissue. *J. Neurochem.* 16:355–361, 1969.
78. Desha, C. McF. and Fuerst, R.: Chemical and enzymatic synthesis of γ-pantothenate. *Biochim. Biophys. Acta* 86:33–38, 1964.
79. Dreifuss, J. J., Kelly, J. S., and Krnjevic, K.: Cortical inhibition and γ-aminobutyric acid. *Exp. Brain Res.* 9:137–154, 1969.
80. Editorial: Advantages of an antagonist. *Nature* 226:1199–1200, 1970.
81. Elliott, K. A. C.: γ-Aminobutyric acid and other inhibitory substances. *Br. Med. Bull.* 21:70–75, 1965.
82. Elliott, K. A. C. and van Gelder, N. M.: Occlusion and metabolism of γ-aminobutyric acid by brain tissue. *J. Neurochem.* 3:28–40, 1958.
83. Engberg, I. and Thaller, S.: On the interaction of picrotoxin with GABA and glycine in the spinal cord. *Brain Res.* 19:151–154, 1970.
84. Eugster, C. H.: Isolation, structure, and synthesis of centrally active compounds from *Amanita muscaria* (L. ex Fr.) Hooker. In Efron, D. H., Holmstedt, B., and Kline, N. S. (Eds.): *Ethnopharmacologic Search for Psychoactive Drugs.* Washington: U. S. Public Health Service publication no. 1645, 1967, pp. 416–418.
85. Fahn, S. and Côté, L. J.: Regional distribution of γ-aminobutyric acid (GABA) in brain of Rhesus monkey. *J. Neurochem.* 15:209–213, 1968.
86. Falchetto, S., Kato, G., and Provini, L.: The action of carnitines on cortical neurones. *Can. J. Physiol. Pharmacol.* 49:1–7, 1971.
87. Fedinec, A. A. and Shank, R. P.: Effect of tetanus toxin on the content of glycine, gamma-aminobutyric acid, glutamate, glutamine and aspartate in the rat spinal cord. *J. Neurochem.* 18:2229–2234, 1971.
88. Fonnum, F.: The distribution of glutamate decarboxylase and aspartate transaminase in subcellular fractions of rat and guinea-pig brain. *Biochem. J.* 106:401–412, 1968.
89. Fonnum, F., Storm-Mathisen, J., and Walberg, F.: Glutamate decarboxylase in inhibitory neurons. A study of the enzyme in Purkinje cell axons and boutons in the cat. *Brain Res.* 20:259–275, 1970.
90. Gaitonde, M. K.: Sulphur amino acids. *Handbook Neurochem.* 3:225–287, 1970.
91. Galindo, A.: GABA-picrotoxin interaction in the mammalian central nervous system. *Brain Res.* 14:763–767, 1969.
92. Galindo, A., Krnjević, K., and Schwartz, S.: Microiontophoretic studies on neurones in the cuneate nucleus. *J. Physiol.* (Lond.) 192:359–377, 1967.
93. Gayon, G. R. and Davies, D. D.: Inhibition of the Krebs cycle by glyoxylic acid and by γ-hydroxy-α-ketoglutaric acid. *C. R. Acad. Sci.* (*Paris*) 265D:7111–713, 1967.

94. Giurgea, C., Dauby, J., Moeyersoons, F., et al.: The neuropharmacology of a new antidepressant chemically related to gamma aminobutyric acid. In Garattini, S. and Dukes, M. N. G. (Eds.): *Antidepressant Drugs*. Amsterdam: Excerpta Medica Foundation, 1967, pp. 222–232.
95. Giurgea, C. E., Moeyersoon, F., and Evraerd, A. C.: A GABA-related hypothesis on the mechanism of action of the antimotion sickness drugs. *Arch. Int. Pharmacodyn. 166*:238–251, 1967.
96. Gottesfeld, Z. and Elliott, K. A. C.: Factors that affect the binding and uptake of GABA by brain tissue. *J. Neurochem. 18*:683–690, 1971.
97. Gottesfeld, Z., Kelly, J. S., and Renaud, L. P.: The *in vivo* neuropharmacology of amino-oxyacetic acid in the cerebral cortex of the cat. *Brain Res. 42*:319–335, 1972.
98. Graham, L. T. and Aprison, M. H.: Distribution of some enzymes associated with the metabolism of glutamate, aspartate, γ-aminobutyrate and glutamine in cat spinal cord. *J. Neurochem. 16*:559–566, 1969.
99. Graham, L. T., Shank, R. P., Werman, R., et al.: Distribution of some synaptic transmitter suspects in cat spinal cord: glutamic acid, aspartic acid, γ-aminobutyric acid, glycine and glutamine. *J. Neurochem. 14*:465–472, 1967.
100. Guidotti, A., Badiani, G., and Pepeu, G.: Taurine distribution in cat brain. *J. Neurochem. 19*:431–435, 1972.
101. Haber, B., Kuriyama, K., and Roberts, E.: L-Glutamic acid decarboxylase: A new type in glial cells and human brain gliomas. *Science 168*:598–599, 1970.
102. Haber, B., Sze, P. Y., Kuriyama, K., et al.: GABA as a repressor of L-glutamic acid decarboxylase (GAD) in developing chick embryo optic lobes. *Brain Res. 18*:545–547, 1970.
103. Hammer, W., Holmstedt, B., Karlen, B., et al.: The metabolism of tremorine. Identification of a new biologically active metabolite, N-(4-pyrrolidino-2-butynyl)-γ-aminobutyric acid. *Biochem. Pharmacol. 17*:1931–1941, 1968.
104. Hammerstad, J. P., Murray, J. E., and Cutler, R. W. P.: Efflux of amino acid neurotransmitters from rat spinal cord slices. II. Factors influencing the electrically induced efflux of ^{14}C-glycine and ^{3}H-GABA. *Brain Res. 35*:357–367, 1971.
105. Hebb, C.: CNS at the cellular level: Identity of transmitter agents. *Ann. Rev. Physiol. 32*:165–192, 1970.
106. Hendrick, J. L. and Sallach, H. J.: The nonoxidative decarboxylation of hydroxypyruvate in mammalian systems. *Arch. Biochem. Biophys. 105*:261–269, 1964.
107. Hill, R. G., Simmonds, M. A., and Straughan, D. W.: Antagonism of GABA by picrotoxin in the feline cerebral cortex. *Br. J. Pharmacol. 44*:807–809, 1972.
108. Hökfelt, T. and Ljungdahl, A.: Light and electron microscopic autoradiography on spinal cord slices after incubation with labeled glycine. *Brain Res. 32*:189–194, 1971.

109. Hökfelt, T. and Ljungdahl, A.: Uptake of (3H)noradrenaline and γ-(^{3}H)aminobutyric acid in isolated tissues of rat: An autoradiographic and fluorescence microscopic study. *Prog. Brain Res.* 34:87–102, 1971.
110. Hökfelt, T. and Ljungdahl, A.: Autoradiographic identification of cerebral and cerebellar cortical neurons accumulating labeled gamma-aminobutyric acid (^{3}H-GABA). *Exp. Brain Res.* 14:354–362, 1972.
111. Hope, D. B.: Cystathionine accumulation in the brains of pyridoxine deficient rats. *J. Neurochem.* 11:327–337, 1964.
112. Hopkin, J. and Neal, M. J.: Effect of electrical stimulation and high potassium concentrations on the efflux of ^{14}C-glycine from slices of spinal cord. *Br. J. Pharmacol.* 42:215–223, 1971.
113. Hosein, E. A., Kato, A., Vine, E., et al.: The identification of acetyl-1-carnitylcholine in rat brain extracts and the comparison of its cholinomimetic properties with acetylcholine. *Can. J. Physiol. Pharmacol.* 48:709–722, 1970.
114. Hösli, L., Andres, P. F., and Hösli, E.: Effects of glycine on spinal neurones grown in tissue culture. *Brain Res.* 34:399–402, 1971.
115. Hösli, L. and Haas, H. L.: Effects of histamine, histidine, and imidazole acetic acid on neurones of the medulla oblongata of the cat. *Experientia* 27:1311–1312, 1971.
116. Hösli, L. and Tebēcis, A. K.: Actions of amino acids and convulsants on bulbar reticular neurones. *Exp. Brain Res.* 11:111–127, 1970.
117. Hösli, L., Tebēcis, A. K., and Haas, H. L.: Depressant amino acids and possible antagonists on medullary reticular neurones. *Experientia* 27:732–733, 1971.
118. Howells, D. J.: The inhibitory transmitter: GABA or its choline ester? *J. Pharm. Pharmacol.* 23:794–795, 1971.
119. Hubig, W.: The three forms of glycine. *Z. Naturforsch.* 13b:633–638. 1958.
120. Huffman, R. D. and McFadin, L. S.: Suppression of presynaptic inhibition and cerebellar disfacilitation by bicuculline. *Life Sci. Part I,* 11:113–121, 1972.
121. Iitaka, Y.: Crystal structure of β-glycine. *Acta Cryst.* 13:35–45, 1960.
122. Iitaka, Y.: The crystal structure of γ-glycine. *Acta Cryst.* 14:1–10, 1961.
123. Ito, M. and Yoshida, M.: The origin of cerebellar-induced inhibition of Deiters' neurones. I. Monosynaptic inhibition of the inhibitory postsynaptic potentials. *Exp. Brain Res.* 2:330–349, 1966.
124. Iversen, L. L.: The uptake, storage, release, and metabolism of GABA in inhibitory nerves. In Snyder, S. H. (Ed.): *Perspectives in Neuropharmacology.* New York: Oxford University Press, 1972, pp. 75–111.
125. Iversen, L. L. and Bloom, F. E.: Studies of the uptake of ^{3}H-GABA and ^{3}H-glycine in slices and homogenates of rat brain and spinal cord by electron microscopic autoradiography. *Brain Res.* 41:131–143, 1972.

126. Iversen, L. L. and Johnston, G. A. R.: GABA uptake in rat central nervous system: comparison of uptake in slices and homogenates and the effects of some inhibitors. *J. Neurochem.* 18:1939–1950, 1971.
127. Iversen, L. L., Mitchell, J. F., and Srinivasan, V.: The release of γ-aminobutyric acid during inhibition in the cat visual cortex. *J. Physiol.* 212:519–534, 1971.
128. Iversen, L. L. and Neal, M. J.: The uptake of ^{3}H-GABA by slices of rat cerebral cortex. *J. Neurochem.* 15:1141–1149, 1968.
129. Iversen, L. L. and Snyder, S. H.: Synaptosomes: different populations storing catecholamines and γ-aminobutyric acid in homogenates of rat brain. *Nature* 220:797–798, 1968.
130. Jacobsen, J. G. and Smith, L. H.: Biochemistry and physiology of taurine. *Physiol. Rev.* 48:424–511, 1968.
131. Janssen, P. A. J.: The pharmacology of haloperidol. *Int. J. Neuropsychiatr.* 3:S10–18, 1967.
132. Jarboe, C. H. and Porter, L. A.: The preparative column chromatographic separation of picrotoxin. *J. Chromatogr.* 19:427–428, 1965.
133. Jarboe, C. H., Porter, L. A., and Buckler, R. T.: Structural aspects of picrotoxinin action. *J. Med. Chem.* 11:729–731, 1968.
134. Jasper, H. H. and Koyama, I.: Rate of release of amino acids from the cerebral cortex in the cat as affected by brainstem and thalamic stimulation. *Can. J. Physiol. Pharmacol.* 47:889–905, 1969.
135. Jenny, E. and Solberg, R.: Biochemische Eigenschaften partiell gereinigter Glutaminsäurecarboxylase (L-glutamate-1-carboxylase, E.C. 4.1.1.15) aus Kalbshirnrinde. *Helv. Physiol. Pharmacol. Acta* 26:270–277, 1969.
136. Jinnai, D., Sawai, A., and Mori, A.: γ-Guanidinobutyric acid as a convulsive substance. *Nature* 212:617, 1966.
137. Johnson, E. S., Roberts, M. H. T., and Straughan, D. W.: Amino acid induced depression of cortical neurones. *Br. J. Pharmacol.* 38:659–666, 1970.
138. Johnston, G. A. R.: The intraspinal distribution of some depressant amino acids. *J. Neurochem.* 15:1013–1017, 1968.
139. Johnston, G. A. R.: Muscimol and the uptake of γ-aminobutyric acid by rat brain slices. *Psychopharmacol.* 22:230–233, 1971.
140. Johnston, G. A. R., Beart, P. M., Curtis, D. R., et al.: Bicuculline methochloride, a GABA antagonist. *Nature* 240:219–220, 1972.
141. Johnston, G. A. R. and Curtis, D. R.: γ-Aminobutyrylcholine and central inhibition. *J. Pharm. Pharmacol.* 24:251–252, 1972.
142. Johnston, G. A. R., Curtis, D. R., de Groat, W. C., et al.: Central actions of ibotenic acid and muscimol. *Biochem. Pharmacol.* 17:2488–2489, 1968.
143. Johnston, G. A. R., de Groat, W. C., and Curtis, D. R.: Tetanus toxin and amino acid levels in cat spinal cord. *J. Neurochem.* 16:797–800, 1969.
144. Johnston, G. A. R. and Iversen, L. L.: Glycine uptake in rat central nervous system slices and homogenates: Evidence for different

uptake systems in spinal cord and cerebral cortex. *J. Neurochem.* *18*:1951–1961, 1971.
145. Johnston, G. A. R. and Mitchell, J. F.: The effect of bicuculline, metrazol, picrotoxin and strychnine on the release of ^{3}H-GABA from rat brain slices. *J. Neurochem* *18*:2441–2446, 1971.
146. Johnston, G. A. R. and Vitali, M. V.: Glycine-producing transaminase activity in extracts of spinal cord. *Brain Res.* *12*:471–472, 1969.
147. Johnston, G. A. R. and Vitali, M. V.: Glycine: 2-oxoglutarate transaminase in rat cerebral cortex. *Brain Res.* *14*:201–208, 1969.
148. Johnston, G. A. R., Vitali, M. V., and Alexander, H. M.: Regional and subcellular distribution studies on glycine: 2-oxoglutarate transaminase activity in cat spinal cord. *Brain Res.* *20*:361–367, 1970.
149. Jordan, C. C. and Webster, R. A.: Release of acetylcholine and ^{14}C-glycine from the cat spinal cord *in vivo*. *Br. J. Pharmacol.* *43*:441P, 1971.
150. Kaczmarek, L. K., Agrawal, H. C., and Davison, A. N.: The biochemistry of taurine in developing rat brain. In Carson, N. A. J. and Raine, D. N. (Eds.): *Inherited Disorders of Sulphur Metabolism*. Baltimore: Williams and Wilkins, 1971, pp. 63–69.
151. Deleted in proof.
152. Kandera, J., Levi, G., and Lajtha, A.: Control of cerebral metabolite levels. II. Amino acid uptake and levels in various areas of the rat brain. *Arch. Biochem. Biophys.* *126*:242–260, 1968.
153. Kellerth, J. -O. and Szumski, A. J.: Effects of picrotoxin on stretch activated postsynaptic inhibitions in spinal motoneurones. *Acta Physiol. Scand.* *66*:146–156, 1966.
154. Kelly, J. S. and Krnjevic, K.: The action of glycine on cortical neurones. *Exp. Brain Res.* *9*:155–163, 1969.
155. Key, B. J. and White, P. R.: Neuropharmacological comparison of cystathionine, cystine, homoserine and alpha-ketobutyric acid in cats. *Neuropharmacol.* *9*:349–357, 1970.
156. Kier, L. B. and Truitt, E. B.: Molecular orbital studies on the conformation of γ-aminobutyric acid and muscimol. *Experientia* *29*:988–989, 1970.
157. King, E. J.: The ionization constants of glycine and the effect of sodium chloride upon its second ionization. *J. Am. Chem. Soc.* *73*:155–159, 1951.
158. King, E. J.: The ionization constants of taurine and its activity coefficient in hydrochloric acid solutions from electromotive force measurements. *J. Am. Chem Soc.* *75*:2204–2209, 1953.
159. King, E. J.: The thermodynamics of ionization of amino acids. I. The ionization constants of γ-aminobutyric acid. *J. Am. Chem. Soc.* *76*:1006–1008, 1954.
160. King, J. S., Goodman, H. O., and Thomas, J. J.: Urinary amino acid excretion in mongolism. *Acta Genet. Statist. Med.* *16*:132–154, 1966.
161. Koster, R.: Effects of glycine on electroshock and barbiturates in mice. *Pharmacologist*, *9*:225, 1967.
162. Kraus, P.: Substrate induction of mouse brain L-glutamate decarboxylase. *Z. Physiol. Chem.* *349*:1425–1427, 1968.

163. Krnjevic, K. and Schwartz, S.: The action of γ-aminobuytric acid on cortical neurones. *Exp. Brain Res. 3*:320–326, 1967.
164. Kuriyama, K., Haber, B., and Roberts, E.: Occurrence of a new L-glutamic acid decarboxylase in several blood vessels of the rabbit. *Brain Res. 23*:121–123, 1970.
165. Kuriyama, K., Haber, B., Sisken, B., et al.: The γ-aminobutyric acid system in rabbit cerebellum. *Proc. Nat. Acad. Sci. U.S.A. 55*:846–852, 1966.
166. Kuriyama, K., Roberts, E., and Kakefuda, T.: Association of the γ-aminobutyric acid system with a synaptic vesicle fraction from mouse brain. *Brain Res. 8*:132–152, 1968.
167. LaBella, F., Vivian, S. and Queen, G.: Abundance of cystathionine in the pineal body. Free amino acids and related compounds of bovine pineal, anterior and posterior pituitary and brain. *Biochim Biophys. Acta 158*:286–288, 1968.
168. Laborit, H., Baron, C., London, A., et al.: Central nervous activity and comparative general pharmacology of glyoxylate, glycolate and glycolaldehyde. *Agressologie 12*:187–211, 1971.
169. Levi, G.: Regional differences in cerebral amino acid transport. *Prog. Brain Res. 29*:219–228, 1968.
170. Levi, G., Amaldi, P. and Morisi, G.: Gamma-aminobutyric acid (GABA) uptake by the developing mouse brain *in vivo*. *Brain Res. 41*:435–451, 1972.
171. Levy, R. A. and Anderson, E. G.: The effect of the GABA antagonists bicuculline and picrotoxin on primary afferent terminal excitability. *Brain Res. 43*:171–180, 1972.
172. Liang, C.: Studies on experimental thiamine deficiency. Trends of keto acid formation and detection of glyoxylic acid. *Biochem J. 82*:429–434, 1962.
173. Logan, W. J. and Snyder, S. H.: High affinity uptake systems for glycine, glutamic and aspartic acids in synaptosomes of rat central nervous tissues. *Brain Res.* 42:413–431, 1972.
174. McGeer, P. L., McGeer, E. G., and Wada, J. A.: Glutamic acid decarboxylase in Parkinson's disease and epilepsy. *Neurology 21*:1000–1007, 1971.
175. Mangan, L. L. and Whittaker, V. P.: The distribution of free amino acids in subcellular fractions of guinea-pig brain. *Biochem J. 98*:128–137, 1966.
176. Manske, R. H. F.: The alkaloids of fumaraceous plants. II. Dicentra cucullaria (L.) Bernh. *Can. J. Res. 7*:265–269, 1932.
177. Manske, R. H. F.: The alkaloids of fumaraceous plants. III. A new alkaloid, bicuculline, and its constitution. *Can. J. Res. 8*:142–146, 1933.
178. Marcus, R. J., Winters, W. D., Roberts, E., et al.: Neuropharmacological studies of imidazole-4-acetic acid actions in the mouse and rat. *Neuropharmacology 10*:203–215, 1971.
179. Marks, N.: Peptide hydrolases. *Handbook Neurochem. 3*:133–171, 1970.

180. Matus, A. I. and Dennison, M. E.: Autoradiographic localization of tritiated glycine at 'flat-vesicle' synapses in spinal cord. *Brain Res.* 32:195–197, 1971.
181. Maxwell, R. and Roth, R. H.: Conversion of 1,4-butanediol to γ-hydroxybutyric acid in rat brain and in peripheral tissue. *Biochem. Pharmacol.* 21:1521–1533, 1972.
182. Mercier, F. and Delphaut, J: Sur les convulsions produites par la laudanosine. *C.R. Soc. Biol.* (Paris) 118:168–170, 1935.
183. Mitchell, J. F. and Roberts, P. J: Evoked release of amino acids from the intact spinal cord. *Br. J. Pharmacol.* 45:175–176P, 1972.
184. Mitchell, J. F. and Srinivasan, V.: Release of ^{3}H-γ-aminobutyric acid from the brain during synaptic inhibition. *Nature* 224:663–666, 1969.
185. Mitoma, C. and Neubauer, S. E.: Gamma-hydroxybutyric acid and sleep. *Experientia* 24:12–13, 1968.
186. Miyata, Y. and Otsuka, M.: Distribution of γ-aminobutyric acid in cat spinal cord and the alteration produced by local ischaemia. *J. Neurochem.* 19:1833–1834, 1972.
187. Murray, J. E. and Cutler, R. W. P.: Clearance of glycine from cat cerebrospinal fluid: Faster clearance from spinal subarachnoid than from ventricular compartment. *J. Neurochem.* 17:703–704, 1970.
188. Nakajima, T., Kakimoto, Y., Kumon, A., et al.: α-(γ-Aminobutyryl)-lysine in mammalian brain: Its identification and distribution. *J. Neurochem.* 16:417–422, 1969.
189. Nakajima, T., Wolfgram, F., and Clark, W. G.: The isolation of homoanserine from bovine brain. *J. Neurochem.* 14:1107–1112, 1967.
190. Nakajima, T., Wolfgram, F., and Clark, W. G.: Identification of 1,4-methylhistamine, 1,3-diaminopropane and 2,4-diaminobutyric acid in bovine brain. *J. Neurochem.* 14:1113–1118, 1967.
191. Neal, M. J.: The uptake of ^{14}C-glycine by slices of mammalian spinal cord. *J. Physiol.* (Lond.) 215:103–117, 1971.
192. Neal, M. J. and Iversen, L. L.: Subcellular distribution of endogenous and ^{3}H-γ-aminobutyric acid in rat cerebral cortex. *J. Neurochem.* 16:1245–1252, 1969.
193. Neal, M. J. and Iversen, L. L.: Autoradiographic localization of ^{3}H-GABA in rat retina. *Nature (New Biol.)* 235:217–218, 1972.
194. Obata, K.: Pharmacological study on postsynaptic inhibition of Deiters' neurones. *Abstr. XXIII Internat. Physiol., Congr.,* Tokyo, 406, 1965.
195. Obata, K., Ito, M., Ochi, R., et al.: Pharmacological properties of the postsynaptic inhibition by Purkinje cell axons and the action of γ-aminobutyric acid on Deiters' neurones. *Exp. Brain Res.* 4:43–57, 1967.
196. Obata, K. and Takeda, K.: Release of γ-aminobutyric acid into fourth ventricle induced by stimulation of the cat's cerebellum. *J. Neurochem.* 16:1043–1047, 1969.
197. Obata, K., Takeda, K., and Shinozaki, H.: Further study on pharmacological properties of the cerebellar-induced inhibition of Deiters' neurones. *Exp. Brain Res.* 11:327–342, 1970.

198. Okada, Y., Nitsch-Hassler, C., Kim, J. S., et al.: Role of γ-aminobutyric acid (GABA) in extrapyramidal motor system. I. Regional distribution of GABA in rabbit, rat, guinea pig and baboon CNS. *Exp. Brain Res. 13*:514–518. 1971.
199. Okaya, Y.: Refinement of the crystal structure of taurine. An example of computer-controlled experimentation. *Acta Cryst. 21*:726–735, 1966.
200. O'Neal, R. M., Chen, C., Reynolds, C. S., et al.: The "neurotoxicity" of L-2,4-diaminobutyric acid. *Biochem. J. 106*:699–706, 1968.
201. Otsuka, M.: γ-Aminobutyric acid in the nervous system. In Bourne, G. H. (Ed.): *The Structure and Function of Nervous Tissue.* New York: Academic Press, 1972, 4:249–289.
202. Otsuka, M., Obata, K., Miyata, Y., et al.: Measurement of γ-aminobutyric acid in isolated nerve cells of cat central nervous system *J. Neurochem. 18*:287–295, 1971.
203. Parry Jones, G., Roberts, R. T., and Ahmed, A. I.: A nuclear magnetic resonance investigation of γ-aminobutyric acid (GABA). *Mol. Phys. 22*:547–549, 1971.
204. Peck, E. J. and Awapara, J.: Formation of taurine and isethionic acid in rat brain. *Biochim. Biophys. Acta 141*:499–506, 1967.
205. Perry, T. L., Berry, K., Hansen, S., et al.: Regional distribution of amino acids in human brain obtained at autopsy. *J. Neurochem. 18*:513–519, 1971.
206. Phillis, J. W., Tebēcis, A. K., and York, D. H.: Histamine and some antihistamines: Their actions on cerebral cortical neurones. *Br. J. Pharmacol. 33*:426–440, 1968.
207. Pohle, W. and Matthies, H.: Die Topohistochemie von Transmittersystemen in Kortex und Hippocampus des Kaninchens. *Acta Biol. Med. Germ. 25*:447–454, 1970.
208. Pong, S. F. and Graham, L. T.: N-methylbicuculline, a convulsant more potent than bicuculline. *Brain Res. 43*:486–490, 1972.
209. Ponnuswamy, P. K. and Sasisekharan, V.: Conformation of amino acids. II. Potential energy of conformations of glycine, alanine and aminobutyric acid. *Int. J. Protein Res. 2*:37–45, 1970.
210. Porter, L. A: Picrotoxin and related substances. *Chem. Rev. 67*:441–464, 1967.
211. Rees, R. and Smith, H.: Structure and biological activity of some reduction products of strychnine, brucine and their congeners. *J. Med. Chem. 10*:624–627, 1967.
212. Richter, J. J. and Wainer, A. A.: Evidence for separate systems for the transport of neutral and basic amino acids across the blood-brain barrier. *J. Neurochem. 18*:613–620, 1971.
213. Roberts, E. and Kuriyama, K.: Biochemical-physiological correlation in studies of the γ-aminobutyric acid system. *Brain Res. 8*:1–35, 1968.
214. Roberts, E., Kuriyama, K., and Haber, B.: Biochemistry of synaptic inhibition at the cellular levels: The GABA system. In Costa, E. and Giacobini, E. (Eds.): *Biochemistry of Simple Neuronal Models.* New York: Raven Press, 1970, pp. 139–161.

215. Romano, M. and Cerra, M.: Further studies on the toxicity of glyoxylate in the rat. *Gazz. Biochem. 16*:354–360, 1967.
216. Roth, R. H.: Formation and regional distribution of γ-hydroxybutyric acid in mammalian brain. *Biochem. Pharmacol. 19*:3013–3019, 1970.
217. Roth, R. H. and Suhr, Y.: Mechanism of the γ-hydroxybutyrate-induced increase in brain dopamine and its relationship to "sleep." *Biochem. Pharmacol. 19*:3001–3002, 1970.
218. Ruffo, A., Testa, E., Adinolfi, A., et al.: Control of the citric acid cycle by glyoxylate. *Biochem. J. 103*:19–23, 1967.
219. Runyan, T. J. and Gershoff, S. N.: The effect of vitamin B_6 deficiency in rats on the metabolism of oxalic acid precursors. *J. Biol. Chem. 240*:1889–1892, 1965.
220. Ryall, R. W.: The subcellular distributions of acetylcholine, substance P, 5-hydroxytryptamine, γ-aminobutyric acid and glutamic acid in brain homogenates. *J. Neurochem. 11*:131–145, 1964.
221. Salganicoff, L. and de Robertis, E.: Subcellular distribution of the enzymes of the glutamic acid, glutamine, and γ-aminobutyric acid cycles in rat brain. *J. Neurochem. 12*:287–309, 1965.
222. Schmidt, R. F.: Presynaptic inhibition in the vertebrate central nervous system. *Ergeb. Physiol. 63*:20–101, 1971.
223. Scotti de Carolis, A., Lipparini, F., and Longo, V. G.: Neuropharmacological investigations on muscimol, a psychotropic drug extracted from *Amanita muscaria*. *Psychopharmacol. 15*:186–195, 1969.
224. Seiler, N., Wiechmann, M., Fischer, H. A., et al.: Incorporation of putrescine carbon into γ-aminobutyric acid in rat liver and brain in vivo. *Brain Res. 28*:317–325, 1971.
225. Semba, T. and Kano, M.: Glycine in the spinal cord of cats with local tetanus rigidity. *Science 164*:571–572, 1969.
226. Shank, R. P. and Aprison, M. H.: The metabolism *in vivo* of glycine and serine in eight areas of the rat central nervous system. *J. Neurochem. 17*:1461–1475, 1970.
227. Shaw, R. K. and Heine, J. D.: Ninhydrin positive substances present in different areas of normal rat brain. *J. Neurochem. 12*:151–155, 1965.
228. Shaw, K. N. F., Lieberman, E., Koch, R., et al.: Cystathionuria. *Am. J. Dis. Child. 113*:119–127, 1967.
229. Shimizu, H., Kakimoto, Y., Nakajima, T., et al.: Isolation and identification of 2-aminoethylphosphonic acid brain. *Nature 207*:1197–1198, 1965.
230. Shimizu, H., Kakimoto, Y., and Sano, I.: A method of determination of cystathionine and its distribution in human brain. *J. Neurochem. 13*:65–73, 1966.
231. Siggins, G. R., Oliver, A. P., Hoffer, B. J., et al.: Cyclic adenosine monophosphate and norepinephrine: Effects on transmembrane properties of cerebellar Purkinje cells. *Science 171*:192–194, 1971.
232. Small, N. A., Holton, J. B., and Ancill, R. J.: *In vitro* inhibition of serotonin and γ-aminobutyric acid synthesis in rat brain by histidine metabolites. *Brain Res. 21*:55–62, 1970.

233. Smith, S. E.: Kinetics of neutral amino acid transport in rat brain in vitro. *J. Neurochem. 14*:291–300, 1967.
234. Smythies, J. R.: The chemical anatomy of synaptic mechanisms: receptors. *Int. Rev. Neurobiol. 14*:233–331, 1971.
235. Snatzke, G., Wollenberg, G., Hrbek, J., et al.: The optical rotatory dispersion and circular dichroism of the phthalideisoquinoline alkaloids and of their γ-hydroxybenzyltetrahydroisoquinoline derivatives. *Tetrahedron 25*:5059–5086, 1969.
236. Sprince, H., Parker, C. M., and Josephs, J. A.: Homocysteine-induced convulsions in the rat: Protection by homoserine, serine, betain, glycine and glucose. *Agents Actions 1*:9–13, 1969.
237. Srinivasan, V., Neal, M. J., and Mitchell, J. F.: The effect of electrical stimulation and high potassium concentrations on the efflux of ^{3}H-γ-aminobutyric acid from brain slices. *J. Neurochem. 16*:1235–1244, 1969.
238. Stern, P. and Hadzovic, S.: Effect of glycine on experimental hindlimb rigidity in rats. *Life Sci. Part I, 9*:955–959, 1970.
239. Steward, E. G., Player, R., Quilliam, J. P., et al.: Molecular conformation of GABA. *Nature* (New Biol.) *233*:87–88, 1971.
240. Storm-Mathisen, J.: Glutamate decarboxylase in the rat hippocampal region after lesions of the afferent fibre systems. Evidence that the enzyme is localized in intrinsic neurones. *Brain Res. 40*:215–235, 1972.
241. Susz, J. P., Haber, B., and Roberts, E.: Purification and some properties of mouse brain L-glutamic decarboxylase. *Biochemistry 5*:2870–2876, 1966.
242. Sverdlov, Yu. S., and Kozhechkin, S. N.: On presynaptic inhibitory mechanisms of monosynaptic reflexes. In Birjukova, D. A. and Zimkina, N. V. (Eds.): *Mechanisms of the Nervous System*. Leningrad: Nauka, 1969, pp. 136–143.
243. Sytinsky, I. A. and Vasilijev, V. Y.: Some catalytic properties of purified γ-aminobutyrate-α-oxoglutarate transaminase from rat brain. *Enzymologia 39*:1–11, 1970.
244. Sze, P. Y.: Possible repression of L-glutamic acid decarboxylase by gamma-aminobutyric acid in developing mouse brain. *Brain Res. 19*:322–325, 1970.
245. Sze, P. Y., Kuriyama, K., and Roberts, E.: Thiosemicarbazide and γ-aminobutyric acid. *Brain Res. 25*:387–396, 1971.
246. Sze, P. Y. and Lovell, R. A.: Reduction of level of L-glutamate decarboxylase by γ-aminobutyric acid in mouse brain. *J. Neurochem. 17*:1657–1664, 1970.
247. Taberner, P. V., Rick, J. T., and Kerkut, G. A.: The action of gamma-hydroxybutyric acid on cerebral glucose metabolism. *J. Neurochem. 19*:245–254, 1972.
248. Tachiki, K. H., de Feudis, F. V., and Aprison, M. H.: Studies on the subcellular distribution of γ-aminobutyric acid in slices of rat cerebral cortex. *Brain Res. 36*:215–217, 1972.
249. Takeuchi, A. and Takeuchi, N.: A study of the action of picrotoxin on the inhibitory neuromuscular junction of the crayfish. *J. Physiol. 205*:377–391, 1969.

250. Tallan, H. H.: A survey of amino acids and related compounds in nervous tissue. In Holden, J. T. (Ed.): *Amino Acid Pools.* Amsterdam: Elsevier, 1962, pp. 471–585.
251. Tallan, H. H., Moore, S., and Stein, W. H.: L-Cystathionine in human brain. *J. Biol. Chem. 230*:707–716, 1958.
252. Tebēcis, A. K.: Effects of monoamines and amino acids on medial geniculate neurones of the cat. *Neuropharmacology 9*:381–390, 1970.
253. Theobald, W., Buch, O., Kunz, H. A., et al.: Pharmakologische und experimental-psychologische Untersuchungen mit 2 Inhaltsstoffen des Fliegenpilzes (*Amanita muscaria*). *Arzneim. Forsch. 18*:311–315, 1968.
254. Thomas, J. J., Goodman, H. O., King, J. S., et al.: Taurine excretion and intelligence in mongolism. *Proc. Soc. Exp. Biol. Med. 119*:832–833, 1965.
255. Tomita, K.: Crystal data and some structural features of γ-aminobutyric acid, 3-aminopropane sulfonic acid and their derivatives. *Tetrahedron Letters* 2587–2588, 1971.
256. Tsunoo, S., Horisaka, K., and Yamaguchi, A.: ω-Aminosulfonic acids, with special reference to the pharmacology of taurine. *Showa Igakkai Zasshi 28*:301–316, 1968.
257. Tureen, L. L.: Effect of experimental temporary vascular occlusion on the spinal cord. I. Correlation between structural functional changes. *Arch. Neurol. Psychiatr. 35*:789–807, 1936.
258. Turnbull, M. J., Slater, P. and Briggs, I.: An investigation of the pharmacological properties of homocarnosine. *Arch. Int. Pharmacodyn. 196*:127–132, 1972.
259. Tursky, T.: Inhibition of brain glutamate decarboxylase by adenosine triphosphate. *Eur. J. Biochem. 12*:544–549, 1970.
260. Tursky, T.: Inhibition of brain pyridoxal kinase by gamma-aminobutyric acid. *Biologia 27*:187–191, 1972.
261. Uhr, M. L. and Sneddon, M. K.: Glycine and serine inhibition of D-glycerate dehydrogenase and 3-phosphoglycerate dehydrogenase of rat brain. *FEBS Lett. 17*:137–140, 1971.
262. Uhr, M. L. and Sneddon, M. K.: The regional distribution of D-glycerate dehydrogenase and 3-phosphoglycerate dehydrogenase in the cat central nervous system: correlation with glycine levels. *J. Neurochem. 19*:1495–1500, 1972.
263. Valdes, F. and Orrego, F.: Strychnine inhibits the binding of glycine to rat brain-cortex membrane. *Nature 226*:761–762, 1970.
264. van Gelder, N. M.: The effect of aminooxyacetic acid on the metabolism of γ-aminobutyric acid in the brain. *Biochem. Pharmacol. 15*:533–539, 1966.
265. van Gelder, N. M.: The action *in vivo* of a structural analogue of GABA: Hydrazinopropionic acid. *J. Neurochem. 16*:1355–1360, 1969.
266. van Regemorter, N., Mardens, Y., Lowenthal, A., et al.: Distribution of two peptides of γ-aminobutyric acid in human brain and CSF. *Clin. Chim. Acta 38*:59–65, 1972.

267. von Seiler, N. and Wiechmann, M.: The occurrence of γ-aminobutyric acid and γ-amino-β-hydroxybutyric acid in animal tissues. *Z. Physiol. Chem. 350*:1493–1500, 1969.
268. Waksman, A. and Bloch, M.: Identification of multiple forms of aminobutyrate transaminase in mouse and rat brain: Subcellular localization. *J. Neurochem. 15*:99–105, 1968.
269. Waksman, A. and Roberts, E.: Purification and some properties of mouse brain γ-aminobutyric-α-ketoglutaric acid transaminase. *Biochemistry 4*:2132–2139, 1965.
270. Walsh, D. A. and Sallach, H. J.: Comparative studies on the pathways for serine biosynthesis in animal tissues. *J. Biol. Chem. 241*:4068–4076, 1966.
271. Waser, P. G.: The pharmacology of *Amanita muscaria*. In Efron, D. H. Holmstedt, B., and Kline, N. S. (Eds.): *Ethnopharmacologic Search for Psychoactive Drugs*. Washington: U.S. Public Health Service Publication no. 1645, 1967, pp. 419–439.
272. Welch, A. D. and Henderson, V. E.: A comparative study of hydrastine, bicuculline and adlumine. *J. Pharmacol. Exp. Therap. 51*:482–491, 1934.
273. Werman, R.: Amino acids as central neurotransmitters. *Res. Publ. Assoc. Res. Nerv. Ment. Dis. 50*:147–180, 1972.
274. Werman, R., Davidoff, R. A., and Aprison, M. H.: The inhibitory action of cystathionine. *Life Sci. 5*:1431–1440, 1966.
275. Werman, R., Davidoff, R. A., and Aprison, M. H.: Inhibition of motoneurones by iontophoresis of glycine. *Nature 214*:681–683, 1967.
276. Werman, R., Davidoff, R. A., and Aprison, M. H.: Inhibitory action of glycine on spinal neurones in the cat. *J. Neurophysiol. 31*:81–95, 1968.
277. Wofsey, A. R., Kuhar, M. J., and Snyder, S. H.: A unique synaptosomal fraction, which accumulates glutamic and aspartic acids in brain tissue. *Proc. Nat. Acad. Sci. U.S.A. 68*:1102–1106, 1971.
278. Wood, J. D., Watson, W. J., and Stacey, N. E.: A comparative study of hyperbaric oxygen-induced and drug-induced convulsions with particular reference to γ-aminobutyric acid metabolism. *J. Neurochem. 13*:361–370, 1966.
279. Yessaian, N. H., Armenian, A. R., and Buniatian, H. Ch.: Effect of γ-aminobutyric acid on brain serotonin and catecholamines. *J. Neurochem. 16*:1425–1433, 1969.
280. Yoshida, T. and Kikuchi, G.: Major pathways of glycine and serine catabolism in rat liver. *Arch. Biochem. Biophys. 139*:380–392, 1970.
281. Yoshino, Y., de Feudis, F. V., and Elliott, K. A. C.: Omega-amino acids in rat brain. *Can. J. Biochem. 48*:147–148, 1970.

CHAPTER THREE
Excitatory Amino Acids

Peter N. R. Usherwood

The ability of certain acidic amino acids to depolarize neurons and thereby enhance or depress their activity has led to considerable speculation about the role of these substances either as mediators or as modulators of synaptic transmission within the mammalian central nervous system (CNS). This conjecture is reinforced by 1) the discovery of specific neuronal uptake systems for these amino acids, 2) the demonstration that some of these substances can be released by neural stimulation, and 3) the nonuniformity of their distribution in the mammalian brain and spinal cord. Although most of this interest has centered around L-glutamate, L-aspartate and, to a lesser extent, L-cysteate have received some attention as putative transmitters.

L-Glutamate plays a central role in tissue metabolism and is ubiquitously distributed throughout mammalian nervous tissue. For this reason, and because of its predominant cytoplasmic localization, it seemed unacceptable as a neurotransmitter candidate at mammalian central synapses. During the past few years, however, a gradual shift in opinion has occurred owing primarily to discoveries during the 1960s that led to the concept of metabolic compartmentation of amino acids within nervous tissue. The overwhelming evidence favoring the candidacy of L-glutamate as a transmitter at central synapses in lower vertebrates, as well as at central and peripheral synapses in some invertebrates, has not had as much impact on the vertebrate scene as might have been expected. This is indeed sad but not entirely surprising.

DISTRIBUTION OF EXCITANT AMINO ACIDS

Although the free amino acid content of mammalian nervous tissue is significantly higher than that of most other tissues among the excitant acidic amino acids, only L-glutamate and L-aspartate have been found in appreciable amounts in the brain and spinal cord (*40*). In fact L-glutamate and L-aspartate account for approximately 75% of the total free amino acid content of these regions, with L-glutamate reaching concentrations as high as 10^{-2} M (*182, 232*). Consequently, the relationship of these compounds to brain function has been intensively studied both in vivo and in vitro, although not always with synaptic transmission in mind. The ratios of L-glutamate to L-aspartate vary in different parts of the nervous system as shown in Tables 1 and 2. These regional differences may be correlated with the transmitter roles that these two amino acids are assumed to serve, although it is recognized that they play many other roles that could equally account for the differences.

The pattern of distribution of L-glutamate in various parts of the brain is highly consistent when studied in diverse mammals. There is a stepwise increase in L-glutamate content when progressing from lower to higher levels of the brain, the highest concentration being found in the gray matter of the cerebral hemispheres and the lowest in the cerebral peduncle of the midbrain (Table 1). It would be unwise, of course, to infer from these data that higher levels of the CNS are more richly endowed with L-glutamate synapses, since the techniques for obtaining the data on which these generalizations are based do not differentiate between metabolic L-glutamate and supposed transmitter L-glutamate. Furthermore, the data do not agree with known numbers of cell bodies of either glia or neurons in the various parts of the brain, e.g., the cat (*127*), except for the cerebellum, since the DNA contents of the various brain regions (both white and gray) are similar (*161, 180*).

The DNA content of the cerebellum is abnormally high because of the large number of cells packed into the granular layer. The potentially important relationships between L-glutamate content of different brain regions, the sensitivity of neurons in these regions to this amino acid, and the number of synapses in these regions have yet to be established. The distribution of L-aspartate is relatively uniform throughout the mammalian brain (Table 1). Certainly the caudorostral concentration gradient seen with L-glutamate is not readily apparent with L-aspartate. A major part of the nonprotein L-aspartate of the brain is present as the *N*-acetyl derivative, a compound that normally occurs only in nervous tissue (*123, 236*). L-cysteate (Table 1) and cysteine sulfate (*16*) are the other excitant acidic amino acids found in the mammalian brain, but information on their distribution in different areas is very sparse.

The pattern of distribution of L-glutamate in the spinal cord provides convincing support for its presumed role as an excitatory transmitter in

Table 1. Distribution of acidic amino acids in mammalian brain[a]

Site	L-Glutamate	L-Glutamine	L-Aspartate	L-Cysteate	L-Serine[c]	L-Lysine[c]
Forebrain						
Telencephalon	11.2R (218)[e]	4.4R (218)	2.4R (218)	—	1.0R (218)	—
Diencephalon	8.6R (218)	4.2R (218)	2.4R (218)	—	0.6R (218)	—
Mesodiencephalon	11.0C (22)	4.6C (22)	2.0C (22)	—	—	—
Cerebral hemispheres	11.6R (130), 9.3R (219), 10.2–13.0C (196), 6.0C (196), 8.9C (195)	2.6R (219), 4.0C (196)	2.4R (130), 2.4R (219), 1.8C (196)		1.2R (219), 0.6C (196)	0.2R (130), 0.1C (196), 0.2R (219)
Gray matter	8.6D (191), 8.7B (217), 19.9H (192), 10.5S (235), 8.0D (79), 9.6–10.2C (190), 11.6R (6), 11.3K (21), 7.6–11.4C (20)	4.7B (217), 3.8S (235), 4.2–7.2C (20), 4.5D (79), 4.8K (21)	2.2D (191), 2.6–3.6C (20), 2.5D (79), 3.4K (21)	—	0.4D (191)	0.1D (191)
White matter	5.8D (191)	—	1.3D (191)			
Frontal lobe	8.9H (196a)[b]	3.8H (196a)[b]	2.7M (221), 1.9H (196a)[b]	—	0.4D (191)	0.1D (191)
Temporal lobe	12.9C (11), 10.9H (196a)[b]	4.4H (196a)[b]	3.1C (11), 1.9H (196a)[b], 2.7M (221)	—	0.6H (196a)[b], 0.8H (196a)[b]	0.2H (196a)[b], 0.2C (11), 0.3H (196a)[b], 0.3C (11)
Hippocampus	9.5D (79), 12.5C (22)	5.0D (79), 5.7C (22)	2.5D (79), 2.0C (22), 2.9M (221)	—	—	—
Corpus callosum	5.9C (125), 10.6C (11), 5.5H (196a)[b]	2.6H (196a)[b]	1.4C (11), 1.2H (196a)[b]	—	0.7H (196a)[b]	0.3H (196a)[b]
Corpus striatum						
Caudate nucleus	8.5D (191), 11.8C (128), 12.0D (79), 8.6C (196), 10.5H (196a)[b]	7.5D (79), 6.2C (196), 3.2H (196a)[b]	1.4D (191), 1.2C (196), 1.3H (196a)[b], 3.0D (79)	—	0.7C (196), 0.9H (196a)[b], 0.4D (191)	0.2C (196), 0.4H (196a)[b]
Internal capsule	5.6C (128)	—	—	—	—	—
Amygdala	9.4H (196a)[b]	4.7H (196a)[b]	2.7M (221), 1.7H (196a)[b]	—	1.3H (196a)[b]	0.5H (196a)[b]

continued

Table 1. continued

Site	L-Glutamate	L-Glutamine	L-Aspartate	L-Cysteate	L-Serine[c]	L-Lysine[c]
Thalamus	12.4C (11), 9.3D (191), 11.0D (79), 8.9C (128), 8.4H (196a)[b]	5.5D (79), 3.7H (196a)[b]	2.7C (11), 2.2D (191), 3.0D (79), 2.5H (196a)[b]	—	1.3H (196a)[b], 0.4D (191)	0.3C (11), 0.1D (191), 0.6H (196a)[b]
Hypothalamus	6.1D (191), 5.8H (196a)[b]	3.2H (196a)[b]	2.4D (191), 2.8M (221), 2.3H (196a)[b]	—	1.1H (196a)[b], 0.7D (191)	0.5H (196a)[b]
Optic nerve	3.5C (128)	—	—	—	—	—
Midbrain	7.5R (130), 6.4R (219), 7.1R (218), 9.7C (11)	2.1R (219), 3.5R (218)	2.5R (130), 2.5R (219), 4.1C (11), 2.5R (215)	0.2R (219)	0.4R (218), 1.1R (219)	0.4R (130), 0.4C (11), 0.4R (219)
Cerebral peduncle	3.7C (128)	—	—	—	—	—
Colliculus	6.0D (79), 5.9–8.0C (128)	4.0D (79)	3.5D (79)	—	—	—
Hindbrain						
Cerebellum	6.9C (127), 10.3R (130), 12.6C (11), 10.0C (22), 9.9M (221), 8.7D (87), 10.8C (128), 7.5D (79), 8.5C (196), 9.7R (218), 8.4R (219), 9.6H (196a)[b]	3.7D (87), 5.0D (79), 2.6R (219), 6.0C (22), 4.7R (7), 7.8C (196), 5.2H (196a)[b], 2.0R (218)	2.0R (130), 2.9C (11), 2.0D (79), 2.2R (219), 2.0C (22), 2.0R (199), 1.8C (196), 2.3H (196a)[b], 3.1M (221), 2.0R (218)	0.2R (219)	1.2R (219), 0.6R (218), 0.6C (196), 1.2H (196a)[b]	0.4R (130), 0.2C (196), 0.6H (196a)[b], 0.2C (11)
Hemisphere	8.7D (191)	—	2.0D (191)	—	0.4D (191)	0.2D (191)
Vermis	12.0D (191)	—	2.3D (191)	—	0.4D (191)	0.4D (130)
Pons-medulla	6.9R (130), 7.0D (79), 4.7R (219), 7.0C (22), 6.0C (236)	3.5D (79), 1.5R (219), 4.0C (22)	2.7R (130), 1.5D (79), 2.6R (219), 2.0C (22)	0.2R (219)	1.3R (219)	0.5R (219)
Pons	6.2R (218)	2.8R (218)	2.5R (218)	—	0.4R (218)	—
Medulla	5.3D (191), 6.2R (218)	3.1R (218)	2.5D (191), 2.7R (218)	—	0.4R (218), 0.5D (191)	0.1D (191)

[a] Concentrations given in μmol/g of (wet weight) tissue.
[b] Data obtained at autopsy from patients who died suddenly without previous brain disease.
[c] Included for comparison.
[d] Key: B, cattle; C, cat; D, dog; H, human being; K, kitten; M, monkey; R, rat; S, sheep.
[e] Reference number in "Literature Cited" Section.
—, Data not reported or relevant.

this part of the mammalian CNS. Free L-glutamate levels are highest in cord regions containing endings of primary afferent fibers, i.e., dorsal, and to a lesser extent, ventral gray matter (Table 2). This suggests that this amino acid may be the transmitter released from endings of primary afferent fibers, a hypothesis consistent with the high L-glutamate content of primary sensory nuclei of the medulla oblongata (*127*), since many primary afferent fibers from dorsal root sources terminate in these regions which contain neurons with a high sensitivity to L-glutamate. If one accepts the concept that some primary afferent terminals release L-glutamate, then a lower but still elevated L-glutamate level would be expected for the ventral gray matter of the spinal cord, since some of the dorsal root afferent fibers continue into the ventral gray to form monosynaptic connections with motoneurons.

A glance at Table 2 confirms this expectation and also confirms that the L-glutamate level of spinal white matter is, as would be expected, lower than that of gray matter (*70, 98*). However, because collaterals of primary afferent fibers travel up and down one or two spinal segments in dorsal white matter before terminating in gray matter, one might reasonably expect that this region, i.e., dorsal white, will have a higher L-glutamate concentration than ventral white matter, assuming of course that transmitter is manufactured in somata and transported to axon terminals. The data in Table 2 apparently support this contention as well.

These recent studies of L-glutamate distribution in spinal cord regions perhaps provide the most compelling support for a central transmitter role for this amino acid. If it is accepted that L-glutamate is the transmitter released by some primary afferent neurons, then it may seem reasonable to expect higher free L-glutamate levels in that part of the dorsal root which projects toward the spinal cord than in the dorsal root ganglion and distal part of the dorsal root. Data meet this expectation, for in both the dog (*82*) and cat (*81, 82, 127*), L-glutamate is the only amino acid found in uniformly high concentrations in dorsal roots between the ganglia and cord, all other amino acids being found in much higher concentrations in ganglia rather than in dorsal roots.

In the spinal cord, the concentration of L-glutamate becomes progressively higher as one moves from the dorsal root ganglion toward the region containing the terminals of primary afferent fibers, where presumably the population density of synaptic vesicles—the supposed organelles for neurotransmitter storage—is highest. It is important to remember that in sympathetic ganglia, where acetylcholine is the presynaptic transmitter (*104*), norepinephrine (NE) the putative postsynaptic transmitter (*9*), and where the free L-glutamate level is lower than in any region of the spinal cord

Table 2. Concentrations (μmol/g wet weight) of acidic amino acids in mammalian spinal cord and nerves[a]

Site	L-Glutamate	L-Glutamine	L-Aspartate	L-Cysteate	L-Serine[c]	L-Lysine[c]
Spinal cord						
Gray matter	6.4R (218)[d], 6.9C (127)	3.3R (218)	3.5R (218)	--	0.4R (218)	0.4R (235)[d]
Dorsal	6.5C (70, 98)	5.3C (70, 98)	2.1C (70, 98)	--	--	--
Ventral	5.4C (70, 98)	5.4C (70, 98)	3.1C (98)	--	--	--
White matter	3.7R (218)	3.2R (218)	1.3R (218)	--	0.5R (218)	0.5R (235)
Dorsal	4.8C (70, 98), 5.9C (127)	3.6C (70, 98)	1.1C (70, 98)	--	--	--
Ventral	3.9C (70)	3.8C (70, 98)	1.3C (70, 98)	--	--	--
Dorsal root	4.2–4.4D (82), 4.3C (127), 2.8C (127), 4.5C (127)[e], 3.4–4.5C (81), 4.6C (98), 3.8C (82), 2.8R (82)	2.4–2.5C (81), 1.9C (98), 1.6C (82), 3.0R (82), 2.3C (127), 1.9–2.2D (82), 2.1–2.4C (81)	1.2–1.5C (81), 1.0C (98), 1.5C (82), 2.0–2.2D (82), 1.1R (82), 0.7C (127), 1.1–1.2C (81)	--	0.5C (127)	0.1C (127)
Dorsal root ganglion	4.5C (81), 4.3C (127), 3.2C (127), 4.5C (127)	5.1C (81), 6.0C (127)	4.4C (81), 2.3C (127)	--	1.8C (127)	0.2C (127)

Ventral root	1.5R (82), 2.6C (127), 1.7C (127), 2.2C (81), 3.1C (98), 2.2C (82), 3.0–3.2D (82)	1.5C (81), 1.9C (98), 1.5C (82), 2.0–2.1D (82), 2.8R (82), 1.4C (127)	1.2C (81), 1.3C (98), 2.0–2.2D (82), 1.1R (82), 0.7C (127), 1.2C (82)	—	0.3C (127)	0.1C (127)
Spinal fluid	0.03C (125)	0.003–0.008C (125)	0.003–0.008C (125)	—	—	0.02–0.05 (125)
Peripheral nerve						
Afferent	2.0C (81), 3.0C (127), 2.1C (127)	2.1C (81), 1.6C (127)	1.1C (81), 1.0C (127)	—	0.7C (127)	0.1C (127)
Efferent	—	—	—	—	—	—
Mixed	1.0–1.5R (167), 1.8–2.1C (81), 2.2L (94), 2.3R (94, 235)	2.2R (94, 235), 1.8–2.3C (81), 1.8L (94), 4.2L (94), 0.1R (167)	0.7–1.2C (81), 0.2–0.5R (67)	0.02R (167)	0.3R (167)	0.3–0.4R (167)
Sympathetic ganglion	3.0C (127)	—	—	—	—	—

[a] Concentrations rounded off to first decimal place; standard errors have been omitted.
[b] *Key*: C, cat; R, rat; D, dog; L, rabbit.
[c] Included for comparison.
[d] Reference number in "Literature Cited" section.
[e] Terminations of dorsal root in spinal cord.
—, Data not reported or relevant.

(Table 2), synaptic terminals and postsynaptic neurons abound. The mere presence of neuronal somata and nerve terminals alone is, therefore, seemingly insufficient to account for high L-glutamate levels (*127*).

If one assumes that L-glutamate flows from soma to axon terminals in afferent neurons, it follows that dorsal sensory neurons must have a functional polarity with a differential flow mechanism resulting in more free L-glutamate directed centrally than peripherally, and it has been pointed out that this could result from the bipolar origin of dorsal root neurons (*127*). Axonal transport of L-glutamate and other amino acids has been demonstrated (*148, 189, 220*), and in some vertebrate and invertebrate nerve fibers flow rates as high as 720 mm/day have been observed, the flow proceeding proximodistally (*131a*). Although little is known about the flow of L-glutamate along dorsal root afferents, estimates for the flow of labeled L-leucine from dorsal root ganglion down the axonal extensions reveal a slow flow rate of 1.3 mm/day (*153a*) and rapid flow rates of 401 to 410 mm/day (*189a, 189b*) or 500 mm/day (*153a*).

If L-glutamate is the transmitter for dorsal root neurons, and this idea is consistent with their high glutamate context, it seems reasonable to expect that the L-glutamate levels of motor axons in ventral roots and peripheral nerve will be lower than those of primary afferent axons. Unfortunately, data on L-glutamate content of efferent fibers are lacking, although mixed nerves and ventral roots do contain significantly lower levels of L-glutamate than do dorsal roots. Perhaps the approximately 2 μmol/g (wet weight) of L-glutamate found in mixed nerve and ventral roots represents a metabolic pool of L-glutamate common to all neurons, whereas the additional L-glutamate found in the dorsal roots represents *transmitter* L-glutamate (*128*). It is perhaps significant that the components of the cholinergic transmitter system are more concentrated in ventral than in dorsal roots.

The concentrations of L-aspartate in dorsal and ventral roots are identical, but the concentration of this amino acid in ventral gray is higher than in dorsal gray matter (*98*). It seems possible, therefore, that the distribution of L-aspartate parallels that of terminals of spinal interneurons that synapse with motoneurons. This possibility was examined by attempting to produce a relatively selective degeneration of interneurons in the cat spinal cord to determine if L-aspartate and L-glutamate levels in dorsal and ventral gray matter were changed accordingly (*70*). Using temporary aortic occlusion, the investigators found that there was a loss of more than 50% of the small neurons, i.e., in both dorsal and ventral gray matter. The L-glutamate levels in dorsal and ventral gray and white matter fell by about 30% with the loss of these neurons, but there was no correlation between neuron count and L-glutamate concentration. However, the correlation was significant between the fall in concentration of L-aspartate in gray matter and the loss of small neurons. This finding not only implicates interneurons as a major source of L-aspartate, but also favors the latter as a transmitter candidate in these cells. The post-ischemic loss of synapses between primary afferent terminals and degenerated interneurons may account for the lowered L-glutamate levels observed in regions where these synapses normally occur.

The monoamide L-glutamine is not a candidate for excitatory synaptic transmission in mammals, although it always seems to be intimately associated with nervous system L-glutamate and may, therefore, be a precursor of this putative transmitter. It is present in much larger amounts in dorsal root ganglia than is any other amino acid, but its concentration approaches that of L-glutamate in spinal and ventral roots and mixed nerve (Table 2). Since the L-glutamine concentration of spinal gray matter is not significantly reduced following interneuronal degeneration (70), it is suggested that this substance resides primarily in glial cells. The latter are relatively resistant to anoxia (96), and their numbers do not change following aortic occlusion (70). The high ratio of L-glutamine to L-glutamate in ventral gray matter of the spinal cord is unexpected and does not at present fit into a scheme in which L-glutamine is a precursor of transmitter glutamate. L-glutamine is also found in markedly high concentrations in the brain, but its distribution pattern does not reflect that of L-glutamate (Table 2).

The total "free" amino acid content of invertebrate nervous tissue, both central and peripheral, greatly exceeds that of mammalian nervous tissue, L-aspartate being particularly abundant in the former (Table 3). This is apparently because amino acids in invertebrates perform an osmoregulatory role in addition to their metabolic and—in the case of L-glutamate at least—transmitter roles.

DEVELOPMENTAL ASPECTS OF EXCITANT AMINO ACID DISTRIBUTIONS

Concomitant with the development from fetal or neonatal state to that of adult, there is an increase in the concentration of L-glutamate and other free amino acids in the mammalian brain (14, 21, 79, 112–114, 137, 190, 208, 251, 253). In the mouse brain, e.g., the free amino acid content shows significant differences between the newborn and adult in most amino acids (Figure 1). In this mammal, most components of the free amino acid pool decrease during development except for one or two amino acids including L-glutamate and L-aspartate. In the dog (79) and cat (21) there

Table 3. Concentrations (μmol/g wet weight) of excitant acidic amino acids in vertebrate (excluding mammals) and invertebrate nervous systems[a]

Division	L-Glutamate	L-Glutamine	L-Aspartate	L-Cysteate	L-Serine[c]	L-Lysine[c]
Vertebrates						
Fish	5.0B (190)[d]	6.2B (190)	0.3B (190)	—	—	—
Frog	4.4B (190), 1.9N (259)	5.5B (190)	0.7B (190), 0.4N (259)	—	0.2N (259)	0.6N (259)
Tortoise	4.6B (190)	2.1B (190)	0.5B (190)	—	—	—
Hen	9.9B (190), 10.8B (203)	1.9B (190), 4.1B (203)	2.5–3.8B (190, 203)	—	—	—
Pigeon	6.2–14.0B (203, 253)	4.3–7.0B (203, 253)	—	—	—	—
Invertebrates						
Lobster	25.0N (160), 10.1–10.6N (167), 9.6B (167), 9.6–12.2T (167)	—	112.0N (160), 76.0–88.0N (167), 14.7B (167), 16.0–38.0T (167)	0.9–1.4N (167), 0.1B (167), 0.06–0.2T (167)	2.4–2.5N (167), 2.4B (167)	0.2–0.3N (167), 0.5B (167), 0.5T (167)
Crab	35.0N (160), 8.2–10.6N (167), 56.0N (86)[e]	37.0N (86)[e]	138.0N (160), 50.0–54.0N (167), 309.0N (86)[e]	—	0.6–1.0N (167), 4.7N (86)[e]	0.3–0.4N (167), 0.9N (86)[e]
Squid	21.2A (135)	—	62.5A (135), 79.1A (135)	—	—	—
Cuttlefish	39.0C (160)	—	82.0C (160)	—	—	—

[a] Concentrations rounded off to 1st decimal place; standard errors have been omitted.
[b] *Key:* A, axoplasm; B, brain; C, axon; N, nerve; T, thoracic ganglia.
[c] Included for comparison.
[d] Reference number in "Literature Cited" section.
[e] μmol/g of cell water.
—, Data not reported or relevant.

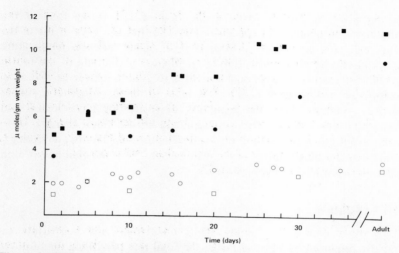

Figure 1. Changes in concentration (μmol/g wet weight) of L-glutamate and L-aspartate in developing cortex of cat and dog. Data are mean values from samples from 1 to 10 animals. Key: ■, L-glutamate in cat cortex; □, L-aspartate in dog cortex; ●, L-glutamate in dog cortex; ○, L-aspartate in cat cortex. (From Berl, S., and Purpura, D. P.: J. Neurochem. 10:237–240, 1963. Dravid, A. R., Himwich, W. A., and Davis, J. M.: J. Neurochem. 12:901–906, 1965.)

is an increase in the concentration of glutamate and L-glutamine during brain development (Figure 1). The concentration of the L-aspartate precursor, acetyl-aspartate, also increases during development of the rat (234), rabbit (123), and human (123, 192) brain. In dogs the L-glutamate concentration in cerebral cortex increases significantly during the first few days after birth, but then falls for five to ten days before increasing once again to reach the adult level (79). Unlike L-glutamate, L-aspartate increases consistently throughout development of the dog brain (79).

In newborn kittens, L-glutamate is approximately 40%, L-glutamine 100%, and L-aspartate 60% of the levels found in adult animals (or cats) (21). The greatest increase in feline brain L-glutamate and L-aspartate occurs between the fourth and 15th postnatal days, all three compounds reaching their adult levels within three to four postnatal weeks, although at different times during this interval. It is, of course, tempting to equate changes in free amino acid levels with development of neural elements that might utilize L-glutamate and L-aspartate as transmitters. That the concentration of the L-aspartate precursor, acetylaspartate, also increases during development of the rat (123, 234) and human (190) brain is also of interest. Marked behavioral changes (89) as well as profound structural changes are taking place in various parts of the brain (*252*) at the time of the alterations in amino acid concentration. The first five days of life in the dog, e.g., are marked by development of stable reflexes, temperature control, and better coordination (*89*), whereas at the end of the fourth week more stable reflexes and responses to stimuli appear as well as an essentially mature EEG (*197*). The value of these ontogenetic studies derives primarily from the possibility of correlating physiological and behavioral development with structural and histochemical changes. However, much more information on the developmental anatomy, physiology, behavior, and biochemistry of the mammalian CNS is required before even speculative correlations are justified.

METABOLISM

Like other nonessential amino acids, L-glutamate and L-aspartate are rapidly produced by nervous tissue, the total rate paralleling the turnover of members of the tricarboxylic acid cycle. The compound L-glutamate occupies a central position in the intermediary metabolism of all tissues but particularly in the nervous system and, because of this, has been for some time and indeed still is regarded by many workers as unlikely to serve a transmitter role. The brain uses mainly glucose as its source of energy most of which is used to maintain its functional activity. Since the metabolic pathways leading to the production of energy are probably under direct control of this activity, it would not be surprising if the metabolism of compounds not directly involved in energy production were separated from the reaction pathways involved in energy production. This idea of metabolic compartmentation has received growing acceptance during the past decade.

Amino Acid Pools

The possibility of separate pools of amino acids for synthesis and storage purposes was first suggested by Cowie and Walton (*47*), and subsequent studies of uptake and metabolism of L-glutamate indicate that nervous tissue contains two pools of this amino acid, one large and one small, with the large pool being relatively inaccessible to the small pool and both having different relations to the tricarboxylic acid cycle. The compartmentation of L-glutamate occurs as the brain matures to the adult form (*22*), the interval of the development of compartmentation varying in different regions of the mammalian CNS (*22*). Attempts have been made to interpret metabolic compartmentation of L-glutamate in terms of particular morphological components such as neurons, glial cells, and, of course, synapses, but these attempts have met with little success to date.

The small L-glutamate pool was first described by Waelsch and his colleagues (*19, 149, 150, 254*) who found that following systemic injection of labeled L-glutamate there is an almost immediate conversion of this labeled material to L-glutamine in the rat brain before it is mixed with whole tissue L-glutamate, the result being that the specific activity of labeled L-glutamine in the brain is four to five times higher than that of L-glutamate after just three min. This occurs only when either L-glutamate, L-glutamine, or a ketogenic precursor is used (*193, 238*). When a glucogenic precursor is used the reverse is true, since carbon from such a source passes through the entire L-glutamate pool before arriving in L-glutamine (*244*). These observations indicate that only a small fraction of tissue L-glutamate normally takes part in the conversion to L-glutamine. In fact, Waelsch calculated that the small L-glutamate pool cannot be larger than 20% of the total pool of this amino acid.

The enzyme system generating L-glutamine from L-glutamate, i.e., glutamine synthetase, is present not only in mammalian brain but also in invertebrate nervous systems such as those of locust, snail, and octopus (*34, 211*). Apart from glutamine synthetase, the small pool of L-glutamate is also associated with glutamate dehydrogenase (*193, 245*) and what has been called a "synthetic tricarboxylic acid cycle", the primary precursors being acetate, butyrate, phenylalanine, proline, tyrosine, bicarbonate, pyruvate, and glucose. The larger pool of L-glutamate is associated with transaminases (*17, 193*), and a "tricarboxylic acid energy cycle." The transaminases presumably have the function of converting their respective amino acids to the corresponding oxoacids in order to keep the tricarboxylic acid cycle intermediates regulated at proper levels (*126*).

Little evidence at present links Waelsch's small L-glutamate pool with a transmitter pool of this amino acid. Since the L-glutamine level in the kitten cerebral cortex does not change significantly during postnatal development, whereas the level of L-glutamate increases markedly (*21*), it seems likely that the small L-glutamate pool is present at birth; therefore this small pool might be localized in cell bodies rather than in axon terminals. This would then account for the low activity of glutamine synthetase found in synaptic endings obtained from the rat brain (except cerebellum) (*215*).

An alternative idea is that the active L-glutamate pool is located in glia (*210*). However, the interval of glial maturation seemingly lags behind that of the development of L-glutamate compartmentation (*47*), although there is a correlation of compartment development with the morphophysiological development of neurons that may represent fine structure differentiation (*47*). It would be very unwise at this time to identify the small, active L-glutamate pool with a transmitter pool, although the concept of metabolic and transmitter pools is seemingly basic to the concept of any proposed transmitter function for L-glutamate. Presumably the transmitter pool is found only in neurons that use this substance as such, the metabolic pool(s) being common to all neurons.

In order to define the sites of the postulated L-glutamate pools, information about the subcellular distribution of enzymes related to the metabolism of this amino acid would be of value. Glutamine synthetase is mainly concentrated in the microsomal fraction (*217*), although some is bound to synaptic membranes (*75a*), whereas the oxidation of glutamate takes place in mitochondria (*9a, 9b*). It is possible that glutamine synthetase activity predominates at sites relatively poor in mitochondria. This seemingly excludes axon terminals; and the observation that incorporation of L-glutamate into protein is less than 1% of that into L-glutamine also suggests that glutamine synthetase in the compartment containing the small L-glutamate pool is associated with membranes that are not the sites of active protein synthesis (*163*). The exclusion of large populations of mitochondria from the small L-glutamate pool is not accepted by Van den Berg and his colleagues (*185, 244, 245*) who, along with others (*18, 193*), have evidence to support the hypothesis that at least two tricarboxylic acid cycles, localized in two populations of mitochondria, are present in brain tissue. One population contains glutamate dehydrogenase, and it is suggested that the mitochondria that it contains are the sites of synthesis of the small L-glutamate pool (*245*).

The localization of enzymes for L-glutamate, L-glutamine, and L-aspartate in the brain and spinal cord has been examined but the findings are difficult to interpret (*97*), and the relationship of the findings to transmitter production is far from being understood. In spinal cord tissue, aspartate aminotransferase has the highest activity, followed in descending order by glutamate dehydrogenase, glutaminase, glutamine synthetase, GABA aminotransferase, and glutamate decarboxylase. For L-glutamate and L-aspartate none of the enzyme activities measured is related to the distribution patterns of these amino acids in the spinal cord. It might thus be concluded that the local high concentrations of these supposed excitatory transmitters in various regions of the spinal cord are consistent with the hypothesis that they are accumulated for synthetic purposes in a specific functional compartment, the patterns for enzyme activity merely reflecting the general metabolic requirements of spinal cord tissue (*97*).

The results of initial studies of the subcellular distribution of L-glutamate in brain homogenates were thought to indicate a uniform distribution of this amino acid throughout the neuronal cytoplasm. This conclusion led investigators to dismiss the effects of L-glutamate on neurons as being unrelated to transmitter function (*214*). There seems to be little doubt that L-glutamate is sequestered in some axon endings in the mammalian CNS, since suspensions of cortical nerve endings excite cortical cells in situ (*146*), and the L-glutamate content of these synaptosomes is sufficient to account for their excitant action (*165*). Furthermore, much of the L-glutamate sequestered by brain slices finds its way into axon terminals (*146b*) where it forms in situ (*146*) an active and readily accessible pool (*36*).

TRANSPORT

We have already seen that some of the excitatory amino acids exist in considerably higher concentrations in neural than in most other mammalian tissues. The distinctive composition of the free amino acid pool of brain tissue in comparison to that of plasma and other tissues was at first thought to be attributable to a barrier limiting the passage of these metabolites between brain and plasma. Apparently such a barrier cannot account for the regional variations in concentrations of brain amino acids referred to earlier in this chapter. It now appears that there are rapid exchanges of amino acids between the plasma and central nervous tissue (*147, 151*), and that the uptake of amino acids by neurons conforms to the criteria of an active transport process, suggesting the participation of carrier-mediated and energy-dependent processes such as those found in other mammalian tissues.

If, as seems likely, the different amino acids are carried by a variety of carriers, it is possible to account for regional heterogeneities of the amino acid pool by having a nonuniform distribution of these carriers. There is no evidence for outward transport of amino acids from resting nervous tissue (*241a*), although passive in-out diffusion occurs. This means that as the intracellular concentration of the amino acid increases, its efflux rate also increases, since efflux, like passive influx, is dependent on concentration (*152, 153*), equilibrium being established when an intracellular concentration is attained at which influx (active plus passive) equals efflux. Much of the recent work on amino acid uptake by brain tissue has been done by using brain slices. Studies of this type demonstrate a substrate specificity similar to that found in such systems as Ehrlich ascites tumor cells (*194*), the intestine (*262*), and kidney (*212*). Instead of a distinct carrier for each amino acid, several transport classes can be distinguished, each specific for a group of amino acids with similar structure and with considerable overlap between classes (27, 28). Interaction between amino acids may occur during transport in brain slices, which may be one of competition for a carrier as seen in other transport systems (27, 28).

L-Glutamate is actively sequestered into brain, spinal cord, and peripheral nerve against considerable concentration gradients for this amino acid (Table 4). Examination of the specificity of the L-glutamate carrier points to a strong interaction between closely related excitatory acidic amino acids, only a slight interaction with neutral amino acids, and none with other substances (27, 28). The carrier appears to be stereoselective, the influx of L-glutamate being much higher than that of its D-enantiomorph (Table 4). L-Glutamine does not compete with L-glutamate for the L-glutamate transport sites in either brain or peripheral nerve (Table 5).

The requirements for binding to the active site for L-glutamate uptake in frog sciatic nerve have been determined by examining the competition between this amino acid and its structural analogues (259). An α-amino

Table 4. Kinetics of uptake of acidic amino acids by nervous tissue

Site	Amino acid	V_m (mol/g/min)[a]	K_m (mol/liter)[b]
Frog sciatic nerve (259)[d]	L-Glutamate	3.7–4.35 × 10^{-9}	2.97–3.48 × 10^{-5}
Rat spinal cord (100)	L-Glutamate	——	5.0 × 10^{-5}
Mouse cerebral hemisphere slices (26)	L-Glutamate	4.0 × 10^{-9}	4.8 × 10^{-4}
	D-Glutamate	1.0 × 10^{-8}	3.0 × 10^{-3}
	L-Aspartate	4.4 × 10^{-9}	4.9 × 10^{-4}
	L-Glutamine	3.3 × 10^{-9}	2.5 × 10^{-3}
	L-Lysine	6.7 × 10^{-10}	1.0 × 10^{-3}
Rat cerebral cortex homogenates (162)	L-Glutamate	——	3.62 × $10^{-5\,c}$ 1.5 × 10^{-3}
	L-Aspartate	——	1.69 × $10^{-5\,c}$ 3.68 × 10^{-4}
	L-Serine	——	2.16 × 10^{-4}
Rat spinal cord homogenates (162)	L-Glutamate	——	1.89 × $10^{-5\,c}$ 1.67 × 10^{-3}
	L-Aspartate	——	2.15 × $10^{-5\,c}$ 3.76 × 10^{-3}
	L-Serine	——	1.08 × 10^{-3}

[a] Maximal velocity of uptake; weights refer to intracellular water.
[b] Michaelis-Menton consant.
[c] High-affinity transport.
[d] Reference number in "Literature Cited" section.
——, Data not reported or relevant.

Table 5. Inhibition of uptake of L-glutamate

Site	Inhibitor compound (B)	Ratio of concentration of L-glutamate to concentration of (B)	% Inhibition	Concentration of L-glutamate K_i^a (mol/liter)
Frog				
Peripheral nerve (259)[b]	L-Aspartate	0.2	24–27	—
		0.1	33–35	—
		0.01	100	—
		0.001	100	—
	L-Glutamine	0.01	(−1)–(−2)	—
	L-Lysine	0.001	25–29	—
	-Aminobutyrate	0.001	(−1)–(−3)	—
Rat				
Spinal cord slices (100)	L-Aspartate	—	—	2.0×10^{-5}
	L-Cysteate	—	—	2.0×10^{-5}
	N-Methyl-DL-aspartate	—	0	—
	DL-Homocysteate	—	0	—
Rat				
Cerebral cortex slices (241)	L-Aspartate	1.0	33.8	—
	L-Glycine	1.0	28.5	—
Mouse	L-Glycine	0.2	8.0–13.0	—
Cerebral hemisphere slices (28)	L-Glycine-methylester	0.2	0	—
	N-Acetylglycine	0.2	0	—
	L-Alanine	0.2	9.0	—
	L-α-Aminobutyrate	0.2	2.0	—
	L-Leucine	0.2	0	—
	L-Serine	0.2	14.0	—
	L-Proline	0.2	12.0	—
	L-Histidine	0.2	1.0	—
	L-Histidine methylester	0.2	0	—
	L-Phenylalanine	0.2	0	—
	N-Acetyl-L-histidine	0.2	1	—
	L-Sarcosine	0.2	12.0	—
	N,N-Dimethylglycine	0.2	0	—

continued

Table 5. *continued*

Site	Inhibitor compound (B)	Ratio of concentration of L-glutamate to concentration of (B)	% Inhibition	K_i^a (mol/liter)	Concentration of L-glutamate
	Betaine	0.2	0	—	—
	β-Alanine	0.2	8.0	—	—
	γ-Aminobutyrate	0.2	7.0	—	—
	δ-Aminovalerate	0.2	0	—	—
	L-2,3-Diaminopropionate	0.2	15	—	—
	L-2,4-Diaminobutyrate	0.2	0	—	—
	L-Lysine	0.2	0	—	—
	L-Arginine	0.2	0	—	—
	L-Aspartate	0.2	64.0	—	—
	D-Aspartate	0.2	59.0	—	—
	N-Acetyl-L-aspartate	0.2	9.0	—	—
	L-Glutamate	0.2	50.0	—	—
	D-Glutamate	0.2	2.0	—	—
	L-Glutamine	0.2	12.0	—	—
	L-Glutamic acid	0.2	23.0	—	—
	-methylester	0.05	19.0	—	—
	L-Glutamic acid	0.2	23.0	—	—
	γ-dimethylester	0.05	2.0	—	—
	N-Acetyl-L-glutamate	0.2	0	—	—
		0.05	0	—	—
	Dicarboxylic acids (succinate and glutarate)	0.2	4	—	—
		0.05	0	—	—
	L-Asparagin	0.2	10.0	—	—
Synaptosomes					
Spinal cord (*162*)	L-Aspartate	—	—	3×10^{-5}	
Cerebral cortex (*162*)	L-Aspartate	—	—	3×10^{-5}	

^{a}Michaelis-Menten inhibitor constant.
^{b}Reference number in "Literature Cited" section.
—, Data not reported or relevant.

group, a terminal carboxyl group, and a carboxyl group adjacent to the amino group are necessary. The carboxyl groups and the α-amino group must be in a particular arrangement, since D-glutamate does not compete with L-glutamate for uptake (Table 5). On the basis of these studies, a three-point attachment of L-glutamate to the carrier via carboxyl and amino groups is suggested and, since these groups are both ionized at physiological pH (41, 222), this attachment may be ionic (259). In many respects the binding properties of the L-glutamate carrier bear a striking resemblance to those of the postulated amino acid receptor on mammalian central neurons except, of course, for the moot point concerning stereospecificity.

L-Cysteate is as effective as L-aspartate as an inhibitor of L-glutamate uptake, whereas N-methyl-DL-aspartate and DL-homocysteate, compounds that are sometimes more potent as excitors than is L-glutamate, have no effect on active uptake of L-glutamate by spinal cord slices (100). The initial influx of many amino acids into brain cells, including L-glutamate, can be described phenomenologically by Michaelis-Menten kinetics (26). The Michaelis-Menten constant (K_m) and the maximal rate (V_m) for influx of L-glutamate and L-aspartate for vertebrate brain, spinal cord, and peripheral nerve are presented in Table 4, whereas some values for inhibition of L-glutamate uptake by L-aspartate and other compounds are presented in Table 5.

When the rat sciatic nerve was incubated in a medium with 1 mM of labeled L-glutamate, the level of the latter in the nerve gradually increased until it was above that of the medium (264). Although this result is qualitatively similar to those obtained with the mammalian brain and spinal cord slices, there are important quantitative differences between the peripheral and central systems. Uptake by nerve, e.g., proceeded at a much slower rate, so that after a three-hr incubation the rat sciatic nerve had sequestered about 18 times less L-glutamate/ml of tissue water than rat cortex slices incubated for one hr. That cortex tissue contains about eight times as much L-glutamate/ml of tissue water as sciatic nerve may partly account for this difference, the sciatic nerve having a corresponding lower transport capacity than have brain slices (264). Although the affinity of the carrier for L-glutamate is seemingly lower than that for cortical or spinal cord slices, mediated transport processes in mammalian sciatic nerve seem possible, especially since this amino acid is not significantly metabolized or incorporated into protein after its entry (264).

The accumulation of amino acids against a concentration gradient is necessarily dependent on a source of energy (1, 230). Uptake is therefore suppressed by metabolic inhibitors and does not occur at low temperatures

(*241a*). Although the available energy supply has a direct influence on transport of L-glutamate and other amino acids, it is unlikely to be a determinant of amino acid levels and distribution, since a decrease in available energy results in a proportionate decrease in transport of all amino acids.

In some tissues amino acid uptake appears to be potassium dependent but in others a marked sodium-dependence has been demonstrated. Margolis and Lajtha (*166*) found that L-glutamate uptake by slices of mouse brain was lowered by reducing the external sodium concentration ($[Na^+]_o$), but more markedly lowered by reducing the external potassium concentration. Their results clearly demonstrate that an electrochemical gradient for sodium is not essential for L-glutamate and L-aspartate uptake, since uptake still occurs when the ratio $[Na^+]_o/[Na^+]_i$[1] is either unity or less than unity. Perhaps sodium affects the affinity of the carriers for these amino acids (*166*). Reduction of $[Na^+]_o$ to zero affects uptake of L-glutamate by rat sciatic nerve only to a very limited extent (*264*).

It seems likely that the differences in the rates of transport of different amino acids found in various areas of the brain may account for the heterogeneous distribution of free amino acids in the mammalian CNS (*11, 158, 159*), although elucidation of the relationship between regional characteristics of transport processes and function remains difficult. Subcellular studies of uptake of excitant acidic amino acids by the mammalian brain and spinal cord are one means of elucidating this possible relationship, particularly with respect to synaptic function.

Although Krnjević and Whittaker (*146*) clearly demonstrated a high L-glutamate concentration in synaptosomes from cortical tissue, subcellular fractionation studies of guinea-pig brain (*165*) failed to show a unique synaptosomal localization for any amino acid, all amino acids examined having the same subcellular distribution as cytoplasmic markers. However, recent amino acid uptake studies support the contention that specific populations of synaptosomes from mammalian central nervous tissue sequester L-glutamate and L-aspartate (*146b, 198, 263*). Synaptosomal fractions from the rat cortex, e.g., possess a high affinity transport system for L-glutamate and L-aspartate (Figure 2).

The efficiency of the high affinity uptake system for L-aspartate in the rat cortex apparently increases during development, accumulation of L-aspartate by synaptosomal fractions from this tissue being three- to four-fold higher in adult material (40 days) than in immature material (seven to nine days) (*198*). Low-affinity amino acid uptake systems,

---

[1] $[Na^+]_i$, internal sodium concentration.

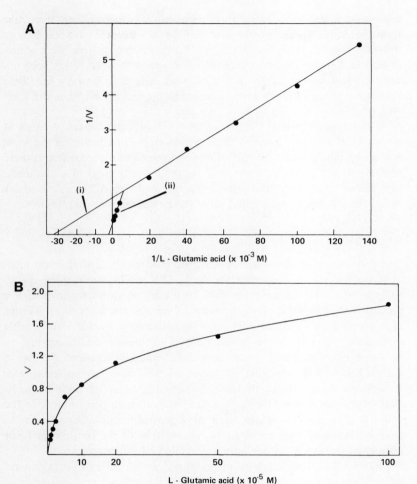

Figure 2. A. Double-reciprocal plot of velocity (v) of L-glutamate accumulation in homogenates of rat cerebral cortex at different L-glutamate concentrations. Each point is the mean of four determinations. Velocity is expressed as μmol of L-glutamate/g/4 min. (i) High affinity system (K_m = 3.6 ± 1.3 × 10^{-5} M); (ii) low affinity system (K_m 1.5 ± 0.9 × 10^{-3} M). B. Computed least squares fit of the same velocity of L-glutamate uptake data as in A, assuming transport by a two-component system. Analysis of these data assuming a one-component system gave a poor fit. (From Logan, W. J., and Snyder, S. H.: *Nature* 234:297–299, 1971.)

apparently common to all neurons, presumably sequester their respective amino acids for the metabolic pool. Although it has not yet been demonstrated unequivocally that the high-affinity uptake systems for L-glutamate and L-aspartate are restricted to neurons that are electrophysiologically and biochemically identifiable as cells that actually use these substances as neurotransmitters, it seems likely that the evidence will soon be forthcoming.

The demonstration of a high-affinity L-glutamate uptake system at terminals of primary afferent fibers would go far in convincing the most sceptical about the possible role of this compound as a transmitter at these sites. Such a system might not only maintain a high level of L-glutamate near presynaptic transmitter release sites but might also play a role in inactivating the transmitter L-glutamate after its release into the synaptic cleft.

RELEASE

For a substance to function as a synaptic transmitter, a mechanism must exist for its release from axon terminals. This mechanism may be specifically localized to the axon terminal, or it may be an intensification of a process that is found in all of the neuronal membrane. Since most neurons have a very high tissue:medium ratio for acidic amino acids, one possible mechanism for ensuring synaptic transfer of information from pre- to postsynaptic cells would be to allow for a dramatic transient increase in permeability of the presynaptic terminal membrane to the transmitter amino acid concomitant with the arrival of either an action potential or its electrotonic derivative. Of course, a specific increase in permeability to the amino acid in question is not essential; a general permeability increase to amino acids is all that is necessary, recognition of the transmitter being assured by postsynaptic membranes programmed for this purpose.

Release of amino acids from peripheral nerve during excitation may serve as a model of the way in which these compounds are released from axon terminals, although this approach may seemingly exclude a transmitter storage and release role for synaptic vesicles. Wheeler, Boyarsky, and Brooks (260) studied the release of a variety of amino acids from desheathed sciatic nerves of bullfrogs with this principle in mind. They found that these nerves release only L-glutamate at greatly increased (> 200%) rates (2.5×10^{-2} pM/cm^2/sec) during stimulation, the increased rate of release of other related substances, like L-aspartate, with stimulation being much less (about 14% above the resting rate; Figure 3). The extra release of L-glutamate during stimulation is not affected by ouabain although it is eliminated by sodium azide (0.2 to 1.0 mM) (Figure 3) and choline (10 mM). They suggested that azide prevents oxidative phosphorylation, which in turn prevents the supply of energy necessary for active extrusion of L-glutamate and other compounds from nerve during stimulation. Since the nerve action potential was unaffected by azide in these experiments it seems unlikely that the additional L-glutamate released during stimulation was linked to the sodium and potassium permeability changes associated with the action potential.

Studies with L-glutamate analogues lead to the conclusion that there is an active extrusion process for L-glutamate, or another substance which carries L-glutamate, with a three-point attachment site for release. The inhibition of release by choline could be explained by assuming that a site present on the carrier molecule can be saturated by choline so that L-glutamate is prevented from attaching and being released during stimulation (260). These initial findings on L-glutamate release from stimulated peripheral nerve are generally confirmed by recent studies of De Feudis (75), who also obtained additional data that led him to conclude that the additional release of L-glutamate during stimulation is in some way linked to depolarization of the axon membrane, the rate of release for L-glutamate being increased, e.g., when peripheral nerve is depolarized by 100 mM of potassium (75).

The release of biogenic amino acids from the mammalian brain after electrical stimulation has been demonstrated both in vivo and in vitro. Acidic amino acids, including L-glutamate and L-aspartate, are released from the rat brain by procedures capable of inducing neuronal depolarization, i.e., electrical stimulation (131) and high potassium concentrations (32, 164), although Katz, Chase, and Kopin (131) found that the release from the stimulated rat striatum is not specific for those amino acids thought to subserve a transmitter role. Roberts and Mitchell (209) obtained a calcium-dependent increased rate of release of L-glutamate, L-aspartate, glycine, and GABA (but not of other amino acids) from isolated frog and toad hemicords when they directly stimulated the rostral cord, but obtained no increase when they stimulated the spinal roots. They suggested that their results support the possibility that these compounds may be acting as neurotransmitters in the spinal cord.

Similar results were obtained in two other laboratories (101, 184). Roberts and Mitchell have suggested that Katz et al. (131) obtained a nonselective release of amino acids because they used a "serum substitute" medium containing L-glutamate (4.9 mM), which enhances the efflux of L-glutamine, L-leucine, and L-lysine from rat cortical slices (6).

Amino acids are also released either from parts of the intact mammalian brain (124, 125) or from isolated parts of this organ (6), and

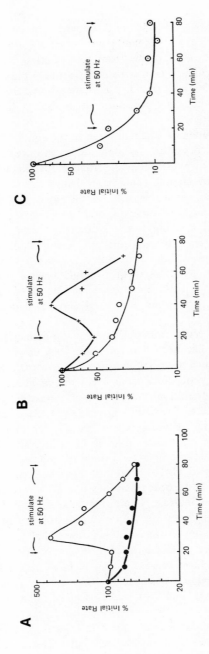

Figure 3. Effect of A, electrical stimulation on rate of release of labeled L-glutamate from frog sciatic nerve. *Key:* •, control; ○, stimulated. B, 40 to 50 min exposure to ouabain (0.1 mM) and sodium azide (0.2–1.0 mM) on rate of release of labeled L-glutamate during stimulation of frog sciatic nerve. *Key:* +, ouabain; ○, azide C, stimulation on rate of release of labeled L-glutamate from frog sciatic nerve soaked for 24 hr in 10 mM of choline (nonisotopic). Nerves were incubated for 24 hr in labeled L-glutamate before stimulation. (From Wheeler, D. D., Boyarsky, L. L., and Brooks, W. H.: *J. Cell. Physiol.* 67:141–148, 1966.)

evidence exists that the type and amount of these amino acids found in whole brain perfusates are dependent on the physiological state of the experimental animal. A significantly greater amount of L-glutamate, e.g., is released from the cat brain during arousal than during rest (*124*), which could be regarded as consistent with the supposed excitatory action of this substance. In the quiescent cat, L-glutamate is released at rates of 7 to 9 μg/hr/cm^2 from the perforated pial surface of cerebral cortex, and it decreases to ca. 70% of this amount during sleep or following lesions of the mesencephalic reticular formation. The rate of release of L-aspartate and L-glutamine does not change (*124*).

A recent extension of these studies demonstrates that an increased liberation of acidic amino acids, particularly L-glutamate, can be obtained by stimulating the reticulocortical system (*125*). Electrical stimulation of the mesencephalic reticular formation at the level of the superior colliculus caused a marked increase in liberation of L-glutamate, whereas no such change could be induced by stimulation of the mesial thalamic recruiting system after lesions of the mesencephalic reticular formation, even though liberation of acetylcholine was increased at this time (*125*). These data suggest that stimulation of specific neural pathways is required to release L-glutamate which, of course, agrees closely with the suspected transmitter role of this substance.

L-Glutamate is released from the isolated chick retina during glutaraldehyde fixation (*249*), a finding that supports other neurochemical (*196a*) and electrophysiological (*188, 241*) evidence that this amino acid functions as a retinal transmitter. This compound is also released from rabbit brain during "spreading depression" although other amino acids are released at this time (*248, 250*). Snodgrass and Iversen (*221a*) studied release of L-glutamate (and GABA) from chopped rat cerebral tissue and found that this amino acid, together with its decarboxylated end product, was the major amino acid released during electrical stimulation. This result was obtained in seven of ten experiments despite a considerable variation in spontaneous release between experiments. There was a smaller increase in release of L-glutamate in response to potassium depolarization than to electrical stimulation.

Finally, when suspensions of synaptosomes from the mammalian CNS are stimulated either electrically or by potassium, they release L-glutamate and L-aspartate (*33, 74*). In fact L-glutamate and L-aspartate (and GABA) are preferentially released by stimulation of rat cortical (*33, 74*) and sheep hypothalamic synaptosomes (*85*). The release of amino acids from brain synaptosomes is calcium dependent (*33, 74, 85*). Since in their studies Snodgrass and Iversen (*221a*) did not obtain release of L-aspartate with

L-glutamate, they suggested that the brain may release L-glutamate independent of L-aspartate.

NEUROPHYSIOLOGY AND PHARMACOLOGY

Responses to Topical Application

The first descriptions of the excitatory actions of acidic amino acids on the mammalian brain were reported by Hyashi in the 1950s (*116–121*). Performed before World War II, these studies dealt with the effects of sodium glutamate injections into the gray matter of the canine cerebral cortex. Initially Hyashi's observations received scant attention and, at the time of their publication, there was certainly no reason to consider L-glutamate as a possible transmitter at cerebral synapses. Since the appearance of Hyashi's papers, additional investigations along this line revealed that L-glutamate and a number of other acidic amino acids cause "spreading depression" when applied in vivo to the cerebral cortices of rabbits (*246, 247*) and cats (*67, 134, 206, 256*), as well as in vitro to isolated slabs of cat cortex (*90*), the electrical silence being preceded by a phase of intense neuronal activity. Thus it seems likely that the action of these substances is related in some way to neuronal excitation. With this in mind van Harreveld (*247*) and Curtis and Watkins (*67*) tested the ability of several compounds either chemically or metabolically related to L-glutamate to produce "spreading depression" in the rabbit cortex.

The threshold concentrations of those compounds that initiate "spreading depression" are given in Table 6. The initiation of this phenomenon is not specific to L-glutamate; D-glutamate is slightly more effective than its stereoisomer (*247*), and *N*-methylaspartate (*67*) is more effective than either D- or L-glutamate, but this finding may indicate that D-glutamate and *N*-methylaspartate are neither metabolized nor sequestered rapidly by nervous tissue and, as a result, they achieve a relatively high concentration at the site of action (*67*). The effectiveness of L-asparagin and L-glutamine in causing "spreading depression" is somewhat difficult to explain since these substances are ineffective when applied iontophoretically to central neurons. It is suggested that they may be converted enzymatically to active substances when they diffuse into brain tissue (*66*). Pentylenetetrazol (Metrazol) enhances the sensitivity of the rabbit cortex to L-glutamate and L-aspartate. Although van Harreveld (*247*) suggested that Metrazol inhibits the enzyme reactions that use L-glutamate and L-aspartate, thus allowing for the buildup of a higher concentration of these amino acids in the cortex, it would nevertheless be of interest to determine whether or not

Table 6. Threshold concentrations of amino acids causing "spreading depression"

Compound	Threshold concentration (mmol/liter)
N-Methyl-D-aspartate	7.6×10^{-2} (256)[b]
D-Homocysteate	2.5×10^{-1} (256)
DL-Homocysteate	5×10^{-1} (256)
N-Methyl-L-aspartate	7×10^{-1} (256)
D-Glutamate	3.6 (246)
D-Aspartate	$>1 \times 10^{1}$ (246)
L-Asparagin	1.3×10^{1} (246)
L-Glutamate	1.5×10^{1} (246)
L-Aspartate	1.7×10^{1} (246)
L-α-Aminoadipate	3.5×10^{1} (246)
L-Glutamine	5.0×10^{1} (246)
L-Proline	1×10^{2} (246)
Succinate	2×10^{2} [a] (246)
α-Ketoglutarate	NM (246)
DL-α-Aminobutyrate	NM (246, 256)
Glutathione	NM (246, 256)
L-Ornithine	NM (246)
γ-Aminobutyrate	NM (246)
Glutarate	NM (246)

[a]Hypertonic solution.
[b]Reference number in "Literature Cited" section.
NM, threshold so high that it was not measurable.

this substance affects uptake of glutamate and aspartate by mammalian central nervous tissue.

Following extracellular studies on the effects of topically applied ω-amino acids on surface potentials of the mammalian brain, it was suggested that such acids of a six-carbon chain length or less and ω-guanidino amino acids of two- or three-carbon chain length selectively block excitatory postsynaptic potentials and eventually cause spreading depression (*116–119, 205, 206*). However, the interpretation of such extracellular studies is difficult, and for more conclusive evidence, one must look to studies using the iontophoretic technique for the application of these amino acids to central neurons, together with intracellular recordings of neuronal events.

Studies on ion and water movement between extra- and intracellular compartments of cortical brain slices show that, when first applied, L-glutamate increases intracellular sodium but decreases intracellular potassium without concomitant movement of water (*103*), changes that accord closely with the concept of neuronal depolarization by this amino acid. Indeed, it has been convincingly demonstrated that isolated guinea-pig cerebral tissues are depolarized by topically applied acidic amino acids (*35, 95, 110*), a five-fold increase in sodium permeability relative to potassium permeability occurring at this time (*35*). The later stages of the action of topically applied L-glutamate are characterized by increased intracellular potassium and water and a notable increase in metabolic expenditure (*103*). Relatively low concentrations (0.5 to 5.0 mM/liter) of L-glutamate also increase intracellular water and chloride of the rat retina (*2, 3*).

Iontophoretic Studies

Direct applications of excitatory amino acids to the surface of either the intact brain or brain slices do not give particularly definitive answers on site and mode of action of these substances because neuronal excitability is often abolished by an excessive excitation, and also because of the complexity inherent in the organization of the mammalian brain. The iontophoretic technique obviates some of the difficulties of interpretation associated with topical applications of acidic amino acids, although we shall find that it has not yet supplied unequivocal information on the site of action of these substances on central neurons.

Iontophoretic studies with L-glutamate and other excitatory amino acids demonstrate the remarkably widespread distribution of sensitivity to these compounds throughout the mammalian CNS (Table 7). Indeed D- and L-glutamate excite neurons in almost all of the nervous system regions that have been so far examined, although not all neurons in these regions are equally responsive to these amino acids, and some fail to respond.

When L-glutamate is applied iontophoretically to brain neurons, the pattern of response recorded from the somata of these cells depends on the amount of drug applied. Small subliminal doses "enhance" responses evoked by other inputs either through presynaptic facilitation (*65, 71, 72*) or through depolarization of either the postsynaptic or nonsynaptic membrane (or both) of the postsynaptic cell, i.e., summation. Larger doses, producing an extracellular concentration of about 10^{-4} M, cause an outright discharge at a frequency related to the amount of glutamate released (Figure 4), and maintained application at this concentration causes prolonged firing without any obvious sign of desensitization. Very high doses have a predominantly depressant action, usually preceded by a phase of strong excitation (see section on "Spreading Depression" in this chapter).

Perhaps one of the most striking features of the action of L-glutamate when applied iontophoretically is its short latency of onset (rarely > 1 sec)

Table 7. Positive responses to iontophoretically applied natural acidic amino acids

Site	Model	L-Glutamate	L-Aspartate	L-Cysteate
Forebrain				
Cerebral cortex	Guinea pig	(146)[a]	—	—
	Cat	(48–49, 59, 61, 67, 73, 138, 139, 141–145, 157, 169, 175, 176, 201, 202, 224, 225)	(48–50, 59, 67, 73, 142, 144)	(67, 142, 144)
	Rabbit	(142, 144)	(142, 144)	(142, 144)
	Monkey	(144)	(144)	
	Rat	(227–229)		
Hippocampus	Cat	(24, 25, 107, 223, 225)	—	—
	Rat	(227–229)	—	—
Thalamus	Cat	(4, 5, 54, 59, 65, 76, 99, 99a, 172, 173, 176, 199)	(59, 99a)	(99a)
Olfactory bulb	Rabbit	(12)	—	—
Corpus striatum	Cat	(265)	—	—
Putamen	Rabbit	(54, 265)	—	—
Subthalamic nucleus	Rabbit	(105, 108)	—	—
	Rat	(227–229)	—	—
Caudate nucleus	Cat	(29, 105)	—	—
	Rabbit	(108)	—	—
Geniculate nucleus	Cat	(54–56, 59, 68, 181, 200, 239, 240)	(59, 68, 181)	(68)

continued

Table 7. continued

Site	Model	L-Glutamate	L-Aspartate	L-Cysteate
Amygdala	Rabbit	(106)	—	—
	Cat	(231)	—	—
Midbrain colliculus	Cat	(61)		
Hindbrain				
Pons-medulla	Cat	(30, 37, 39, 54–56, 59, 61, 68, 115, 181, 239, 240)	(115)	—
Cuneate nucleus	Cat		(99a)	(99a)
	Monkey	(91)		
Gracilis nucleus	Cat	(179, 226)	—	—
Cerebellum	Cat	(42, 142, 144, 169, 170, 171)	(142)	(142)
	Rabbit	(142, 144)	(142)	(142)
	Monkey	(144)		
Spinal cord				
Afferent neurons	Cat	(65)	—	—
Efferent neurons	Cat	(23, 53, 59, 62–64, 66–68)	(59, 62–64, 66, 68)	(64, 66, 68)
Interneurons	Cat	(23, 58, 59, 63, 64, 66–68, 266, 99a)	(58, 59, 62–64, 66–68, 99a)	(64, 66–68, 99a)
Renshaw cells	Cat	(58, 59, 62–64, 66, 68)	(58, 59, 62–64, 66, 68)	(64, 66, 68)

[a]Reference numbers in "Literature Cited" section.
——, Data not reported or relevant.

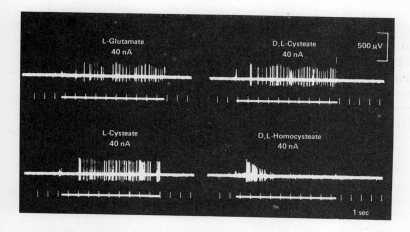

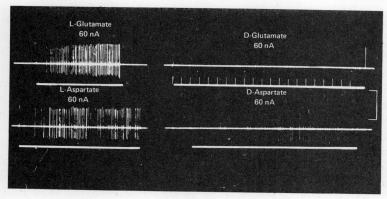

Figure 4. Effect of A, L-glutamate and sulfonated derivatives applied iontophoretically with equal ejection currents near unit in cat's cortex under Dial. Note block of spike production presumably owing to excessive depolarization by DL-homocysteic acid. B, Comparing actions of L- and D-isomers of glutamate and aspartate on unit in cat's cortex under Dial. Substances were similar, near-saturated solutions, at pH 8.5; iontophoresis was by identical inward currents. Spikes recorded extracellularly. (From Krnjević, K., and Phillis, J. W.: *J. Physiol.* (Lond.) *165*:274–304, 1963.)

and even more rapid termination (Figure 4). Postresponse depression following iontophoretic application of this amino acid can occur when the spontaneous firing frequency is lower than normal. When spike blockage is achieved with high concentrations—presumably by excessive depolarization—recovery is achieved within a few seconds following termination of the iontophoretic current.

Although L-glutamate is active in all regions of brain studied so far (Table 7), and although most central neurons respond to the application of this amino acid with the production of spikes, it is not equally effective on all cells in these regions. Some quiescent cells in the cortex, e.g., are only partially depolarized by L-glutamate and do not generate spikes. The suggestion that these are glial cells (138) seems unfounded since the latter are apparently not depolarized by L-glutamate (145). Neurons in different parts of the thalamus vary markedly in their responsiveness to L-glutamate, those in superficial regions being relatively insensitive whereas those in deep areas are strongly excited by this amino acid (176). In the olfactory bulb very few (about 18%) neurons are sensitive to L-glutamate (13). That some neurons in this region exhibit a decrease in the rate of spontaneous activity without gross depolarization when exposed to this amino acid suggests the possibility of either desensitization of excitatory receptors on these neurons by L-glutamate or a nonsynaptic action for this amino acid.

Alternatively, L-glutamate may act as an inhibitory transmitter at some central synapses. This would also explain the reduction in discharge rate of cells in dorsal column nuclei when treated with this amino acid (226). There are differences in sensitivity to L-glutamate and L-aspartate in cat lateral geniculate neurons receiving inputs from different parts of the retina (181). For neurons in the inferior field, there appears to be a gradient of sensitivity to glutamate and aspartate that is correlated with the distance of the receptive field from the center. Cells in the center of the field are more sensitive to L-glutamate than to L-aspartate, whereas the inverse applies to cells in the inferior field. Those in the intermediate field are equally sensitive to these amino acids.

Differences in sensitivity of cat medullary neurons, i.e., those in the dorsal column nuclei (nucleus cuneatus and nucleus gracilis), to L-glutamate and other amino acids have also been reported (226). In this part of the brain, however, the variability in cell sensitivity cannot be related to the location of the cell in the brainstem. Differences in sensitivity of cortical neurons have been observed, although some reports suggest that there are no significant differences in the sensitivity of particular classes of cortical neurons to excitant amino acids (57, 69).

The only concrete and possibly significant generalization that can be derived so far from these fragmentary data is that neurons synapsing with primary afferent terminations appear to exhibit highly developed sensitivities to natural acidic amino acids.

The relative potencies of acidic amino acids on a generalized vertebrate central neuron are described in Table 8, whereas Table 9 contains a list of

Table 8. Hierarchical arrangement of acidic amino acids according to potency[a] on generalized vertebrate neuron

N-Methyl-D-aspartic acid (68)[b]
D-Homocysteic acid (68)
N-Methyl-DL-aspartic acid (68)
DL-Homocysteic acid (68)
DL-Ibotenic acid (129)
N-Iminomethyl-D-aspartic acid (68)
N-Ethyl-D-aspartic acid (68)
N-Ethyl-DL-aspartic acid (68)
DL-2-Amino-4-sulfino-N-butyric acid (68)
L-Cysteic acid (66, 68)
L-Cysteine sulfinic acid (66)
L-2-Amino-3-sulfinopropionic acid (68)
N,N-Propyl-D-aspartic acid (68)
L-Homocysteic acid (68)
γ-Methylene-L-glutamic acid (59)
γ-Fluoro-L-glutamic acid (59)
L-Glutamic acid (66, 68)
L-Aspartic acid (66, 68)
N-Methyl-DL-glutamic acid (68)

N-Methyl-L-glutamic acid (68)
N-Methyl-D-glutamic acid (68)
N-Methyl-L-aspartic acid (66, 68)
DL-2-Amino-5-sulfo-N-valeric acid (68)
N,N-Dimethyl-DL-aspartic acid (66, 68)
D-Glutamic acid (66, 68)
β-Aminoglutaric acid (66)

D-Aspartic acid (66, 68)
D-Cysteic acid (68)
N-Methyl-DL-cysteic acid (68)
β-Hydroxyglutamic acid (68)
N-Methyl-DL-homocysteic acid (68)
N,N-Dimethyl-L-glutamic acid (66)
N-Iminomethyl-L-aspartic acid (68)
DL-2-(Aminomethyl) succinic acid (66)
N-Ethyl-L-aspartic acid (68)
DL-α-Aminopimelic acid (66)
N-Methyl-L-cysteic acid (66, 68)
Aminomalonic acid (66)
DL-α-Aminoadipic acid (66, 68)
α-Methyl-DL-aspartic acid (59)[c]
DL-N-Allyl-aspartic acid (59)[c]
DL-N-2-Hydroxyethylaspartic acid (59)[c]
DL-N-Isopropylaspartic acid (59)[c]
DL-threo-β-hydroxyaspartic acid (59)[c]

γ-Methyl-L-glutamic acid (59)[c]
γ-Hydroxymethyl-L-glutamic acid (59)[c]
ς-Carboxymethyl-L cysteine (59)[c]
DL-Serine-O-sulfate (59)[c]
DL-Guanidinosuccinic acid (59)[c]
meso-α,β-Diaminosuccinic acid (59)[c],[d]

[a]Potencies determined by effect of iontophoretic application on extracellular spike frequency.
[b]Reference number in "Literature Cited" section.
[c]Active but potency not specified; potencies determined by effect of iontophoretic application on spike frequency.
[d]See Table 9.

Table 9. Inactive acidic amino acids and structurally related nonamino compounds that do not excite mammalian central neurons

N,N-Propyl-L-aspartic acid (68)[a]	L-Glutamic acid-γ-methyl ester (66)
O-Phosphoryl-L-serine (66)	N-Phthalyl-DL-glutamine (59)
DL-O-Tyrosine (66)	L-Glutamine (66)
DL-β-Aminoadipic acid (66)	N-Phthalyl-DL-glutamic acid (59)
γ-Aminopimelic acid (66)	L-Glutamic acid-γ-methylamide (66)
2-Aminoethylmalonic acid (66)	L-Glutamic acid-γ-dimethylamide (66)
Iminodiacetic acid (66)	γ-L-Glutamylglycine (66)
α-Methyl-DL-glutamic acid (66)	L-Glutamic acid-γ-hydrazide (66)
meso-α,β-Diaminosuccinic acid (66)	L-Isoglutamine (66)
L-α-ε-Diaminopimelic acid (66)	L-Glutamic acid-α-methyl ester (66)
N-Methyl-DL-glutamic acid (66)	α-L-Glutamylglycine (66)
α-Glutarobetaine (66)	Pyroglutamic acid (66)
N-Acetyl-L-aspartic acid (59, 66)	Malonic acid (66)
N-Carbamyl-DL-aspartic acid (66)	Succinic acid (66)
N-Acetyl-L-glutamic acid (59, 66)	Glutaric acid (66)
N-Carbamoyl-L-glutamic acid (66)	L-Malic acid (66)
L-Asparagin (66)	Ethylamine (66)
D-Asparagin (49)	Propylamine (66)

[a]Reference number in "Literature Cited" section.

inactive acidic amino acids and structurally related substances. Although the relative positions of neighboring compounds in Table 8 are somewhat arbitrary, the only contentious feature of the hierarchical arrangement of the amino acids in this table concerns the relative positions of D- and L-glutamate. Curtis and Watkins (68) found that D-glutamate was approximately one-half to two-thirds as potent as the L-isomer, and this was also found by Andersen and Curtis (5) to be the case for thalamic neurons. However, according to Krnjević (139) and Krnjević and Phillis (142, 144), cat, rabbit, and monkey cortical cells are more sensitive to L-glutamate than to its steroisomer (Figure 4), the usual sequence for these cells in descending order being L-cysteate>L-glutamate>L-aspartate>D-aspartate> D-glutamate.

This profile is further clouded by the finding that cells of the cat pericruciate cortex do not behave very differently toward iontophoretically administered D- and L-glutamate, L-glutamate being only 1.25 to 2.0 times more effective than D-glutamate (49). This accords with data obtained from lateral geniculate neurons (68), neurons of the ventrobasal thalamus (52), and cells of the cat cuneate nucleus (226). According to Crawford and Curtis (49), some of the discrepancies between their results and those of Krnjević and Phillis (142, 144) are attributable to differences in technique, although this explanation has obviously failed to satisfy Krnjević (140). Possibly the stereospecificity of glutamate receptors on mammalian central neurons differs in different parts of the CNS.

Alternatively, mammalian neuron receptors may form subpopulations with varying stereospecificities. Differences in the relative densities of these populations may conceivably account for the recorded differences in sensitivity to L- and D-glutamate. The occurrence of two populations of amino acid receptors on central neurons, i.e., a population of synaptic receptors with high affinity for one or more of the natural acidic amino acids and a population of nonsynaptic receptors with high affinity for substances like DL-homocysteate, may also account for the recorded regional differences in relative sensitivity to different acidic amino acids, e.g., to L-glutamate and DL-homocysteate, although DL-homocysteate, like some other unnatural acidic acids, is usually much more potent than any of the natural acidic amino acids (Table 8; Figure 4A).

The suggestion that some central neurons may have two types of acidic amino acid receptors was previously advanced by McLennan, Huffman, and Marshall (176). This observation accords closely with the concept that insect muscle fibers may have two types of acidic amino acid receptors, namely, highly specific synaptic glutamate receptors coupled to ionophores for sodium and possibly one or more other ionic species (*243*), and less specific nonsynaptic receptors (*155*) coupled to ionophores for chloride (*154, 156*). It may be significant that the excitatory synaptic receptors on insect and on crustacean muscle fibers have a much higher (× 100) affinity for L- than for D-glutamate (*233, 243*).

The suggestion that differences in the properties and distribution of central amino acid receptors may account for the fact that surface stimulation of the cortex inhibits firing of pyramidal cells excited by L-glutamate for much longer periods than firing at the same frequency induced by DL-homocysteate (*115*) has been challenged, because substances like DL-homocysteate are unlikely to be inactivated as effectively as L-glutamate. As a result there may well be differences in the distribution of L-glutamate and DL-homocysteate in cortical tissues after iontophoretic application of these compounds and, therefore, differences in the area and location of membranes excited by each of these substances in relation to the location of the inhibitory synapses activated by surface stimulation. Nevertheless, L-aspartate, a natural amino acid like L-glutamate, excites neurons throughout the CNS, yet regional differences in the relative sensitivities of neurons to these two amino acids occur. Possibly these differences truly reflect differences in properties of synaptic acidic amino acid receptors on these cells.

The excitant action of L-glutamate on mammalian neurons is simulated by a variety of structurally related compounds (Table 8). The major requirements for excitant activity are an amino group and two acidic groups. The acidic group may be carbonyl, sulfonic or sulfinic, and the optimal separation between the α-amino group and one of the acidic groups is two to ten C-atoms (*66*). Assuming some stereospecificity, these observations would be consistent with a three-point receptor for the excitatory amino acids (*66*) similar to that postulated for acidic amino acid receptors on arthropod muscle fibers (*243*). The longer chain ω-amino monocarboxylic acids like ϵ-aminocaproic and ω-aminocaprylic acids, which might either excite brain excitatory synapses (*117, 119*) or selectively block brain inhibitory synapses, thereby causing enhancement of ongoing excitatory activity when applied topically (*206*), give somewhat ambiguous results when applied iontophoretically (*49, 66, 138, 142*), sometimes showing both excitation and inhibition. For example, ω-aminocaproic acid has a slight depressant effect on cortical neurons but, after a latency of 15 to 90 sec, a paroxysmal high frequency firing is produced by this compound and by ω-aminocaprylic acid, which continues for as long as 25 sec after termination of the iontophoretic ejection current (*49*).

These observations are not consistent with the excitation experienced with L-glutamate and are not observed when neutral forms of ω-aminocaprylic acid and ϵ-aminocaproic acid are applied on cortical neurons, although the latter depress the firing of cortical neurons induced by DL-homocysteate (49). Perhaps large pH changes occur when the cationic form of the ejected amino acid releases its extra proton on reaching the extracellular medium around the micropipette (49).

In some respects, studies of the effects of acidic amino acids on mammalian spinal neurons have been more instructive than those on mammalian brain cells. Spinal motoneurons, Renshaw cells, and interneurons are all excited by iontophoretically applied L-glutamate (Table 7). Intracellular application of this amino acid to motoneurons does not significantly affect the membrane potential of these cells (8, 43); neither does extracellular application of L-glutamate to somata and axon processes of spinal afferent neurons (187). Furthermore, the excitatory amino acids are effective at concentrations that are similar to those of acetylcholine required to excite cholinoceptive neurons (69). It may seem reasonable to assume therefore that there are receptors for L-glutamate at synapses on spinal neurons.

Nevertheless, in view of the widespread occurrence of acidic amino acid sensitivity throughout the mammalian CNS, it is suggested that this sensitivity is unrelated to synaptic function (49, 52, 56, 61, 63, 67–69). However, there is little support for this concept at the present time, even though the exact site(s) of action of the excitatory amino acids has not yet been unequivocally established. There is, however, a growing awareness of the possibility that the distribution of sensitivity to acidic amino acids on spinal neurons may be more complex than was first imagined and, although a discrete population of acidic amino acid receptors may well be present at excitatory synapses on these cells, one or more populations of receptors for these substances may occur nonsynaptically (23).

High concentrations of acidic amino acids, particularly the more potent ones, cause rapid depolarization block of spinal neurons. As is also the case for brain neurons, this effect does not apparently result from desensitization but is mainly a consequence of maintained alteration in neuronal membrane conductance and potential. At arthropod somatic excitatory nerve-muscle synapses, both bath-applied and iontophoretically applied L-glutamate, but not L-aspartate, cause rapid desensitization to both the amino acid and the natural transmitter (243, 255). However, the rapid desensitization of these arthropod synaptic receptors may perhaps reflect their essentially phasic control function.

Spinal neurons, like brain neurons, respond to many different acidic amino acids (*64, 69*). Differences in the excitant amino acid sensitivity of different neurons are not very marked, although Renshaw cells seem to be relatively more sensitive to L-aspartate than to L-glutamate when compared with a population of dorsal interneurons (*80*). Peak sensitivity is usually associated, as it is for brain neurons, with separation of the acidic groups by two to three carbon atoms with the amino group optimally situated in the position with respect to one of the acidic groups (*69*). The D-form of glutamate is almost as effective on spinal neurons as the L-form; however, this lack of enantiomeric difference is not general.

However, the relative potencies of amino acid excitants are presumably determined to a great extent by the rate of removal of these compounds from the extracellular environment of the activated cells—L-glutamate, L-aspartate, and L-cysteate being quickly inactivated by uptake, whereas N-methyl-D-aspartate and DL-homocysteate are "inactivated" much less effectively (*58, 59*). If D-glutamate is "inactivated" less effectively than L-glutamate but has the same potency when applied iontophoretically, it may then be reasonable to assume that the excitatory amino acid receptors on spinal neurons also exhibit stereospecificity. The suggested slow rate of "inactivation" of N-methyl-D-aspartate may account for the prolonged aftereffects that this substance has on spinal neurons, with action continuing ten to 20 sec after iontophoretic current termination, even with low doses of this compound (*68*).

In general, substances that require a longer time to evoke a certain spike frequency in spinal interneurons also have the longer postapplication duration of action (*68*). The isoxazole, DL-ibotenic acid, has an excitant action on cat spinal interneurons and Renshaw cells (*129*), and the action of this compound is of slower onset and of longer duration than that of L-glutamate. Somewhat surprisingly, in view of the structural similarities between ibotenate and glutamate, DL-ibotenate does not activate the acidic amino acid receptors at excitatory synapses on insect somatic muscle fibers, although it readily activates receptors on the nonsynaptic membrane of these fibers (*154, 155*) where it increases chloride permeability (*154, 156*).

Mode of Action of Excitant Amino Acids

The enhanced excitability of vertebrate central neurons caused by topical (*35, 95, 110, 111*) and iontophoretic (*53, 59, 138, 145, 168*) application of amino acids is atributable to a reversible depolarization accompanied by an increase in membrane conductance (*145, 168*). With high drug concentrations EPSPs and sometimes also IPSPs are much reduced in magnitude owing in part to the shunting effect of membrane activation produced by the amino acid, although IPSPs may at other times be enhanced by the change in membrane potential. A highly specific test of the principle that the receptor activated by a putative transmitter is identical to that activated by the natural transmitter substance is to determine that they both activate ionophores for the same species of ions and cause the membrane potential to migrate toward the same equilibrium point.

The predominantly dendritic location of excitatory synapses on vertebrate central neurons (*122, 207*) has hitherto prevented an accurate measurement of the equilibrium potential for L-glutamate, and therefore the ionic mechanisms underlying the action of this amino acid are still a matter of conjecture. With recording and current microelectrodes normally located in somata, meaningful comparisons of synaptic and L-glutamate excitations are not possible. An additional problem arises when attempts are made to compare accurately the reversal potentials of synaptic- and drug-induced events, owing to the nonlinear voltage-current characteristics of parts or the whole of the neuronal membrane (*186*).

Curtis (*53*) showed that although passive hyperpolarization of the motoneuronal membrane enhanced the depolarizing effect of L-glutamate whereas depolarization depressed it, the reversal potential for the L-glutamate action was at a lower depolarized level than that resulting from excitatory synaptic action. Proponents of a role for this amino acid as an excitatory transmitter argue that this result may be expected if L-glutamate and synaptic action are identical, in view of the dendritic location of the excitatory synapses. Opponents of the idea argue that the different reversal potentials result from activation of different membrane components involving different ionophores. The reversal potentials for L-glutamate action on arthropod somatic muscle fibers have negative values (*15, 233*) and do not seemingly always correspond with those for EPSPs (but see *237*). However, this may be in part owing to the nonsynaptic glutamate receptors that, when activated, increase chloride permeability (*154–156*).

The plateau depolarization obtained during the action of N-methyl-D-aspartate on mammalian central neurons is about -20 to -30 mV from the resting value of -60 to -70 mV (*59*). This may well represent the reversal potential for the action of this excitant, in which case it is markedly different from that obtained with L-glutamate on lamprey spinal neurons, which was at near zero potential (*168*). One must be cautious, however, in interpreting the results for N-methyl-D-aspartate in view of its unknown extracellular distribution. When ejected iontophoretically from a drug electrode, it may affect the inhibitory inputs of the neurons under examination as well as act directly on the membrane of that cell. It is also possible that the receptors for substances like N-methyl-D-aspartate and DL-homocysteate are distinct and less specific than the postulated synaptic receptors for natural substances like L-glutamate.

Many of the problems that obviate a meaningful comparison between iontophoretically generated and synaptically driven EPSPs also complicate the interpretation of such events when modified by intracellular injection of ions. When chloride ions are injected into spinal motoneurons, the depolarizing action of DL-homocysteate on these cells is unaffected (Figure 5). Apparently chloride ions do not contribute to the depolarization induced by this substance, a finding that comports with earlier results on the effect of this anion on the motoneuron EPSP (44). However,

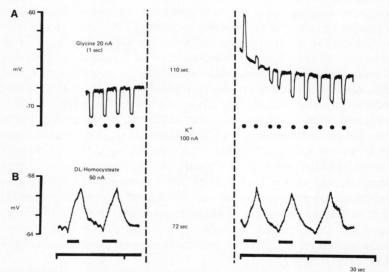

Figure 5. Effects of intracellular injection of potassium ions on potentials recorded intracellularly from cat spinal motoneuron. Peroneal motoneuron, potassium citrate recording electrode. A, Extracellular iontophoretic administration of glycine, 20 nA for 1 sec, dot beneath record. Potassium ions injected intracellularly with current of 100 nA for 110 sec, between vertical broken lines. B, Extracellular iontophoretic administration of DL-homocysteate, 50 nA, 6 sec, black bar beneath records. Potassium ions injected intracellularly with current of 100 nA for 72 sec. Ordinates: membrane potential, mV. Abscissae: time 30 sec. (From Curtis, D. R., Duggan, A. W., Felix, et al.: *Brain Res. 41*:283–301, 1972.)

injections of chloride ions into motoneuron somata may not profoundly influence the distribution of chloride across distant excitatory synaptic membrane in the short term (Figure 5).

The absence of a correlation between reversal potentials for L-glutamate and the EPSP, together with the almost ubiquitous action of this amino acid on mammalian central neurons, dissuaded many investigators from accepting L-glutamate as a transmitter substance and led them to suggest that its action was nonspecific and exerted generally on the neuronal membrane *(49, 52, 56, 61, 67–69)*. This belief was reinforced by the finding that tetrodotoxin reduces the influx of sodium ions into incubated cerebral tissue, i.e., an influx induced by high concentrations of L-glutamate *(174, 266)*, since it was implied that this amino acid depolarizes mammalian neurons by increasing their permeability to sodium, presumably by utilizing the sodium channels used for spike generation *(35, 103, 183)*. It was later shown that tetrodotoxin does not completely block the glutamate-induced influx of sodium ions *(174)*; neither does it block the depolarization of central neurons induced by DL-homocysteate *(59)*. However, it remains to be demonstrated that the tetrodotoxin-resistant sodium influx bears any relation to glutamate receptor-mediated activation of synaptic sodium ionophores.

Since amino acids are transported across neuronal membranes by carrier-mediated systems, it is suggested that the depolarizations of central neurons induced by substances like L-glutamate are a consequence of these membrane transport processes. However, organic mercurials, which inhibit such transport processes, enhance rather than diminish the sensitivity of nerve cells to L-glutamate and L-aspartate *(38)*. Furthermore in both vertebrate and invertebrate systems, wherein L-glutamate is a putative transmitter, the specific sites for uptake of this amino acid are separable from the presumptive specific receptor sites for acidic amino acids. For example, *N*-methyl-DL-aspartate, DL-homocysteate, and DL-ibotenate— among the most powerful known excitants of mammalian central neurons—are presumably not transported by glutamate carrier systems in the mammalian brain, since they do not interfere with glutamate uptake in brain slices *(10)*.

Acidic amino acids may act by chelating calcium *(62)*. Such a mechanism of action would exclude a direct excitatory synaptic role for these substances on mammalian neurons. There is no doubt that the concentration of calcium ions in the extracellular neuronal environment influences the permeability of neuronal membranes and that calcium chelating agents like ethylenediaminetetraacetic acid (EDTA) and ATP *(92)* can depolarize cells and cause firing. However, the excitant acidic amino acids complex calcium ions only weakly, and the long latency (five to 20 sec compared with one of three sec for excitant amino acids) of excitatory action of strong chelating agents on neurons is clearly very different from that of acidic amino acids like L-glutamate (62). When a series of chelating agents was applied iontophoretically to cat spinal neurons, the relative effectiveness of each was found to be related to its calcium chelating potency (62). Some of these excitant chelators do not contain amino groups, e.g., pyrophosphoric acid and citric acid. While these findings do not preclude a chelating action for the amino acids, they do raise doubts about the adequacy of such an explanation.

Antagonists

The involvement of acidic amino acids in transmission at central excitatory synapses would be substantiated by the demonstration that substances that block excitatory transmission also block the response to these amino acids, although this would not exclude the possibility that transmitter and amino acids act at different sites. It appears to be almost axiomatic that the structure of a putative transmitter should be ignored when searching for antagonists. That substances like ergometrine, lysergic acid diethylamide (LSD 25), methysergide maleate, 2-methoxyaporphine (2-MA), 1-methionine-DL-sulfoximine (MSO), and 1-hydroxy-3-amino pyrrolidone-2 (HA-966) have been suggested as possible L-glutamate antagonists seemingly adds some weight to this contention.

The search for antagonists of L-glutamate action on central neurons has not, however, been exclusively restricted to chemicals with exotic structures. For example, diverse glutamate derivatives, which have little or no excitatory action, have been tested on single neurons of the cat thalamic nuclei (ventralis lateralis and ventralis posterolateralis) by examining their effects on the responsiveness of these cells to synaptic activation, excitatory amino acids, and acetylcholine (Table 10). Two of these substances, α-methyl-DL-glutamate (α-MG) and the diethylester of L-glutamate (GDEE), reversibly blocked the depolarizing action of L-glutamate without markedly affecting the response to acetylcholine (59, 99, 177). The response to DL-homocysteate was either less affected than that to L-glutamate or unaltered by these compounds, suggesting once again the possibility that DL-homocysteate and L-glutamate react with different receptor populations.

Synaptic activation of neurons in the nucleus ventralis posterolateralis, elicited by stimulation of peripheral nerves, was also antagonized by these substances, although not in every cell where the antagonist had abolished

Table 10. Effects of possible antagonists on pharmacological excitation of cat central neurons

Type	Excitant	α-MG[a]	GDEE[b]	MSO[c]	2-MA[d]	LSD 25[e]	UML 491[f]	BOL 148[g]	HA-966[h]
Spinal interneuron	L-Aspartate	0/7[i] (59)[m]		4/7 (59)					
	L-Glutamate	0/7 (59)	5/14 (99a)	9/12 (59)					
	DL-Homocysteate	0/6 (59)	11/18 (99a)	10/12 (59)					
Spinal Renshaw cell	Acetylcholine	0/7 (59)	5/5 (59)	0/8 (59)	0/13 (59)				
	L-Aspartate	0/7 (59)	5/5 (59)	5/8 (59)	7/8 (59)				
	L-Glutamate	0/7 (59)	3/16 (99a)	6/8 (59)	10/11 (59)				
	DL-Homocysteate	0/7 (59)		5/7 (59)	7/8 (59)				
Ventrobasal thalamo-cortical neuron	Acetylcholine	0/7 (59)	0/9 (59)	0/17 (59)	0/6 (59)				
	L-Aspartate	0/5 (59)	2/4 (59)	3/5 (59)					
	L-Glutamate	0/7 (59)	6/11 (99a)	7/17 (59)	5/6 (59)				
			4/9 (59)						
	DL-Homocysteate	0/7 (59)	41/49 (99a)	3/14 (59)	1/6 (59)				
			20/40 (99a)						
	L-Cysteate		3/4 (99a)						
Lateral geniculo-cortical neuron	Acetylcholine	8/9 (diminished) (59)	0/8 (59)	0/9 (59)					
	L-Aspartate	8/9 (enhanced) (59)	8/8 (59)	6/7 (59)					
	L-Glutamate	8/9 (enhanced) (59)	8/8 (59)	8/9 (59)					
	DL-Homocysteate	8/9 (enhanced) (59)	8/8 (59)	2/4 (59)					
Pyramidal tract neuron	Acetylcholine			0/20 (59)	0/5 (59)				
	L-Aspartate	0/7[i] (59)	4/7 (59)	13/13 (59)	5/5 (59)				
	L-Glutamate	0/7[i] (59)	4/7 (59)	20/20 (59)	4/5 (59)				
	DL-Homocysteate			19/20 (59)	4/5 (59)				
Cuneate nucleus neuron	L-Aspartate		10/17 (99a)						
	L-Glutamate		14/19 (99a)						
	DL-Homocysteate		6/18 (99a)						
	L-Cysteate		3/7 (99a)						
Brainstem neuron	5-Hydroxytryptamine					31/32 (30)	23/47 (30)	6/33 (30)	
	Noradrenaline					1/10 (30)	0/10 (30)	1/6 (30)	

continued

Table 10. continued

Type	Excitant	α-MG[a]	GDEE[b]	MSO[c]	2-MA[d]	LSD 25[e]	UML 491[f]	BOL 148[g]	HA-966[h]
	Acetylcholine					0/15 (30)	0/9 (30)	0/9 (30)	
	L-Glutamate					20/27 (30)	12/21 (30)	1/19 (30)	
	DL-Homocysteate					0/8 (30)	2/11 (30)	0/5 (30)	
Thalamic nuclei neuron	Acetylcholine	(99)[k]	(99)[k]						
	L-Glutamate	41/76 (99)	26/38 (99)						
	DL-Homocysteate	(99)[l]	(99)[l]						
Pericruciate neuron	Acetylcholine								11/21 (73)
	L-Glutamate								38/40 (73)
	L-Aspartate								7/7 (73)

[a] α-Methyl-DL-glutamate.
[b] L-Glutamic acid diethyl ester.
[c] L-Methionine-DL-sulfoximine.
[d] 2-Methoxyaporphine hydrochloride.
[e] D-Lysergic acid diethylamide tartrate.
[f] Methysergide maleate.
[g] 2-Bromo-lysergic acid diethylamide.
[h] 1-Hydroxy-3-aminopyrrolidone-2.
[i] Each set of figures indicates number of cells for which antagonism was demonstrated, as ratio of total number of cells tested; for data from (59), apart from spinal interneurons, cells are only included when there was no antagonism of excitation by acetylcholine.
[j] Four other cells were excited by α-MG.
[k] Rarely antagonized.
[l] Less antagonized than L-glutamate.
[m] Reference number in "Literature Cited" section.

the response to L-glutamate. Possible antagonism between L-glutamate and GDEE was first suggested by the work of Dési, Farkas, Sós, et al. (76), who found that GDEE impairs learning and behavior in the rat and changes the EEG, effects that can be counteracted by intraperitoneal injections of L-glutamate.

The results of more recent studies on the effectiveness of L-glutamate diethylester as a glutamate antagonist have been less convincing (59). In some cells of the cat ventrobasal thalamus, e.g., it reduced the effectiveness of acetylcholine without modifying the effect of L-glutamate and DL-homocysteate (Table 10), although it did reduce the sensitivity of spinal neurons to both L-glutamate and DL-homocysteate. The effect of α-MG on the sensitivity of central neurons to acidic amino acids has also been reexamined (59). The results of this study failed to confirm earlier findings of glutamate antagonism in either the ventrobasal thalamus or in other regions of the cat CNS (Table 10).

The compounds 2-MA and MSO have been found to diminish the actions of L-glutamate, L-aspartate, and DL-homocysteate on neurons in many different parts of the cat CNS without reducing their sensitivity to acetylcholine (Figure 6). In general the effects of L-aspartate and L-glutamate were reduced to the same extent, whereas the effect of DL-homocysteate was antagonized to a lower degree. Although 2-MA and MSO failed to reduce the spontaneous firing of central neurons, they did cause a small but significant reduction in the number of action potentials of some spinal interneurons evoked through cutaneous afferents. Marked regional differences in the effectiveness of these substances were found (Table 10).

LSD 25 antagonizes the excitatory actions of L-glutamate on cat brainstem neurons, which are excited by 5-hydroxytryptamine (5-HT) but not on neurons that are inhibited by this indolalkylamine (30) (Table 10; see also Ch. 7). Results from attempts to determine if this antagonism was competitive were not entirely satisfactory, though suggestive of such a possibility (30). Somewhat surprisingly, LSD 25 does not influence excitation by DL-homocysteate. The excitatory action of L-glutamate on brainstem neurons was also antagonized by methysergide but once again only in neurons excited by 5-HT. The antagonism to L-glutamate excitation by methysergide is weaker than that by LSD 25 (Table 10). The compound 2-bromo-lysergic acid diethylamide (BOL 148) is a much weaker antagonist than either LSD 25 or methysergide.

In the mammalian cortex and hippocampus, 5-HT depresses both synaptically induced firing and the response to L-glutamate but since it also depresses the response to acetylcholine, it is unlikely that the site of action of 5-HT is at "glutamatergic" synapses (31). It is of interest to note,

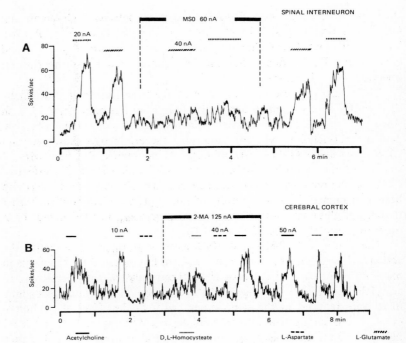
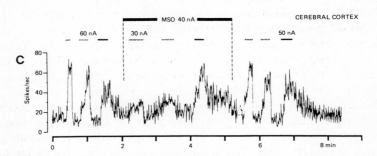

Figure 6. Effect of A, L-methionine-DL-sulfoximine (MSO) on firing of feline spinal dorsal horn interneuron. B, 2-methoxyaporphine HCl (2-MA) on firing of feline pyramidal tract neuron. C, MSO on firing of feline pyramidal tract neuron. Firing of neurons caused by acetylcholine (solid lines), L-glutamate (cross-hatched lines), L-aspartate (broken lines), and DL-homocysteate (dotted lines). MSO and 2-MA applied iontophoretically at times indicated by bar and vertical broken lines. Ordinates: spikes/sec. Abscissae: time in min. Spikes recorded extracellularly. (From Curtis, D. R., Duggan, A. W., Felix, D.: *Brain Res. 10*:447–462, 1970.)

however, that 5-HT blocks transmission at insect excitatory nerve-muscle synapses, i.e., synapses that are probably "glutamatergic" (*109, 243*).

The compound HA-966 causes a variety of centrally-mediated depressant effects in several mammalian species. When applied iontophoretically to cat cortical neurons, it antagonized L-glutamate and L-aspartate excitation in most cells tested (*73*). It had a smaller but nevertheless significant effect on the responses of these cells to acetylcholine. In neurons of the cuneate nucleus, HA-966 also antagonized the excitatory responses evoked by stimulation of primary afferent fibers (*73*). However, more critical studies of the mode of action of HA-966 are required before a decision can be made about its supposed antagonism at excitatory synapses on mammalian neurons.

INACTIVATION

After application of a chemical substance to the surface of a neuron, the fall in concentration of active material at any one point depends not only on free diffusion (*83*) but also on the rate of absorption on inert macromolecular surfaces, the efficacy of transport across adjacent membranes, and (possibly) enzymatic inactivation. If the substance reacts with receptors on the neuron, the decay of this action will have a time course dependent on both the rate of fall of the concentration of the substance and the stability of the receptor-substance complex (*64*). The ideas implicit in these statements are drawn, to some extent, from studies of some cholinergic systems in which the enzyme acetylcholinesterase appears to play an important role in limiting the time course of the postsynaptic action of acetylcholine.

Following the initial studies of iontophoretic application of L-glutamate to mammalian central neurons, it was decided that since there is rapid cessation of activation of spinal motoneurons and Renshaw cells by diverse dicarboxylic amino acids, it seemed improbable that any one enzyme system could remove all of these amino acids at a similar rate. The prolongation of EPSPs and L-glutamate potentials by inhibition of an enzyme for which L-glutamate is known to be a substrate would indicate that the naturally occurring transmitter and L-glutamate are possibly identical, especially if the enzyme has a high specificity for L-glutamate. However, the initial studies on enzyme inhibitors were very unpromising in this respect (*64*) (Table 11).

Most of the inhibitors that were tested in these studies showed a low potency when compared with acetylcholinesterase inhibitors. Some of the

Table 11. Effect of supposed enzyme inhibitors on response of mammalian central neurons to general and iontophoretic application of excitatory amino acids

Compound	Animal[a]	Site	Effect of iontophoretic application[d]	Animal	Site	Effect of general application[d]
Sodium-p-chloromercuribenzoate		Spinal cord (64)[e]	0			
Iproniazid		Spinal cord (64)	0			
Isoniazid		Spinal cord (64)	0			
Phenylhydrazine HCl		Spinal cord (64)	0			
Hydroxylamine HCl		Spinal cord (64)	0			
Semicarbazide HCl		Spinal cord (58, 64)	0			
Thiosemicarbazide HCl	Rat	Cortex (227–229)	0	Rat	Cortex (227–229)	+ve
	Rat	Thalamus (227–229)	0	Rat	Thalamus (227–229)	+ve
	Rat	Subthalamus (227–229)	0	Rat	Subthalamus (227–229)	+ve
	Rat	Hippocampus (227–229)	0	Rat	Hippocampus (227–229)	+ve
Pyridoxal-5-phosphate + Thiosemicarbazide HCl[b]	Rat	Cortex (227–229)	0 or –ve			
	Rat	Thalamus (227–229)	0 or –ve			

Thiocarbohydrazide HCl	Rat	Subthalamus (227–229)	0 or −ve
Sodium benzoate	Rat	Hippocampus (227–229)	0 or −ve
Quinine HCl		Spinal cord (64)	0
Ephedrine HCl		Spinal cord (64)	0
p-Phenylenediamine HCl		Spinal cord (64)	0
Sodium fluoride		Spinal cord (64)	0
Sodium cyanide		Spinal cord (64)	0
DL-Methionine sulfoxide		Spinal cord (64)	0
DL-Methionine sulfoximine		Spinal cord (64)	0
p-Chloromercuriphenyl sulfonate		Spinal cord (58)	+ve
DL-C-Allyglycine		Spinal cord (58)	−ve[c]
Hydrazinoacetic acid		Spinal cord (58)	−ve[c]
Hydrazinopropionic acid		Spinal cord (58)	−ve[c]
p-Hydroxymercuribenzoate		Spinal cord (58)	+ve

[a] Cat, unless otherwise indicated.
[b] Applied parenterally.
[c] Depressed "spontaneous" firing of spinal neurons.
[d] 0 = no effect; +ve = potentiation of glutamate response measured as an increase in spike frequency; −ve = depression of glutamate response measured as a decrease in spike frequency.
[e] Reference number in "Literature Cited" section.

substances included in Table 11, namely, the hydrazides, have been known for some time to cause convulsive activity in mammals when injected systemically. It is suggested that they inhibit the conversion of L-glutamate in the brain to γ-aminobutyrate by glutamate decarboxylase *(132, 133, 261)*. This is thought to occur through interference with the prosthetic group of this enzyme formed by a vitamin (B_6) of the B-complex, pyridoxal-5-phosphate. Systemic injection of thiosemicarbazide is followed by increased neuronal activity in subcortical structures *(204)*, and convulsive activity occurs under conditions of vitamin B deficiency *(45, 46)*.

Although the first studies of the effects of enzyme inhibitors on the action of iontophoretically applied excitant amino acids gave negative results, additional investigation in this field is far from being discouraged. Steiner and Ruf *(227–229)* have studied the effects of L-glutamate on neurons of the rat cortex, thalamus, subthalamus, striatum, and hippocampus before, during, and after iontophoretic and systemic administration of thiosemicarbazide and pyridoxal-5-phosphate. The response to L-glutamate, i.e., change in spike frequency, was potentiated with thiosemicarbazide, the increased excitability of the treated neurons outlasting the iontophoretic application of L-glutamate by three sec (Figures 7 and 8).

Following the application of thiosemicarbazide, the rate of discharge of spontaneously active rat brain neurons increased as early as one min after the start of the injection and was sustained for more than 30 min. Pyridoxal-5-phosphate partially reverses the excitability increase to L-glutamate induced by iontophoretically applied thiosemicarbazide, provided that the delivery of pyridoxal-5'-phosphate precedes that of L-glutamate by at least 30 sec, and that the thiosemicarbazide-induced aftereffect of L-glutamate is completely abolished by concomitant iontophoresis of pyridoxal-5-phosphate. However, after the application of thiosemicarbazide, single unit activity was not changed by iontophoretically applied pyridoxal-5-phosphate *(229)*. When applied alone, pyridoxal-5-phosphate either has no effect or results in slight activation of central neurons *(229)*.

It is suggested that these changes induced by thiosemicarbazide are caused by impairment of glutamic decarboxylase, which would result in the prolongation of the biological half-life of L-glutamate and, in consequence, increase the total amount of this substance in the neuronal environment *(228)*. Following studies on cat spinal interneurons with semicarbazide *(58)*, it was suggested that the enhanced sensitivity of supraspinal neurons to L-glutamate *(227–229)* after administration of

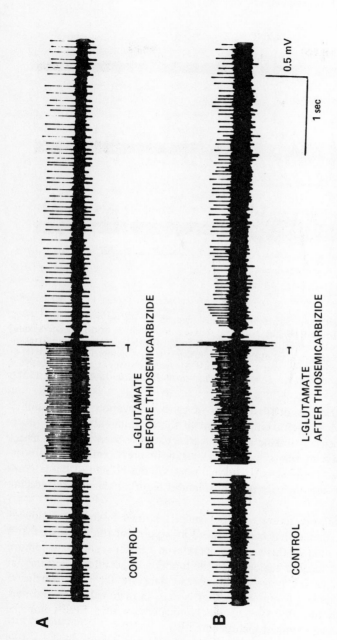

Figure 1. Activation of rat thalamic neuron by L-glutamate (A) before and (B) after systemic administration of thiosemicarbazide (100 mg/kg). After thiosemicarbazide administration L-glutamate effect was enhanced; on termination of glutamate iontophoresis spike frequency slowly (2.5 sec) returned to control level. Calibration: 1 sec, 0.5 mV. Spikes recorded extracellularly. (From Steiner, F. A., and Ruf, K.: *Schweiz. Arch. Neurol. Psychiatr.* 100:310–320, 1967.)

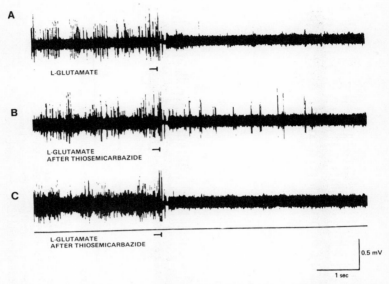

Figure 8. Activation of silent hippocampal neuron of rat by L-glutamate iontophoresis (A) before and (B and C) after systemic administration of thiosemicarbazide. Excitation outlasting time of actual iontophoresis (B) after systemic administration of thiosemicarbazide was abolished by concomitant iontophoresis of pyridoxal-5-phosphate (_____). Calibration: 1 sec, 0.5 mV. Spikes recorded extracellularly. (From Steiner, F. A., and Ruf, K.: *Schweiz. Arch. Neurol. Psychiatr. 100*:310–320, 1967.)

thiosemicarbazide is attributable to a general elevation of excitability rather than to a specific enhancement of the excitable action of L-glutamate (*58*). Phenylhydrazine and semicarbazide have been tested on insect excitatory nerve-muscle synapses and act both presynaptically and postsynaptically at these sites. However, none of the effects of these compounds can be convincingly related to inhibition of a transmitter-degrading enzyme (*78*).

The convulsant action of MSO and its known inhibitory action of glutamine synthetase has been adduced in support of the proposal that a particular fraction of this enzyme operates at certain synapses to remove glutamate in a manner analogous to the function of acetylcholinesterase at cholinergic synapses; however, supportive data are far from conclusive (*213*). A compelling counterargument is that, in mammalian central neurons, iontophoretically applied MSO enhances neither EPSPs nor the response to locally applied L-glutamate (*59*).

The mercurial, *p*-chloromercuriphenylsulfonate (pMs), when applied iontophoretically (with large ejection currents) over prolonged periods (ten min) to cat Renshaw cells, causes a progressive enhancement of the responses of these cells to L-glutamate, L-aspartate, D-glutamate, and DL-homocysteate without affecting the responses to acetylcholine (58). At the same time an increase in spontaneous background activity is noted. The excitant effects of DL-homocysteate and D-glutamate were usually enhanced less than those of L-aspartate and L-glutamate (Figure 9). Furthermore, preliminary results indicate that the action of the powerful excitant N-methyl-D-aspartate is relatively unaffected by concentrations of pMs that are adequate to enhance the action of L-glutamate and L-aspartate.

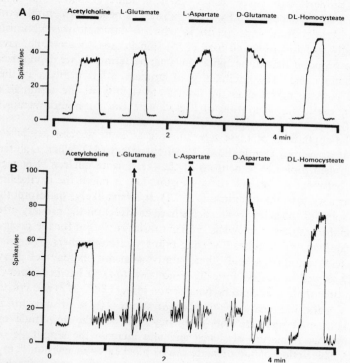

Figure 9. Effect of iontophoresis of p-mercuriphenylsulfonate on excitation of feline Renshaw cell. Cell was excited by iontophoretic administration of acetylcholine, L-glutamate, D-glutamate, L-aspartate, and DL-homocysteate during periods indicated by horizontal bars. A. Control. B. Immediately after administration of p-mercuriphenylsulfonate for 10 min. Vertical arrows indicate frequencies of order of 150 spikes/sec. Ordinates: spikes/sec. Spikes recorded extracellularly. Abscissae: Time in min. (From Curtis, D. R., Duggan, A. W., Felix, D., et al.: *Exp. Brain Res.* *41*:283–301, 1972.)

It is unlikely but not disproven that the potentiation of the responses to natural acidic amino acids by pMs is due to enzyme inhibition in synaptic clefts. It is more likely to result from inhibition of uptake of these substances, since uptake of DL-aspartate in rat cortical slices is 37% inhibited by 10^{-4} M pMs (58). The lack of effect of pMs on DL-homocysteate and N-methyl-D-aspartate may be attributable to the absence of uptake systems for these substances at central nervous synapses.

PRESYNAPTIC ACTION OF ACIDIC AMINO ACIDS

Although the excitatory amino acids seem to have no effect on the electrical properties of nerve trunks, there is some evidence that afferent fibers may be depolarized by these substances near their central terminations. Results from experiments in which L-glutamate was administered topically to feline (84) and toad (216) spinal cords, for example, have led to the proposal that some spinal presynaptic terminals are depolarized by this amino acid. The excitability of primary afferent fibers in the cat spinal cord is increased in the immediate vicinity of the terminals after iontophoretic administration of L-glutamate and DL-homocysteate (Figure 10).

The application of these substances to afferent fibers at regions other than their central terminations is without effect (65). When L-glutamate was applied topically to neurons of the rat cuneate nucleus at concentrations between 10^{-4} M and 10^{-2} M (Figure 11), it raised the excitability of the primary afferent terminals (71, 72). It seems likely, therefore, that at least some of the acidic amino acids have an effect on the primary afferent terminal membrane somewhat similar to that envisaged for the postsynaptic membrane of synapses involving primary afferent fibers. The principle that acidic amino acid and other putative transmitters may act presynaptically as well as postsynaptically is not new (84). It has been known for some time that, at excitatory synapses on insect (242, 243) and crustacean (88) somatic musculature, low concentrations (10^{-8} M in some instances) of L-glutamate increase the frequency but not the amplitude of the spontaneous miniature discharge and also increase the amplitude of the EPSP. This action of L-glutamate can be blocked by magnesium ions (77).

IDENTIFICATION OF CNS TRANSMITTERS

In his valuable review of criteria for the identification of central nervous transmitters, Werman (257) listed those factors that he considered fundamental to the identification of a drug with a natural transmitter. They include: an inactivating enzyme, the putative transmitter in the presynap-

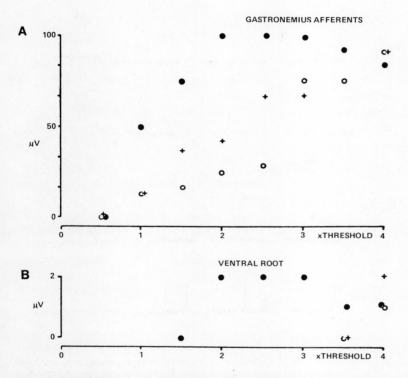

Figure 10. Excitability of terminations of feline gastrocnemius group 1a afferents and motoneurons before (○), during (●), and after (+) iontophoretic administration of DL-homocysteate (100 μA). Ordinates: amplitude (μV) of potentials recorded monophasically from (A) gastrocnemius nerve and (B) ventral root. Abscissae: multiples of just-above-threshold stimulus for afferent fibers. (From Curtis, D. R., and Ryall, R. W.: *Exp. Brain Res.* 1:195–204, 1966.)

tic neuron, collection of the putative transmitter in perfusates, synthesizing enzymes for the putative transmitter, precursors of the putative transmitter in the presynaptic neuron or associated glial cells, a specific release mechanism for the putative transmitter, identical action of putative transmitter and natural transmitter, and pharmacological identity of putative transmitter and natural transmitter. In a more recent review Werman (*258*) suggests that there are in fact only two criteria that need to be established in order to identify a transmitter: identity of action and collectibility.

The first criterion is perhaps not essential and probably not applicable to any synaptic systems that utilize acidic amino acids as transmitters,

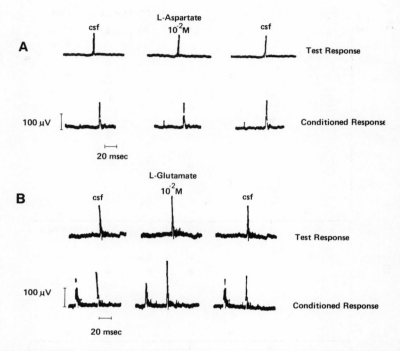

Figure 11. Effect of topical application of L-apartate and L-glutamate to restricted area of rat cuneate nucleus on primary afferent terminal excitability. A, Effect of these as amino acids on activity in ulnar nerve induced by brief test stimuli passed from extracellular electrode close to terminals. B, Effect on antidromic response of ulnar nerve conditioned by supramaximal stimulus applied to ipsilateral median nerve which caused presynaptic inhibition in cuneate nucleus. Both test and conditioned responses were enhanced by L-glutamate but were unaffected by L-aspartate. CSF, artificial cerebrospinal fluid. (From Davidson, N., and Southwick, C. A. P.: *J. Physiol.* (Lond.) *219*:689–708, 1971.)

since these substances could be readily removed from the synaptic cleft by simple diffusion and by sequestration by nerve terminals and other adjacent structures. Indeed enzymatic inactivation seems improbable since the time course of action of L- and D-glutamate are very similar (*64, 66*). The demonstrable effects of enzyme inhibitors like hydrazides on glutamate-sensitive neurons are difficult to interpret since their precise site of action is uncertain. It will be necessary to demonstrate clearly that these inhibitors act specifically at excitatory synapses by prolonging transmitter action, before suggestions of enzymatic inactivation of transmitter at these sites can be taken seriously.

It is not sufficient to use changes in spike parameters as criteria for supposed synaptic action of a substance. It may be instructive to mention that, after exhaustive studies of the effects of amino acid enzyme inhibitors at insect "L-glutamate" synapses, the data suggest that enzymatic inactivation of transmitter does not occur at these sites (*78*).

Since acidic amino acids like L-glutamate and L-aspartate are ubiquitous throughout the mammalian CNS, these substances seemingly satisfy the second criterion, since all that is required in these circumstances is a specific release mechanism for the appropriate amino acid in the axon terminal and a postsynaptic membrane programmed to identify this substance when it is released into the synaptic cleft. That L-glutamate is more concentrated in the terminals of primary afferent fibers than elsewhere in these sensory structures adds weight to arguments that it is the transmitter released by at least some of these fibers. It is somewhat disturbing to hear comments to the effect that L-cysteate is an unlikely transmitter candidate since it is found in the mammalian CNS in much lower concentrations than L-glutamate. Only a few years ago the trend was to reject L-glutamate as a transmitter because of its *high* central nervous concentration.

Studies on the collectibility of putative acidic amino acid transmitters have given equivocal results, possibly because of the high affinity uptake systems for these substances at axon terminals, which may in some cases reduce or even eliminate "spillover" of transmitter. This may also account for the somewhat unpredictable results obtained with invertebrate nerve-muscle preparations (*136, 180a*). In some cases it has been possible to demonstrate that L-glutamate is released upon stimulation of certain neural pathways in the mammalian CNS considered to utilize this amino acid as a transmitter; but in other cases amino acids without a known neurophysiological role, such as leucine, lysine, and cycloleucine, have also been released during neural stimulation (*131*). Studies on the release of amino acids from bullfrog sciatic nerves demonstrate that either the axonal or glial membranes or both of these structures have built-in mechanisms for release of L-glutamate and to a more limited extent, L-aspartate (*260*). The absence of synaptic vesicles at the release sites in these studies poses certain conceptual difficulties when attempts are made to relate these findings to transmitter release for axon terminals.

The demonstration in central nervous tissue of enzymes capable of synthesizing substances such as L-glutamate, L-aspartate, and L-cysteate, together with the precursors of these amino acids, is not particularly difficult, since all of these substances are so intimately involved in many aspects of cell metabolism. It is not essential to demonstrate that such enzymes and precursors are specifically located in axon terminals, since longitudinal transport of amino acids clearly occurs in all nerve cells.

The problems of convincingly demonstrating that a putative acidic amino acid transmitter mimics the action of natural central transmitter are considerable. In the CNS, unlike at the periphery, it has not been possible to determine if the action of an iontophoretically administered substance is localized to particular synapses or, indeed, if it is even restricted to synaptic regions. There seems to be little doubt that the effect of excitant amino acids is to cause membrane depolarization through reaction with specific receptor sites located postsynaptically and possibly also presynaptically. Differences in neuronal sensitivity to certain natural acidic amino acids seemingly make it unlikely that their depolarizing action involves binding with receptors common to the general neuronal membrane, although it is difficult to decide on the basis of available (but conflicting) data whether or not these differences are in all cases real or apparent owing to problems of technique.

The remoteness of excitatory synapses on dendrites of central neurons calls into question comparative studies of reversal potentials of natural and putative transmitter and, related to this, studies of the ionic aspects of synaptic activation and putative transmitter action. Werman (*257*) described a quantitative procedure for determining whether or not the molecular mechanism evoked by a putative transmitter was identical with that produced by the natural transmitter, which he considered circumvented this problem. However, his proposals do not solve the problems associated with the possible presence of two (or more) different receptor populations for an excitant acidic amino acid—one synaptic, the other nonsynaptic—activating different ionic channels and producing different reversal potentials.

It would not be easy to differentiate clearly between these two sets of receptors with present in vitro iontophoretic application techniques. However, with the development of tissue culture techniques for central neurons perhaps it may be possible to apply acidic amino acids to clearly identifiable sites on these cells, and L-glutamate- and L-aspartate-sensitive electrodes (*12, 258*) may improve resolution of any possible synapses. using these substances as transmitters. It remains true, of course, that L-glutamate mimics the action of the transmitter at some invertebrate central and peripheral synapses. Unfortunately mammalian physiologists and pharmacologists are often reluctant to admit that their experimental material has evolutionary links with the invertebrate kingdom.

Studies of the pharmacology of excitant amino acid receptors on mammalian central neurons are still in their infancy, especially with regard to the search for suitable antagonists of L-glutamate and L-aspartate (60). In fact the role of antagonists in identifying transmitters is the subject of some debate, and Werman (258) has gone so far as to suggest that "transmitters . . . tell us more about blocking agents than the latter tell us about transmitters." There is an element of truth in this statement, but since it is possible to demonstrate that a substance has a specific postsynaptic effect without having previously identified the transmitter, it seems likely that antagonists will play a major role in establishing the identity of transmitters in the future as they have done in the past. Admittedly investigations of the site of action of possible antagonists is a lengthy, time-consuming process, and in the CNS such investigations are confronted with problems similar to those experienced when testing putative transmitters. Undoubtedly antagonists that selectively block the action of just one or a small group of acidic amino acids, while leaving the action of other excitant amino acids unimpaired, would be of great value in differentiating between, e.g., possible L-glutamate and L-aspartate synapses and synaptic and nonsynaptic receptors for acidic amino acids.

A recent editorial in *Nature* (Vol. 238, p. 70, 1972) implied that the role of L-glutamate as an excitatory transmitter in vertebrates has not yet been clarified. This seems to be a decidedly gloomy and misleading conclusion. I am optimistic that in the next few years speculation concerning the possible excitatory transmitter role of this amino acid and other natural acidic amino acids will be fully rewarded, and that the role of L-glutamate as a transmitter, at least at primary afferent synapses, will be convincingly demonstrated. Perhaps by that time vertebrate synaptic physiopharmacology will be fully cognizant of trends in the invertebrate world and will be ready to consider the possibility that acidic amino acids may also function as inhibitory transmitters in the vertebrate brain and spinal cord.

Finally, I wonder if there is any evolutionary significance to the fact that vertebrates have adopted acetylcholine as their chemical mediator at somatic nerve-muscle synapses, with L-glutamate possibly serving this function at primary sensory endings, whereas in arthropods the inverse situation seems to apply.

LITERATURE CITED

1. Abadom, P. N. and Scholefield, P. G.: Amino acid transport in brain cortex slices. III. The utilization of energy for transport. *Can. J. Biochem. Physiol.* 40:1603–1618, 1962.
2. Ames, A.: Effect of glutamate and glutamine on the intracellular electrolytes of nervous tissue. *Neurology* 8 (Suppl. 64): 1958.

3. Ames, A., Tsukada, Y., and Nesbett, F. B.: Intracellular Cl^-, Na^+, K^+, Ca^{++}, Mg^{++}, and P in nervous tissue; response to glutamate and to changes in extracellular calcium. *J. Neurochem.* 14:145–159, 1967.
4. Andersen, P. and Curtis, D. R.: The excitation of thalamic neurones by acetylcholine. *Acta Physiol. Scand.* 61:85–99, 1964.
5. Andersen, P. and Curtis, D. R.: The pharmacology of the synaptic and acetylcholine-induced excitation of ventrobasal thalamic neurones. *Acta Physiol. Scand.* 61:100–120, 1964.
6. Arnfred, T. and Hertz, L.: Effects of potassium and glutamate on brain cortex slices: Uptake and release of glutamic and other amino acids. *J. Neurochem.* 18:259–265, 1971.
7. Aprison, M. H., Graham, L. T., Jr., Livengood, D. R., et al.: Distribution of glutamic acid in the cat spinal cord and roots. *Fed. Proc.* 24:462, 1965.
8. Araki, T., Ito, M., and Oscarsson, O.: Anion permeability of the synaptic and nonsynaptic motoneurone membrane. *J. Physiol.* (Lond.) 159:410–435, 1961.
9. Axelrod, J. and Kopin, I. J.: The uptake, release and metabolism of noradrenaline in sympathetic nerves. In Akert, K. and Waser, P. G. (Eds.): *Mechanisms of Synaptic Transmission, Progress in Brain Research.* Amsterdam: Elsevier, 1969.
9a. Bálazs, R. and Haslam, J.: Exchange transamination and the metabolism of glutamate in brain. *Biochem. J.* 94:131–141, 1965.
9b. Bálazs, R., Machiyama, Y., Hammond, B. J., et al.: The operation of the γ-aminobutyrate bypath of the tricarboxylic acid cycle in brain tissue *in vitro*. *Biochem. J.* 116:445–467, 1970.
10. Balcar, V. and Johnston, G. A. R.: Glutamate uptake by brain slices and its relation to the depolarization of neurones by acidic amino acids. *J. Neurobiol.* 3:295–301, 1972.
11. Battistin, L., Grynbaum, A., and Lajtha, A.: Distribution and uptake of amino acids in various regions of the cat brain *in vitro*. *J. Neurochem.* 16:1459–1468, 1969.
12. Baum, G.: Determination of acetylcholinesterase by an organic substrate selective electrode. *Ann. Biochem.* 39:65–72, 1971.
13. Baumgarten, R. von, Bloom, F. E., Oliver, A. P., et al.: Response of individual olfactory nerve cells to microelectrophoretically administered chemical substances. *Pflügers Arch.* 277:125–140, 1963.
14. Baxter, C. F., Schade, J. P., and Roberts, E.: Maturational changes in cerebral cortex. II. Levels of glutamic acid decarboxylase, γ-aminobutyric acid and some related amino acids. In Roberts, E. (Ed.): *Inhibition in the Nervous System and Gamma-aminobutyric Acid.* Oxford: Pergamon Press, 1969, pp. 214–218.
15. Beránek, R. and Miller, P. L.: The action of iontophoretically applied glutamate on insect muscle fibers. *J. Exp. Biol.* 49:83–93, 1968.
16. Bergeret, B. and Chatagner, F.: Sur la présence d'acide cystéinesulfinique dans le cerveau du rat normal. *Biochim. Biophys. Acta* 14:297, 1954.
17. Berl, S., Clarke, D. D., and Nicklas, W. J.: Compartmentation of citric acid cycle metabolism in brain: Effect of aminooxyacetic acid,

ouabain and Ca^{2+} on the labelling of glutamate, glutamine, aspartate and GABA by [1-^{14}C]acetate, [U-^{14}C]glutamate and [U-^{14}C]aspartate. *J. Neurochem. 17*:999–1007, 1970.
18. Berl, S. and Clarke, D. D.: Compartmentation of amino acid metabolism. In Lajtha, A. (Ed.): *Handbook of Neurochemistry* (vol. II). New York: Plenum Press, 1969, pp. 447–472.
19. Berl, S., Lajtha, A., and Waelsch, H.: Amino acid and protein metabolism–VI. Cerebral compartments of glutamic acid metabolism. *J. Neurochem. 7*:186–197, 1961.
20. Berl, S. and McMurtry, J. G.: Isolated cerebral cortex: Changes in levels of glutamic acid, glutamine, aspartic acid and γ-aminobutyric acid. *Arch. Biochem. Biophys. 118*:645–648, 1967.
21. Berl, S. and Purpura, D. P.: Postnatal changes in amino acid content of kitten cerebral cortex. *J. Neurochem. 10*:237–240, 1963.
22. Berl, S. and Purpura, D. P.: Regional development of glutamic acid compartmentation in immature brain. *J. Neurochem. 13*:293–304, 1966.
23. Bernardi, G., Zieglgänsberger, W., Herz, A., et al.: Intracellular studies on the action of L-glutamic acid on spinal neurones of the cat. *Brain Res. 39*:523–525, 1972.
24. Biscoe, T. J. and Straughan, D. W.: The pharmacology of hippocampal neurones. *J. Pharm. Pharmacol. 17*:60–61, 1965.
25. Biscoe, T. J. and Straughan, D. W.: Micro-electrophoretic studies of neurones in the cat hippocampus. *J. Physiol.* (Lond.) *183*:341–359, 1966.
26. Blasberg, R. G.: Specificity of cerebral amino acid transport: A kinetic analysis. *Prog. Brain Res. 29*:245–256, 1968.
27. Blasberg, R. and Lajtha, A.: Substrate specificity of steady-state amino acid transport in mouse brain slices. *Arch. Biochem. Biophys. 112*:361–377, 1965.
28. Blasberg, R. and Lajtha, A.: Heterogeneity of the mediated transport systems of amino acid uptake in brain. *Brain Res. 1*:86–104, 1966.
29. Bloom, F. E., Costa, E., and Salmoiraghi, G. C.: Anesthesia and the responsiveness of individual neurons of the caudate nucleus of the cat to acetylcholine, norepinephrine and dopamine administered by microelectrophoresis. *J. Pharmacol. Exp. Ther. 150*:244–252, 1965.
30. Boakes, R. J., Bradley, P. B., Briggs, I., et al.: Antagonism of 5-hydroxytryptamine by LSD 25 in the central nervous system: A possible neuronal basis for the actions of LSD 25. *Br. J. Pharmacol. 40*:202–218, 1970.
31. Bonta, I. L., De Vos, C. J., Grijsen, H., et al.: 1-Hydroxy-3-aminopyrrolidone-2(HA-966): A new GABA-like compound, with potential use in extrapyramidal diseases. *Br. J. Pharmacol. 43*:514–535, 1971.
32. Bradford, H. F.: Response of synaptosomes to electrical stimulation. *Biochem. J. 117*:36P, 1970.
33. Bradford, H. F.: Metabolic response of synaptosomes to electrical stimulation: Release of amino acids. *Brain Res. 19*:239–247, 1970.

34. Bradford, H. F., Chain, E. B., Cory, H. T., et al.: Glucose and amino acid metabolism in some invertebrate nervous systems. *J. Neurochem.* 16:969–978, 1969.
35. Bradford, H. F. and McIlwain, H.: Ionic basis for the depolarization of cerebral tissues by excitatory acidic amino acids. *J. Neurochem.* 13:1163–1177, 1966.
36. Bradford, H. F. and Thomas, A. J.: Metabolism of glucose and glutamate by synaptosomes from mammalian cerebral cortex. *J. Neurochem.* 16:1495–1504, 1969.
37. Bradley, P. B., Dhawan, B. N., and Wolstencroft, J. H.: Pharmacological properties of cholinoceptive neurones in the medulla and pons of the cat. *J. Physiol.* (Lond.) 183:658–674, 1966.
38. Bradley, P. B. and Wolstencroft, J. H.: The action of drugs on single neurones in the brain stem. In Bradley, P. B., Flugel, F., and Hoch, P. (Eds.): *Neuropsychopharmacology* (Vol. 3). Amsterdam: Elsevier, 1964, pp. 237–240.
39. Bradley, P. B. and Wolstencroft, J. H.: Actions of drugs on single neurones in the brain-stem. *Br. Med. Bull.* 21:15–18, 1965.
40. Brady, R. O. and Tower, D. B.: *The Neurochemistry of Nucleotides and Amino Acids.* New York: Wiley, 1960.
41. Caldwell, P. C.: Studies on the internal pH of large muscles and nerve fibres. *J. Physiol.* (Lond.) 142:22–62, 1958.
42. Chapman, J. B. and McCance, I.: Acetylcholine-sensitive cells in intra-cerebellar nuclei of the cat. *Brain Res.* 5:535–538, 1967.
43. Coombs, J. S., Eccles, J. C., and Fatt, P.: The specific ionic conductances and the ionic movements across the motoneuronal membrane that produce the inhibitory postsynaptic potential. *J. Physiol.* (Lond.) 130:326–373, 1955.
44. Coombs, J. S., Eccles, J. C., and Fatt, P.: Excitatory synaptic action in motoneurones. *J. Physiol.* (Lond.) 130:374–395, 1955.
45. Coursin, D. B.: Seizures in Vitamin B_6 deficiency. In Roberts, E. (Ed.): *Inhibition in the Nervous System and Gamma-Aminobutyric Acid.* Oxford: Pergamon Press, 1960, pp. 294–301.
46. Coursin, D. B.: Vitamin B_6 metabolism in infants and children. *Vitam. Horm.* 22:755–786, 1964.
47. Cowie, D. B. and Walton, B. R.: Kinetics of formation and utilisation of metabolic pools in the biosynthesis of protein and nucleic acid. *Biochem. Biophys. Acta* 21:211–226, 1956.
48. Crawford, J. M.: The sensitivity of cortical neurones to acidic amino acids and acetylcholine. *Brain Res.* 17:287–296, 1970.
49. Crawford, J. M. and Curtis, D. R.: The excitation and depression of mammalian cortical neurones by amino acids. *Br. J. Pharmacol.* 23:313–329, 1964.
50. Crawford, J. M. and Curtis, D. R.: Pharmacological studies on feline Betz cells. *J. Physiol.* (Lond.) 186:121–138, 1966.
51. Crowshaw, K., Jessup, S. J., and Ramwell, P. W.: Thin-layer chromatography of 1-dimethylaminonaphthalene-5-sulphonyl derivaties of amino acids present in superfusates of cat cerebral cortex. *Biochem. J.* 103:79–85, 1967.

52. Curtis, D. R.: Actions of drugs on single neurones in the spinal cord and thalamus. *Br. Med. Bull. 21*:5–9, 1963.
53. Curtis, D. R.: The action of amino acids upon mammalian neurones. In Curtis, D. R. and McIntyre, A. K. (Eds.): *Studies in Physiology, Presented to J. C. Eccles.* Heidelberg: Springer-Verlag, 1965, pp. 34–43.
54. Curtis, D. R. and Andersen, P.: Acetylcholine—central transmitter? *Nature 195*:1105–1106, 1962.
55. Curtis, D. R. and Davis, R.: Pharmacological studies upon neurones of the lateral geniculate nucleus of the cat. *Br. J. Pharmacol. 18*:217–246, 1962.
56. Curtis, D. R. and Davis, R.: The excitation of the lateral geniculate neurones by quaternary ammonium derivatives. *J. Physiol.* (Lond.) *165*:62–82, 1963.
57. Curtis, D. R. and Crawford, J. M.: Central synaptic transmission—microelectrophoretic studies. *Ann. Rev. Pharmacol. 9*:209–240, 1969.
58. Curtis, D. R., Duggan, A. W., and Johnston, G. A. R.: The inactivation of extracellularly administered amino acids in the feline spinal cord. *Exp. Brain Res. 10*:447–462, 1970.
59. Curtis, D. R., Duggan, A. W., Felix, D., et al.: Excitation of mammalian central neurones by acidic amino acids. *Brain Res. 41*:283–301, 1972.
60. Curtis, D. R. and Johnston, G. A. R.: Amino acid transmitters. In Lajtha, A. (Ed.): *Handbook of Neurochemistry* (Vol. 4). New York: Plenum Press, 1970, pp. 115–134.
61. Curtis, D. R. and Koizumi, K.: Chemical transmitter substances in brain stem of cat. *J. Neurophysiol. 24*:80–90, 1961.
62. Curtis, D. R., Perrin, D. D. and Watkins, J. C.: The excitation of spinal neurones by the iontophoretic application of agents which chelate calcium. *J. Neurochem. 6*:1–20, 1960.
63. Curtis, D. R., Phillis, J. W., and Watkins, J. C.: Chemical excitation of spinal neurones. *Nature 183*:611–613, 1959.
64. Curtis, D. R., Phillis, J. W., and Watkins, J. C.: The chemical excitation of spinal neurones by certain acidic amino acids. *J. Physiol.* (Lond.) *150*:656–682, 1960.
65. Curtis, D. R. and Ryall, R. W.: Pharmacological studies upon spinal presynaptic fibers. *Exp. Brain Res. 1*:195–204, 1966.
66. Curtis, D. R. and Watkins, J. C.: The excitation and depression of spinal neurones by structurally related amino acids. *J. Neurochem. 6*:117–141, 1960.
67. Curtis, D. R. and Watkins, J. C.: Analogues of glutamic and γ-amino-n-butyric acids having potent actions on mammalian neurones. *Nature 191*:1010–1011, 1961.
68. Curtis, D. R. and Watkins, J. C.: Acidic amino acids with strong excitatory actions on mammalian neurones. *J. Physiol.* (Lond.) *166*:1–14, 1963.
69. Curtis, D. R. and Watkins, J. C.: The pharmacology of amino acids related to gamma-aminobutyric acid. *Pharmacol. Rev. 17*:347–392, 1965.

70. Davidoff, R. A., Graham, L. T., Shank, R. P., et al.: Changes in amino acid concentrations associated with loss of spinal interneurons. *J. Neurochem.* 14:1025–1031, 1967.
71. Davidson, N. and Southwick, C. A. P.: The effect of topically applied amino acids on primary afferent terminal excitability in the rat cuneate nucleus. *J. Physiol.* (Lond.) 210:172–173, 1970.
72. Davidson, N. and Southwick, C. A. P.: Amino acids and presynaptic inhibition in the rat cuneate nucleus. *J. Physiol.* (Lond.) 219:689–708, 1971.
73. Davies, J. and Watkins, J. C.: Is 1-hydroxy-3-aminopyrrolidione-2 (HA 966) a selective excitatory amino-acid antagonist? *Nature* 238:61–63, 1972.
74. De Belleroche, J. S. and Bradford, H. F.: Metabolism of beds of mammalian cortical synaptosomes: Response to depolarizing influences. *J. Neurochem.* 19:585–602, 1972.
75. DeFeudis, F. V.: Effects of electrical stimulation of the efflux of L-glutamate from peripheral nerve *in vitro*. *Exp. Neurol.* 30:291–296, 1971.
75a. De Robertis, E., Sellinger, O. Z., Rodriguez, H., and de Lores Arnaiz, G.: Nerve endings in methionine sulphoximine convulsant rats, a neurochemical and ultrastructural study. *J. Neurochem.* 14:81–89, 1967.
76. Dési, I., Farkas, I., Sós, J., et al.: Neurophysiological effects of glutamic acid ethylester. *Acta Physiol. Acad. Sci. Hung.* 32:323–335, 1967.
77. Dowson, R. J. and Usherwood, P. N. R.: The effect of low concentrations of L-glutamate and L-aspartate on transmitter release at the locust excitatory nerve-muscle synapse. *J. Physiol.* (Lond.) 229:13–14P, 1972.
78. Dowson, R. J. and Usherwood, P. N. R.: The mode of action of phenylhydrazine hydrochloride on the locust neuromuscular system. *J. Insect Physiol.* 19:355–368, 1973.
79. Dravid, A. R., Himwich, W. A., and Davis, J. M.: Some free amino acids in dog brain during development. *J. Neurochem.* 12:901–906, 1965.
80. Duggan, A. W.: Amino Acids as Transmitters. Ph.D. Thesis, Aust. National Univ., 1971.
81. Duggan, A. W. and Johnston, G. A. R.: Glutamate and related amino acids in cat spinal roots, dorsal root ganglia and peripheral nerves. *J. Neurochem.* 17:1205–1208, 1970.
82. Duggan, A. W. and Johnston, G. A. R.: Glutamate and related amino acids in cat, dog, and rat spinal roots. *Comp. Gen. Pharmacol.* 1:127–128, 1970.
83. Eccles, J. C. and Jaeger, J. C.: The relationship between the mode of operation and the dimensions of the junctional regions at synapses and motor end-organs. *Proc. R. Soc.* Lond. (Biol.) 148:38–56, 1958.
84. Eccles, J. C., Schmidt, R. F., and Willis, W. D.: Pharmacological studies on presynaptic inhibition. *J. Physiol.* (Lond.) 168:500–530, 1963.

85. Edwardson, J. A., Bennett, G. W., and Bradford, H. F.: Release of amino acids and neurosecretory substances after stimulation of nerve-endings (synaptosomes) isolated from the hypothalamus. *Nature* 240:554–556, 1972.
86. Evans, P. D.: The free amino acid pool of the haemocytes of *Carcinus maenas* (L.). *J. Exp. Biol.* 56:501–507, 1972.
87. Flock, E. W., Block, M. A., Grindlay, J. H., et al.: Changes in free amino acids of brain and muscle after total hepatectomy. *J. Biol. Chem.* 200:529–536, 1953.
88. Florey, E. and Woodcock, B.: Presynaptic excitatory action of glutamate applied to crab nerve-muscle preparations. *Comp. Biochem. Physiol.* 26:651–661, 1968.
89. Fox, M. W.: The ontogeny of behaviour and neurological responses in the dog. *Anim. Behav.* 12:301–310, 1964.
90. Frank, G. B. and Jhamandas, K.: Effects of general stimulant drugs on the electrical responses of isolated slabs of cat's cerebral cortex. *Br. J. Pharmacol.* 39:716–723, 1970.
91. Galindo, A., Krnjević, K., and Schwartz, S.: Patterns of firing in the cuneate nucleus. *Proc. Can. Fed. Biol. Soc.* 10:140, 1967.
92. Galindo, A., Krnjević, K., and Schwartz, S.: Micro-iontophoretic studies on neurones in the cuneate nucleus. *J. Physiol.* (Lond.) 192:359–377, 1967.
93. Galindo, A., Krnjević, K., and Schwartz, S.: Patterns of firing in cuneate neurones and some effects of flaxedil. *Exp. Brain Res.* 5:87–101, 1968.
94. Gerard, R. W.: Metabolism and function in the nervous system. In Elliot, K. A. C., Page I. H., and Quastel, J. H. (Eds.): *Neurochemistry*. Springfield, (Ill.): C.C. Thomas, 1955, pp. 458–498.
95. Gibson, I. M. and McIlwain, H.: Continuous recording of changes in membrane potential in mammalian cerebral tissues *in vitro*: recovery after depolarization by added substances. *J. Physiol.* (Lond.) 176:261–283, 1965.
96. Glees, P.: *Neuroglia, Morphology and Function*. Springfield, (Ill.): C.C. Thomas, 1953.
97. Graham, L. T., Jr. and Aprison, M. H.: Distribution of some enzymes associated with the metabolism of glutamate, aspartate, γ-aminobutyrate and glutamine in cat spinal cord. *J. Neurochem.* 16:559–566, 1969.
98. Graham, L. T., Jr., Shank, R. P., Werman, R., et al.: Distribution of some synaptic transmitter suspects in cat spinal cord: Glutamic acid, aspartic acid, γ-aminobutyric acid, glycine, and glutamine. *J. Neurochem.* 14:465–472, 1967.
99. Haldeman, S., Huffman, R. D., Marshall, K. C., et al.: The antagonism of the glutamate-induced and synaptic excitations of thalamic neurones. *Brain Res.* 39:419–425, 1972.
99a. Haldeman, S. and McLennan, H.: The antagonistic action of glutamic acid diethylester towards amino acid-induced and synaptic excitations of central neurones. *Brain Res.* 45:393–400, 1972.
100. Hammerschlag, R., Potter, L. T., and Vinci, J.: *In vitro* uptake of

^{14}C-glutamate by rat spinal cord. *Trans. Am. Soc. Neurochem.* 2:78, 1971.

101. Hammerstad, J. P., Murray, J. E., and Cutler, R. W. P.: Efflux of amino acid neurotransmitters from rat spinal cord slices. II. Factors influencing the electrically induced effects of [^{14}C]glycine and ^{3}H-GABA. *Brain Res. 35*:357–367, 1971.
102. Hansson, H.-A.: Ultrastructural studies on the long term effects of sodium glutamate on the rat retina. *Virchows Arch.* (Pathol. Anat.) 6:1–11, 1970.
103. Harvey, J. A. and McIlwain, H.: Excitatory acidic amino acids and the cation content and sodium ion flux of isolated tissues from the brain. *Biochem. J. 108*:269–274, 1968.
104. Hebb, C. O. and Krnjević, K.: The physiological significance of acetylcholine. In Elliott, K. A. C., Page, I. H., and Quastel, J. H. (Eds.): *Neurochemistry.* Springfield, (Ill.): C.C. Thomas, 1963.
105. Herz, A. and Freytag-Loringhoven, V. H.J.: Über die synaptische Erregung im Corpus Striatum und deren antagonistische Beeinflussung durch mikroelektrophoretisch verabfolgte Glutaminsäure und Gamma-Aminobuttersäure. *Pflügers Arch. 299*:167–184, 1968.
106. Herz, A. and Gogolák, G.: Mikroelektrophoretische Untersuchungen am Septum des Kaninchens. *Pflügers Arch. 285*:317–330, 1965.
107. Herz, A. and Nacimiento, A. C.: Über die Wirkung von Pharmaka auf Neurone des Hippocampus nach mikroelektrophoretischer Verabfolgung. *Arch. Exp. Path. Pharmacol. 251*:295–314, 1965.
108. Herz, A. and Zieglgänsberger, W.: Synaptic excitation in the corpus striatum inhibited by microelectrophoretically administered dopamine. *Experientia 22*:839–840, 1966.
109. Hill, R. B. and Usherwood, P. N. R.: The action of 5-hydroxytryptamine and related compounds on neuromuscular transmission in the locust *Schistocerca gregaria. J. Physiol.* (Lond.) *157*:393–401, 1961.
110. Hillman, H. H., Campbell, J. W., and McIlwain, H.: Membrane potentials in isolated and electrically stimulated mammalian cerebral cortex. Effects of chlorpromazine, cocaine, phenobarbitone and protamine on the tissues' electrical and chemical responses to stimulation. *J. Neurochem. 10*:325–339, 1963.
111. Hillman, H. H. and McIlwain, H.: Membrane potentials in mammalian cerebral tissues *in vitro:* Dependence on ionic environment. *J. Physiol.* (Lond.) *157*:263–278, 1961.
112. Himwich, H. E. and Himwich, W. A.: Brain metabolism in relation to ageing. *Res. Pub. Assoc. Nerv. Ment. Dis. 35*:19–30, 1956.
113. Himwich, W. A. and Peterson, J. C.: Correlation of chemical maturation of the brain in various species with neurologic behaviour. In Masserman, J. H. (Ed.): *Biological Psychiatry* (Vol. 1). New York: Grune and Stratton, 1959, pp. 2–16.
114. Himwich, W. A., Sullivan, W. T., Kelley, B., et al.: Chemical constituents of human brain. *J. Nerv. Ment. Dis. 122*:441–447, 1955.
115. Hösli, L. and Tebēcis, A. K.: Actions of amino acids and convulsants on bulbar reticular neurones. *Exp. Brain Res. 11*:111–127, 1970.

116. Hyashi, T.: A physiological study of epileptic seizures following cortical stimulation in animals and its application to human clinics. *Jap. J. Physiol. 3*:46–64, 1952.
117. Hyashi, T.: *Neurophysiology and Neurochemistry of Convulsion.* Tokoyo Dainihon - Tosho, 1954.
118. Hyashi, T.: Effects of sodium glutamate on the nervous system. *Keio, J. Med. 3*:183–192, 1954.
119. Hyashi, T.: *Chemical Physiology of Excitation in Muscle and Nerve.* Tokyo: Nakayama-Shoten, 1956, pp. 152–157.
120. Hyashi, T. and Nagai, K.: Action of ω-amino acids on the motor cortex of higher animals, especially γ-amino-β-oxy-butyric acid as the real inhibitory principle in brain. *XX Int. Physiol. Cong. 2*:410, 1956.
121. Hyashi, T. and Shuhara, R.: Substances which produce epileptic seizure when applied on the motor cortex of dogs, and substances which inhibit the seizure directly. *XX Int. Physiol. Cong. 2*:410–411, 1956.
122. Jack, J. J. B., Miller, S., Porter, R. et al.: The distribution of Group 2a synapses on lumbosacral spinal motoneurones in the cat. In Andersen, P. and Jansen, J. K. S. (Eds.): *Excitatory Synaptic Mechanisms.* Oslo: Universitetsforlaget, 1970, pp. 199–205.
123. Jacobson, K. B.: Studies on the role of N-acetyl aspartic acid in mammalian brain. *J. Gen. Physiol. 43*:323–333, 1959.
124. Jasper, H. H., Khan, R. T., and Elliott, K. A. C.: Amino acids released from the cerebral cortex in relation to its state of activation. *Science 147*:1448–1449, 1965.
125. Jasper, H. H. and Koyama, I.: Rate of release of amino acids from the cerebral cortex in the cat as affected by brainstem and thalamic stimulation. *Can. J. Physiol. Pharmacol. 47*:889–905, 1969.
126. Johnson, J. L.: Glutamic acid as a synaptic transmitter in the nervous system. A review. *Brain Res. 37*:1–19, 1972.
127. Johnson, J. L. and Aprison, M. H.: The distribution of glutamic acid, a transmitter candidate, and other amino acids in the dorsal sensory neuron of the cat. *Brain Res. 24*:285–292, 1970.
128. Johnson, J. L. and Aprison, M. H.: The distribution of glutamate and total free amino acids in thirteen specific regions of the cat central nervous system. *Brain Res. 26*:141–148, 1971.
129. Johnston, G. A. R., Curtis, D. R., De Groat, W. C., et al.: Central actions of ibotenic acid and muscimol. *Biochem. Pharmacol. 17*: 2488–2489, 1968.
130. Kandera, J., Levi, G., and Lajtha, A.: Control of cerebral metabolite levels. II. Amino acid uptake and levels in various areas of the rat brain. *Arch. Biochem. Biophys. 126*:249–260, 1968.
131. Katz, R. I., Chase, T. N., and Kopin, I. J.: Effect of ions on stimulus-induced release of amino acids from mammalian brain slices. *J. Neurochem. 16*:961–967, 1969.
131a. Kerkut, G. A., Shapira, A., and Walker, R. J.: The transport of ^{14}C-labelled material from CNS muscle along a nerve trunk. *Comp. Biochem. Physiol. 23*:729–748, 1967.

132. Killam, K. F.: Convulsant hydrazides II. Comparison of electrical changes and enzyme inhibition induced by the administration of thiosemicarbazide. *J. Pharmacol. Exp. Ther.* *119*:263—271, 1957.
133. Killam, K. F. and Bain, J. A.: Convulsant hydrazides I. *In vitro* and *in vivo* inhibition of vitamin B_6 enzymes by convulsant hydrazides. *J. Pharmacol. Exp. Ther.* *119*:255—262, 1957.
134. Knaape, H. H. and Wiechert, P.: Krampfaktivität nach intracerebraler Injektion von L-Glutamat. *J. Neurochem.* *17*:1171—1175, 1970.
135. Koechlin, B. A.: On the chemical composition of the axoplasm of squid giant nerve fibers with particular reference to its ion pattern. *J. Biophys. Biochem. Cytol.* *1*:511—529, 1955.
136. Kravitz, E. A., Slater, C. R., Takahashi, K., et al.: Excitatory transmission in invertebrates—glutamate as a potential neuromuscular transmitter compound. In Andersen, P. and Jansen, J. K. S. (Eds.): *Excitatory Synaptic Mechanisms*. Oslo: Universitetsforlaget, 1970, pp. 85—94.
137. Krebs, H. R., Eggleston, L. V., and Hems, R.: Distribution of glutamine and glutamic acid in animal tissues. *Biochem. J.* *44*:159—163, 1949.
138. Krnjević, K.: Micro-iontophoretic studies on cortical neurons. *Int. Rev. Neurobiol.* *7*:41—98, 1964.
139. Krnjević, K.: Actions of drugs on single neurones in cerebral cortex. *Br. Med. Bull.* *21*:10—14, 1965.
140. Krnjević, K.: Central transmitters in vertebrates. In Andersen, P. and Jansen, J. K. S. (Eds.): *Excitatory Synaptic Mechanisms*. Oslo: Universitetsforlaget, 1970, pp. 95—117.
141. Krnjević, K., Mitchell, J. F., and Szerb, J. C.: Determination of iontophoretic release of acetylcholine from micropipettes. *J. Physiol.* (Lond.) *165*:421—426, 1963.
142. Krnjević, K. and Phillis, J. W.: The actions of certain amino acids on cortical neurones. *J. Physiol.* (Lond.) *159*:62—63P, 1961.
143. Krnjević, K. and Phillis, J. W.: Actions of certain amines on cerebral cortical neurones. *Br. J. Pharmacol.* *20*:471—491, 1963.
144. Krnjević, K. and Phillis, J. W.: Iontophoretic studies of neurones in the mammalian cerebral cortex. *J. Physiol.* (Lond.) *165*:274—304, 1963.
145. Krnjević, K. and Schwartz, S.: Some properties of unresponsive cells in the cerebral cortex. *Exp. Brain Res.* *3*:306—319, 1967.
146. Krnjević, K. and Whittaker, V. P.: Excitation and depression of cortical neurones by brain fractions released from micropipettes. *J. Physiol.* (Lond.) *179*:298—322, 1965.
146a. Kubíček, R. and Dolének, A.: Taurine et acides amines dans la rétine des animaux. *J. Chromatogr.* *1*:266—268, 1958.
146b. Kuhar, M. J. and Snyder, S. H.: The subcellular distribution of free H^3-glutamic acid in rat cerebral cortical slices. *J. Pharmacol. Exp. Ther.* *171*:141—152, 1970.
147. Lajtha, A.: Amino acid and protein metabolism of the brain. V. Turnover of leucine in mouse tissues. *J. Neurochem.* *3*:358—365, 1959.

148. Lajtha, A.: Protein metabolism in nerve. In Folch-Pi, J. (Ed.): *Chemical Pathology of the Nervous System.* Oxford: Pergamon Press, 1971, pp. 268–276.
149. Lajtha, A., Berl, S., and Waelsch, H.: Amino acid and protein metabolism of the brain. IV. The metabolism of glutamic acid. *J. Neurochem.* 3:332, 1959.
150. Lajtha, A., Berl, S., and Waelsch, H.: Compartmentalization of glutamic acid metabolism in the central nervous system. In Roberts, E. (Ed.): *Inhibition of the Nervous System and GABA.* Oxford: Pergamon Press, 1960, pp. 460–467.
151. Lajtha, A. and Mela, P.: The brain barrier system. I. The exchange of free amino acids between plasma and brain. *J. Neurochem.* 7:210–217, 1961.
152. Lajtha, A. and Toth, J.: The brain barrier system. V. Stereospecificity of amino acid uptake, exchange and efflux. *J. Neurochem.* 10:909–929, 1963.
153. Lajtha, A. and Toth, J.: The effects of drugs on uptake and exit of cerebral amino acids. *Biochem. Pharmacol.* 14:729–738, 1965.
153a. Lasek, R.: Axoplasmic transport in cat dorsal root ganglion cells: As studied with [1-^{3}H] leucine. *Brain Res.* 7:360–377, 1968.
154. Lea, T. J. and Usherwood, P. N. R.: Increased chloride permeability of insect muscle fibres on exposure to ibotenic acid. *J. Physiol.* (Lond.) 211:32–33P, 1970.
155. Lea, T. J. and Usherwood, P. N. R.: The site of action of ibotenic acid and the identification of two populations of glutamate receptors on insect-muscle fibres. *Comp. Gen. Pharmacol.* 4:333–350, 1973.
156. Lea, T. J. and Usherwood, P. N. R.: Effect of ibotenic acid on chloride permeability of insect-muscle fibres. *Comp. Gen. Pharmacol.* 4:351–363, 1973.
157. Legge, K. F., Randic, M., and Straughan, D. W.: The pharmacology of the pyriform cortex. *Br. J. Pharmacol.* 26:87–107, 1966.
158. Levi, G., Kandera, J., and Lajtha, A.: Control of cerebral metabolite levels. I. Amino acid uptake and levels in various species. *Arch. Biochem. Biophys.* 119:303–311, 1967.
159. Levi, G. and Lajtha, A.: Cerebral amino acid transport *in vitro* II. Regional differences in amino acid uptake by slices from the central nervous system of the rat. *J. Neurochem.* 12:639–648, 1965.
160. Lewis, P. R.: The free amino-acids of invertebrate nerve. *Biochem. J.* 52:330–338, 1952.
161. Logan, J. E., Mannell, W. A., and Rossiter, R. J.: Estimation of nucleic acids in tissue from the nervous system. *Biochem. J. 51:* 470–482, 1952.
162. Logan, W. J. and Snyder, S. H.: Unique high affinity uptake systems for glycine, glutamic and aspartic acids in central nervous tissue of the rat. *Nature* 234:297–299, 1971.
163. Machiyama, Y., Bálazs, R., and Mérei, T.: Incorporation of [^{14}C] glutamate into glutathione in rat brain. *J. Neurochem.* 17:449–453, 1970.
164. Machiyama, Y., Bálazs, R., and Richter, D.: Effect of K$^+$ stimulation

on GABA metabolism in brain slices *in vitro*. *J. Neurochem.* 14: 591–594, 1967.
165. Mangan, J. L. and Whittaker, V. P.: The distribution of free amino acids in subcellular fractions of guinea-pig brain. *Biochem. J.* 98: 128–137, 1960.
166. Margolis, R. K. and Lajtha, A.: Ion dependence of amino acid uptake in brain slices. *Biochim. Biophys. Acta* 163:374–385, 1968.
167. Marks, N., Datta, R. K., and Lajtha, A.: Distribution of amino acids and of exo- and endopeptidases along vertebrate and invertebrate nerves. *J. Neurochem.* 17:53–63, 1970.
168. Martin, A. R., Wickelgren, W. O., and Beránek, R.: Effects of iontophoretically applied drugs on spinal interneurones of the lamprey. *J. Physiol.* (Lond.) 207:653–665, 1970.
169. McCance, I. and Phillis, J. W.: Discharge patterns of elements in cat cerebellar cortex, and their responses to iontophoretically applied drugs. *Nature* 204:844–846, 1964.
170. McCance, I. and Phillis, J. W.: The action of acetylcholine on cells in cat cerebellar cortex. *Experientia* 20:217–218, 1964.
171. McCance, I. and Phillis, J. W.: Cholinergic mechanisms in the cerebellar cortex. *Int. J. Neuropharmacol.* 7:447–462, 1968.
172. McCance, I., Phillis, J. W., Tebēcis, A. K., et al.: The pharmacology of acetylcholine-excitation of thalamic neurones. *Br. J. Pharmacol. Chemother.* 32:652–662, 1968.
173. McCance, I., Phillis, J. W., and Westerman, R. A.: Acetylcholine-sensitivity of thalamic neurones: Its relationship to synaptic transmission. *Br. J. Pharmacol. Chemother,* 32:635–651, 1968.
174. McIlwain, H., Harvey, J. A., and Rodriguez, G.: Tetrodotoxin on the sodium and other ions of cerebral tissues, excited electrically and with glutamate. *J. Neurochem.* 16:363–370, 1969.
175. McLennan, H.: Inhibitions of long duration in the cerebral cortex. A quantitative difference between excitatory amino acids. *Exp. Brain Res.* 10:417–426, 1970.
176. McLennan, H., Huffman, R. D., and Marshall, K. C.: Patterns of excitation of thalamic neurones by amino-acids and by acetylcholine. *Nature* 219:387–388, 1968.
177. McLennan, H., Marshall, K. C., and Huffman, R. D.: The antagonism of glutamate action at central neurones. *Experientia* 27:1116, 1971.
178. McLennan, H. and York, D. H.: The action of dopamine on neurones of the caudate nucleus. *J. Physiol.* (Lond.) 189:393–402, 1967.
179. Meyer, M.: Die Wirkung von Acetylcholin, L-Glutaminsäure und Dopamin auf Neurone im Gebiet der Nuclei cuneatus und gracilis der Katze. *Helv. Physiol. Pharmacol. Acta* 23:325–340, 1965.
180. Mihailovic, L., Jankovic, R. D. Petrovic, M., et al.: Distribution of DNA and RNA in different regions of cat's brain. *Experientia* 14:9–10, 1958.
180a. Miller, R., and Daoud, A.: Release of amino acids from resting and stimulated superfused insect nerve-muscle preparations. In preparation.

181. Morgan, R., Vrvoba, G., and Wolstencroft, J. H.: Correlation between the retinal input to lateral geniculate neurones and their relation response to glutamate and aspartate. *J. Physiol.* (Lond.) *224*:41–42P, 1972.
182. Muscini, E. and Marcucci, F.: Free amino acids in brain after treatment with psychotropic drugs. In Holden, J. T. (Ed.): *Amino Acid Pools.* Amsterdam: Elsevier, 1962, p. 486–492.
183. Narahasi, T., Moore, J. W., and Scott, W. R.: Tetrodotoxin blockage of sodium conductance increase in lobster giant axons. *J. Gen. Physiol. 47*:965–974.
184. Neal, M. J. and Hopkin, J.: Effect of electrical stimulation and high potassium concentrations on the efflux of [^{14}C] glycine from slices of spinal cord. *Br. J. Pharmacol. Chemother. 42*::215–223, 1971.
185. Neidle, A., van den Berg, C. J., and Grynbaum, A.: The heterogeneity of rat brain mitochondria isolated on continuous sucrose gradients. *J. Neurochem. 16*:225–234, 1969.
186. Nelson, P. G. and Frank, K.: Anomalous rectification in cat spinal motoneurones and effect of polarizing currents on excitatory postsynaptic potentials. *J. Neurophysiol. 30*:1097–1113, 1967.
187. Nishi, S., Soeda, H., and Koketsu, K.: Effect of alkali-earth cations on frog spinal ganglion cell. *J. Neurophysiol. 28*:457–471, 1965.
188. Noell, W. K. and Lasansky, A.: Effects of electrophoretically applied drugs and electrical currents on the ganglion cell of the retina. *Fed. Proc. 18*:15, 1959.
189. Ochs, S. and Johnson, J. L.: Fast and slow phases of axoplasmic flow in ventral root nerve fibers. *J. Neurochem. 16*:845–853, 1969.
189a. Ochs, S. and Ranish, N.: Characteristics of the fast transport system in mammalian nerve fibers. *J. Neurochem. 2*:247–261, 1969.
189b. Ochs, S., Sabri, M. I., and Johnson, J. L.: Fast transport system of materials in mammalian nerve fibers. *Science N.Y., 163*:686–687, 1969.
190. Okumura, N., Otsuki, S., and Aoyama, T.: Studies on the free amino acids and related compounds in the brains of fish, amphibia, reptiles, aves and mammals by ion exchange chromatography. *J. Biochem. (Tokyo) 46*:207, 1959.
191. Okumura, N., Otsuki, S., and Fukai, N.: Amino acid concentration in different parts of the dog brain. *Acta Med. Okayama 13*:27–30, 1959.
192. Okumura, N., Otsuki, S., and Kameyama, A.: Studies on free amino acids in human brain. *J. Biochem. (Tokyo) 47*:315–320, 1960.
193. O'Neal, R. M. and Koeppe, R. E.: Precursors *in vivo* of glutamate, aspartate and their derivatives of rat brain. *J. Neurochem. 13*:835–847, 1966.
194. Oxender, D. L. and Christensen, H. N.: Distinct mediating systems for the transport of neutral amino acids by the Ehrlich cell. *J. Biol. Chem. 238*:3686–3699, 1963.
195. Pepeu, G., Bartolini, A., and Bartolini, R.: Differences of GABA

content in the cerebral cortex of cats transected at various midbrain levels. *Biochem. Pharmacol. 19*:1007–1013, 1970.
196. Perry, T. L., Sander, H. D., Hansen, S., et al.: Free amino acids and related compounds in five regions of biopsied cat brain. *J. Neurochem. 19*:2651–2656, 1972.
196a. Perry, T. L., Berry, K., Hansen, S., et al.: Regional distribution of amino acids in human brain obtained at autopsy. *J. Neurochem. 18*:513–519, 1971.
197. Petersen, J., Di Perri, R., and Himwich, W. A.: Comparative development of the EEG in rabbit, cat and dog. *Electroencephalogr. Clin. Neurophysiol. 17*:557–563, 1964.
198. Peterson, N. A. and Raghupathy, E.: Characteristics of amino acid accumulation by synaptosomal particles isolated from rat brain. *J. Neurochem. 19*:1423–1438, 1972.
199. Phillis, J. W. and Tebēcis, A. K.: The responses of thalamic neurones to iontophoretically applied monoamines. *J. Physiol.* (Lond.) *192*: 715–745, 1967.
200. Phillis, J. W., Tebēcis, A. K., and York, D. H.: A study of cholinoceptive cells in the lateral geniculate nucleus. *J. Physiol.* (Lond.) *192*:695–713, 1967.
201. Phillis, J. W. and York, D. H.: Cholinergic inhibition in the cerebral cortex. *Brain Res. 5*:517–520, 1967.
202. Phillis, J. W. and York, D. H.: Pharmacological studies on a cholinergic inhibition in the cerebral cortex. *Brain Res. 10*:297–306, 1968.
203. Porcellati, G. and Thompson, R. H. S.: The effect of nerve section on the free amino acids of nervous tissue. *J. Neurochem. 1*:340–347, 1957.
204. Preston, J. B.: The influence of thiosemicarbazide on electrical activity recorded in the anterior brain stem of the cat. *J. Pharmacol. Exp. Ther. 115*:39–45, 1955.
205. Purpura, D. P.: Pharmacological actions of omega-amino acid drugs on different cortical synaptic organizations. In Roberts, E., (Ed.): *Inhibition in the Nervous System and Gamma-Aminobutyric Acid.* Oxford: Pergamon Press, 1960, pp. 495–514.
206. Purpura, D. P., Girado, M., Smith, T. G., et al.: Structure-activity determinants of pharmacological effects of amino acids and related compounds on central synapses. *J. Neurochem. 3*:238–268, 1959.
207. Rall, W., Burke, R. E., Smith, T. G., et al.: Dendritic location of synapses and possible mechanisms for the monosynaptic EPSP in motoneurons. *J. Neurophysiol. 30*:1169–1113, 1967.
208. Roberts, E., Frankel, S., and Harman, P. J.: Amino acids of nervous tissue. *Proc. Soc. Exp. Biol. Med. 74*:383–387, 1950.
209. Roberts, P. J. and Mitchell, J. F.: The release of amino acids from the hemisected spinal cord during stimulation. *J. Neurochem. 19*: 2473–2481, 1972.
210. Rose, S. P. R.: The compartmentation of glutamate and its metabolites in fractions of neuron cell bodies and neuropil; studied by intraventricular injection of [U-^{14}C] glutamate. *J. Neurochem. 17*:809–816, 1970.

211. Rose, S. P. R. and Cory, H. T.: Glutamate metabolism in octopus brain *in vivo*; absence of a Waelsch effect. *J. Neurochem. 17*:817–820, 1970.
212. Rosenberg, L. E., Downing, S. J., and Sagal, S.: Competitive inhibition of dibasic amino acid transport in rat kidney. *J. Biol. Chem. 237*:2265–2270, 1962.
213. Rowe, W. B. and Meister, A.: Identification of L-methionine-S-sulfoximine as the convulsant isomer of methionine sulphoximine. *Proc. Nat Acad. Sci. 66*:500–506, 1970.
214. Ryall, R. W.: The subcellular distributions of acetylcholine, substance P, 5-hydroxytryptamine, γ-aminobutyric acid and glutamic acid in brain homogenates. *J. Neurochem. 11*:131–145, 1964.
215. Salganicoff, L. and De Robertis, E.: Subcellular distribution of the enzymes of the glutamic acid, glutamine and γ-aminobutyric acid cycles in rat brain. *J. Neurochem. 12*:287–309, 1965.
216. Schmidt, R. F.: Pharmacological studies in the primary afferent depolarization of the toad spinal cord. *Pflügers Arch. 277*:325–346, 1963.
217. Sellinger, O. Z. and De Balbian Verster, F.: Glutamine synthetase of rat cerebral cortex: Intracellular distribution and structural latency. *J. Biol. Chem. 237*:2836–2844, 1962.
218. Shank, R. P. and Aprison, M. H.: The metabolism *in vivo* of glycine and serine in eight areas of the rat central nervous system. *J. Neurochem. 17*:1461–1475, 1970.
219. Shaw, R. K. and Heine, J. D.: Ninhydrin positive substances present in different areas of normal rat brain. *J. Neurochem. 12*:151–155. 1965.
220. Singer, M. and Salpeter, M. M.: The transport of ^{3}H-1-histidine through the Schwann and myelin sheath into the axon, including a re-evaluation of myelin function. *J. Morphol. 120*:281–316, 1966.
221. Singh, S. I. and Malhotra, C. L.: Amino acid content of monkey brain-1. General pattern and quantitative value of glutamic acid/glutamine, γ-aminobutyric acid and aspartic acid. *J. Neurochem. 9*:37–42, 1962.
221a. Snodgrass, S. R. and Iversen, L. L.: A sensitive double isotope derivative assay to measure release of amino acids from brain *in vitro*. *Nature 241*:155–156, 1973.
222. Spyropoulous, C. S.: Cytoplasmic pH of nerve fibers. *J. Neurochem. 5*:185–194, 1960.
223. Stefanis, C.: Hippocampal neurons: Their responsiveness to microelectrophoretically administered endogenous amines. *Pharmacologist 6*:171, 1964.
224. Stefanis, C.: Electrophysiological properties of cortical motoneurons during iontophoretic application of chemical substances. *Physiologist 7*:263, 1964.
225. Stefanis, C.: Discussion of the serotonin effects on central neurons. *Adv. Pharmacol. 6*:414–418, 1968.
226. Steiner, F. A. and Meyer, M.: Actions of L-glutamate, acetylcholine and dopamine on single neurones in the nuclei cuneatus and gracilus of the cat. *Experientia 22*:58–59, 1966.

227. Steiner, F. A. and Ruf, K.: Interactions of L-glutamic acid, thiosemicarbazide and pyridoxal-5'-phosphate at single unit level in rat brain. *Brain Res. 3*:214–216, 1966.
228. Steiner, F. A. and Ruf, K.: Excitatory effects of L-glutamic acid upon single unit activity in rat brain and their modification by thiosemicarbazide and pyridoxal-5'-phosphate. *Helv. Physiol. Acta 24*:181–192, 1966.
229. Steiner, F. A. and Ruf, K.: Interactions of L-glutamic acid, gamma-aminobutyric acid and pyridoxal-5'-phosphate at the neuronal level. *Schweiz. Arch. Neurol. Psychiatr. 100*:310–320, 1967.
230. Stern, J. R., Eggleston, L. V., Hems, R., et al. Accumulation of glutamic acid in isolated brain tissue. *Biochem. J. 44*:410–418, 1949.
231. Straughan, D. W. and Legge, K. F.: The pharmacology of amygdaloid neurones. *J. Pharm. Pharmacol. 17*:675–677, 1965.
232. Strecker, H. J.: Glutamic acid and glutamine. In Richter, D. (Ed.): *Metabolism of the Nervous System*. London: Pergamon Press, 1957, pp. 459–474.
233. Takeuchi, A. and Takeuchi, N.: The effect on crayfish muscle of iontophoretically applied glutamate. *J. Physiol.* (Lond.) *170*:296–317, 1964.
234. Tallan, H. H.: Studies of the distribution of N-acetic-L-aspartic acid in brain. *J. Biol. Chem. 224*:41–45, 1957.
235. Tallan, H. H.: A survey of the amino acids and related compounds in nervous tissue. In Holden, J. T. (Ed.): *Amino Acid Pools*. Amsterdam: Elsevier, 1962, pp. 471–485.
236. Tallan, H. H., Moore, S., and Stein, W. H.: N-acetyl-L-aspartic acid in brain. *J. Biol Chem. 219*:257–264, 1956.
237. Taraskevich, P. S.: Reversal potentials of L-glutamate and the excitatory transmitter at the neuromuscular junction of the crayfish. *Biochim. Biophys. Acta 241*:700–703, 1971.
238. Tarkowski, S. and Cremer, J. E.: Metabolism of glucose and free amino acids in brain, studied with ^{14}C-labelled glucose and butyrate in rats intoxicated with carbon disulphide. *J. Neurochem. 19*:2631–2640, 1972.
239. Tebēcis, A. K.: Are 5-hydroxytryptamine and noradrenaline inhibitory transmitters in the medial geniculate nucleus? *Brain Res. 6*:780–782, 1967.
240. Tebēcis, A. K.: Acetylcholine and medial geniculate neurones. *Aust. J. Exp. Biol. Med. Sci. 46*:3P, 1968.
241. Trubatch, R., Verhulst, C., and van Harreveld, A.: Glutamate as a transmitter: Comparison between the crustacean neuromuscular junction and chicken retina. *Comp. Biochem. Physiol. 45A*:183–194, 1973.
241a. Tsukada, Y., Nagata, Y., Hirano, S., et al.: Active transport of amino acid into cerebral cortex slices. *J. Neurochem. 10*:241–256, 1963.
242. Usherwood, P. N. R.: Insect neuromuscular mechanisms. *Am. Zool. 7*:553–582, 1967.
243. Usherwood, P. N. R. and Machili, P.: Pharmacological properties of excitatory neuromuscular synapses in the locust. *J. Exp. Biol. 49*:341–361, 1968.

244. van den Berg, C. J.: Compartmentation of glutamate metabolism in the developing brain; experiments with labelled glucose, acetate, phenylalanine, tyrosine, and proline. *J. Neurochem.* 17:973–983, 1970.
245. van den Berg, C. J., Kržalić, Lj., Mela, P., et al.: Compartmentation of glutamate metabolism in brain. Evidence for the existence of two different tricarboxylic acid cycles in brain. *Biochem. J. 113*: 281–290, 1969.
246. van Harreveld, A.: Compounds in brain extracts causing spreading depression of cerebral cortical activity and contraction of crustacean muscle. *J. Neurochem. 3*:300–315, 1959.
247. van Harreveld, A.: Physiological effects of glutamate and some related compounds. In Roberts, E. (Ed.): *Inhibition in the Nervous System and Gamma-aminobutyric Acid.* Oxford: Pergamon Press, 1960, pp. 454–459.
248. van Harreveld, A. and Fifkova, E.: Glutamate release from the retina during spreading depression. *J. Neurobiol.* 2:13–29, 1970.
249. van Harreveld, A. and Fifkova, E.: Release of glutamate from the retina during glutaraldehyde fixation. J. Neurochem. 19:237–241, 1972.
250. van Harreveld, A. and Kooiman, M.: Amino acid release from the cerebral cortex during spreading depression and asphyxiation. *J. Neurochem. 12*:431–439, 1965.
251. Vernadakis, A. and Woodbury, D. M.: Electrolyte and amino acid changes in rat brain during maturation. *Am. J. Physiol. 203*:748–752, 1962.
252. Voeller, K., Pappas, G. D., and Purpura, D. P.: Electron microscope study of development of cat superficial neocortex. *Exp. Neurol.* 7:107–130, 1963.
253. Waelsch, H.: Glutamic acid and cerebral function. In Anson, M. L., Edsall, J. T., and Vailey, K. (Eds.): *Advances in Protein Chemistry* (Vol. 6). New York: Academic Press, 1951, pp. 299–341.
254. Waelsch, H.: Metabolism of proteins and amino acids In Richter, D. (Ed.): *Metabolism of the Nervous System.* London: Pergamon Press, 1957, pp. 431–448.
255. Walther, C. and Usherwood, P. N. R.: Characterization of the glutamate receptors at the locust excitatory neuromuscular junction. *Verh. Dtsch. Ges. Helgoland 65*:309–312, 1972.
256. Watkins, J. C.: The synthesis of some acidic amino acids possessing neuropharmacological activity. *J. Med. Pharm. Chem.* 5:1187–1199, 1962.
257. Werman, R.: Criteria for identification of a central nervous system transmitter. *Comp. Biochem. Physiol. 18*:745–766, 1966.
258. Werman, R.: CNS cellular level: Membranes. *Ann. Rev. Physiol.* 34:337–374, 1972.
259. Wheeler, D. D. and Boyarsky, L. L.: Influx of glutamic acid in peripheral nerve—characteristics of influx. *J. Neurochem. 15*: 1019–1031, 1968.
260. Wheeler, D. D., Boyarsky, L. L., and Brooks, W. H.: The release of

amino acids from nerve during stimulation. *J. Cell. Physiol. 67*: 141–148, 1966.
261. Williams, H. L. and Bain, J. A.: Convulsive effects of hydrazides: Relationship to pyridoxine. *Int. Rev. Neurobiol. 3*:319–348, 1961.
262. Wilson, T. H.: *Intestinal Absorption*. Philadelphia: Saunders, 1962.
263. Wofsey, A. R., Kuhar, M. J., and Snyder, S. H.: A unique synaptosomal fraction, which accumulates glutamic and aspartic acids, in brain tissue. *Proc. Nat. Acad. Sci. 68*:1102–1106, 1971.
264. Yamaguchi, M., Yanos, T., Yamaguchi, T., et al.: Amino acid uptake in the peripheral nerve of the rat. *J. Neurobiol. 1*:419–433, 1970.
265. York, D. H.: A microiontophoretic study of neurones in the globus pallidus-putamen complex. *Aust. J. Exp. Biol. Med. Sci. 46*:3–6P, 1968.
266. Zieglgänsberger, W. and Puil, E. A.: Tetrodotoxin interference of CNS excitation by glutamic acid. *Nature 239*:204–205, 1972.

CHAPTER FOUR
Acetylcholine and Synaptic Transmission in the Central Nervous System

John W. Phillis

More than 25 years have elapsed since W. S. Feldberg (93) published his classic paper entitled "Present Views on the Mode of Action of Acetylcholine in the Central Nervous System." At the end of a decade that saw acetylcholine (ACh) established as the humoral transmitter released at ganglionic and neuromuscular synapses, he was optimistic about the acceptance of ACh as a central transmitter and was able to state: "The present position of the theory of acetylcholine as central transmitter is all but settled." Although many authors currently display greater reserve when they approach this subject, it is reasonable to claim that the passage of time has added substance to Feldberg's conviction that ACh is indeed a synaptic transmitter in the central nervous system (CNS). Even so, it is now apparent that cholinergic synapses often merely modulate the activity of nerve cells, conditioning their state of excitability so that nervous impulses conveyed by other transmitters, possibly the amino acids, will have an appropriate effect.

BACKGROUND AND HISTORICAL REVIEW

Many of the more salient features of the metabolism of ACh were already known in 1945, and in his review Feldberg discusses its synthesis, release, and hydrolysis. Thus in that year it was shown that ACh is synthesized in vertebrate tissues by the enzyme choline acetyltransferase (93), which is both present in cholinergic neurons and capable of transferring acetyl groups from coenzyme A (CoA) to choline (95, 163, 192). The requirement of choline and acetate for ACh synthesis was recognized, as was the need for adenosine triphosphate (ATP) as the source of chemical energy for the acetylation reaction.

The intervening years have seen an expansion and consolidation of knowledge about the synthesis and hydrolysis of ACh. Two major reports on the distribution of choline acetyltransferase in the CNS demonstrated that the levels fluctuate extensively from one area to another (*96, 115*). Reliance on the levels of ACh and choline acetyltransferase as indicators of cholinergic synapses has, however, proved to be misleading in some instances. Evidence will be presented herein, e.g., that ACh is an excitatory transmitter in the cerebellar cortex, even though this structure has very low levels of ACh and its synthesizing enzyme.

The most notable advances within the area of this review have resulted from the development of histochemical techniques for the localization of cholinesterase and choline acetyltransferase, and especially from the refinement of techniques for analysis of the actions of ACh at a cellular level. Techniques for the study of ACh release from central nervous structures have been developed, and the results of these experiments have also contributed to our understanding of the role of ACh.

When viewed together, the results from these various endeavors add support to the contention that ACh fulfills some of the criteria of an important transmitter in the brain. This chapter will be especially concerned with the actions of ACh on single neurons in different regions of the CNS, for only with the recent development of microtechniques for the application of drugs directly into the extracellular environment of single nerve cells has this been possible. The techniques are those of Nastuk (*194*), the substances being applied by iontophoresis from five-barrel glass micropipettes. Multiple-barrel micropipettes permit the recording of potentials from neurons while substances are ejected into their immediate extracellular environment. The details of this method have been described by Curtis (*60*), who pioneered the application of the technique to the CNS.

Although the identification of ACh as the active cholinergic principle in the brain is generally accepted, it is suggested that other choline esters or related substances may also contribute to the activity of brain extracts (*122, 125, 126*). The possibility that propionylcholine and butyrylcholine may be present in fresh brain extracts is militated against by gas chromatographic analysis, which readily separates these esters from ACh (*112, 232*). The ACh contents of brain extracts, as determined by this technique, are comparable to the estimates obtained by bioassay, and no traces of propionylcholine or of butyrylcholine have been detected.

The ACh and choline acetyltransferase of brain homogenates are largely associated with the particulate fraction—and on subcellular fractionation by centrifugation in a density-gradient column of sucrose solutions, the bulk of the ACh and choline acetyltransferase is found in the synaptosome or nerve ending fraction. In most species ACh is a component of the synaptic vesicles within the synaptosomes, whereas the choline acetyltransferase is associated with the vesicular or synaptosomal membrane (*167, 276, 288*). The biosynthesis of ACh in nervous tissue has recently been reviewed in detail by Hebb (*113*) and the interested reader is referred to this article.

HISTOCHEMICAL STUDIES

Cholinesterase

Until the recent development of a specific histochemical method for demonstrating the distribution of choline acetyltransferase (*40*), emphasis was placed on studies of the localization of acetylcholinesterase in the CNS (see reviews by Gerebtzoff [*105*]; Koelle [*146, 147*]; Silver [*241*]; and Ishii and Friede [*129*]). The significance of the distribution of acetylcholinesterase as an indicator of cholinergic pathways has been evaluated in a limited number of situations either by studying the correlation between the relative levels of choline acetyltransferase and acetylcholinesterase (*107, 161*) or by a pharmacological analysis of the synaptic effects of stimulation of enzyme-containing pathways (*170, 171*).

Shute and Lewis (*236, 237*) have emphasized that acetylcholinesterase may be present in the cell bodies of neurons that are unlikely to be cholinergic. They consider the presence of the enzyme in the axon and presynaptic terminals to be an essential criterion of a cholinergic neuron. However, the demonstration of acetylcholinesterase in noradrenaline-containing nerve fibers in sympathetically innervated organs (*88, 131, 132*) throws considerable doubt on the validity of this assumption. Although a high level of acetylcholinesterase activity in the somal and axonal processes may be indicative of a cholinergic neuron, this need not always be so.

By combining histochemistry with operative procedures, Shute and Lewis (*236*) developed a method for determining the polarity of cholinesterase-staining fibers in the CNS. When such fibers are transected, there is an accumulation of acetylcholinesterase in the severed axons on the cell-body side, and a decrease in the level of the enzymatic activity on the distal side of the lesion, where the axons are now separated from their cells of origin. An extensive survey of the presence and polarity of acetylcholinesterase-containing fibers enabled these authors to trace two major ascending pathways from the reticular and tegmental nuclei of the brain stem of the rat. The pathways, which project to virtually all cortical and subcortical structures, are the "dorsal" and "ventral tegmental pathways" (161, 239).

The dangers inherent in the hypothesis that the histochemically demonstrable acetylcholinesterase at synaptic junctions is indicative of cholinergic transmission have recently been stressed by Karczmar (138) and can be illustrated by recent studies on the cerebellar cortex. The widespread presence of the enzyme at synapses in the guinea-pig cerebellar cortex led Kása and Csillik (140) to conclude that ACh is the universal transmitter in this structure. This conclusion is difficult to reconcile both with the low levels of choline acetyltransferase in the cerebellar cortex of this species (107) and with the insensitivity of many of the granule and Purkinje cells in the guinea-pig cerebellar cortex to ACh (168). There is, moreover, no apparent correlation between the levels of choline acetyltransferase and acetylcholinesterase in the various layers of the cerebellar cortices of different species (107).

The functional importance of pseudocholinesterase in the CNS is even less clearly defined (241). Pseudocholinesterase is present in some neurons (236) where it may have a role similar to that of acetylcholinesterase. The pseudocholinesterase in glial cells may be involved in the control of activity in adjacent neurons (78) by influencing the permeability of glial cells to various ions or by protecting the neurons from exogenously released ACh.

The difficulties associated with the interpretation of histochemically determined cholinesterase distribution have been stressed, and in summary it can be said that acetylcholinesterase in nerve fibers, their terminals, and the membranes of associated postsynaptic elements may be a useful indicator of potential cholinergic synapses. In this respect the technique serves the valuable function of focusing attention on these junctions and—in association with data on the distribution, release, and pharmacology of ACh—assists in the identification of cholinergic synapses. For this reason relevant details on cholinesterase distribution will be included in the sections on cholinergic responses of different areas of the CNS.

Choline Acetyltransferase

Because of the accepted limitations of cholinesterase localization as a solution to the problem of defining cholinergic synapses, efforts have been made to study the distribution of choline acetyltransferase, the ACh-synthesizing enzyme. From all that is known of the distribution of choline acetyltransferase, it would be expected to give more certain indication of the distribution of ACh and hence of cholinergic neurons and synapses (*204*). A histochemical method for localization of sites of ACh synthesis was recently proposed by Burt (*40*) in which sections of tissue are incubated in acetyl-CoA, choline, and a lead salt. Free CoA, released during the synthesis of ACh, is precipitated by the lead. The precipitate so formed is then converted to lead sulfide, which can readily be visualized. The technique has now been used to stain sections from the spinal cord and cerebellum (*40, 141, 142*). Reaction product was found in the cell body, dendrites, and axons of motoneurons (*40, 114, 115*). In preliminary studies of the cerebellar cortex and deep nuclei of the rat (*40*), the deep nuclei were heavily stained, the molecular layer was nearly devoid of activity, the soma of Purkinje cells were unstained, and the granular layer was moderately stained.

Csillik (*59*) has attempted to use carbon-14 labeled hemicholinium, which interferes with the choline transporting mechanism of nerve terminals, to localize by autoradiography the sites of ACh synthesis. In sections of the spinal cord, silver grains are concentrated over the motoneuronal somata and in the neuropil of the gray matter. Activity in the cerebellum is confined to the glomerular islands of the granular layer, the parallel fibers of the molecular layer, and the somata of Purkinje cells. In the cerebral cortex and caudate nucleus activity is confined to the neuropil.

RELEASE OF ACETYLCHOLINE

Physiological Studies

The desirability of being able to demonstrate release of a transmitter from those junctions at which it is a suspected mediator was stressed in the introduction to this review. Acceptance of ACh as the mediator at synapses in ganglia and skeletal muscle was greatly facilitated when release of this agent from ganglia and muscle was shown to occur during stimulation of the presynaptic nerve fibers (*71, 94*).

Experimental methods for revealing a release of ACh from the brain were originally developed by MacIntosh and Oborin (*166*). Their technique of placing small, open-ended cylinders, filled with saline, on the surface of the cortex is still currently employed in the majority of experiments on the cerebral cortex.

Use has been made of implanted cannulas to detect release in areas of the brain that are not as conveniently exposed as the cortical hemispheres.

Such cannulas have been placed in the ventricular system of the brain to detect an efflux of ACh from adjacent structures. Of especial significance was the development of the Gaddum push-pull cannula, an arrangement of two fine concentrically positioned tubes, the tips of which can be inserted into individual nuclei or other areas of the brain and spinal cord. Acetylcholine from the surrounding tissue diffuses into the saline stream flowing through the tubes and can thus be collected and measured.

Release From Cerebral Cortex A considerable volume of literature on ACh release from the cerebral cortex has accumulated since MacIntosh and Oborin (*166*) published their original report. Most of the investigators have used anesthetized animals, although in some instances unanesthetized *cerveau isolé* or *encéphale isolé* preparations have been used (*12, 42, 206, 240*). More recently chronic conscious preparations have been developed (*15, 48, 134*). Identification of the active principle in perfusates as ACh (or a pharmacologically identical agent) has been provided by several investigators, including Mitchell (*186*) and Szerb (*266*).

The rate of release of ACh from the cerebral cortex is influenced by the state of arousal of the animal. Spontaneous release from the surface of the brain of anesthetized animals is reduced as the depth of anesthesia is increased (*48, 166, 186*), and Collier and Mitchell (*47*) were able to demonstrate that release from the cerebral cortex of conscious, freely moving rabbits increases during periods of activity.

More recently Jasper and Tessier (*134*) correlated ACh release with electroencephalographic potentials recorded from the cortex of intact, freely moving cats. Release rates of ACh into a cylinder over the postcruciate and anterior suprasylvian cortices were lowest during slow wave-sleep (average rate 1.2 ng/min/cm^2) and increased during paradoxical (REM) sleep (2.2 ng/min/cm^2) and during waking (2.1 ng/min/cm^2). These results indicate that desynchronization of the EEG, whether occurring during arousal or paradoxical sleep, is associated with an increase in the rate of release of ACh. This fact suggests that liberation of cortical ACh is related to activation of the EEG rather than to the animal's behavioral responsiveness.

The relationship between cortical ACh release and electroencephalographic arousal has also been investigated by Szerb (*268*). He concluded that increases in ACh release occur in conjunction with EEG activation, but that the pathways involved are not identical, since the two phenomena do not vary in a parallel fashion when the reticular formation is stimulated at different frequencies or when different subcortical areas are stimulated. Stimulation of the septum in the cat, e.g., markedly increased ACh output with only moderate EEG activation (*268*).

The spontaneous release of ACh from chronically undercut cortical slabs is reduced to one-fifth of that of the control cortex (*48*). This finding implies that the majority of cholinergic nerve endings are extracortical although approximately 20% must originate from local intracortical neurons. This conclusion is substantiated by the observation that direct stimulation of undercut slabs produces an increase in ACh release that is only 25% of that resulting from stimulation of intact control cortex. The persistence of appreciable (35% of control value) levels of choline acetyltransferase in isolated cortical slabs (*109*) is also consistent with the concept of intracortical cholinergic neurons.

There is general agreement that a nonspecific, reticulocortical pathway, involved in the cortical arousal response, is responsible for the widespread, evoked release of ACh from the cerebral cortex that is observed following a variety of sensory stimuli. The finding that stimulation of audiovisual or forepaw receptors produced roughly comparable bilateral increases in the release of ACh from the sensorimotor, parietal, and audiovisual cortices of the cat (*202*) is consistent with the concept of a diffuse projection system. Stimulation of the reticular formation itself also increased release to a comparable extent in the various cortical areas (*202*). A widely distributed reticulocortical system of cholinergic fibers, supplemented by local intracortical cholinergic circuits could account for these experimental findings, and it is unnecessary, therefore, to propose the existence of a second group of more specific cholinergic projection pathways to the cortex.

This second proposal was initiated by Mitchell (*186*) to account for his observation that during periods of forepaw stimulation the increase in release from the contralateral somatosensory cortex was considerably greater than that from the ipsilateral cortex. Subsequent experiments by the Mitchell group (*47, 48, 115*) in which the medial and lateral geniculate nuclei were stimulated directly, apparently showed that cholinergic neurons formed part of other specific pathways to the cortex. Unilateral stimulation of the lateral geniculate body of the rabbit produced a large increase in release of ACh from the ipsilateral primary visual area of the cortex and a widespread but smaller increase from other areas of both cortical hemispheres. Constant retinal illumination of the rabbit also released a large amount of ACh from the primary visual cortices and less from the somatosensory cortex (*47*).

Two-Pathway Hypothesis Collier and Mitchell concluded that two ascending cholinergic pathways were activated by retinal or lateral geniculate stimulation. One presumptive pathway was the ascending reticular arousal system previously mentioned, and the second was a pathway specifically associated with the visual system, activation of which would cause a localized release from the ipsilateral visual cortex. This second system was identified as the pathway responsible for recruiting and augmenting responses in the cortex. The effects of vertical or horizontal lesions, which separated the lateral geniculate nucleus from midline or more caudal structures in the brain, were described in a subsequent report (*48*). Their finding essentially was that such lesions abolished the increases in ACh release from the contralateral, but not from the ipsilateral, visual cortices during unilateral stimulation of the lateral geniculate nucleus. The conclusion was that geniculocortical cholinergic fibers are responsible for the localized release from the visual cortex.

The proposal that a cholinergic pathway, identical with the augmenting and recruiting pathways, is partially responsible for release in the primary receiving area for a specific modality of stimulation is largely based on the evidence that after undercutting or isolation of the lateral geniculate nucleus, release from the ipsilateral primary visual area evoked by stimulation of the nucleus was unaffected, whereas release from other cortical areas was abolished. An alternative explanation for this finding, admittedly speculative, follows from the suggestion that fibers of the cholinergic arousal system projecting to the primary receiving areas of the cerebral cortex may send collaterals to the related relay nuclei in the thalamus and metathalamus. If such an organizational structure exists, stimulation in the relay nuclei would be expected to activate cholinergic projections to the associated cortical receiving area.

Hemsworth and Mitchell (*116*) have shown that stimulation of the medial geniculate nucleus increases the release of ACh from the ipsilateral auditory cortex more than that from the visual cortex, and they endorsed the two-pathway hypothesis previously formulated by Collier and Mitchell. They suggest that the relatively uniform increases in evoked ACh output from the different receiving areas of the cortex, described by Phillis (*202*), were a result of the short duration (15 min) of the stimulations employed by the latter investigator. Apparently, peripheral stimulation produces a slow increase in ACh release and the maximum is often reached only after 60 min (*47, 116*). Thus, although adequate for direct thalamic or geniculate stimulation, a 15-min stimulation of a peripheral receptor may produce a relatively small response, and it would be difficult to demonstrate the separate effect of specific pathways.

Extent of Stimulation It has always been difficult for this author to accept the proposal that a 45- to 60-min stimulation is necessary to reveal the ACh release resulting from activation of the augmenting and recruiting pathways. Alternative explanations for the gradual increase in release during prolonged stimulation may have to be found. Anesthetized animals were used in these experiments, and it is more than likely that the protracted stimulation initiated a slow but progressive arousal from the anesthesia, with a concurrent increase in the release of ACh. A similar progressive increase in release of ACh from the cortex of unstimulated, Dial-anesthetized animals has been described by Jhamandas, Phillis, and Pinsky (*135*). Stimulation would merely enhance the response.

The effects of photic stimulation on ACh release from the cortex may be complicated by the existing levels of EEG activation (*13*). Unanesthetized animals, with midpontine transections and an activated EEG, failed to respond to photic stimulation with an increased release of ACh. In contrast animals with a prepontine transection, synchronized EEG, and low cortical ACh release rates, showed an increase in basal release during 15-min periods of photic stimulation. Furthermore, in these experiments simultaneous increases of comparable magnitude were observed from both visual and sensorimotor cortices, in agreement with the results of Phillis (*202*) on anesthetized cats; ACh output from the visual cortex of four of the five cats with a midpontine transection actually declined during the photic stimulation (*13*). Specification of the experimental conditions and simultaneous EEG recording may, therefore, be necessary before valid comparisons of sets of experimental data can be made.

Release from Other Areas of Brain The cerebellar cortex contains approximately one-tenth the amounts of ACh and choline acetyltransferase that are in the cerebral cortex (*206*), which conclusion is consistent with the finding that ACh release into cups on the cerebellar cortex is considerably lower than that from the cerebral hemispheres. A release of ACh from the unstimulated cortex, which may be enhanced by direct stimulation, has been demonstrated (*185, 206*).

When the cerebral ventricles of anesthetized cats are perfused with an anticholinesterase-containing solution, ACh appears in the effluent collected from the cannulated aqueduct (*23*). The greatest amount of this ACh comes from structures lining the anterior horn of the lateral ventricle, including: the caudate nucleus, the olfactory gray matter, and perhaps the septum. Less comes from the structures lining the ventral one-half of the third ventricle, the walls of which contain the nuclei of the hypothalamus (*18*). Using the same techniques of perfusion of the lateral cerebral ventricle, Portig and Vogt (*219*) demonstrated that stimulation of the substantia nigra or of the contralateral caudate nucleus, as well as cutaneous and auditory stimuli, increased the release of ACh from the feline caudate nucleus. Experiments with push-pull cannulas confirm that the caudate nucleus is a source of the ACh appearing in the lateral ventricle

(*174, 188*). A spontaneous release of ACh from the feline caudate nucleus was observed, which could be augmented by stimulation of the thalamic nucleus ventralis anterior as well as of a portion of the anterior sigmoid cortex. Release from the hippocampus may also contribute to ventricular ACh. Smith (*244*) placed small cups on the dorsal surface of the rabbit hippocampus and demonstrated an efflux of ACh which could be enhanced by septal stimulation.

Push-pull cannulas have also showed that ACh is released from the thalamus, hypothalamus, and superior and inferior colliculi. A spontaneous efflux of ACh, which could be augmented by a variety of stimuli, has been observed with the cannula tip located in either the dorsal or ventral thalamic masses of cats (*210*). Release from the monkey thalamus was greatest in the N. ventralis lateralis posterior and lowest in the N. lateralis dorsalis (*191*). Liberation of ACh from the hypothalamus occurred principally in the caudal portions (*191*). The same investigators also demonstrated a release of ACh from the superior and inferior colliculi (*19*).

Metz (*180, 181*) has studied the effects of hypercapnia and hypoxia on ACh release from the medulla in an attempt to demonstrate an involvement of cholinergic synapses in the respiratory control systems. Hypercapnia, but not hypoxia, increased the rate of release of ACh. The increases in ACh release were more pronounced in "respiratory-responsive areas" of the medulla than they were in "nonresponsive areas" and correlated closely with the increase of integrated electrical activity in the responsive areas.

Bülbring and Burn (*38*) initially demonstrated that sciatic nerve stimulation caused a release of ACh from the dog spinal cord. Release of ACh has also been observed in amphibian and feline spinal cords (*157, 187*). Antidromic stimulation of the motor axons increased the release of ACh into the perfusate or venous effluent from spinal cords of both species. Release following stimulation of this pathway probably occurs at the terminals of motor axon collaterals that, in the mammalian spinal cord, synapse with Renshaw cells (*84*). Dorsal root stimulation evoked a comparable release of ACh from the feline spinal cord. Although some of this ACh may have been released at motor axon collateral terminals subsequent to reflex excitation of motoneurons, it is also possible that it may have originated in part at other unidentified cholinergic synapses in the lumbosacral cord.

A possible complication of the vascular perfusion techniques employed by both Bülbring and Burn (*38*) and Kuno and Rudomin (*157*) is that part of the ACh in the perfusates may have been derived from the paravertebral muscles. Edery and Levinger (*86*) have since developed a method of perfusing the intermeningeal spaces of the spinal cord to guard against this possibility. With this technique they confirmed that ACh is released from the cat spinal cord by sciatic nerve stimulation and concluded that release is from motor axon collaterals to Renshaw cells and from ventral roots.

Pharmacological Studies

Central nervous system stimulants, such as strychnine, picrotoxin, leptazol, and amphetamine, cause increases in release of ACh from the cerebral cortex, associated with desynchronization of the electroencephalogram (*20, 117, 186, 199*). Anesthetics and sedatives decrease ACh release (*186, 202, 207*). The effects of narcotic analgesics are more variable. In unanesthetized rabbits (*15*) and barbiturate-anesthetized cats (*135*), morphine reduced the release of ACh from various regions of the cortex. This effect was reversed in the cat by administration of the narcotic antagonists levallorphan, naloxone, and nalorphine. In unanesthetized cats, however, morphine (0.5 and 5 mg/kg) caused an increased release of ACh, which feature correlated closely with signs of behavioral arousal (*208*). Since morphine causes EEG synchronization and general sedation in the rabbit (*15*) and EEG desynchronization and arousal in the cat (*85*), the different effects of this substance on cortical ACh release from these two species were not altogether unanticipated. Indeed, these results add support to the concept that ACh is involved in arousal.

Use has been made of the effect of pharmacological agents to study the relationship between the ACh content of the cerebral cortex and its rate of release. The results from both their own and studies by others have been summarized by Szerb, Malik, and Hunter (*269*) in a report on the effects of different groups of drugs and midbrain lesions on the ACh content of and release from cerebral cortex. Pentobarbital decreased release of ACh whereas it increased content. Picrotoxin increased release and decreased content. Atropine, in low doses (1 mg/kg), provided the only exception to the general conclusion that increased release is associated with a decreased content and vice versa, in that it increased both release and content. Very large doses of atropine (25 mg/kg) caused an even greater enhancement of release, a finding that was associated with a decreased content of cortical ACh.

The enhancement of release of ACh produced by antimuscarinic agents such as atropine and hyoscine has been the subject of many investigations, and several explanations of this effect have been proposed. A decreased rate of inactivation of ACh, for instance by blocking reuptake as proposed by Polak (*217*) and Schuberth and Sundwall (*234*), or by the occupation of specific or nonspecific receptors (*12, 267*), may constitute a partial explanation for the phenomenon. An increased synthesis and release of ACh by cortical slices in atropine has been demonstrated by Bertels-Meeuws and Polak (*21*). As atropine has no influence on the synthesis of ACh by cell-free choline acetyltransferase systems, some investigators suggest that the apparent stimulation of synthesis observed in brain slices may be secondary to an enhancement of release (*106, 189, 233*).

Two possibilities have been considered by Polak (*218*) to explain the stimulatory action of atropine on ACh release from brain slices. One hypothesis is that atropine may displace ACh from the storage sites in the nerve ending, and the other that atropine may act by a form of disinhibition in preventing an inhibitory effect of the released ACh on further release. The second possibility gains support from studies at peripheral junctions, which have demonstrated that ACh applied to nerve terminals in ganglia and at nerve-muscle synapses reduces ACh release by nerve impulses (*45, 127, 216, 270*). Szerb and Somogy (*270*) have recently drawn attention to the fact that ACh release in vivo can be measured only after inhibition of the cholinesterases, which would otherwise rapidly hydrolyze released ACh, and they suggest that the resultant unphysiologically high extracellular concentration of ACh may in itself depress additional release of ACh.

To circumvent the difficulties associated with cholinesterase inhibitors, they developed a technique for studying ACh release by measuring the efflux of labeled choline from cerebral cortical slices (*247*). Choline release was greatly depressed in the presence of eserine but was restored to control levels when atropine was added to the incubation medium. Atropine by itself did not potentiate the evoked release of ACh. On the basis of these observations, the enhancing effect of atropine on ACh release in vitro and partially in vivo can be explained by its overcoming the artifactual depression caused by inhibition by cholinesterase.

MacIntosh (*165*) made another suggestion to explain the action of atropine. According to his explanation, atropine blocks cholinergic synapses which form part of inhibitory neuronal circuits controlling the activity of neurons from which ACh is released. Recent findings (*81*) appear to support this hypothesis, since tetrodotoxin, which blocks nervous conduction, abolished the stimulating action of atropine when the former was applied topically to the exposed cortex. Dudar and Szerb (*81*) suggest that an inhibitory feedback circuit, by which cortical cholinoceptive neurons inhibit cholinergic neurons in the brainstem, is interrupted by atropine. This explanation fails to account for the effects of atropine on brain slices or on isolated cortical slabs, and possibly several mechanisms are brought into operation simultaneously. The enhancement in ACh release initiated by atropine was not attributable to an inhibition of cholinesterase, for in all of these experiments specific cholinesterase inhibitors had been applied prior to the atropine.

Calcium ions in the solution bathing the cortex appear to be necessary for optimal release of ACh. Both spontaneous and evoked release are appreciably reduced in the absence of calcium (*116, 222*). Release was increased slightly when the calcium concentration was raised to twice its normal level. Small amounts of magnesium ions facilitate, and large amounts depress, ACh release (*116*). Perfusion of the cortical surface with a Ringer's solution in which all sodium ions were replaced by lithium ions blocked ACh release evoked by stimulation of the contralateral forepaw in cats but did not affect resting release (*29*). The effects of sodium-free Ringer's solution were attributed in part to interference with the conduction of impulses in cortical neurons as well as to interference with ACh release or synthesis, or both. Birks (*25*) had previously shown that low-sodium solutions reduce ACh synthesis and release from sympathetic ganglia, possibly by interfering with ACh uptake which, in brain slices, is partly dependent on the operation of a sodium pump (*162*).

Hemicholinium and triethylcholine applied to the surface of the cortex depress both the spontaneous and the evoked release of ACh (*116, 269*); this effect is reversed by choline. The reduction in spontaneous release of ACh by triethylcholine and hemicholinium supports the suggestion that cortical cholinergic nerve endings are spontaneously active since these agents, both of which block the synthesis of ACh by competition with choline transport across the presynaptic cell membrane (*204*), are effective only at continuously stimulated peripheral cholinergic junctions.

Tetrodotoxin, which abolishes conduction in nerve fibers, also reduces release of ACh from the cortex (*28, 81*), a finding that lends support to the concept that activity in the cholinergic nerve endings is associated with release.

SINGLE NEURONS OF BRAIN AND SPINAL CORD

Neuropharmacological investigations can be carried out at several levels, depending on the type of information required. For some purposes the drug may be administered orally to an intact animal, and its responses to the environment can be observed. Alternatively the effects of drugs administered by diverse routes can be assessed on spontaneous or evoked activity from different areas of the nervous system. Although these techniques may yield useful information about the effects of drugs on whole populations of neurons, they rarely permit a determination of the actual site and mode of action of a particular substance. For this reason the wealth of observations obtained with systemically administered or topically applied ACh will not be dealt with in this review. Although such experiments provide extensive basic information about the effects of ACh on the nervous system, they have largely failed to yield convincing evidence that it functions as a synaptic transmitter.

With the development of more sophisticated electrophysiological techniques for the analysis of synaptic physiology at a cellular level, there have been rapid advances in the refinement of pharmacological microtechniques. In particular the development of multiple-barrel micropipettes from which drugs can be injected into the immediate vicinity of nerve cells has led to dramatic advances in our understanding of the actions of ACh in the CNS. Even with this method, however, caution must be exercised in the interpretation of results. Studies, summarized in Table 1, show that some central neurons are excited and others depressed by ACh.

As in the periphery, both nicotinic and muscarinic receptors have been detected on central neurons, but it is questionable whether or not an ACh receptor on the neuronal membrane necessarily indicates a direct relationship with synaptic mechanisms.

A variety of drugs are available for the study of the cholinergic system, including inhibitors of synthesis and hydrolysis of ACh, and several types of anticholinergic agents, all of which can be employed in the identification of cholinergic synapses. In some instances it is possible to block the synaptic action exerted by impulses in a particular synaptic pathway with ACh antagonists while simultaneously abolishing the action of iontophoretically applied ACh. Hemicholinium, which blocks ACh synthesis, reduces the excitation of Renshaw cells by orthodromic impulses in motor axon collaterals (221). Observations such as these undoubtedly help to establish ACh as a transmitter at specific synaptic junctions.

Difficulties may exist in demonstrating specific interactions between the synaptically released transmitter and drugs that modify the synthesis or action of ACh. Substances administered iontophoretically probably influence only a limited area of the neuronal membrane and may fail to reach the activated synapses, even though their concentrations are adequate to affect locally released ACh. If, on the other hand, the antagonists are administered systemically to achieve a more uniform concentration, there may be complications associated with diffusional barriers, or with localizing the site of their action if multisynaptic pathways are involved in the afferent pathway being tested.

Finally, the failure of iotophoretically applied ACh to produce a demonstrable effect does not necessarily indicate that it is not a trans-

Table 1. Responses of neurons in the CNS to iontophoretically applied acetylcholine

Region	Cell type	Animal[a]	Preparation or anesthetic procedure	Effect[b]	Response blocked by	Investigators[c]
Spinal Cord	Motoneuron		Barbiturate	0		64
			Decerebrate	0		285
	Interneuron		Barbiturate	+,-		64
			Decerebrate	+,(+)		284
	Renshaw cell		Barbiturate	-,(+)	Atropine	68
			Barbiturate	+	DHE	62
			Barbiturate	+	DHE	65–67
				+	Atropine	
			Barbiturate	-		26
			Decerebrate	+		26
	Sympathetic preganglionic		Barbiturate	0		76
			Decerebrate	0		123
	Parasympathetic preganglionic		Chloralose	0,-		228
Medulla			Decerebrate	+	DHE	34
				-	Atropine	
			Decerebrate	+	DHE	33
Paramedian nucleus			Decerebrate	+,-	Atropine	230
Lateral vestibular nucleus			Decerebrate	+	DHE	6
Vestibular nucleus			Decerebrate	+	Gallamine	291
Cochlear nucleus			Barbiturate	+,-	Atropine	255
			Barbiturate	+	Atropine, DHE	50

continued

Table 1. continued

Region	Cell type	Animal[a]	Preparation or anesthetic procedure	Effect[b]	Response blocked by	Investigators[c]
Cuneate & gracile nuclei			Barbiturate	+,−		182
			Barbiturate	(+),−		103
			Barbiturate	+,−		254
Raphé nuclei		Rat	Barbiturate	+,−	Atropine, DHE	53
Pons			Decerebrate	+,−		34
			Decerebrate	0		63
Raphé nuclei		Rat	Barbiturate	+,−		53
Formatio reticularis		Rat	Local anesthetic & gallamine	0		262
			Gallamine	+,−		156
Red nucleus			Barbiturate or decerebrate	−		74
Cerebellum						
Cortex	Granule cell		Barbiturate	0		57
			Barbiturate	+	DHE, Atropine	170
	Purkinje cell		Barbiturate	+	Atropine	57
			Barbiturate	+	DHE, Atropine	170
	Basket cell		Barbiturate	0		57
Deep nuclei			Decerebrate	+	Atropine	43
Inferior colliculus			Decerebrate	+		63
Superior colliculus			Local anesthetic & gallamine	+,−		262
Hypothalamus			Barbiturate	+,−		32
			Ether	+,−	Atropine	197
Preoptic area	Thermosensitive	Rat	Urethane	+		17
	Thermosensitive	Rat	Urethane	+		17
Supraoptic nucleus	Neurosecretory	Rat	Barbiturate & urethane	+	Atropine	80
	Neurosecretory		Decerebrate, halothane, or urethane	−	Atropine	7
Paraventricular nucleus	Neurosecretory	Rabbit	Urethane	+	DHE	58
	Neurosecretory	Rabbit	Urethane	+		190
	Nonneurosecretory	Rabbit	Urethane	−		190

Region		Preparation	Effect	Antagonist	Ref.
Medial geniculate		Halothane, decerebrate or methoxyflurane	+,−	Atropine, DHE	271–273
	Geniculocortical neurons	Halothane, decerebrate or methoxyflurane	+	Atropine, DHE	271–273
Lateral geniculate		Barbiturate	(+)		61
		Halothane or methoxyflurane	+,−	Mytolon	209
		Decerebrate	+,−		231
		Chloralose-urethane	+		253
	Geniculocortical neurons	Halothane or methoxyflurane	+	Mytolon	209
Thalamus		Decerebrate	+	DHE, atropine	231
		Barbiturate	+,−	DHE, atropine	2, 3
		Barbiturate or methoxyflurane	+	DHE, atropine	171, 172
		Barbiturate	+		73
		Barbiturate	+	DHE, atropine	2, 3
	Thalamocortical neurons	Barbiturate or methoxyflurane	+	DHE, atropine	171, 172
Caudate nucleus		Decerebrate	+,−		31
		Decerebrate	+,−	Atropine	176
Globus pallidus/putamen	Rabbit	Local anesthetic & gallamine	+,−		120
	Rabbit	Local anesthetic & gallamine	+,−		120
		Methoxyflurane	0	Atropine, DHE	292
Septum	Rabbit	Local anesthetic & gallamine	(+),−		118
Olfactory bulb	Rabbit	Barbiturate or decerebrate	+	DHE, atropine	14
Amygdala		Barbiturate or chloralose	+		263
Hippocampus		Barbiturate or chloralose-urethane	+,−	Atropine	27
		Barbiturate	+,−		119
Dentate gyrus		Chloralose-urethane	+		252
Piriform cortex		Barbiturate or chloralose & urethane	+,−	Atropine	252
					253
					159

continued

Table 1. continued.

Region	Cell type	Animal[a]	Preparation or anesthetic procedure	Effect[b]	Response blocked by	Investigators[c]
Cerebral cortex						
Pericruciate		Rat	Barbiturate or urethane	+,−	Atropine	256
	Betz		Barbiturate	+	Atropine	150, 151
	Betz		Barbiturate	+	Atropine	56
	Deep pyramidal		Cerveau isolé	+	Atropine	150, 151
	Deep pyramidal		Cerveau isolé	+	Atropine	86
			Barbiturate & ether	+,−	Atropine	223
			Methoxyflurane	+,−	Atropine	211–215
			Methoxyflurane	+,−	Atropine	136
Visual cortex			Barbiturate	+	Atropine	150, 151
			Encéphale isolé	+		249
			Methoxyflurane	−		213–215
Auditory cortex			Encéphale isolé	+	Atropine	249
			Barbiturate	+	Atropine	150
			Methoxyflurane	+,−	Atropine	213–215

[a]Cat, unless otherwise noted.
[b]Key: 0, no effect; −, depression; +, excitation; ( ), effect was weak and inconstant.
[c]Numbers are reference numbers in "Literature Cited" section.

mitter for neurons in a particular structure. Reactive neurons may be overlooked on account of their small size or limited occurrence, and it is clearly established that many of the anesthetic agents commonly employed in neuropharmacological studies may reduce or abolish the actions of ACh.

Spinal Cord

Acetylcholine and choline acetyltransferase are present in the spinal cord and ventral spinal roots. The levels in dorsal roots are very low (*204*); ACh levels are apparently higher in the dorsal than in the ventral gray matter. Acetylcholinesterase activity is concentrated in the ventral horn, intermediomedial and intermediolateral columns and the substantia gelatinosa of the dorsal gray matter (*105, 129, 195*). Acetylcholinesterase has been identified at the surface of physiologically identified Renshaw cells where it is presumably associated with the terminals of the motor axon collaterals (*90*). Histochemical studies on the distribution of the enzyme in the white matter suggest that both ascending and descending cholinergic tracts are present (*111, 280, 281*).

The most extensively investigated cholinergic synapses in the CNS are those on Renshaw cells. These neurons are readily excited by ACh, the predominant receptor having nicotinic properties, and the evidence clearly shows that ACh is the excitatory transmitter released by collateral branches of spinal motor axons (*62, 65, 67, 82, 83*). The excitation initiated by iontophoretically applied ACh reaches a peak frequency within a few seconds of the commencement of application and ceases within seconds of its termination. Excitation is abolished by dihydro-β-erythroidine, which also blocks the synaptic excitation of Renshaw cells by orthodromic impulses in the motor axon collaterals. Anticholinesterases potentiate both synaptic excitation and the action of ACh. Hemicholinium-3 progressively reduces and finally abolishes the excitation induced by prolonged repetitive excitation of motor axon collaterals. Its action is presumably owing to a diminished release of ACh from the presynaptic terminals, for the sensitivity of Renshaw cells either to applied ACh or to the firing produced by noncholinergic afferents is unaffected (*221*).

Antidromic stimulation of the ventral root fibers increases the amount of ACh in the effluent of perfused cat spinal cords (*157*). Cholinomimetics also excite Renshaw cells by interaction with muscarinic receptors which can be blocked by atropine. The late phase of the discharge evoked by impulses in motor axon collaterals may be mediated by these muscarinic receptors (*67*). In the presence of dihydro-β-erythroidine and atropine,

ACh depresses Renshaw cells. Curtis and Ryall (67) suggested that the pause between the early and late phases of the synaptically evoked discharge may result from a desensitization of the nicotinic excitatory receptors with an unmasking of the inhibitory effects of synaptically released ACh. Activation of the muscarinic receptors follows, initiating the late synaptic excitant response which may last for hundreds of milliseconds. Ryall (227) subsequently provided evidence to show that the pause is owing to mutual inhibitory interaction between Renshaw cells.

Identification of Renshaw elements is the subject of recent controversy. These elements, localized histologically by dye injection after electrophysiological identification, were found to be: 1) of small size, 2) devoid of dendrites, 3) and in close proximity to motoneurons (90). The elements were tentatively identified as terminal boutons on the motoneuron membrane, an interpretation that was consistent with the proposal (282) that the Renshaw element is the terminal bulb of the motor axon collateral itself. Definite morphological identification of Renshaw cells as small interneurons by Procion yellow dye injection has now been made by Jankowska and Lindström (133). A possible explanation for the disparity between these findings is that Erulkar and his colleagues were recording from the terminal boutons of Renshaw cell axons on the motoneuron membrane.

A similar cholinoceptive Renshaw-type cell, involved in inhibition of a different nature, may exist in the amphibian spinal cord. Antidromic volleys in the ventral roots of these animals do not induce inhibition of motoneurons but initiate rather a depolarization of the dorsal root fiber terminals with the generation of a dorsal root potential (8, 148). This dorsal root potential is readily blocked by dihydro-β-erythroidine and other ACh-antagonists (143, 148, 187), suggesting that the motor axon collaterals (242) activate a cholinoceptive interneuron that in turn depolarizes afferent nerve terminals by release of an unidentified excitatory transmitter. The physiological importance of this pathway may be in its activation of presynaptic inhibition of afferent volleys in the dorsal roots.

Acetylcholine has relatively weak excitant or depressant effects on some spinal interneurons (68, 284), but cholinergic pathways have not been identified to account for these ACh receptors. Histochemical studies of the distribution of acetylcholinesterase in fibers within the spinal cords of cats or rats suggest that both ascending and descending cholinergic pathways are present (111, 280). It may be relevant that dorsal root stimulation enhances release of ACh from the feline spinal cord to the same extent as does ventral root stimulation (157).

Little attention has been paid to the actions of ACh on motoneurons in nonanesthetized animals. The few neurons that have been tested were unaffected (285). Histochemical studies which have demonstrated choline acetyltransferase and acetylcholinesterase-containing synaptic endings on the motoneuron membrane, suggest that ACh receptors may be present (40, 245, 282). Weight (282) considers that muscarinic receptors on the motorneuron membrane may be involved in the generation of recurrent inhibition of motoneurons, and further studies of the action of ACh on the motoneuron membrane appear to be warranted.

Medulla and Pons

The pharmacology and distribution of ACh-sensitive neurons in the medulla and pons have been studied by several groups of investigators (Table 1). Most of these investigations were carried out on physiologically unidentified neurons. Neurons in the various brainstem nuclei differed markedly in their sensitivity to ACh. Approximately one-third of the "reticular" neurons were excited by ACh, and one-tenth were depressed (33, 34, 230). Approximately one-half of the neurons in the lateral vestibular (Deiters') nucleus (291) and almost all of those in the paramedian reticular nucleus (6) were excited by ACh. Excitatory receptors in the medulla and pons had a combination of nicotinic and muscarinic properties, whereas the inhibitory receptors for ACh were solely muscarinic (33).

Evidence exists that the excitatory pathway from the superior olive to the ipsilateral cochlear nucleus of the brainstem is cholinergic. Acetylcholine excites neurons of the cochlear nucleus as does olivary stimulation, and the effects of both are prevented by atropine or dihydro-β-erythroidine (50). Injection of hemicholinium-3 into the ventricular system of the brain depresses the excitation of cochlear neurons evoked by olivary stimulation (49). Also consistent with the suggestion that certain olivocochlear fibers are cholinergic is the presence of acetylcholinesterase in some axons in the tract (225).

The olivocochlear bundle, which can be traced from the superior olivary nucleus to the organ of Corti (224), may be identical with the centrifugal pathway from the superior olivary nucleus to the cochlear nucleus. Within the cochlea of the inner ear, only the terminals of this olivocochlear bundle show staining for acetylcholinesterase (198, 235, 237). In normal cats Schuknecht et al. (235) found a marked precipitation of the reaction product beneath the hair cells, indicating the site of the enzyme. This staining reaction was greatly reduced after section of the olivocochlear bundle. On the assumption that these fibers were cholinergic, Desmedt and Monaco (79) studied the effects of several ACh antagonists on the suppression of transmission from cochlear hair cells to the afferent auditory axons induced by activation of the olivocochlear bundle. None of the antagonists showed an effect, casting considerable doubt on the assumption that the cholinesterase in olivocochlear fibers is related to a cholinergic function. Strychnine and brucine, in relatively low doses, did block the inhibitory effects of stimulation of the olivocochlear bundle.

Early experiments (246) failed to reveal an action of ACh on cochlear responses, but Daigneault and Brown (71) subsequently showed that intraarterial administration of ACh depressed the neural response (N1) of the cochlea to acoustic stimulation. Strychnine depresses this action of ACh as well as the effects of stimulation of the olivocochlear bundle. After application of hemicholinium-3 the effects of olivocochlear bundle stimulation are also reduced (36, 110). Hypersensitivity to ACh develops after chronic section of the olivocochlear bundle (70), which finding is also consistent with the concept of a cholinergic innervation. Thus with the exception of the failure of anticholinergic drugs to block the neurally induced depression of the cochlear response, the evidence favors the existence of a cholinergic junction between olivocochlear bundle fibers and the hair cells. Further investigation may yield effective ACh antagonists. An analogous situation may exist with the ACh receptors on the hearts of molluscs of the *Veneridae* group, which are unaffected by most ACh antagonists. At these receptors only mytolon (benzoquinonium) and mytelase are known effective ACh antagonists (204).

Cerebellum

Acetylcholine-excited neurons have been observed in both the cerebellar deep nuclei and cortex. About 80% of the neurons in the fastigial, interpositus, and dentate nuclei were demonstrably excited by ACh (43). Cholinomimetic agents, including ACh, usually facilitated firing induced by L-glutamate, but some neurons were excited directly after a short latency. Atropine and hyoscine, but not dihydro-β-erythroidine, readily antagonized the actions of ACh. The deep nuclei have a high level of choline acetyltransferase (107), and many of the cells and fibers in the nuclei contain acetylcholinesterase (203, 210). Cholinesterase-containing fibers leaving the cerebellum in the superior and inferior peduncles (Phillis [203]) probably originate from neurons in the deep nuclei, and collateral branches of these axons may form cholinergic synapses on other deep nuclear cells. Cholinesterase-containing fibers enter the cerebellum of the cat and rat in all three peduncles (203, 237), and some of them may synapse with deep nuclear cells. Small but definite amounts of choline acetyltransferase are present in the cerebellar peduncles (*96, 115*), which finding is consonant with the concept that they contain cholinergic fibers.

Iontophoretic studies have established ACh-excited neurons in the cerebellar cortex, although some disagreement exists over the identification of the neurons. According to Crawford, Curtis, Voorhoeve, et al. (*57*), these are exclusively the Purkinje cells, whereas McCance and Phillis (*169, 170*) identified them as being both granule layer and Purkinje cells. In the investigations of McCance and Phillis, ACh-excited neurons were found with increasing frequency in the deeper layers of the cortex. The receptors in these cells were "nicotinic" and dihydro-β-erythroidine was the most effective antagonist.

Some of the disparities between the findings of the two groups of investigators can be readily explained. The Canberra team recorded for the most part from cells near the surface of the cortex and used carbachol extensively to reveal cholinergic receptors. Carbachol is resistant to hydrolysis by acetylcholinesterase and will therefore diffuse more readily than ACh through the granular layer of the cortex, which contains a high concentration of acetylcholinesterase. The high proportion of superficial Purkinje cells excited by carbachol may have resulted from the ability of this compound to activate a sufficient number of the scattered cholinoceptive granule cells to initiate relayed firing of the Purkinje cells. McCance and Phillis, using ACh, would have been unable to excite more than a few of the superficially located Purkinje cells.

With two micropipettes in one folium—one pipette to excite chemically cells synapsing with a Purkinje cell and the other to record from the latter—an attempt was made to ascertain whether or not ACh receptors are present on basket or granule cells. Granule and basket cells were shown to be excited by DL-homocysteic acid but to be insensitive to ACh and carbachol (*57*). However, these experiments were restricted to the superficial folia of the cerebellar vermis, and cholinergic receptors may be more prevalent on granule cells of the deeply located folia. Purkinje cells, especially those of the deeper folia, are excited during the application of ACh into the vicinity of their cell bodies but do not respond when ACh is applied in the molecular layer. The question arises, therefore, is Purkinje cell excitation the result of a direct action of ACh on the cell soma or is it secondary to the excitation of cells in the adjacent granule cell layer, with subsequent synaptic excitation of Purkinje cells by parallel fibers? The question is currently unresolved. Purkinje cell bodies of the adult cat are almost completely devoid of acetylcholinesterase (*203, 241*), although choline acetyltransferase staining boutons terminaux have been described on the surface of Purkinje cells (*40*).

The cerebella of many, but not all, mammalian species are particularly rich in acetylcholinesterase (5). By contrast the cerebellar content of ACh and choline acetyltransferase is among the lowest of the entire brain. The ACh is known to be located in the synaptosomal fraction (130). The stratification of the cerebellar cortex makes it particularly suitable for ultramicroanalytical studies of the enzyme content of its particular layers. Goldberg and McCaman (107) compared the cholinesterase and choline acetyltransferase contents of the molecular and granular layers and underlying white matter of the cerebella of five species, but were unable to demonstrate a correlation between the levels of the two enzymes (Table 2).

Histochemical studies show that cholinesterase-containing mossy fibers enter the cerebellum in the middle peduncle and to a less marked extent in the superior and inferior peduncles (200, 202). Many of the cells within the granular layer contain acetylcholinesterase which persists after peduncular transection or isolation of the cortex, and is therefore not likely to be associated with the mossy afferent terminals. The significance of these cholinesterase-containing granular cells and of the high content of the enzyme in the molecular layer has yet to be determined, since pharmacological studies indicate that transmission between parallel fibers and Purkinje cells is noncholinergic (57, 170).

The inferior olive, from which the climbing fibers originate (265), contains cholinesterase (129), but these fibers are unlikely to be cholinergic as neither dihydro-β-erythroidine nor atropine alters the configuration of the negative-positive potential evoked at the surface of the cerebellar cortex by olivary stimulation (57, 170).

Further evidence that some mossy afferent fibers are cholinergic was obtained from experiments in which various brainstem nuclei were stimulated. Increases in the discharge of ACh-sensitive granular neurons induced by stimulation of the nucleus reticularis pontis oralis complex of the mesencephalic reticular formation were prevented by locally applied dihydro-β-erythroidine (170). According to Shute and Lewis (237) cholinergic mossy afferents should also arise in the cochlear, vestibular, cuneate and gracile nuclei, and lateral and tegmental reticular nuclei. Stimulation of the latter two nuclei elicited discharges from granular cells, but they could not be prevented or reduced by dihydro-β-erythroidine (170).

McCance (168) evaluated the effects of ACh on neurons in the cerebellar cortex of rats, guinea pigs, and monkeys. Considerable differences were found between species, the level of response being considerably lower in monkeys and rats than in guinea pigs and cats. Twelve and 19% of the neurons tested in the monkey and rat, respectively, were excited by ACh, whereas the level in both cats and guinea pigs approached 40%. These

Table 2. Cholineacetyltransferase[a] and acetylcholinesterase[b] activity in cerebella of several species

	Rat		Pigeon		Guinea pig		Rabbit		Cat	
	Choline acetyl-transferase	Cholin-esterase	Choline acetyl-transferase	Cholin-esterase	Choline acetyl-transferase	Cholin-esterase	Choline acetyl-transferase	Cholin-esterase	Choline acetyl-transferase	Cholin-esterase
Molecular Layer	0.98 ± 0.27	1,000 ± 123	1.55 ± 0.30	4,293 ± 316	0.62 ± 0.15	3,757 ± 285	<0.1	2,087 ± 216	0.71 ± 0.19	2,330 ± 288
Granular layer	1.92 ± 0.37	1,472 ± 89	1.36 ± 0.19	2,399 ± 190	1.60 ± 0.22	2,390 ± 131	1.52 ± 0.14	3,702 ± 643	1.23 ± 0.18	3,694 ± 282
White matter	2.75 ± 0.48	1,620 ± 223	1.38 ± 0.22	775 ± 132	2.60 ± 0.08	1,192 ± 69	1.23 ± 0.10	1,270 ± 126	0.44 ± 0.10	1,193 ± 211
Deep nuclei	6.90 ± 0.51	1,525 ± 232	10.5 ± 1.5	610 ± 58	10.6 ± 1.5	1,354 ± 189			3.78 ± 1.08	1,790 ± 276

Data from (107).
[a] Activity expressed as μmol of acetylcholine synthesized/g dry weight/hr.
[b] Activity expressed as μmol of acetylthiocholine hydrolyzed/g dry weight/hr; each value represents mean ± standard error.

results accord reasonably well with the levels of histochemically and manometrically demonstrable acetylcholinesterase in the cerebellar cortex (5, 201). Whereas the molecular layer of the guinea-pig cortex stains intensively for acetylcholinesterase (99, 201), only half of the Purkinje cells tested were excited by ACh.

Hypothalamus

The hypothalamus contains low to moderate amounts of ACh and choline acetyltransferase (115, 164); ACh is released principally in the caudal hypothalamus (191). The hypothalamic gray matter shows a uniform distribution of moderate acetylcholinesterase activity, the nerve cells showing relatively more staining than the neuropil. The paraventricular nucleus, in particular, stands out for the intensity of its staining (129).

Acetylcholine has both excitant and depressant effects on neurons in the feline hypothalamus. They are most pronounced in the paraventricular, ventral median, and lateral nuclei (32, 197). Atropine reduces ACh excitation but not depression (197). ACh increases the firing of warm-sensitive cells in the preoptic and anterior hypothalamic regions of cats and rats but has no effect on cool-sensitive neurons (17). These effects are in agreement with what would have been predicted from the results of microinjection studies in which ACh produced coordinated behavioral and physiological heat-dissipation responses (16).

Acetylcholine excites neurosecretory cells in the supraoptic nucleus through nicotinic receptors and inhibits with muscarinic receptors (7, 80). Neurosecretory cells of the paraventricular nucleus are also excited (57, 190).

Thalamic and Geniculate Nuclei

The thalamus contains moderate amounts of ACh (93, 163). Studies of the amount of thalamic choline acetyltransferase have been conducted on a variety of species (96, 115, 293). The regional distribution of choline acetyltransferase has been studied in the rat and monkey thalami (92, 107, 205), and activity in both species was lowest in the anterior thalamus. Acetylcholinesterase activity is highest in the nuclei of the internal medullary lamina and in the reticular thalamic nucleus (129). According to their schema, Shute and Lewis (239) envisage an innervation of thalamic and geniculate nuclei by cholinergic fibers of the dorsal tegmental pathway. Cholinesterase-containing efferents in the superior cerebellar peduncle may connect cerebellar deep nuclear cells to neurons in the ventrolateral nucleus of the thalamus (203).

Levels of ACh in the lateral geniculate nucleus are comparable to those in the thalamus (46, 75). Some controversy surrounds the possible presence of cholinergic retinogeniculate fibers in the optic nerve. The ACh content of optic nerves is low as is their choline acetyltransferase activity (75, 96, 115, 164). The contribution of retinogeniculate fibers to the ACh content of the lateral geniculate nucleus has been studied by bilateral ophthalmectomy. Five days after the surgical procedure there was no change in the lateral geniculate (cat) levels of ACh, choline acetyltransferase, or cholinesterase, even though synaptic transmission had failed (11). Three to four weeks after unilateral enucleation, the ACh content of the contralateral lateral geniculate nucleus of the rabbit in which there is nearly complete crossing of the optic nerve had been reduced by more than 20% (183).

Ablation of the visual cortex caused a comparable loss of ACh in the ipsilateral lateral geniculate nucleus of the rabbit. The difference between the effects of eye removal in these two species is not immediately apparent. DeRoetth (77) reports that appreciable levels of choline acetyltransferase are present in the rabbit optic nerve (but see [115]), and it is therefore conceivable that a higher proportion of the fibers are cholinergic than in the cat. It is also possible that the longer period allowed for degeneration in the rabbit experiments may have been sufficient for transynaptic degeneration to occur. The results of both groups of experiments indicate that a substantial proportion of the cholinergic fibers in the lateral geniculate nucleus must originate from regions other than the retina.

Deffenu et al. (75) have correlated the ACh content of the lateral geniculate bodies of normal and bilaterally enucleated cats with the degree of synchronization of the cortical EEG produced by different levels of brain transection. A relationship was postulated between the degree of EEG activation and the level of ACh in the lateral geniculate, and it was suggested that the ACh content of the geniculate was largely dependent on the level of activity of fibers originating in the reticular formation.

Choline acetyltransferase is present in the medial geniculate nucleus in amounts comparable to those in the thalamus and lateral geniculate nucleus (96, 115).

Neurons that are excited by iontophoretically applied ACh are found throughout the thalamus (2, 73, 171, 172, 175), lateral geniculate nucleus (61, 209, 231, 253), and medial geniculate nucleus (271, 272). A large majority of the thalamocortical and geniculocortical relay neurons are excited by ACh, the sensitivity of some of these cells being almost equal to that of Renshaw cells. The ACh receptors on these cells have mixed nicotinic-muscarinic properties. Neurons that are depressed by ACh are also found in all three areas.

The absence, on the one hand, of effects of ACh antagonists on the excitation of thalamic and geniculate neurons by afferent volleys in spinothalamic, optic, auditory, and corticothalamic fibers indicates that these pathways are unlikely to be cholinergic. On the other hand, the firing evoked in all three areas by stimulation of the mesencephalic reticular formation is abolished by ACh antagonists, suggesting the presence of a cholinergic pathway (the dorsal tegmental pathway of Shute and Lewis?). Release of ACh from the thalamus is enhanced by stimulation of the mesencephalic reticular formation (*210*). The finding that the excitation of some ventrolateral thalamic neurons by impulses in the superior cerebellar peduncle is depressed by atropine (*100, 172, 177*) supports the suggestion that cholinesterase-containing fibers in the peduncle may be cholinergic (*203*).

Basal Ganglia

The caudate nucleus contains particularly high concentrations of ACh, choline acetyltransferase, and acetylcholinesterase (*39, 96, 115, 164*). A high rate of ACh turnover is indicated by the rapid reduction in ACh levels following intracaudate injection of hemicholinium (*114*). Acetylcholinesterase activity is very high in the caudate nucleus and putamen and somewhat less in the pallidum (*129*); the enzyme is diffusely distributed in the neuropil of all three areas (see also section on "Release from Other Areas of Brain" in this chapter).

A proportion of caudate nucleus neurons are excited by ACh, whereas others are depressed (*31*), and a remarkable correlation has been observed between the effects of ACh and those of stimulation of the nucleus ventralis anterior of the thalamus (*176*). Both types of response to ACh and the responses to nucleus ventralis anterior stimulation are readily prevented by atropine, suggesting that the final synapse in this thalamocaudate pathway may be cholinergic. A survey of the globus pallidus-putamen complex (*292*) reveals that ACh excites 60% and depresses 13% of the neurons in this area. The receptors for both effects had a mixture of nicotinic and muscarinic properties.

Cholinergic synapses in the striatal complex, possibly concerned with both excitation and inhibition, may provide an explanation for the beneficial effects of atropine in Parkinson's disease.

Hippocampus

The distribution of acetylcholinesterase in the hippocampus has been the subject of several investigations (*105, 236, 238, 259, 261*). According to Shute and Lewis (*239*) acetylcholinesterase-containing afferents to the hippocampus and dentate area travel through the fimbria from both the medial septal nucleus and the nucleus of the diagonal band and serve as a connection with the ascending reticular system. Hippocampal acetylcholinesterase seems to be a reliable indicator of cholinergic nervous elements. Its laminar distribution coincides with that of choline acetyltransferase (98), and the two enzymes disappear concomitantly following interruption of the septohippocampal pathway (161, 173, 259, 260).

The pharmacological evidence is at present less convincing but indicates a cholinergic excitation of hippocampal cells (27, 119, 252). Most of the cholinoceptive units were less than 0.5 mm below the dorsal surface of the hippocampus and were tentatively identified as pyramidal cells (27). The receptors on these neurons were muscarinic.

Amygdala and Piriform Cortex

Both of these areas contain choline acetyltransferase (115). Great variation is found for acetylcholinesterase activity in the amygdaloid subnuclei. Very weak activity occurs in the nucleus amygdalae lateralis, which is separated by a sharp borderline from the very high activity in the nucleus amygdalae basalis pars medialis and lateralis. Activity is less pronounced in the nuclei corticalis and medialis and in the lobus piriformis (129, 154).

Acetylcholine excited or facilitated the action of glutamate on about 16% of cells in the amygdala (263) and from 30 to 50% of piriform cortical neurons (159). The receptors on neurons in both areas were muscarinic, with excitation developing and terminating slowly. A few neurons that were depressed by ACh were encountered in the piriform cortex.

Olfactory Bulb

This part of the rhinencephalon also contains ACh and choline acetyltransferase (107, 115, 164). Well-defined zones of cholinesterase staining are found above and below the unstained mitral cells (156), and some stained fibers are present in the lateral olfactory tract (144).

In contrast with the remainder of the rhinencephalon, the predominant action of ACh on all cells in the olfactory bulb of the unanesthetized rabbit was depression (14). Atropine and dihydro-β-erythroidine blocked the depression of these neurons by ACh, without affecting the inhibition of mitral and tufted cells evoked by lateral olfactory tract stimulation (229).

Cerebral Cortex

The considerable volume of indirect evidence favoring a role for ACh as a transmitter in the cerebral cortex has recently gained support from iontophoretic studies that show that it excites some cortical neurons and depresses others (*56, 150, 151, 211, 213–215, 223, 249, 256, 257*). These results, in conjunction with those from studies on the metabolism of ACh, have led to the proposal that both excitatory and inhibitory cholinergic pathways are present in the cerebral cortex.

All of the components of the ACh metabolic system are found in the cortex: ACh (*88, 164*) being particularly concentrated in nerve endings and synaptic vesicles (*289*), choline acetyltransferase (*115*), and acetylcholinesterase (*39*). Release of ACh from the cerebral cortex has been discussed in detail in the section on "Release of Acetylcholine" where evidence was presented that the release was associated with a cortical arousal system.

Acetylcholinesterase-staining fibers are found ramifying throughout the various cortical layers, although the majority occur deeply in relation to the pyramidal cells of layer V and the polymorph cells of layer VI. These branches originate from fibers that form a predominantly tangential system running beneath the cortex (*154*). The main subcortical connections of this system can be traced to the corpus striatum and septal area (*154*) and may be identical to the system of ascending fibers described by Shute and Lewis (*239*). Septal lesions in the rat cause a decrease in cortical ACh (*200*).

Although no cerebral cortical neurons in the cat show intense staining, some of the spindle or polymorph cells in the deepest layer of the cortex often have relatively large amounts of intracellular enzyme and may therefore give rise to some of the fibers already described. Many large pyramidal cells in layer V of the cortex are lightly but definitely stained.

Studies on the development of acetylcholinesterase staining in embryos show that the primary cortical elements of the cat are devoid of the enzyme (*155*). In the primitive forebrain, acetylcholinesterase is found only in the developing lenticular nucleus and septum. Fibers from these areas spread out to innervate the rest of the hemisphere.

After isolation the choline acetyltransferase and acetylcholinesterase contents of cat cortical slabs fall rapidly for three days and then reach fairly stable levels (choline acetyltransferase at 35% of control value; acetylcholinesterase at 43% of control) which are sustained for at least another two weeks (the duration of isolation in these particular experiments [*109*]). Histochemical observation of large isolated slabs reveals that there is substantial preservation of the normal staining pattern with a clearly visible system of fibers running beneath and invading the cortex. Small slabs, with damage to the deeper layers of the cortex, were almost devoid of histochemically demonstrable acetylcholinesterase activity (*153*). These observations on isolated slabs all strongly suggest that there are both intrinsic and extrinsic components of the cortical cholinergic system: an intracortical system, which may originate in the polymorph cells of layer VI, and an extracortical system from the corpus striatum and septal area.

Neurons that are excited by ACh have been observed in all regions of the neocortex, although they seem to be more common in or near the primary afferent areas. It may be significant that ACh release from the primary afferent areas is greater than that from the associational cortex (202). Few ACh-excited cells are found in the upper layers of the cortex, most of them being concentrated within the middle depth range. Since most of the cells activated by antidromic stimulation of the medullary pyramids are excited by ACh, it is possible that ACh-excited units in general are deep pyramidal neurons.

ACh-excited neurons in the deeper layers of the cortex characteristically display an irregular spontaneous discharge. The receptors on these cells have well-defined muscarinic properties (56, 151, 257); ACh excitation was slow in onset and maximum firing rates were attained after a latent period of ten to 60 sec during which spontaneous activity was often depressed. Acetylcholine also caused some excitation of superficial cells (above 600 μ) in the rat cortex which was mimicked by nicotine and blocked by (+)-tubocurarine (257), suggesting that the receptors had nicotinic properties. The latency of excitation of these cells tended to be shorter than that for the deeper neurons.

Intracellular studies show that ACh initiates a slow depolarization of cortical neurons, which may or may not lead to an outright discharge, but which potentiates the effects of other excitatory inputs. There is a tendency for repetitive firing apparently attributable to a slowing of the repolarization after action potentials (152). During the depolarization membrane resistance is increased, and studies on the voltage-current relationship indicate a decrease in membrane conductance for an ion with a highly negative equilibrium potential. Krnjević et al. (152) suggest that ACh depolarizes by decreasing both the potassium conductance of the membrane and the delayed potassium currents associated with the repolarizing phase of the action potential. A similar mechanism is postulated to account for the generation of the cholinergic slow excitatory postsynaptic potential in frog sympathetic ganglion cells (285a).

The identity of the postulated excitatory cholinergic afferent pathways to the cortex has yet to be established by electrophysiological methods. Systemically administered atropine depresses the spontaneous activity of ACh-excited neurons and reduces the repetitive firing (after discharge) evoked by single pulse stimulation of peripheral nerves or thalamic relay nuclei (*150*). Otherwise attempts to relate the excitant actions of ACh to the operation of specific afferent pathways to the neocortex have not been successful. Transcallosal fibers and short latency pathways from the thalamus are probably not cholinergic, since many of the synaptically driven cells are unaffected by ACh and responses evoked by thalamic or transcallosal stimulation are not affected by atropine. Iontophoretically administered atropine did depress the repetitive firing of neurons in the visual cortex evoked by stimulation of the mesencephalic reticular formation (*250*). Atropine also suppresses the desynchronization of the electroencephalogram (cortical arousal) induced by stimulation of the mesencephalic reticular formation (*42, 137*).

Considered in its entirety, the evidence presented strongly suggests that the excitatory system in the cerebral cortex forms the final link in an ascending cholinergic pathway from the striatum and mesencephalic reticular formation. The level of cortical arousal may thus be determined, at least in part, by activity in ascending cholinergic fibers that would control the excitability of deep pyramidal cells in particular. The relatively slow action of ACh makes it especially suitable for such a modulatory role.

In addition to excitation, ACh depresses some cortical neurons, particularly those located in the more superficial layers of the cortex (*136, 211, 213–215, 256, 257*). Neurons that are depressed by ACh have been located in all primary cortical areas as well as associational areas. The receptors on these cells have predominantly muscarinic properties, atropine and hyoscine being the most effective ACh antagonists. Strychnine also antagonizes the depressant action of ACh.

Repetitive stimulation of the adjacent cortex, the mesencephalic reticular formation, or lateral hypothalamus inhibits the responses of many of these ACh-depressed neurons, and this synaptically evoked inhibition can also be abolished by atropine and strychnine (*136, 212, 214*). Inhibitions of this type may have a duration of more than one minute and can be potentiated by anticholinesterases. Comparable inhibitions can be evoked by direct stimulation of the surface of both chronically and acutely isolated cortical slabs, suggesting that cholinergic inhibitory interneurons may be present in the cerebral cortex. Histochemical observations would suggest that these may be the acetylcholinesterase-containing polymorph cells in cortical layer VI.

Ilyutchenok and Gilinsky (*128*) have confirmed an inhibitory cholinergic synapse in the cerebral cortex by demonstrating that topical application of the ACh antagonists, benactizine and atropine, abolishes the inhibition of spontaneously active cortical neurons induced by stimulation of the reticular formation. The electroencephalographic arousal reaction produced by the reticular stimulation was also abolished. An intracortical inhibitory cholinergic interneuron has also been postulated by Vazquez and his colleagues (*149, 278*) to account for the effects of cholinergic drugs on the duration of the epileptiform afterdischarge evoked by surface stimulation of chronically isolated cortical slabs. Cholinergic drugs such as pilocarpine, arecoline, eserine, and oxotremorine significantly decrease the duration of the afterdischarge. Atropine and hyoscine antagonize this cholinergic anticonvulsant effect and increase the duration of the afterdischarge.

The most likely explanation for these findings is that cholinergic inhibitory interneurons in the isolated cortical slab suppress the excitability of the neuronal circuits that generate the afterdischarge. Atropine and tubocurarine (which also antagonizes cholinergic inhibition of cortical neurons) are both convulsants when applied topically to the cerebral cortex. An abolition of cortical cholinergic synaptic inhibition is likely to be an important factor in the generation of these effects.

Weight and Padjen (*283*) recently reported that the slow inhibitory hyperpolarizing postsynaptic potential generated in sympathetic ganglion cells of the bullfrog by ACh are associated with a marked increase in membrane resistance. Electrical depolarization of the membrane decreased the amplitude of the IPSP and could reverse it to a depolarizing potential. Conversely, hyperpolarization of the membrane increased the amplitude of the IPSP. The reversal potential for the slow IPSP was in the region of the sodium equilibrium potential, and it was suggested that ACh generates the potential by inactivating resting sodium conductance.

ACETYLCHOLINE IN CENTRAL DISEASE PROCESSES

Parkinson's Disease

Feldberg (*93*) suggested that the alleviation of parkinsonian rigidity and tremor by atropine might result from a central atropine-ACh antagonism, and many of the drugs that have subsequently been introduced for the treatment of this disease exhibit anticholinergic properties. Most of them are antimuscarinic agents, but this property and the antitremor action are not closely correlated (*52*). An involvement of ACh in Parkinson tremor, but not in other types of tremor, is suggested by Marshall and Schnieden (*176a*), who found that nicotine depressed parkinsonian but not essential, cerebral, or anxiety tremors. Abnormalities in ACh metabolism may be a factor in parkinsonism, although total brain cholinesterase remains within normal limits (*22*).

A deficiency of dopamine (DA) in the caudate nucleus, putamen, and neurons of the substantia nigra is a consistent pathological finding in Parkinson's disease *(124)*, and lesions of the nigrostriatal tracts can induce symptoms of parkinsonism, together with a loss of DA in the corpus striatum *(248)*. It is hypothesized that the cholinergic and dopaminergic systems maintain a balance by exerting mutually antagonistic actions, a decrease in dopaminergic tone or an increase in cholinergic tone resulting in tremor.

Intravenously administered L-dopa causes a brief but distinct amelioration of the major Parkinson symptoms *(124)*, whereas tremorine and oxotremorine induce a tremor that is not unlike that in parkinsonism *(121)*. Direct injection of tremorine into the corpus striatum produces tremor in rats *(54)*. The cholinergic effects of oxotremorine may be attributable either to its substantial muscarinic properties *(44)* or to the increase in ACh content which it induces in the corpus striatum *(9, 55)*. The effects of tremorine are antagonized by the ACh antagonists used as anti-Parkinson drugs *(91)*. The tremor induced by local injection of carbachol into the feline caudate nucleus can be suppressed by intraventricularly administered DA *(51)*.

Development of a parkinsonian syndrome in patients who have received high doses of tranquilizing agents over long periods can also be explained by postulating a reduction in adrenergic tone which may be attributable to a depletion of DA *(1)*, α-adrenergic blockage *(179)*, or dopaminergic blockade *(277)*. An interesting observation is that some of the drugs used to treat Parkinson's disease, such as benztropine and diphenylpyraline, cause a reduction in the accumulation of DA induced by administration of L-dopa, whereas atropine and hyoscine do not have this effect *(101)*. (See Ch. 5 for further discussion on DA and DA-ACh balance.)

Huntington's Chorea

The main pathological finding in Huntington's chorea is a pronounced degeneration of neurons in the corpus striatum. The concentration of DA in the striatum and pallidum of patients with Huntington's chorea remains within normal limits *(87)*. Drugs like reserpine, the phenothiazines, and the butyrophenones, all of which reduce monoaminergic activity, diminish the hyperkinesia of Huntington's chorea *(37)*. These findings all suggest that in this disorder, unlike Parkinson's disease, the dopaminergic neurons of the nigrostriatal system are intact. The hyperkinesia and hypotonia of Huntington's chorea are mirror images of the symptoms of Parkinson's disease. Anticholinergic drugs lack a therapeutic effect in Huntington's chorea and have even been claimed to accentuate the hyperkinesia *(24,*

144). Physostigmine, which accentuates parkinsonian tremor (*82*), alleviates the symptoms of Huntington's disease (*4, 144*).

These observations suggest that there is a reduction in cholinergic tone and an accentuation of dopaminergic activity in Huntington's chorea—a reversal of the abnormality postulated for Parkinson's disease. Aquilonius and Sjöström (*4*) suggest that the administration of oxotremorine or a similar muscarinic agent may be an effective mode of therapy for Huntington's chorea.

Wilson's Disease

Abnormally high concentrations of copper are found in the tissues in this condition, and the symptoms of rigidity, intention tremor, and athetosis may be a result of the damage that occurs from deposition of copper in the brain (*178*). Removal of copper from the body by chelating agents has a beneficial effect on the movement disorder (*69*).

An involvement of ACh in Wilson's disease is suggested by the finding that copper potentiates both the tremor induced by tremorine injected into mice and ACh-induced contractions of the guinea-pig ileum (*41*). The means by which copper ions cause Wilson's disease and potentiate the action of cholinergic agents remain a subject for speculation, but it may be relevant to point out that copper is an active inhibitor of choline acetyltransferase (*220, 243*).

Epilepsy

There was a time, during the nineteenth century and earlier, when atropine was commonly used for the treatment of epilepsy (*108*), and interest in the involvement of ACh in epileptic seizures continues to this day. Williams (*290*) found that the frequency and duration of "petit mal" attacks in patients could be increased by the administration of certain cholinergic drugs and that atropine counteracted these effects. Acetylcholine is frequently detectable in the cerebrospinal fluid (CSF) of epileptics, whereas it seldom appears in the CSF of nonepileptics (*275*). Acetylcholine was found in every specimen taken during or within six hr after a seizure, but its presence could not be attributed to release during excessive neuronal activity since it occurred in a high percentage of cases during seizure-free intervals. Furthermore, in four of six nonepileptics subjected to electroshock therapy, no ACh could be detected in the CSF taken immediately after a generalized seizure.

The ACh content of brain slices from cortical epileptogenic lesions is less than that of normal slices (*274*), and cholinesterase activity in such slices is increased. The ability of anticholinesterases to induce convulsions which can be suppressed by atropine has been extensively documented (*97, 258*).

Studies on the induction of seizures by high concentrations of topically applied ACh (*35, 184*) have raised some doubts about the possibility that epileptiform discharges may be induced with more physiological concentrations. The aforementioned experiments with cholinesterase inhibitors show that this is possible and that cholinergic mechanisms may be involved in certain naturally occurring epileptiform seizures.

CONCLUSIONS

Thus far in this discussion various sections of the ACh system in the CNS have been treated. An important area that has been omitted is concerned with the role of the cholinergic system in behavioral responses. Studies on these relationships include the correlation of brain ACh or acetylcholinesterase levels with the intelligence or responsiveness of the particular strain of animal, and similarly an assessment of the effects of cholinergic and anticholinergic drugs on the animal's performance. The results from several such studies have recently been summarized and discussed in some detail in other reviews (*138, 139, 279, 286*), and the interested reader may consult them.

The evidence presented in this review strongly supports the contention (*93*) that ACh is indeed a transmitter in the CNS. It is now generally accepted that ACh is the transmitter released by motor axon collateral terminals onto Renshaw cells within the spinal cord, although it may only be one of the excitatory transmitters that influence Renshaw cells (*64, 66*). Furthermore it is apparent that ACh is probably not the excitatory transmitter released at synapses in many of the major afferent pathways in the brain.

In those circumstances in which ACh has an excitatory action, its effects are frequently slow in onset and of lengthy duration. Such an action appears to be more consistent with that of an agent which modulates the level of cell excitability than with one involved in the rapid transfer of information. Most of the evidence favors a role for ACh as the transmitter released at terminals of an ascending reticular arousal system, projecting widely to areas of the forebrain, midbrain, and cerebellum. The nature of its effects is consistent with the anticipated functions of such a system in controlling the state of excitability of neurons.

The receptors for ACh range from those that have classic nicotinic properties, such as the major group of cholinergic receptors on the Renshaw cell membrane, to those with pure muscarinic properties, typically the receptors on cortical Betz cells. Furthermore it appears that the nicotinic or muscarinic properties of the receptor may determine the nature of the effect of ACh. Although it has not been evaluated experimentally, it is likely that the action of ACh on the nicotinic receptors of Renshaw cells is to initiate an increase in membrane permeability to sodium and potassium ions, as it does at the skeletal neuromuscular junction. There is evidence (*152*) that ACh activation of the muscarinic receptors on Betz cells initiates a depolarization by decreasing the potassium conductance of the membrane. Additional research will be required to elucidate the action of ACh on neurons, such as thalamo- or geniculocortical relay neurons, which appear to have receptors with properties that are intermediate between those of true nicotinic or muscarinic receptors. Some of these neuronal pharmacological properties may be a result of an admixture of both types of receptor on their membrane, or alternatively a third type of receptor with pharmacological properties intermediate between those of nicotinic and muscarinic receptors may be present.

It appears possible that a clear indication of the role of ACh at the level of individual neurons may be gained from an evaluation of the pharmacological properties of the neuronal receptors. The evidence presented in this review indicates that nicotinic receptors are associated with rapid conduction across the synapse, whereas muscarinic receptors have properties that are more consistent with a modulatory function for the synaptically released ACh. Evidence has recently been published that activation of muscarinic, but not nicotinic, receptors in rabbit cerebral cortical slices caused an increase in their guanosine cyclic 3′,5′-monophosphate (cyclic GMP) content which was accompanied by a slight decrease in the levels of adenosine cyclic 3′,5′-monophosphate (cyclic AMP) (*158*). These findings are compatible with the hypothesis that the physiological response to activation of muscarinic receptors is mediated by cyclic GMP.

The action of ACh on nicotinic receptors, on the other hand, apparently does not involve mediation by a cyclic nucleotide since the levels of cyclic AMP and cyclic GMP are unaffected. Lee and his colleagues (*158*) speculate that the explosive physiological responses associated with the action of ACh on nicotinic receptors develop with a rapidity that precludes mediation by a complex series of biochemical reactions. Activation by ACh of muscarinic receptors initiates more slowly developing responses which could reasonably be mediated by a series of biochemical reactions involving cyclic GMP.

Two examples of an inhibitory synaptic function of ACh have been cited—in the caudate nucleus and in the cerebral cortex. In both instances stimulation of adjacent areas of the brain evoked an inhibition of ACh-depressed cells, and anticholinergic drugs blocked the action of both endogenously and synaptically released ACh. In both instances the ACh receptors have predominantly muscarinic properties. The inhibitory receptors on neurons in the brainstem are also muscarinic, and to date no example of an inhibitory nicotinic receptor has been reported.

The evidence for a role of ACh as a synaptic transmitter in the brain and spinal cord is often inadequate if assessed by the various criteria that have been developed for the identification of transmitters (*204, 287*). However, the criteria themselves can be interpreted more or less rigorously, and with the techniques currently available to researchers some leniency must be adopted in their application. Even so it is currently difficult to satisfy many of the criteria in most areas of the brain and spinal cord. In a few areas such as the caudate nucleus, thalamus, and lateral geniculate bodies, where all of the criteria have been satisfied to a greater or less extent, the evidence strongly suggests that ACh is an excitatory transmitter. A role as an inhibitory transmitter in the cerebral cortex and caudate nucleus is also indicated.

Ultimately, however, complete resolution of this problem will demand additional knowledge, and in particular, detailed comparisons of the actions of both iontophoretically applied ACh and synaptically released transmitter on the permeability of the postsynaptic membrane.

In conclusion, although progress has been made since Feldberg published his classic review in 1945, there is still some way to go to ascertain the role of ACh as a synaptic transmitter in the CNS.

LITERATURE CITED

1. Andén, N. E., Roos, B. E., and Werdinius, B.: Effects of chlorpromazine, haloperidol and reserpine on the levels of phenolic acids in rabbit corpus striatum. *Life Sci. 3*:149–158, 1964.
2. Andersen, P. and Curtis, D. R.: The excitation of thalamic neurones by acetylcholine. *Acta Physiol. Scand. 61*:85–99, 1964a.
3. Andersen, P. and Curtis, D. R.: The pharmacology of the synaptic and acetylcholine-induced excitation of ventrobasal thalamic neurones. *Acta Physiol Scand. 61*:100–120, 1964b.
4. Aquilonius, S. M. and Sjöström, R.: Cholinergic and dopaminergic mechanisms in Huntington's chorea. *Life Sci. 10*:Pt. 1, 405–414, 1971.
5. Austin, L. and Phillis, J. W.: The distribution of cerebellar cholinesterases in several species. *J. Neurochem. 12*:709–717, 1965.
6. Avanzino, G. L., Bradley, P. B., and Wolstencroft, J. H.: Pharmacological properties of neurones of the paramedian ventricular nucleus. *Experientia 22*:410, 1966.

7. Barker, J. L., Crayton, J. W., and Nicoll, R. A.: Noradrenaline and acetylcholine responses of supraoptic secretory cells. *J. Physiol.* (Lond.) *218*:19–32, 1971.
8. Barron, D. H. and Matthews, B. H. C.: The interpretation of potential changes in the spinal cord. *J. Physiol.* (Lond.) *92*:276–321, 1938.
9. Bartolini, A., Bartolini, R., and Pepeu, G. C.: The effect of oxotremorine on the acetylcholine content of different parts of the cat brain. *J. Pharm. Pharmacol.* *22*:59–60, 1970.
10. Bartolini, A., Deffenu, G., Nistri, A., et al.: Effect of septal lesions on acetylcholine output from the cerebral cortex in the cat. *Br. J. Pharmacol.* *41*:399, 1971.
11. Bartolini, A. and Domino, E. F.: Cholinergic mechanisms in the cat visual system. *Fed. Proc.* *30*:425, 1971.
12. Bartolini, A. and Pepeu, G.: Investigations into the acetylcholine output from the cerebral cortex of the cat in the presence of hyoscine. *Br. J. Pharmacol.* *31*:66–73, 1967.
13. Bartolini, A., Weisenthal, L. M., and Domino, E. F.: Effect of photic stimulation on acetylcholine release from cat cerebral cortex. *Neuropharmacology* *11*:113–122, 1972.
14. Baumgarten, R. von, Bloom, F. E., Oliver, A. P., et al.: Response of individual olfactory nerve cells to microelectrophoretically administered chemical substances. *Pflügers Arch.* *277*:125–140, 1963.
15. Beani, L., Bianchi, C., Santinoceto, L., et al.: The cerebral acetylcholine release in conscious rabbits with semipermanently implanted epidural cups. *Int. J. Neuropharmacol.* *7*:469–481, 1968.
16. Beckman, A. L. and Carlisle, H. J.: Effect of intrahypothalamic infusion of acetylcholine on behavioural and physiological thermoregulation in the rat. *Nature* *221*:561–562, 1969.
17. Beckman, A. L. and Eisenman, J. S.: Microelectrophoresis of biogenic amines on hypothalamic thermosensitive cells. *Science* *170*:334–336, 1970.
18. Beleslin, D., Carmichael, E. A., and Feldberg, W.: The origin of acetylcholine appearing in the effluent of perfused cerebral ventricles of the cat. *J. Physiol.* (Lond.) *173*:368–376, 1964.
19. Beleslin, D. B. and Myers, R. D.: The release of acetylcholine and 5-hydroxytryptamine from the mesencephalon of the unanesthetized rhesus monkey. *Brain Res.* *23*:437–442, 1970.
20. Beleslin, D., Polak, R. L., and Sproull, D. H.: The effect of leptazol and strychnine on the acetylcholine release from the cat brain. *J. Physiol.* (Lond.) *181*:308–316, 1965.
21. Bertels-Meeuws, M. M. and Polak, R. L.: Influence of antimuscarinic substances on *in vitro* synthesis of acetylcholine by rat cerebral cortex. *Br. J. Pharmacol.* *33*:368–380, 1968.
22. Bertler, A. and Rosengren, E.: Possible role of brain dopamine. *Pharmacol. Rev.* *18*:769–773, 1966.
23. Bhattacharya, B. K. and Feldberg, W.: Perfusion of cerebral ventricles: Effects of drugs on outflow from the cisterna and aqueduct. *Br. J. Pharmacol.* *13*:156–162, 1958.
24. Birkmayer, W. and Mentasti, M.: Weitere experimentelle Unter-

suchungen über den Catecholaminstoffwechsel bei extrapyramidalen Erkrankungen (Parkinson- und Chorea-Syndrom). *Arch. Psychiatr. Nervenkr. 210*:29—35, 1967.
25. Birks, R. I.: The role of sodium ions in the metabolism of acetylcholine. *Can. J. Biochem. Physiol. 41*:2573—2597, 1963.
26. Biscoe, T. J. and Krnjević, K.: Chloralose and the activity of Renshaw cells. *Exp. Neurol. 8*:395—405, 1963.
27. Biscoe, T. J. and Straughan, D. W.: Microelectrophoretic studies of neurones in the cat hippocampus. *J. Physiol.* (Lond.) *183*:341—359, 1966.
28. Bjegović, M., Geber, J., and Randić, M.: Effect of tetrodotoxin on the spontaneous release of acetylcholine from the cerebral cortex. *Jugoslav. Physiol. Pharmacol. Acta 5*:345—348, 1969.
29. Bjegović, M. and Randić, M.: Effect of lithium ions on the release of acetylcholine from the cerebral cortex. *Nature 230*:587—588, 1971.
30. Bloom, F. E., Costa, E., and Salmoiraghi, G. C.: Analysis of individual rabbit olfactory bulb neuron responses to the microelectrophoresis of acetylcholine, norepinephrine and serotonin synergists and antagonists. *J. Pharmacol. Exp. Ther. 146*:16—23, 1964.
31. Bloom, F. E., Costa, E., and Salmoiraghi, G. C.: Anaesthesia and the responsiveness of individual neurons of the caudate nucleus of the cat to acetylcholine, norepinephrine and dopamine administered by microelectrophoresis. *J. Pharmacol. Exp. Ther. 150*:244—252, 1965.
32. Bloom, F. E., Oliver, A. P., and Salmoiraghi, G. C.: The responsiveness of individual hypothalamic neurones to microelectrophoretically administered endogenous amines. *Int. J. Neuropharmacol. 2*:181—193, 1963.
33. Bradley, P. B., Dhawan, B. N., and Wolstencroft, J. H.: Pharmacological properties of cholinoceptive neurones in the medulla and pons of the cat. *J. Physiol.* (Lond.) *183*:658—674, 1966.
34. Bradley, P. B. and Wolstencroft, J. H.: Actions of drugs on single neurones in the brainstem. *Br. Med. Bull. 21*:15—18, 1965.
35. Brenner, C. and Merritt, H. H.: Effect of certain choline derivatives on electrical activity of the cortex. *Arch. Neurol. 48*:382—395, 1942.
36. Brown, R. D., Daigneault, E. A., and Pruett, J. R.: The effects of selected cholinergic drugs and strychnine on cochlear responses and olivo-cochlear inhibition. *J. Pharmacol. Exp. Ther. 165*:300—309, 1969.
37. Bruyn, G. W.: Huntington's chorea. Historical, clinical and laboratory synopsis. In Vinken, P. J. and Bruyn, G. W. (Eds.): *Handbook of Clinical Neurology* (Vol. 6). Amsterdam: North Holland, 1968, pp. 298—378.
38. Bülbring, E. and Burn, J. H.: Observations bearing on synaptic transmission by acetylcholine in the spinal cord. *J. Physiol.* (Lond.) *100*:337—368, 1941.
39. Burgen, A. S. V. and Chipman, L. M.: Cholinesterase and succinic

dehydrogenase in the central nervous system of the dog. *J. Physiol.* (Lond.) *114*:296–305, 1951.
40. Burt, A. M.: A histochemical procedure for the localization of choline acetyltransferase activity. *J. Histochem. Cytochem. 18*:408–415, 1970.
41. Caesar, P. M. and Schnieden, H.: The effect of copper sulphate and manganese sulphate on the toxicity and tremor of tremorine and on some peripheral responses induced by acetylcholine, noradrenaline, dopamine and 5-hydroxytryptamine. *Biochem. Pharmacol. 15*:1691–1700, 1966.
42. Celesia, G. G. and Jasper, H. H.: Acetylcholine released from the cerebral cortex in relation to state of activation. *Neurology* (Minneap.) *16*:1053–1063, 1966.
43. Chapman, J. B. and McCance, I.: Acetylcholine-sensitive cells in the intracerebellar nuclei of the cat. *Brain Res. 5*:535–538, 1967.
44. Cho, A. K., Haslett, W. L., and Jenden, D. J.: The peripheral actions of oxotremorine, a metabolite of tremorine. *J. Pharmacol. Exp. Ther. 138*:249–257, 1962.
45. Ciani, S. and Edwards, C.: The effect of acetylcholine on neuromuscular transmission in the frog. *J. Pharmacol. Exp. Ther. 142*:21–23, 1963.
46. Cobbin, L. B., Leeder, S., and Pollard, J.: Smooth muscle stimulants in extracts of optic nerves, optic tracts and lateral geniculate bodies of sheep. *Br. J. Pharmacol. 25*:295–306, 1965.
47. Collier, B. and Mitchell, J. F.: The central release of acetylcholine during stimulation of the visual pathway. *J. Physiol.* (Lond.) *184*:239–254, 1966.
48. Collier, B. and Mitchell, J. F.: The central release of acetylcholine during consciousness and after brain lesions. *J. Physiol.* (Lond.) *188*:83–98, 1967.
49. Comis, S. D. and Davies, W. E.: Acetylcholine as a transmitter in the cat auditory system. *J. Neurochem. 16*:423–429, 1969.
50. Comis, S. D. and Whitfield, I. C.: Influence of centrifugal pathways on unit activity in the cochlear nucleus. *J. Neurophysiol. 31*:62–68, 1968.
51. Connor, J. D., Rossi, G. V., and Baker, W. W.: Antagonism of intracaudate carbachol tremor by local injections of catecholamines. *J. Pharmacol. Exp. Ther. 155*:545–551, 1967.
52. Corbin, K. B.: Pharmacologic agents used in the treatment of movement disorders. *Res. Publ. Ass. Nerv. Ment. Dis. 37*:86–103, 1959.
53. Couch, J. R.: Responses of neurons in the raphé nuclei to serotonin, norepinephrine and acetylcholine and their correlation with an excitatory synaptic input. *Brain Res. 19*:137–150, 1970.
54. Cox, B. and Potkonjak, D.: An investigation of the tremorgenic effects of oxotremorine and tremorine after stereotaxic injection into rat brain. *Int. J. Neuropharmacol. 8*:291–297, 1969a.
55. Cox, B. and Potkonjak, D.: The relationship between tremor and change in brain acetylcholine concentration produced by injection

of tremorine or oxotremorine in the rat. *Br. J. Pharmacol.* 35:295–303, 1969b.
56. Crawford, J. M. and Curtis, D. R.: Pharmacological studies on feline Betz cells. *J. Physiol.* (Lond.) *186*:121–138, 1966.
57. Crawford, J. M., Curtis, D. R., Voorhoeve, P. E., et al.: Acetylcholine sensitivity of cerebellar neurones in the cat. *J. Physiol.* (Lond.) *186*:139–165, 1966.
58. Cross, B. A., Moss, R. L., and Urban, I.: Effect of iontophoretic application of acetylcholine and noradrenaline to antidromically identified paraventricular neurones. *J. Physiol.* (Lond.) *214*:28–30, 1971.
59. Csillik, B.: Histochemistry of the synaptic region. In Andersen, P. and Jansen, J. K. S. (Eds.): *Excitatory Synaptic Mechanisms.* Oslo: Universitetsforlaget, 1970, pp. 43–55.
60. Curtis, D. R.: Microelectrophoresis. In Nastuk, W. L. (Ed.): *Physical Techniques in Biological Research* (Vol. 5). New York: Academic Press, 1964, pp. 144–190.
61. Curtis, D. R. and Davis, R.: The excitation of lateral geniculate neurones by quaternary ammonium derivatives. *J. Physiol.* (Lond.) *165*:62–82, 1963.
62. Curtis, D. R. and Eccles, R. M.: The excitation of Renshaw cells by pharmacological agents applied electrophoretically. *J. Physiol.* (Lond.) *141*:435–445, 1958.
63. Curtis, D. R. and Koizumi, K.: Chemical transmitter substances in brain stem of cat. *J. Neurophysiol.* 24:80–90, 1961.
64. Curtis, D. R., Phillis, J. W., and Watkins, J. C.: Cholinergic and noncholinergic transmission in the mammalian spinal cord. *J. Physiol.* (Lond.) *158*:296–323, 1961.
65. Curtis, D. R. and Ryall, R. W.: The excitation of Renshaw cells by cholinomimetics. *Exp. Brain Res.* 2:49–65, 1966a.
66. Curtis, D. R. and Ryall, R. W.: The acetylcholine receptors of Renshaw cells. *Exp. Brain Res.* 2:66–80, 1966b.
67. Curtis, D. R. and Ryall, R. W.: The synaptic excitation of Renshaw cells. *Exp. Brain Res.* 2:81–96, 1966c.
68. Curtis, D. R., Ryall, R. W., and Watkins, J. C.: The action of cholinomimetics on spinal interneurones. *Exp. Brain Res.* 2:97–106, 1966.
69. Curzon, G.: The biochemistry of dyskinesias. *Int. Rev. Neurobiol.* *10*:323–370, 1967.
70. Daigneault, E. A. and Blanton, J. F.: Hypersensitivity of the cochlear N_1 response to acetylcholine after chronic efferent denervation. *Fed. Proc.* 30:425, 1971.
71. Daigneault, E. A. and Brown, R. D.: Acetylcholine suppression of the N_1 component of round window recorded cochlear potentials. *Arch. Int. Pharmacodyn. Ther.* *162*:20–29, 1966.
72. Dale, H. H., Feldberg, W., and Vogt, M.: Release of acetylcholine at voluntary motor nerve endings. *J. Physiol.* (Lond.) *86*:353–380, 1936.

73. Davis, R.: Acetylcholine-sensitive neurons in the ventrolateral thalamic nucleus. In Purpura, D. P. and Yahr, M. D. (Eds.): *The Thalamus.* New York: Columbia University Press, 1966, pp. 193–195.
74. Davis, R. and Vaughan, P. C.: Pharmacological properties of feline red nucleus. *Int. J. Neuropharmacol. 8*:475–488, 1969.
75. Deffenu, G., Bertaccini, G., and Pepeu, G.: Acetylcholine and 5-hydroxytryptamine levels of the lateral geniculate bodies and superior colliculus of cats after visual deafferentation. *Exp. Neurol. 17*:203–209, 1967.
76. DeGroat, W. C. and Ryall, R. W.: An excitatory action of 5-hydroxytryptamine on sympathetic preganglionic neurones. *Exp. Brain Res. 3*:299–305, 1967.
77. DeRoetth, A.: Role of acetylcholine in nerve activity. *J. Neurophysiol. 14*:55–57, 1951.
78. Desmedt, J. E. and La Grutta, G.: The effect of selective inhibition of pseudocholinesterase on the spontaneous and evoked activity of the cat's cerebral cortex. *J. Physiol.* (Lond.) *136*:20–40, 1957.
79. Desmedt, J. E. and Monaco, P.: Mode of action of the efferent olivocochlear bundle on the inner ear. *Nature 192*:1263–1265, 1961.
80. Dreifuss, J. J. and Kelly, J. S.: The activity of identified supraoptic neurones and their response to acetylcholine applied by iontophoresis. *J. Physiol.* (Lond.) *220*:105–118, 1972.
81. Dudar, J. D. and Szerb, J. C.: The effect of topically applied atropine on resting and evoked cortical acetylcholine release. *J. Physiol.* (Lond.) *203*:741–762, 1969.
82. Duvoisin, R. C.: Cholinergic-anticholinergic antagonism in Parkinsonism. *Arch. Neurol. 17*:124–136, 1967.
83. Eccles, J. C., Eccles, R. M., and Fatt, P.: Pharmacological investigations on a central synapse operated by acetylcholine. *J. Physiol.* (Lond.) *131*:154–169, 1956.
84. Eccles, J. C., Fatt, P., and Koketsu, K.: Cholinergic and inhibitory synapses in a pathway from motor axon collaterals to motoneurones. *J. Physiol.* (Lond.) *126*:524–562, 1954.
85. Echols, S. D. and Jewett, R. E.: Effects of morphine on sleep in the cat. *Psychopharmacologia 24*:435–448, 1972.
86. Edery, H. and Levinger, I. M.: Acetylcholine release into the perfused intermeningeal spaces of the cat spinal cord. *Neuropharmacology 10*:239–246, 1971.
87. Ehringer, H. and Hornykiewicz, O.: Verteilung von Noradrenalin und Dopamin (3-Hydroxytyramin) im Gehirn des Menschen und ihr Verhalten bei Erkrankungen des extrapyramidalen Systems. *Klin. Wochenschr. 38*:1236–1239, 1960.
88. Elliott, K. A. C., Swank, R. L., and Henderson, N.: Effects of anaesthetics and convulsants on acetylcholine content of brain. *Am. J. Physiol. 162*:469–474, 1950.
89. Eränkö, O.: Demonstration of catecholamines and cholinesterases in the same section. *Pharmacol. Rev. 18*:353–358, 1966.

90. Erulkar, S. D., Nichols, C. W., Popp, M. B., et al.: Renshaw elements: Localization and acetylcholinesterase content. *J. Histochem. Cytochem.* *16*:128–135, 1968.
91. Everett, G. M., Blockus, L. E., and Shepperd, I. M.: Tremor induced by tremorine and its antagonism by anti-Parkinson drugs. *Science* *124*:79, 1956.
92. Fahn, S. and Côté, L. J.: Regional distribution of choline acetylase in the brain of the rhesus monkey. *Brain Res.* *7*:323–325, 1968.
93. Feldberg, W. S.: Present views on the mode of action of acetylcholine in the central nervous system. *Physiol. Rev.* *25*:596–642, 1945.
94. Feldberg, W. and Gaddum, J. H.: The chemical transmitter at synapses in a sympathetic ganglion. *J. Physiol.* (Lond.) *81*:305–319, 1934.
95. Feldberg, W. and Mann, T.: Properties and distribution of the enzyme system which synthesizes acetylcholine in nervous tissue. *J. Physiol.* (Lond.) *104*:411–425, 1946.
96. Feldberg, W. and Vogt, M.: Acetylcholine synthesis in different regions of the central nervous system. *J. Physiol.* (Lond.) *107*:372–381, 1948.
97. Ferguson, J. H. and Jasper, H. H.: Laminar DC studies of acetylcholine-activated epileptiform discharge in cerebral cortex. *Electroencephalogr. Clin. Neurophysiol.* *30*:377–390, 1971.
98. Fonnum, F.: Topographical and subcellular localization of choline acetyltransferase in rat hippocampal region. *J. Neurochem.* *17*:1029–1037, 1970.
99. Friede, R. L. and Fleming, L. M.: A comparison of cholinesterase distribution in the cerebellum of several species. *J. Neurochem.* *11*:1–7, 1964.
100. Frigyesi, T. L. and Purpura, D. P.: Acetylcholine sensitivity of thalamic synaptic organizations activated by brachium conjunctivum stimulation. *Arch. Int. Pharmacodyn. Ther.* *163*:110–132, 1966.
101. Fuxe, K., Goldstein, M., and Ljungdahl, A.: Antiparkinsonian drugs and central dopamine neurones. *Life Sci.* *9*:811–824, 1970.
102. Gaddum, J. H.: Push-pull cannulae. *J. Physiol.* (Lond.) *155*:1–2P, 1961.
103. Galindo, A., Krnjević, K., and Schwartz, S.: Microiontophoretic studies on neurones in the cuneate nucleus. *J. Physiol.* (Lond.) *192*:359–377, 1967.
104. Gastaut, H.: *The Epilepsies*. Springfield, (Ill.): C. C. Thomas, 1954.
105. Gerebtzoff, M. A.: *Cholinesterases, A Histochemical Contribution to the Solution of Some Functional Problems*. London: Pergamon Press, 1959.
106. Giarman, N. J. and Pepeu, G.: The influence of centrally acting cholinolytic drugs on brain acetylcholine levels. *Br. J. Pharmacol. Chemother.* *23*:123–130, 1964.
107. Goldberg, A. M. and McCaman, R. E.: A quantitative microchemical study of choline acetyltransferase and acetylcholinesterase in the cerebellum of several species. *Life Sci.* *6*:1493–1500, 1967.

108. Gowers, W. R.: *Epilepsy and Other Chronic Convulsive Diseases.* New York: Wood, 1885.
109. Green, J. R., Halpern, L. M., and Van Niel, S.: Choline acetylase and acetylcholine esterase changes in chronic isolated cerebral cortex of cat. *Life Sci.* 9:481–488, 1970.
110. Guth, P. S. and Amaro, J.: A possible cholinergic link in olivo-cochlear inhibition. *Int. J. Neuropharmacol.* 8:49–53, 1969.
111. Gwyn, D. G. and Wolstencroft, J. H.: Ascending and descending cholinergic fibers in cat spinal cord: histochemical evidence. *Science* 153:1543–1544, 1966.
112. Hanin, I. and Jenden, D. J.: Estimation of choline esters in brain by a new gas chromatographic procedure. *Biochem. Pharmacol.* 18: 837–845, 1969.
113. Hebb, C.: Biosynthesis of acetylcholine in nervous tissue. *Physiol. Rev.* 52:918–957, 1972.
114. Hebb, C. O., Ling, G. M., McGeer, E. G., et al.: Effect of locally applied hemicholinium on the acetylcholine content of the caudate nucleus. *Nature* 204:1309, 1964.
115. Hebb, C. O. and Silver, A.: Choline acetylase in the central nervous system of man and some other mammals. *J. Physiol.* (Lond.) 134:718–728, 1956.
116. Hemsworth, B. A. and Mitchell, J. F.: The characteristics of acetylcholine release mechanisms in the auditory cortex. *Br. J. Pharmacol.* 36:161–170, 1969.
117. Hemsworth, B. A. and Neal, M. J.: Effect of central stimulant drugs on acetylcholine release from rat cerebral cortex. *Br. J. Pharmacol.* 34:543–550, 1968.
118. Herz, A. and Gogolák, G.: Microelektrophoretische Untersuchungen am Septum des Kaninchens. *Pflügers Arch.* 285:317–330, 1965.
119. Herz, A. and Nacimiento, A. C.: Über die Wirkung von Pharmaka auf Neurone des Hippocampus nach mikroelektrophoretischer Verabfolgung. *Naunyn-Schmiedebergs Arch. Exp. Path. Pharmacol.* 251:295–314, 1965.
120. Herz, A. and Zieglgänsberger, W.: The influence of microelectrophoretically applied biogenic amines, cholinomimetics and procaine on synaptic excitation in the corpus striatum. *Int. J. Neuropharmacol.* 7:221–230, 1968.
121. Holmstedt, B. and Lundgren, G.: Tremorgenic agents and brain acetylcholine. In *Mechanisms of Release of Biogenic Amines.* Oxford: Pergamon Press, 1966.
122. Holtz, P. and Schümann, H. J.: Butyrylcholin in Gehirnextrakten. *Naturwissenschaften* 41:306, 1954.
123. Hongo, T. and Ryall, R. W.: Electrophysiological and microelectrophoretic studies on preganglionic sympathetic neurones in the spinal cord. *Acta Physiol. Scand.* 68:96–104, 1966.
124. Hornykiewicz, O.; Dopamine (3-hydroxytyramine) and brain function. *Pharmacol. Rev.* 18:925–964, 1966.
125. Hosein, E. A., Kato, A., Vine, E., et al.: The identification of

acetyl-L-carnitycholine in rat brain extracts and the comparison of its cholinomimetic properties with acetylcholine. *Can. J. Physiol. Pharmacol. 48*:709–722, 1970.
126. Hosein, E. A. and Orzeck, A.: Some physiological and biochemical properties of acetyl-L-carnitine isolated from brain tissue extracts. *Int. J. Neuropharmacol. 3*:71–76, 1964.
127. Hubbard, J. I., Schmidt, R. F., and Yokota, T.: The effect of acetylcholine upon mammalian motor nerve terminals. *J. Physiol. (Lond.) 181*:810–829, 1965.
128. Ilyutchenok, R. Y. and Gilinsky, M. A.: Anticholinergic drugs and neuronal mechanisms of reticulocortical interaction. *Pharmacol. Res. Comm. 1*:242–248, 1969.
129. Ishii, T. and Friede, R. L.; A comparative histochemical mapping of the distribution of acetylcholinesterase and nicotinamide adenine dinucleotide-diaphorase activities in human brain. *Int. Rev. Neurobiol. 10*:231–275, 1967.
130. Israël, M. and Whittaker, V. P.: The isolation of mossy fibre endings from the granular layer of the cerebellar cortex. *Experientia 21*:325–326, 1965.
131. Jacobowitz, D.: A method for the demonstration of both acetylcholinesterase and catecholamines in the same nerve trunk. *Life Sci. 4*:297–303, 1965.
132. Jacobowitz, D. and Koelle, G. B.: Histochemical correlations of acetylcholinesterase and catecholamines in postganglionic autonomic nerves of the cat, rabbit and guinea pig. *J. Pharmacol. Exp. Ther. 148*:225–237, 1965.
133. Jankowska, E. and Lindström, S.: Morphological identification of Renshaw cells. *Acta Physiol. Scand. 81*:428–430, 1971.
134. Jasper, H. H. and Tessier, J.: Acetylcholine liberation from cerebral cortex during paradoxical (REM) sleep. *Science 172*:601–602, 1971.
135. Jhamandas, K., Phillis, J. W., and Pinsky, C.: Effects of narcotic analgesics and antagonists in the in vivo release of acetylcholine from the cerebral cortex of the cat. *Br. J. Pharmacol. 43*:53–66, 1971.
136. Jordan, L. M. and Phillis, J. W.: Acetylcholine inhibition in the intact and chronically isolated cerebral cortex. *Br. J. Pharmacol. 45*:584–595, 1972.
137. Kanai, T. and Szerb, J. C.: Mesencephalic reticular activating system and cortical acetylcholine output. *Nature 205*:80–82, 1965.
138. Karczmar, A. G.: Is the central cholinergic nervous system overexploited? *Fed. Proc. 28*:147–157, 1969.
139. Karczmar, A. G.: Central cholinergic pathways and their behavioral implications. In *Principles of Psychopharmacology*. New York: Academic Press, 1970.
140. Kása, P. and Csillik, B.: Cholinergic excitation and inhibition in the cerebellar cortex. *Nature 208*:695–696, 1965.
141. Kása, P., Mann, S. P., and Hebb, C.: Localization of choline acetyltransferase. *Nature 226*:812–814, 1970a.

142. Kása, P., Mann, S. P., and Hebb, C.: Ultrastructural localization in spinal neurones. *Nature* 226:814—816, 1970b.
143. Kiraly, J. K. and Phillis, J. W.: The action of some drugs on the dorsal root potentials on the isolated toad spinal cord. *Br. J. Pharmacol. Chemother.* 17:224—231, 1961.
144. Klawans, H. L. and Rubovits, R.: Central cholinergic-anticholinergic antagonism in Huntington's chorea. *Neurology* 22:107—116, 1972.
145. Koelle, G. B.: The histochemical localization of cholinesterases in the nervous system of the rat. *J. Comp. Neurol.* 100:211—235, 1954.
146. Koelle, G. B.: Cytological distributions and physiological functions of cholinesterases. *Handbook Exp. Pharmacol.* 15:187—298, 1963.
147. Koelle, G. B.: Significance of acetylcholinesterase in central synaptic transmission. *Fed. Proc.* 28:95—100, 1969.
148. Koketsu, K.: Intracellular slow potential of dorsal root fibres. *Am. J. Physiol.* 184:338—344, 1956.
149. Krip, G. and Vazquez, A. J.: Effects of diphenylhydantoin and cholinergic agents on the neuronally isolated cerebral cortex. *Electroencephalogr. Clin. Neurophysiol.* 30:391—398, 1971.
150. Krnjević, K. and Phillis, J. W.: Acetylcholine-sensitive cells in the cerebral cortex. *J. Physiol.* (Lond.) 166:296—327, 1963a.
151. Krnjević, K. and Phillis, J. W.: Pharmacological properties of acetylcholine-sensitive cells in the cerebral cortex. *J. Physiol.* (Lond.) 166:328—350, 1963b.
152. Krnjević, K., Pumain, R., and Renaud, L.: The mechanism of excitation by acetylcholine in the cerebral cortex. *J. Physiol.* (Lond.) 215:247—268, 1971.
153. Krnjević, K., Reiffenstein, R. J., and Silver, A.: Chemical sensitivity of neurons in long-isolated slabs of cat cerebral cortex. *Electroencephalogr. Clin. Neurophysiol.* 29:269—282, 1970.
154. Krnjević, K. and Silver, A.: A histochemical study of cholinergic fibres in the cerebral cortex. *J. Anat.* 99:711—759, 1965.
155. Krnjević, K. and Silver, A.: Acetylcholinesterase in the developing forebrain. *J. Anat.* 100:63—89, 1966.
156. Krug, M., Schmidt, J. and Matthies, H.: Beeinflussung des Impulsmusters von spontan tätigen Neuronen der pontinen Formatio reticularis der Ratte durch Noradrenalin, Serotonin und Azetylcholin. *Acta Biol. Med. Ger.* 25:455—467, 1970.
157. Kuno, M. and Rudomin, P.: The release of acetylcholine from the spinal cord of the cat by antidromic stimulation of motor nerves. *J. Physiol.* (Lond.) 187:177—193, 1966.
158. Lee, T-P., Kuo, J. F., and Greengard, P.: Role of muscarinic cholinergic receptors in regulation of guanosine 3',5'-cyclic monophosphate content in mammalian brain, heart muscle and intestinal smooth muscle. *Proc. Nat. Acad. Sci. U.S.A.* 69:3287—3291, 1972.
159. Legge, K. F., Randić, M., and Straughan, D. W.: The pharmacology of neurones in the pyriform cortex. *Br. J. Pharmacol. Chemother.* 26:87—107, 1966.
160. Lewis, P. R. and Shute, C. C. D.: The cholinergic limbic system: Projections to hippocampal formation, medial cortex, nuclei of the

ascending cholinergic reticular system, and the subfornical organ and supra-optic crest. *Brain 90*:521—540, 1967.
161. Lewis, P. R., Shute, C. C. D., and Silver, A.: Confirmation from choline acetylase analyses of a massive cholinergic innervation to the rat hippocampus. *J. Physiol.* (Lond.) *191*:215—224, 1967.
162. Liang, C. C. and Quastel, J. H.: Uptake of acetylcholine in rat brain cortex slices. *Biochem. Pharmacol. 18*:1169—1185, 1969.
163. Lipmann, F. and Kaplan, N. O.: A common factor in the enzymatic acetylation of sulfanilamide and of choline. *J. Biol. Chem. 162*: 743—744, 1946.
164. MacIntosh, F. C.: The distribution of acetylcholine in the peripheral and the central nervous system. *J. Physiol.* (Lond.) *99*:436—442, 1941.
165. MacIntosh, F. C.: Synthesis and storage of acetylcholine in nervous tissue. *Can. J. Biochem. Physiol. 41*:2555—2571, 1963.
166. MacIntosh, F. C. and Oborin, P. E.: Release of acetylcholine from intact cerebral cortex. Abstr. *XIX Int. Physiol. Congr.* 580—581, 1953.
167. McCaman, R. E., Arnaiz, G. R. De L., and De Robertis, E. D. P.: Species differences in subcellular distribution of choline acetylase in the CNS. A study of choline acetylase, acetylcholinesterase, 5-hydroxytryptophan decarboxylase and monoamine oxidase in four species. *J. Neurochem. 12*:927—935, 1965.
168. McCance, I.: Ph.D. Thesis. Melbourne, Monash University, 1969.
169. McCance, I. and Phillis, J. W.: The discharge patterns of elements in cat cerebellar cortex and their responses to iontophoretically applied drugs. *Nature 204*:844—846, 1964.
170. McCance, I. and Phillis, J. W.: Cholinergic mechanisms in the cerebellar cortex. *Int. J. Neuropharmacol. 7*:447—462, 1968.
171. McCance, I., Phillis, J. W., and Westerman, R. A.: Acetylcholine-sensitivity of thalamic neurones: Its relationship to synaptic transmission. *Br. J. Pharmacol. Chemother. 32*:635—651, 1968.
172. McCance, I., Phillis, J. W., Tebēcis, A. K., et al.: The pharmacology of ACh-excitation of thalamic neurones. *Br. J. Pharmacol. Chemother. 32*:652—662, 1968.
173. McGeer, E. G., Wada, J. A., Terao, A., et al.: Amine synthesis in various brain regions with caudate or septal lesions. *Exp. Neurol. 24*:277—284, 1969.
174. McLennan, H.: The release of acetylcholine and of 3-hydroxytyramine from the caudate nucleus. *J. Physiol.* (Lond.) *174*:152—161, 1964.
175. McLennan, H.: *Synaptic Transmission.* Philadelphia: Saunders, 1970.
176. McLennan, H. and York, D. H.: Cholinergic mechanisms in the caudate nucleus. *J. Physiol.* (Lond.) *187*:163—175, 1966.
176a. Marshall, J. and Schnieden, H.: Effects of adrenaline, noradrenaline, atropine, and nicotine on some types of human tremor. *J. Neurol. Neurosurg. Psychiat. 29*:214—218, 1966.
177. Marshall, K. C. and McLennan, H.: The synaptic activation of neu-

rones of the feline ventrolateral thalamic nucleus: Possible cholinergic mechanisms. *Exp. Brain Res.* 15:472–483, 1972.
178. Martin, J. P.: Wilson's disease. In Vinken, P. J. and Bruyn, G. W. (Eds.): *Handbook of Clinical Neurology* (Vol. 6). Amsterdam, North Holland, pp. 267–278.
179. Martin, W. R., Riehl, J. L., and Unna, K. R.: Chlorpromazine. III. The effects of chlorpromazine and chlorpromazine sulphoxide on vascular responses to L-epinephrine and levarterenol. *J. Pharmac. Exp. Ther.* 130:37–45, 1960.
180. Metz, B.: Hypercapnia and acetylcholine release from the cerebral cortex and medulla. *J. Physiol.* (Lond.) 186:321–332, 1966.
181. Metz, B.: Correlation between the electrical activity and acetylcholine release from the cerebral cortex and medulla during hypercapnia. *Can. J. Physiol. Pharmacol.* 49:331–337, 1971.
182. Meyer, M.: Die Wirkung von Acetylcholin, L-Glutaminsäure und Dopamin auf Neurone im Gebiet der Nuclei cuneatus und gracilis der Katze. *Helv. Physiol. Pharmacol. Acta* 23:325–340, 1965.
183. Miller, E., Heller, A., and Moore, R. Y.: Acetylcholine in rabbit visual system nuclei after enucleation and visual cortex ablation. *J. Pharmacol. Exp. Ther.* 165:117–125, 1969.
184. Miller, F. R., Stavraky, G. W., and Woonton, G. A.: Effects of eserine, acetylcholine, and atropine on the electrocorticogram. *J. Neurophysiol.* 3:131–138, 1940.
185. Mitchell, J. F.: Release of acetylcholine from the cerebral cortex and the cerebellum. *J. Physiol.* (Lond.) 155:22P, 1961.
186. Mitchell, J. F.: The spontaneous and evoked release of acetylcholine from the cerebral cortex. *J. Physiol.* (Lond.) 165:98–116, 1963.
187. Mitchell, J. F. and Phillis, J. W.: Cholinergic transmission in the frog spinal cord. *Br. J. Pharmacol. Chemother.* 19:534–543, 1962.
188. Mitchell, J. F. and Szerb, J. C.: The spontaneous and evoked release of acetylcholine from the caudate nucleus. *Abstr. XXII Intern. Physiol. Congr.* No. 819, 1962.
189. Molenaar, P. C. and Polak, R. L.: Stimulation by atropine of acetylcholine release and synthesis in cortical slices from rat brain. *Br. J. Pharmacol.* 40:406–417, 1970.
190. Moss, R. L., Urbana, I., and Cross, B. A.: Microelectrophoresis of cholinergic and aminergic drugs on paraventricular neurons. *Am. J. Physiol.* 223:310–318, 1972.
191. Myers, R. D. and Beleslin, D. B.: The spontaneous release of 5-hydroxytryptamine and acetylcholine within the diencephalon of the unanesthetized Rhesus monkey. *Exp. Brain Res.* 11:539–552, 1970.
192. Nachmansohn, D. and Berman, M.: Studies on choline acetylase. III. On the preparation of the coenzymes and its effect on the enzyme. *J. Biol. Chem.* 165:551–563, 1946.
193. Nachmansohn, D. and Machado, A. L.: The formation of acetylcholine. A new enzyme: "Choline acetylase." *J. Neurophysiol.* 6:397–403, 1943.

194. Nastuk, W. L.: Membrane potential changes at a single muscle endplate produced by transitory application of acetylcholine with an electrically controlled microjet. *Fed. Proc. 12*:102, 1953.
195. Navaratnam, V. and Lewis, P. R.: Cholinesterase-containing neurones in the spinal cord of the rat. *Brain Res. 18*:411–425, 1970.
196. Neal, M. J., Hemsworth, B. A., and Mitchell, J. F.: The excitation of central cholinergic mechanisms by stimulation of the auditory pathway. *Life Sci. 7*:757–763, 1968.
197. Oomura, Y., Ooyama, H., Yamamoto, T., et al.: Behavior of hypothalamic unit activity during electrophoretic application of drugs. *Ann. N. Y. Acad. Sci. 157*:642–665, 1969.
198. Osen, K. K. and Roth, K.: Histochemical localization of cholinesterases in the cochlear nuclei of the cat, with notes on the origin of acetylcholinesterase-positive afferents and the superior olive. *Brain Res. 16*:165–185, 1969.
199. Pepeu, G. and Bartolini, A.: Effect of psychoactive drugs on the output of acetylcholine from the cerebral cortex of the cat. *Eur. J. Pharmacol. 4*:254–263, 1968.
200. Pepeu, G., Mulas, A., Ruffi, A., et al.: Brain acetylcholine levels in rats with septal lesions. *Life Sci. 10*:(Pt. 1) 181–184, 1971.
201. Phillis, J. W.: Cholinergic mechanisms in the cerebellum. *Br. Med. Bull. 21*:26–29, 1965.
202. Phillis, J. W.: Acetylcholine release from the cerebral cortex: Its role in cortical arousal. *Brain Res. 7*:378–389. 1968a.
203. Phillis, J. W.: Acetylcholinesterase in the feline cerebellum. *J. Neurochem. 15*:691–704, 1968b.
204. Phillis, J. W.: *The Pharmacology of Synapses.* Oxford: Pergamon Press, 1970.
205. Phillis, J. W.: The pharmacology of thalamic and geniculate neurones. *Int. Rev. Neurobiol. 14*:1–48, 1971.
206. Phillis, J. W. and Chong, G. C.: Acetylcholine release from the cerebral and cerebellar cortices: Its role in cortical arousal. *Nature 207*:1253–1255, 1965.
207. Phillis, J. W. and Jhamandas, K.: The effects of chlorpromazine and ethanol on *in vivo* release of acetylcholine from the cerebral cortex. *Comp. Gen. Pharmacol. 2*:306–310, 1971.
208. Phillis, J. W., Mullin, W. J., and Pinsky, C.: Morphine enhancement of acetylcholine release into the lateral ventricle and from the cerebral cortex of unanesthetized cats. *Comp. Gen. Pharmacol. 4*:189–200, 1973.
209. Phillis, J. W., Tebēcis, A. K., and York, D. H.: A study of cholinoceptive cells in the lateral geniculate nucleus. *J. Physiol.* (Lond.) *192*:695–713, 1967.
210. Phillis, J. W., Tebēcis, A. K., and York, D. H.: Acetylcholine release from the feline thalamus. *J. Pharm. Pharmacol. 20*:476–478, 1968.
211. Phillis, J. W. and York, D. H.: Cholinergic inhibition in the cerebral cortex. *Brain Res. 5*:517–520, 1967a.
212. Phillis, J. W. and York, D. H.: Strychnine block of neural and drug

induced inhibition in the cerebral cortex. *Nature 216*:922–923, 1967b.
213. Phillis, J. W. and York, D. H.: An intracortical cholinergic inhibitory synapse. *Life Sci. 7*:65–69, 1968a.
214. Phillis, J. W. and York, D. H.: Pharmacological studies on a cholinergic inhibition in the cerebral cortex. *Brain Res. 10*:297–306, 1968b.
215. Phillis, J. W. and York, D. H.: Nicotine, smoking and cortical inhibition. *Nature 219*:89–91, 1968c.
216. Pilar, G.: Effect of ACh on pre- and postsynaptic elements of avian ciliary ganglion synapses. *Fed. Proc. 28*: (2) 670, 1969.
217. Polak, R. L.: Effect of hyoscine on the output of acetylcholine into perfused cerebral ventricles of cats. *J. Physiol.* (Lond.) *181*:317–323, 1965.
218. Polak, R. L.: An Analysis of the Influence of Antimuscarinic Agents on Synthesis, Storage and Release of Acetylcholine by Cortical Slices from Rat Brain. (Doctoral Thesis.) Amsterdam, Drukkerij de Bij, 1–79, 1971.
219. Portig, P. J. and Vogt, M.: Release into the cerebral ventricles of substances with possible transmitter function in the caudate nucleus. *J. Physiol.* (Lond.) *204*:687–715, 1969.
220. Potter, L. T. and Glover, V. A. S.: Choline acetyltransferase from rat brain. *J. Biol. Chem. 243*:3864–3870, 1968.
221. Quastel, D. M. J. and Curtis, D. R.: A central action of hemicholinium. *Nature 208*:192–194, 1965.
222. Randić, M. and Padjen, A.: Effect of calcium ions on the release of acetylcholine from the cerebral cortex. *Nature 215*:990, 1967.
223. Randić, M., Siminoff, R., and Straughan, D. W.: Acetylcholine depression of cortical neurons. *Expl. Neurol. 9*:236–242, 1964.
224. Rasmussen, G. L.: The olivary peduncle and other fiber projections of the superior olivary complex. *J. Comp. Neurol. 84*:141–220, 1946.
225. Rasmussen, G. L.: Anatomic relationships of the ascending and descending auditory systems. In Fields, W. S. and Alford, B. B. (Eds.): *Neurological Aspects of Auditory and Vestibular Disorders.* Springfield, (Ill.): C. C. Thomas, 1964.
226. Russell, R. W.: Behavioral aspects of cholinergic transmission. *Fed. Proc. 28*:121–131, 1969.
227. Ryall, R. W.: Renshaw cell mediated inhibition of Renshaw cells: Patterns of excitation and inhibition from impulses in motor axon collaterals. *J. Neurophysiol. 33*:257–270, 1970.
228. Ryall, R. W. and De Groat, W. C.: The microelectrophoretic administration of noradrenaline, 5-hydroxytryptamine, acetylcholine and glycine to sacral parasympathetic preganglionic neurones. *Brain Res. 37*:345–347, 1972.
229. Salmoiraghi, G. C., Bloom, F. E., and Costa, E.: Adrenergic mechanisms in rabbit olfactory bulb. *Am. J. Physiol. 207*:1417–1424, 1964.

230. Salmoiraghi, G. C. and Steiner, F. A.: Acetylcholine sensitivity of cat's medullary neurones. *J. Neurophysiol.* 26:581–597, 1963.
231. Satinsky, D.: Pharmacological responsiveness of lateral geniculate nucleus neurons. *Int. J. Neuropharmacol.* 6:387–397, 1967.
232. Schmidt, D. E., Szilagyi, P. I. A., Alkon, D. L., et al.: Acetylcholine: Release from neural tissue and identification by pyrolysis–gas chromatography. *Science* 165:1370–1371, 1969.
233. Schuberth, J.: On the biosynthesis of acetyl coenzyme A in the brain. 1. The enzymic formation of acetyl coenzyme A from acetate, adenosine triphosphate and coenzyme A. *Biochim. Biophys. Acta* 98:1–7, 1965.
234. Schuberth, J. and Sundwall, A.: Effects of some drugs on the uptake of acetylcholine in cortex slices of mouse brain. *J. Neurochem.* 14:807–812, 1967.
235. Schuknecht, H., Churchill, J., and Doran, R.: The localization of acetylcholinesterase in the cochlea. *Arch. Otorhinlaryngol.* 69:549–559, 1959.
236. Shute, C. C. D. and Lewis, P. R.: The use of cholinesterase techniques combined with operative procedures to follow nervous pathways in the brain. *Bibl. Anat.* 2:34–49, 1961.
237. Shute, C. C. D. and Lewis, P. R.: Cholinesterase-containing pathways of the hindbrain: Afferent cerebellar and centrifugal cochlear fibers. *Nature* 205:242–246, 1965.
238. Shute, C. C. D. and Lewis, P. R.: Electron microscopy of cholinergic terminals and acetylcholinesterase-containing neurones in the hippocampal formation of the rat. *Z. Zellforsch.* 69:334–343, 1966.
239. Shute, C. C. D. and Lewis, P. R.: The ascending cholinergic reticular system: Neocortical, olfactory and subcortical projections. *Brain* 90:497–520, 1967.
240. Sie, G., Jasper, H. H., and Wolfe, L.: Rate of ACh release from cortical surface in "encephale" and "cerveau isolé" cat preparations in relation to arousal and epileptic activation of the EcoG. *Electroencephalogr. Clin. Neurophysiol.* 18:206, 1965.
241. Silver, A.: Cholinesterases of the central nervous system with special reference to the cerebellum. *Int. Rev. Neurobiol.* 10:57–109, 1967.
242. Silver, M. S.: Motoneurons of the spinal cord of the frog. *J. Comp. Neurol.* 77:1–39, 1942.
243. Slater, P.: The estimation of the "free" and "bound" acetylcholine content of rat brain. *J. Pharm. Pharmacol.* 23:514–518, 1971.
244. Smith, C. M.: The release of acetylcholine from rabbit hippocampus. *Br. J. Pharmacol.* 45:172P, 1972.
245. Söderholm, U.: Histochemical localization of esterases, phosphatases and tetrazolium reductases in the motor neurones of the spinal cord of the rat and the effect of nerve division. *Acta Physiol. Scand.* 65 (Suppl. 256): 1–60, 1965.
246. Sohmer, H. and Feinmesser, M.: Studies on the influence of acetylcholine, eserine, and atropine on cochlear potentials in the guinea pig and cat. *Arch. Int. Pharmacodyn. Ther.* 144:446–453, 1963.

247. Somogyi, G. T. and Szerb, J. C.: Demonstration of acetylcholine release by measuring labelled choline efflux from cerebral cortex slices. *J. Neurochem. 19*:2667–2678, 1972.
248. Sourkes, T. L. and Poirier, L. J.: Neurochemical bases of tremor and other disorders of movement. *Can. Med. Assn. J. 94*:53–60, 1966.
249. Spehlmann, R.: Acetylcholine and prostigmine electrophoresis at visual cortex neurons. *J. Neurophysiol. 26*:127–139, 1963.
250. Spehlmann, R.: Acetylcholine facilitation, atropine block of synaptic excitation of cortical neurons. *Science 165*:404–405, 1969.
251. Spehlmann, R.: Excitability of partially deafferented cortex. II. Microelectrode studies. *Arch. Neurol. 22*:510–514, 1970.
252. Stefanis, C.: Hippocampal neurons: Their responsiveness to microelectrophoretically administered endogenous amines. *Pharmacologist 6*:171, 1964.
253. Steiner, F. A.: Influence of microelectrophoretically applied acetylcholine on the responsiveness of hippocampal and lateral geniculate neurones. *Pflügers Arch. 303*:173–180, 1968.
254. Steiner, F. A. and Meyer, M.: Actions of L-glutamate, acetylcholine and dopamine on single neurones in the nuclei cuneatus and gracilis of the cat. *Experientia 22*:58–59, 1966.
255. Steiner, F. A. and Weber, G.: Die Beeinflussung labyrinthär erregbarer Neurone des Hirnstammes durch Acetylcholin. *Helv. Physiol. Acta 23*:82–89, 1965.
256. Stone, T. W.: Cholinergic mechanisms in the rat cerebral cortex. *J. Physiol.* (Lond.) *222*:155–156P, 1972.
257. Stone, T. W.: Cholinergic mechanisms in the rat somatosensory cortex. *J. Physiol.* (Lond.) *225*:485–499, 1972.
258. Stone, W. E.: The role of acetylcholine in brain metabolism and function. *Am. J. Phys. Med. 36*:222–255, 1957.
259. Storm-Mathisen, J.: Quantitative histochemistry of acetylcholinesterase in rat hippocampal region correlated to histochemical staining. *J. Neurochem. 17*:739–750, 1970.
260. Storm-Mathisen, J.: Glutamate decarboxylase in the rat hippocampal region after lesions of the afferent fiber systems. Evidence that the enzyme is localized in intrinsic neurones. *Brain Res. 40*:215–235, 1972.
261. Storm-Mathisen, J. and Blackstad, T. W.: Cholinesterase in the hippocampal region. Distribution and relation to architectonics and afferent systems. *Acta Anat. 56*:216–253, 1964.
262. Straschill, M. and Perwein, J.: Effect of iontophoretically applied biogenic amines and of cholinomimetic substances upon the activity of neurons in the superior colliculus and mesencephalic reticular formation of the cat. *Pflügers Arch. 324*:43–55, 1971.
263. Straughan, D. W. and Legge, K. F.: The pharmacology of amygdaloid neurones. *J. Pharm. Pharmacol. 17*:675–677, 1965.
264. Szentágothai, J.: Synaptic architecture of the spinal motoneuron pool. *Electroencephalogr. Clin. Neurophysiol.* Suppl. *25*:4–19, 1967.

265. Szentágothai, J. and Rajkovits, K.: Über den Ursprung der Kletterfasern des Kleinhirns. *Z. Anat. Entwicklungsgesch. 121*:130–141, 1959.
266. Szerb, J. C.: Acetylcholine-like activity released from brain "in vivo." *Nature 197*:1016, 1963.
267. Szerb, J. C.: The effect of tertiary and quaternary atropine on cortical acetylcholine output and on the electroencephalogram in cats. *Can. J. Physiol. Pharmacol. 42*:303–314, 1964.
268. Szerb, J. C.: Cortical acetylcholine release and electroencephalographic arousal. *J. Physiol.* (Lond.) *192*:329–343, 1967.
269. Szerb, J. C., Malik, H., and Hunter, E. G.: Relationship between acetylcholine content and release in the cat's cerebral cortex. *Can. J. Physiol. Pharmacol. 48*:780–790, 1970.
270. Szerb, J. C. and Somogyi, G. T.: Depression of acetylcholine release from cerebral cortical slices by cholinesterase inhibition and by oxotremorine. *Nature 241*:121–122, 1973.
271. Tebēcis, A. K.: Properties of cholinoceptive neurones in the medial geniculate nucleus. *Br. J. Pharmacol. 38*:117–137, 1970a.
272. Tebēcis, A. K.: Studies on cholinergic transmission in the medial geniculate nucleus. *Br. J. Pharmacol. 38*:138–147, 1970b.
273. Tebēcis, A. K.: Cholinergic and non-cholinergic transmission in the medial geniculate nucleus of the cat. *J. Physiol.* (Lond.) *226*:153–172, 1972.
274. Tower, D. B.: Nature and extent of the biochemical lesion in human epileptogenic cerebral cortex. An approach to its control in vitro and in vivo. *Neurology 5*:113–130, 1955.
275. Tower, D. B. and McEachern, D.: Acetylcholine and neuronal activity. II. Acetylcholine and cholinesterase activity in the cerebrospinal fluids of patients with epilepsy. *Can. J. Research. Sec. E, 27*:120–131, 1949.
276. Tucek, S.: On subcellular localization and binding of choline acetyltransferase in the cholinergic nerve endings of the brain. *J. Neurochem. 13*:1317–1327, 1966.
277. Van Rossum, J. M.: The significance of dopamine receptor blockage for the mechanism of action of neuroleptic drugs. *Arch. Int. Pharmacodyn. 160*:492–494, 1966.
278. Vazquez, A. J., Krip, G., and Pinsky, C.: Evidence for a muscarinic inhibitory mechanism in the cerebral cortex. *Exp. Neurol. 23*: 318–331, 1969.
279. Votava, Z.: Pharmacology of the central cholinergic synapses. *Ann. Rev. Pharmacol. 7*:223–240, 1967.
280. Waldron, H. A.: Cholinergic fibers in the spinal cord of the rat. *Brain Res. 4*:113–116, 1967.
281. Waldron, H. A.: Cholinergic fibers in the spinal cord of the guinea pig. *Brain Res. 12*:250–252, 1969.
282. Weight, F. F.: Cholinergic mechanisms in recurrent inhibition of motoneurones. In Efron, D. H. (Ed.): *Psychopharmacology: A Review of Progress*. Washington, Govt. Printing Office, 1968.
283. Weight, F. F. and Padjen, A.: Slow postsynaptic inhibition and

sodium inactivation in frog sympathetic ganglion cells. *Abstr. Fifth Int. Pharmacol. Congr.* 1489, 1972.
284. Weight, F. F. and Salmoiraghi, G. C.: Responses of spinal cord interneurons to acetylcholine, norepinephrine and serotonin administered by microelectrophoresis. *J. Pharmacol. Exp. Ther. 153*: 420–427, 1966.
285. Weight, F. F. and Salmoiraghi, G. C.: Motoneurone depression by norepinephrine. *Nature 213*:1229–1230, 1967.
285a. Weight, F. F. and Votava, J.: Slow synaptic excitation in sympathetic ganglion cells: Evidence for synaptic inactivation of potassium conductance. *Science 170*:755–758, 1970.
286. Weiss, B. and Heller, A.: Methodological problems evaluating the role of cholinergic mechanisms in behavior. *Fed. Proc. 28*:135–146, 1969.
287. Werman, R.: A Review–criteria for identification of a central nervous system transmitter. *Comp. Biochem. Physiol. 18*:745–766, 1966.
288. Whittaker, V. P., Michaelson, I. A., and Kirkland, R. J. A.: The separation of synaptic vesicles from nerve-ending particles ("synaptosomes"). *Biochem. J. 90*:293–303, 1964.
289. Whittaker, V. P. and Sheridan, M. N.: The morphology and acetylcholine content of isolated cerebral cortical synaptic vesicles. *J. Neurochem. 12*:363–372, 1965.
290. Williams, D. J.: The effect of choline-like substances on the cerebral electrical discharges in epilepsy. *J. Neurol. Psychiatr. 4*:32–47, 1941.
291. Yamamoto, C.: Pharmacologic studies of norepinephrine, acetylcholine and related compounds on neurons in Deiters' nucleus and the cerebellum. *J. Pharmac. Exp. Ther. 156*:39–47, 1967.
292. York, D. H.: Possible dopaminergic pathway from substantia nigra to putamen. *Brain Res. 20*:233–249, 1970.
293. Zetler, G. and Schlosser, L.: Über die Verteilung von Substanz P und Cholinacetylase in Gehirn. *Arch. Exp. Path. Pharmacol. 224*: 159–175, 1955.

CHAPTER FIVE

On the Physiology and Pharmacology of Cerebral Dopamine Neurons

Detlef Bieger and Charles H. Hockman

As a putative transmitter in the central nervous system (CNS), dopamine (DA) enjoys the distinction of being the primordial catecholamine and of displaying a topological association with phylogenetically older regions of the brain, namely, the corpus striatum and mesolimbic areas which include "paleocortical" structures as well as the limbic cortex. From this evolutionary perspective, one may speculate that DA is a prototype of a neurohumoral transmitter in central nervous systems.

Any account of DA and its physiological significance in CNS function would be incomplete without a historical note on the prominent role played by reserpine in early brain monoamine research. Aware that the alkaloid depleted brain norepinephrine (NE) and serotonin stores, Carlsson and his colleagues, in 1957, reasoned that the reserpine-induced "motor inactivation syndrome" resulted from an absence of these amines at central synapses, and that it ought to be reversed by the administration of their amino acid precursors, the free amines being unable to penetrate the blood-brain barrier. This prediction was borne out when it was shown that DOPA produced a dramatic reversal of reserpine symptoms and signs of overt motor excitation in both mice and rabbits (*108*). Inasmuch as brain levels of NE had fallen to undetectable values in these reserpinized animals, experimenters were surprised to find that they remained low after DOPA administration, despite the conspicuous recovery of motor functions (*108*). Thus the spotlight was focused on DA which, at that time, had just been detected in the mammalian brain (*371, 549*), and for which Blaschko (*74*) had proposed a physiological function in peripheral adrenergic tissues, independent of its ubiquitous role as a biochemical precursor of NE.

With the aid of a new fluorimetric procedure (*109, 112*), the presence of DA in the brain was confirmed, and its unique distribution pattern established. As expected, reserpine induced a marked loss in DA from the corpus striatum, the principal site of occurrence, affirming the involvement of catecholamines in its central actions. Of far greater heuristic significance was the fact that these findings implicated DA as a physiologically active substance in the brain by linking its presence in the corpus striatum to the normal functioning of this region in extrapyramidal motor activity. As reserpine was already known to induce extrapyramidal symptoms in man, indistinguishable from those of Parkinson's disease (*60, 99*), a cause-and-effect relationship between striatal DA deficiency and parkinsonian extrapyramidal dysfunctions was immediately perceived.

These developments triggered the release of a veritable avalanche of studies on the role of DA as a transmitter in the brain, as teams of researchers explored its significance in the pathoneurochemistry of movement disorders. Moreover, the forging of an etiological link between DA and reserpine-induced extrapyramidal sensorimotor dysfunction provided a firm foothold for a pharmacological search and evaluation of drugs that might prove of therapeutic value.

Having drawn the reader's attention to what we believe are eventful landmarks in the unfolding of the "DA story," we shall now examine the role of this amine in the CNS by bringing together data from several disciplines, all of which have contributed immensely to our present understanding of the physiology and pathology of the basal ganglia. This corpus of knowledge firmly buttresses the candidacy of DA as a neurotransmitter.

NEUROCHEMICAL PROPERTIES

Dopamine

Since the scope of this chapter is restricted to mammals, the reader interested in the role of DA in invertebrate systems is referred to several excellent reviews (*271, 304, 557*).

In 1957 Montagu (*371*) identified DA as a constituent of mammalian brains, and shortly afterward Bertler and Rosengren (*60*) showed that its regional distribution in the brain did not parallel that of NE. At that time NE was already considered to be a putative transmitter on the basis of a nonuniform distribution pattern that could not be ascribed to differences in regional vascularity and adrenergic vasomotor innervation (*536, 537*). As the immediate precursor of NE, DA was expected to share its locale; however, dissimilarities in the topochemistry of the two amines suggested a physiological role for DA independent of that for NE. Before long a succession of neurochemical studies demonstrated high amounts of DA in the neostriatum, accounting for approximately 80% of the total content of this amine in the brain (*54, 60, 173, 451*). These observations were extended by several investigators (*85, 274, 275, 332, 424*), and it was subsequently shown that DA occurred in other areas of the brain, including the nucleus accumbens (*28, 212*), the tuberculum olfactorium (*28, 212*), the substantia nigra (SN) (*54, 275*), the pallidum (*274*), and the median eminence (*213*). The persistence of DA in limbic cortical areas after destruction of the noradrenergic ceruleocortical pathway (*72*) suggests a more widespread telencephalic occurrence of the amine than was at first believed. Significant amounts of DA have also been reported in peripheral organs, including autonomic ganglia (*332*), splenic nerves (*182*), carotid bodies (*164*), and the sinoatrial node (*35*). (For a detailed discussion of peripheral distribution, see Hornykiewicz [*271*].)

Historically it is pertinent to recall that biochemical studies on the regional distribution of DA gave birth to the concept of its role as a transmitter in the CNS at a time when these observations lacked morphological support, and when classic neurohistological methods of silver impregnation had failed to provide evidence for ascending fiber projections from DA-containing areas of the midbrain to regions of the basal ganglia. In fact it was not until the early 1960s when Falck and Hillarp's formaldehyde fluorescence technique (*189*) was applied to the study of brain tissues that a specific DA neuron system was identified in the brain. This highly significant advance set the stage for a conceptual integration of biochemical knowledge of regional distribution of DA, and for the study of the functional organization of this neuron system in both neurophysiological and neuropharmacological terms. In its original form the Falck-Hillarp technique did not permit a distinction between the fluorophores formed from NE and DA. Therefore the histochemical differentiation between the two catechol neuron systems in situ necessitated either the application of drugs known to interfere differentially with catecholamine metabolism in vivo or, preferably, direct spectrofluorimetric determination of each catecholamine in purified tissue extracts (*112*).

Recently it has become possible to differentiate between DA and NE at a cellular level by microspectrofluorimetric analysis after pretreatment of tissue sections with HCl (*70, 175*). Thus biochemical data on the localization of brain DA can now be interpreted in terms of a well-defined topography of central DA neuron systems. Although the most systematic and extensive analysis of the anatomical organization of DA neurons has been performed in the rat (*521*), evidence indicates that the same basic neurochemical organization is applicable to other mammals, including man (*395*).

Dopamine Pathways and Cellular Localization Currently four major DA pathways are recognized in the brain, and they may be divided into long- and short-axon projections. The bulk of long DA fibers originates in the ventral tegmentum of the mesencephalon and gives rise to three main projections: 1) the nigroneostriatal fibers that ascend from cell bodies in the pars compacta of the SN and terminate within the caudate-putamen complex (*17, 24, 25, 55, 521*); 2) the mesolimbic DA fibers that originate from cell bodies dorsal to the interpeduncular nucleus of the ventromedial tegmentum and project to the nucleus accumbens, the tuberculum olfactorium, and the dorsolateral part of the nucleus interstitialis striae terminalis and perhaps also to a portion of the nucleus of the diagonal band (*16, 28, 212*); and 3) a smaller contingent of DA cells that innervates the nucleus centralis of the amygdala (*28*). Recent histofluorescent microscopic observations by Nobin and Björklund (*395*) on material obtained from aborted human fetuses have revealed that DA cells in the ventral tegmentum of the mesencephalon can be differentiated into four groups, two of which correspond to the pars compacta of the SN, the other two located between the pars compacta and nucleus ruber; however, no such cells were detectable in the pars reticulata which is known to send a projection to the thalamus (*114*). Projections from the pars compacta "proper," which would include the first two groups, may be assumed to be identical with the nigroneostriatal pathway described in laboratory mammals, whereas we may speculate on the subrubral and juxta-oculomotor aggregates as being the source of DA terminals recently discovered in other telencephalic areas, including the pallidum and claustrum (*395*), as well as in the limbic cortex (*264, 265*). The majority of short-axon DA neurons are located in the hypothalamus and are associated with periventricular structures, the most conspicuous of which is the cell group of the arcuate nucleus which gives rise to the tuberoinfundibular pathway (*71*). This neuronal system is believed to act as an intermediary between forebrain structures and median eminence neurons which elaborate gonadotropin-releasing factors. (For a more detailed discussion, see Knigge et al. [*311*], and Keller and Lichtensteiger [*298*].) Another population of DA neurons is found in the retina coextensive with the amacrine layer and may represent a specialized subgroup of amacrine cells (*315, 316*).

Detailed histofluorescence studies, in conjunction with electron microscopic observations, have revealed several neurochemical features that the long-axon central DA neurons evidently share with other monoamine-containing neurons of the brain and peripheral nervous system (*155, 156*). Thus the perikarya in the pars compacta of the SN exhibit weak to moderate fluorescence, which diminishes progressively with increasing distances from the cell body. As the preterminal axons approach their sites of termination, they ramify extensively and develop intensely fluorescent varicosities that are closely packed with dense-core vesicles. The proximodistal gradient in fluorescence is closely correlated with estimated DA concentrations in the different segments of the neuron as determined by histofluorimetric measurements. On the basis of data from chemical assays of the regional distribution of DA in the mid- and forebrain, in conjunction with fluorescence and electron microscopic estimates of numbers of nigral DA perikarya and striatal DA-containing varicosities, Swedish investigators (*30*) have attempted to calculate intraneuronal amounts of the amine. A summary of their results is presented in Figure 1.

It is noteworthy not only that the terminal arborization of the average neuron contains at least 50 times more amine than the perikaryon, but also that the amine is *concentrated* about 40-fold in the terminal varicosity when compared with the perikaryon. The explanation for this unique distribution pattern comes from axoplasmic transport studies. Although most of them have been performed on peripheral nerves (see review by Dahlström [*154*]), there is strong evidence that the same process also operates in nigrostriatal neurons (*203*). Since the nerve terminal itself appears to be unable to synthesize protein, a major function of this axonal flow is to supply the terminal with the enzymatic machinery necessary for the synthesis of DA. In keeping with this concept, the amount of DA traveling along the preterminal axon would be insignificant when compared with the amounts synthesized locally in the terminals (*221*).

Apart from obvious morphological differences, long- and short-axon DA neurons also appear to differ in their dynamics of transmitter metabolism (*34, 128*). Furthermore, the long-axon nigroneostriatal and mesolimbic DA neurons appear to contain subpopulations that can be distinguished from one another on the basis of ontogenetic as well as neuropharmacological characteristics. It is postulated that in the rat there are two types of DA terminals in both the neostriatum and limbic forebrain (*407*). The neostriatum of the adult animal normally exhibits diffuse fluorescence, making it virtually impossible to visualize single varicosities after freeze-drying and paraformaldehyde treatment (see review by Fuxe and Jonsson [*216*]). Although this diffuseness may be owing to a loose binding of DA to storage sites, observations in newborn rats demonstrate that the typical DA fluorescence first appears in circumscribed areas of the

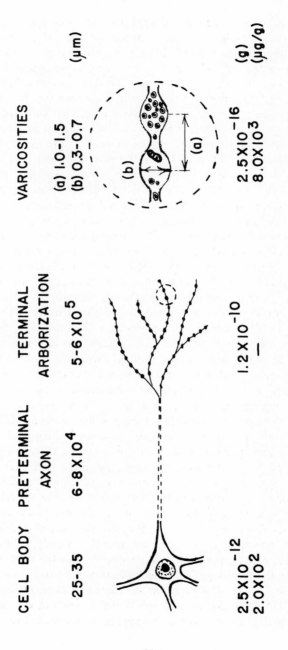

Figure 1. Rat nigroneostriatal dopamine neuron: dimensions and intraneuronal amine distribution. Value below insert refers to *single* varicosity. (After Anden, N.-E., Fuxe, K., Hamberger, B., and Hökfelt, T.: *Acta Physiol. Scand.* 67:306–312, 1966.)

nucleus, single varicosities being visible as individual fluorescent dots. Dotted fluorescence also prevails in limbic areas that, in the adult animal, exhibit additional diffuse fluorescence.

It is noteworthy that the striatal fluorescent islands, which are invisible in the adult rat, reappear during depletion of DA following arrest of its synthesis. Furthermore, under this condition limbic areas lose their diffuse fluorescence more readily than the dotted fluorescence, whereas depletion in the neostriatum is more rapid in its lateral parts than it is medially. It thus appears that neither striatal nor mesolimbic DA neuron groups represent homogeneous populations, but rather separate systems of a more complex functional organization. The possibility of a functional heterogeneity within individual dopaminergic pathways assumes a special significance in view of the present controversy over intraneuronal compartmentation of DA metabolism, a controversy that arose from biochemical turnover studies, all of which are based on the tacit assumption of homogeneity (see section on "Turnover" under "Regulation of Dopamine Synthesis" in this chapter).

Subcellular Localization After differential centrifugation of striatum homogenates, only 30 to 40% of the endogenous DA is recovered as particle-bound, the bulk being found in the cytosol (high-speed supernatant) (*57, 228, 331, 396*). By contrast, exogenous DA has been reported to be mainly particle-bound in homogenates of striatal and hypothalamic tissue. After intraventricular administration of ^{3}H-DL-DOPA, labeled DA and NE can be recovered from synaptosomal fractions of rat striatum and hypothalamus homogenates as prepared by sucrose-density gradient centrifugation (*228*). Although levels of NE in striatal subcellular fractions are too low to permit an accurate estimation, hypothalamic fractions believed to contain synaptic vesicles yield twice as much NE as DA. All of these findings have been taken as an indication that DA has less affinity for storage granules than has NE.

Accordingly, the low proportion of particle-bound endogenous DA is held to reflect the ease with which the amine escapes from binding sites during tissue homogenization. Since particulate:supernatant ratios of *exogenous* DA in various regions of the brain parallel those of *exogenous* NE—and may even surpass those of NE formed from DA introduced exogenously (*228*)—factors other than "low-affinity" binding to storage vesicles may contribute to the differential subcellular distribution pattern of DA when it is compared with NE. Among them, the most salient may be that DA synthesis, unlike that of NE, takes place in the neuronal cytosol.

Evidence for the unique subcellular localization of DA also comes from histofluorescence studies with the Falck-Hillarp method. In peripheral noradrenergic fiber terminals, it has been shown that when the amine is localized predominantly to storage granules, the relationship between intraneuronal amine concentration and fluorescence intensity is linear over a relatively small range (as high as 40% of normal endogenous amounts as determined by chemical analysis [216]). Linearity over a much wider range of intraneuronal amine concentrations is observed when the presumptive granular storage sites are eliminated, i.e., after treatment with reserpine (292). This quenching of fluorescence can be ascribed to the very high intravesicular amine concentration (of the order of magnitude of 1×10^5 µg/g).

After inhibition of DA synthesis in neostriatal fiber terminals, quantitative microfluorimetric measurements of fluorescence intensity suggest a linear relationship between concentration and fluorescence intensity over the entire range of endogenous DA levels (216). Since the concentration of DA in striatal nerve terminals is calculated to be of the same order of magnitude as that of NE in peripheral noradrenergic fiber terminals (30), the low degree of quenching therefore appears to reflect the high axoplasmic extragranular levels of DA. This observation may also explain the normal appearance of striatal DA terminals as a diffuse fluorescent mass. At present, the functional significance of the unique subcellular distribution of DA in neostriatal fiber terminals is unknown. Other peculiarities of striatal DA metabolism make it likely that these terminals possess neurochemical properties different from those of central and peripheral NE fibers, as will be discussed in a subsequent section of this chapter.

Metabolism

The discovery that many centrally acting drugs, such as major tranquilizers and narcotic analgesics, exert powerful effects on DA metabolism of the brain generated a strong impetus for research into the role of this amine in brain function. Interest in brain DA metabolism received an additional boost when it was shown that nigrostriatal DA is severely depleted in brains of parkinsonian patients (173, 275), and that the administration of L-DOPA produces dramatic clinical improvement (79). During the past two decades, many investigators (see review by Hornykiewicz [270]) have attempted to explain the nature of this biochemical defect and to elucidate the mechanisms by which the aforementioned drugs alter basal ganglia function. This research has contributed much to our knowledge about both the biosynthetic pathway of DA formation and the physiological mechanisms participating in its control.

The general reaction sequence that Blaschko (*73, 75, 77, 79*) proposed as the pathway of catecholamine formation is now well established in terms of the various enzymes involved and their intracellular localization in the central and peripheral nervous systems (Figure 2). As shown with striatal slices in vitro, all of the enzymes required for the biosynthesis of DA are present in this tissue (*350*). Tyrosine hydroxylase (EC1.10.31), the initial catalyst of catecholamine synthesis, exhibits a discrete distribution pattern in the mammalian brain (*139, 357, 359, 534*). Unlike the adrenal enzyme, the brain enzyme (*357, 379, 384*) appears to be mainly particle bound (*45, 355*) and is highly localized to nerve-ending particles (synaptosomes) (*45, 188, 355, 378, 382*). In the bovine caudate it is reported to be associated with synaptic vesicles (*188, 355, 382*).

A smaller fraction of the enzyme exists in a soluble form (*318, 383*). The ratio of soluble:particulate enzyme is highly correlated with the ratio of perikarya:nerve terminals in a given brain area. Thus the DA terminal-rich striatum exhibits a higher proportion of the particulate enzyme (30 to 55% in the rat [*318*]; 80% in the bovine caudate nucleus [*383*]) than the midbrain where the soluble form constitutes about 90% of the total activity (*318*). The results of lesion experiments involving the nigrostriatal pathway favor a close association between nigrostriatal DA fibers and striatal tyrosine hydroxylase (*239, 373*).

Subcellular fractionation studies on lysed striatal synaptosomes make it probable that, in its particulate state, the enzyme is associated with synaptosomal membranes rather than with synaptic vesicles (*318*). With preparations of the striatal synaptosomal enzyme, the tyrosine K_m is approximately 0.05 mM and does not differ for both forms of the enzyme (*318*). Other investigators report K_m ten times lower (*356*). Apparent K_m values for rat striatal tissue slices vary between 0.028 mM (tritiated H_2O method) and 0.024 mM (radiolabeled-tyrosine-dopamine conversion) (*63*). It is noteworthy that the particulate form of the enzyme displays a two-fold higher V_{max} (*318*) and a lower K_m (0.15 mM) (*318, 383*) for the synthetic cofactor—6,7-dimethyl-5,6,7,8-tetrahydropterin—than does the soluble form with a K_m of 0.75 mM. Although tyrosine hydroxylation is commonly held to be the rate-limiting step in DA synthesis (*558*), the molecular mechanism underlying induction, short-term modulation, and feedback control of the enzyme are still being actively investigated. (See following section in this chapter.)

Pyridoxal phosphate-dependent L-DOPA decarboxylase (EC 4.1.1.26), the first catecholamine-synthesizing enzyme to be discovered (*266*), catalyzes the transformation of L-DOPA to DA. Although there is still some discussion regarding the subcellular topography of this soluble enzyme,

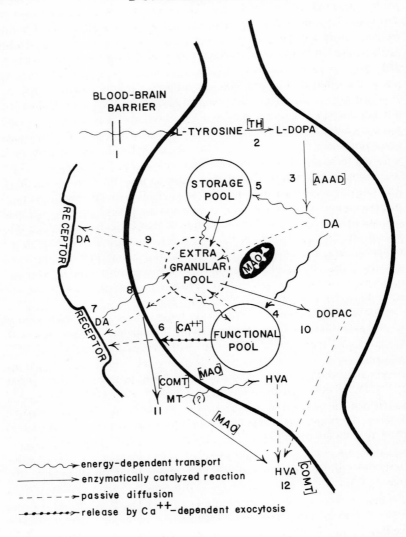

Figure 2. Central DA synapse summarizing fundamental aspects of metabolism and proposed organization of storage, uptake, and release processes. Numerals denote sequential steps in life cycle of DA and potential substrates of pharmacological manipulations (details in text). *Biosynthesis:* 1, L-tyrosine is transported across blood-brain barrier, i.e., vascular-glial AAAD and MAO, into DA perikarya or axon terminals via energy-dependent, stereoselective uptake mechanism for large, neutral L-amino acids, probably operating at level of neuronal cell membrane. 2, L-DOPA

data from axoplasmic transport studies favor a cytoplasmic localization (*157*). Given that the respective K_m values of L-DOPA and tyrosine are of the same order of magnitude, the much larger reaction velocity of the decarboxylase explains why conventional analytical methods fail to detect endogenous L-DOPA in brain tissue under physiological conditions (*55, 101, 102, 161*). Since the enzyme is also known to metabolize various other synthetic as well as naturally occurring aromatic L-amino acids, it is appropriately called aromatic L-amino acid decarboxylase (**AAAD**) (*344*). Unlike other catecholamine-synthesizing enzymes, AAAD occurs not only in adrenergically innervated tissue but also in brain areas containing serotonin perikarya or terminals where it catalyzes the decarboxylation of L-5HTP, the precursor of serotonin (*81, 139, 344, 483*). Accordingly, in human autoptic material its regional distribution pattern in the brain closely parallels that of endogenous catecholamines as well as serotonin (*342*). Most investigators regard this enzyme as a single protein with separate substrate affinity sites; however, this concept has been challenged (*474*).

In summary, available data on the intraneuronal localization of the two biosynthetic enzymes in DA nerve terminals favor the view that the final step of DA synthesis occurs within the neuronal cytoplasm. In catecholamine neurons other than dopaminergic ones, the amine is

---

formed by catalytic action of tyrosine hydroxylase (TH) in presence of reduced pteridine cofactor; this step subject to end-produce inhibition. 3, DA arises from decarboxylation of L-DOPA catalyzed by AAAD in presence of cofactor pyridoxal-5-phosphate. *Storage:* 4 and 5, ATP–Mg^{2+}-dependent, reserpine-sensitive uptake concentrates newly synthesized DA in storage organelles of functional and main storage compartments, former receiving bulk of amine. *Release:* 6, Exocytotic secretion of DA from functional pool triggered by nerve impulses and requires Ca^{2+}. Vesicles of main storage compartment largely inaccessible to nerve impulses. 7, Released into periaxonal space, DA diffuses to and interacts with postsynaptic receptors. *Reuptake:* 8, Neuronal Uptake$_1$ limits the duration of DA receptor interaction. This process satisfies criteria of active transport mechanisms, exhibits low stereochemical selectivity and resistance to reserpine. 9, "Reuptake"-DA is immediately sequestered from axoplasm, i.e., extragranular pool, by vesicular uptake mainly into long-term storage organelles. Release of DA from latter by indirect DA mimetics probably entails passage through extragranular pool and does not require nerve impulses or CA^{2+} for extrusion through axolemma to postsynaptic structures. *Catabolism:* During passage through extragranular pool portion of DA is converted by MAO to DOPAC, freely diffusible catabolite. By keeping axoplasmic concentrations of free DA low, in addition to vesicular uptake, MAO activity promotes effectiveness of Uptake$_1$. Released DA that has escaped Uptake$_1$ is 3-*O*-methylated by extraneuronal COMT to form methoxytyramine (MT). Bulk of MT is converted by extraneuronal MAO to HVA. Smaller portion could be actively (?) transported into axoplasm, deaminated to HVA and passively diffused back into periaxonal interstitial fluid. Not depicted are presynaptic DA receptors presumed to mediate feedback regulation of TH activity.

actively transported into vesicular organelles where it is converted to NE through the catalytic action of dopamine-β-hydroxylase. In DA neurons, however, the amine is actively accumulated in storage vesicles where it is available for release into the synaptic cleft upon depolarization of the nerve terminal membrane.

The catabolism of endogenous DA involves two basic steps in a manner analogous to that in peripheral adrenergic nerves or tissues (Figure 2). These reactions are catalyzed by monoamine oxidase, EC1.4.3.4. (MAO) and catechol-O-methyltransferase, EC 2.1.1.6 (COMT), and both the deaminated and O-methylated derivatives of DA are found in brain tissues. 3,4-Dihydroxyphenylacetic acid (DOPAC) was the first catabolite to be discovered, followed by homovanillic acid (HVA) (*33*) and methoxytyramine (MT) (*111*). The regional distribution of these compounds closely agrees with that of endogenous DA. In the human brain the distinct caudorostral gradient of HVA within the internal capsule (between SN and striatum) was adduced as evidence for the existence of the nigrostriatal DA pathway when this anatomical connection was still being debated (*270*). Enzymatic breakdown of endogenous DA occurs at two principal sites (*239, 441*), one intraneuronal and the other extraneuronal. Oxidative deamination leading to DOPAC is assumed to occur predominantly at the intraneuronal site, in keeping with the localization of MAO in the outer membrane of mitochondria and in microsomes. However, oxidative deamination is by no means restricted to intraneuronal sites, since the bulk of the enzyme is in fact found outside DA neurons; 3-O-methylation by COMT is believed to occur outside the DA nerve terminal.

Since steady-state concentrations of MT and DOPAC in the striatum and cerebrospinal fluid (CSF) of many species are low when compared with HVA, the latter is now regarded as the common end product of DA metabolism. Although the formation of DOPAC usually leads that of HVA, several observations militate against a sequential action of MAO and COMT on the DA molecule. These are: 1) after drug treatments that accelerate the formation of HVA from endogenous DA, inhibition of COMT causes only a small increase in DOPAC in mice (*441*); 2) probenecid, an agent known to block the active transport of organic acids across the blood-brain barrier and the CSF-choroid plexus barrier is much more effective in raising striatal levels of HVA than of DOPAC—the two acidic metabolites may therefore be formed and eliminated at different sites (*441*); and 3) after probenecid treatment, labeled DOPAC and HVA accumulate at rates that add up to the rate of formation of tritiated DA from tritiated tyrosine under steady-state conditions (*89*). These data are consistent with an intraneuronal DOPAC formation and although the acid appears to be present in caudate tissue in a freely diffusible form (*361*), only an insignificant fraction is converted to HVA. By contrast, the bulk of HVA represents DA metabolism outside the DA terminal, its major source being MT which, in turn, appears to be derived primarily from DA released from the terminals (*441*). A minor fraction of HVA may be formed intraneuronally.

Questions have been raised about the functional significance of these two catabolic pathways. Assumed to reflect a *wasting* of nonstorable surplus DA, DOPAC production is attributed by some workers to metabolic neuronal activity (*98*). In essence it is postulated that MAO is concerned with the presynaptic (intraneuronal) metabolism of DA but not with terminating its postsynaptic action. Conversely, while there is no evidence to implicate COMT in the limitation of the postsynaptic action of endogenous DA (*43, 281*), striatal HVA formation has often been used as an index of the functional activity of nigral DA neurons, since electrical stimulation of the SN increases the amount of HVA appearing in ventricular perfusates; however, it is important to note that this is a delayed response (*427*). (For additional discussion, see section on "Regulation of Dopamine Synthesis" in this chapter.)

Only recently have changes in striatal DA catabolism, observed in the presence of various drugs, been correlated with direct measurements of neuronal activity of nigral DA cells, and from such experiments there is an indication that striatal DOPAC levels are closely related to the functional state of nigral pars compacta neurons (*4, 448*). Electrical stimulation or destruction of the nigrostriatal pathway produces rapid increments or decrements of DOPAC, respectively. In addition, several drugs known to raise DOPAC levels accelerate the firing of nigral DA cells. These data suggest that a revision of the concept that presynaptic MAO merely subserves a metabolic disposal function is in order. While there is little doubt that reuptake of released DA into the nerve terminal constitutes the primary mechanism of "inactivation," the process itself appears to depend critically on MAO activity in the terminal, since it is governed by the gradient of free amine across the axolemma (*517*). Accordingly an increase in free axoplasmic DA resulting from MAO blockade would diminish the capacity of the terminal to recapture released DA. If DOPAC were partially derived from DA reentering the terminal, its rate of formation ought to reflect the rate of activity of the reuptake process, and hence the level of functional activity. This idea receives support from numerous in vivo and in vitro studies of striatal DA metabolism (*226*): Most of the DA formed from tyrosine in striatal slices in vitro appears to be retained inside the neuron in a compartment that is protected from MAO. In vivo, however, 50 to 60% of newly formed DA (from tyrosine injected locally into the striatum) is metabolized by MAO. This difference is most readily explained by the influence of nerve impulse activity. The pattern of DA catabolism in DA neurons previously outlined only covers its fundamental aspects, and one should allow for the existence of species differences, as well as for alternative pathways, e.g., the formation of neutral alcoholic metabolites (*510*) and MAO bypassing transamination reactions (*413, 470*). Furthermore, there is renewed interest in condensation products of DA formed with its deaminated catabolite, 3,4-dihydroxyphenylacetaldehyde (*267*). The tetrahydroisoquinoline and papaveroline products formed from this and similar condensation reactions are structurally related to morphine alkaloids and, according to some workers, may play a role both in alcohol addiction (*251*) and in the long-term effects of L-DOPA in the treatment of parkinsonism (*450, 482*).

Regulation of Dopamine Synthesis

As indicated earlier, tyrosine hydroxylation is generally held to limit the rate of catecholamine formation under physiological conditions and thus to be the step at which biological regulation is likely to be most efficient. Indeed there is substantial experimental evidence to support the concept that accelerated brain catecholamine synthesis, induced by a variety of physiological stimuli and pharmacological manipulations, is invariably associated with enhanced tyrosine hydroxylation. Several possible mechanisms by which such regulation might be effected have been identified in peripheral and central catecholaminergic tissues; however, none of them has yet been characterized at a molecular level (*370, 558*). Since there is no direct assay for the quantitative measurements of tyrosine hydroxylase protein, changes in activity of this enzyme have usually been verified by the demonstration of alterations in the rate of catecholamine formation from radiolabeled L-tyrosine or, more recently, by the measurement of L-DOPA accumulation in the presence of AAAD inhibitors (*101, 102, 117*). Results obtained with these methods clearly indicate that, on the one hand, short-term control of amine synthesis involves regulation of the actual enzyme activity, but not of its intraneuronal concentration. On the other hand, long-term control is likely to involve the synthesis of the enzyme itself. For striatal tyrosine hydroxylase, however, direct evidence is derived almost entirely from studies that document the correlation between ontogenetic growth patterns of nigrostriatal DA neurons and striatal tyrosine hydroxylase activity (see review by Coyle [*143*]).

Feedback Regulation At least three mechanisms are postulated to explain short-term regulation of DA synthesis at the tyrosine hydroxylation step in central DA neurons, and the most extensively studied are of a feedback type. One operates inside the nerve terminal (intraneuronal) and the others through an interneuronal link, either at the level of the synapse (intrasynaptic) or via a loop back to the nigral cell (transneuronal). The intraneuronal mechanism fulfills the criteria of end-product inhibition (*107, 288, 358, 379, 389, 486*). In other words, the rate of tyrosine hydroxylation declines as the concentration of free axoplasmic DA rises within the nerve terminal. Kinetic measurements from in vitro studies on the adrenal enzyme (*318, 381, 520*) suggest that this inhibition is noncompetitive with tyrosine but is competitive with both the artificial and the putative natural cofactor of the enzyme in its reduced form, and probably involves competition with the oxidized pteridine for a binding site on a cofactor-regenerating enzyme, the hypothetical pteridine reductase (*279, 377*).

Another feedback mechanism is believed to be triggered by various manipulations all of which ultimately lead to the exclusion of DA from receptors at pre- or postsynaptic sites. For example, cessation of impulse flow in nigrostriatal fiber terminals either after axotomy (hemisection of the forebrain at the level of the caudal hypothalamus) or by agents that inhibit nigral cell firing produces a transient increase in the rate of tyrosine hydroxylation (*297, 488*) and a more prolonged rise in intraneuronal DA levels (*14, 32, 194, 499*), both phenomena being unique to DA neurons. The mediatory role of synaptic DA receptor sites has been deduced from the observations that apomorphine (a presumptive DA receptor stimulant) and amphetamine (an indirect DA mimetic agent) counteract the post-axotomy activation of tyrosine hydroxylase (*297, 448*), and that the former attenuates the rise in intraneuronal DA levels, whereas the latter abolishes it (*32*). Both effects of apomorphine can, in turn, be antagonized by haloperidol (*32, 297*), a DA receptor-blocking agent that, when given alone, does not change the postaxotomy responses. At higher doses, however, the apomorphine effect is apparently obscured by a direct inhibitory action on tyrosine hydroxylase itself (*230*). More precise information about the threshold concentration for this effect in vivo is desirable.

It should also be pointed out that the apomorphine-haloperidol antagonism is demonstrable only during the first half-hour following axotomy in the tyrosine hydroxylation experiments. At a later interval, haloperidol apparently diminishes enzyme activity on the lesioned side. End-product inhibition is questionable at this point or even at longer postaxotomy intervals, since the DA synthesis rates determined during the increased steady-state levels do not differ from synthesis rates observed in the striatum under control conditions (*448*). Nonetheless, end-product inhibition remains operative in transected neurons, i.e., in the absence of nerve impulses, since the postaxotomy rise is completely abolished if free cytoplasmic DA concentrations are increased by treatment with reserpine or MAO inhibitors, or both (*32, 448*).

The aforementioned observations argue for the existence of a synaptic control mechanism that operates in the absence of neuronal activity, and one that is susceptible to drugs presumed to act upon DA receptors. By no means do they preclude that the effects of such agents on DA synthesis and turnover in the intact brain involve a transneuronal impulse–flow-dependent feedback mechanism. For example, only if the nigrostriatal pathway is intact or nigral cells are active (*445, 446*) do neuroleptic DA antagonists accelerate intrastriatal DA synthesis and utilization (*21, 398*). (See section on "Neuroleptics" under "Dopamine Receptor Antagonists" in this chapter.)

Recent observations substantiate the transneuronal feedback concept (*87, 89*). Agents that activate striatal dopaminergic transmission decelerate the firing of nigral neurons, whereas blocking agents have the opposite effect on these cells. In interpreting these observations, one must be cautious not to overlook the possibility that these drugs may be exerting either direct actions on nigral cells or indirect effects secondary to changes in anesthetic state, since activity levels of these neurons vary inversely with the level of anesthesia. Furthermore, many of the agents shown to influence nigral cells are known either to potentiate or to counteract the effects of general anesthetics.

In addition to synaptic feedback control, conformation-dependent alterations in kinetic properties of tyrosine hydroxylase attendant on binding of the enzyme to synaptic membranes may constitute yet another means of short-term regulation of DA synthesis (*318*). Thus the membrane-bound form of the enzyme differs from the soluble form in displaying a higher reaction velocity and greater affinity for the pteridine cofactor. Moreover catecholamine feedback inhibition, in the presence of low cofactor concentrations, may operate with greater sensitivity on the membrane-bound enzyme when compared with its soluble form. Since the soluble form—but not the particulate form—exhibits cooperativity between cofactor binding and DA inhibition, transition from the former state to the latter may entail a shift, already at low cofactor concentrations, from virtual inactivity to relatively high activity. In addition Ca^{2+}

appears to promote binding of the enzyme to synaptic membranes. This raises the possibility that acute fluctuations in axoplasmic Ca^{2+} levels, such as may result from abrupt changes in neuronal activity, may play a significant role in modulating enzymatic activity.

At this point questions might be posed regarding the physiological conditions under which these different feedback mechanisms operate, as well as their functional interrelationships. It is fair to assume that *both* intraneuronal and receptor-mediated regulation of DA synthesis and turnover enable the neuron terminal to adjust its DA output to variable physiological demands. Possibly these two feedback mechanisms differ from one another in their range of operation. Thus the coincidental activation of tyrosine hydroxylase and the accumulation of intraneuronal DA that occurs after cessation of nigrostriatal impulse flow may not be so paradoxical as they appear at first glance. Apparently the accumulating DA is effectively removed from axoplasm by storage granules and thus prevented from exerting its inhibitory action on tyrosine hydroxylase. Only after these storage sites become saturated, can end-product inhibition make itself felt, since the storage capacity of the terminal is apparently unsaturated under steady-state conditions. End-product feedback control of DA synthesis may be unidirectional, i.e., more apt to slow down than speed up synthesis. Conversely, receptor-mediated feedback may provide a bidirectional control of synthesis rate since it is probably independent of the saturation of storage sites. At present our ignorance about the molecular mechanisms underlying receptor-mediated synaptic feedback does not permit more than speculation about the manner by which information of receptor activity is encoded and conveyed to the presynaptic terminal.

Turnover Attempts to measure DA turnover specifically, and its synthesis in the brain, have been reinforced by the belief that the rate of transmitter synthesis within a given neuron is a sensitive indicator of its functional state *(135)*. Methods devised for this purpose *(136)* aim at assessing: 1) conversion rates of radiolabeled precursors into specific amines; 2) utilization rates of endogenous (labeled) amines in the presence of inhibitors of synthesis or catabolism; and 3) disappearance rates of exogenous tracer amines following their introduction into endogenous pools. None of these techniques, however, affords accurate quantitation of the actual rate of DA synthesis in the intact brain.

When striatal slices are incubated with L-tyrosine, large amounts of DA (but not of NE) are synthesized *(226, 228, 350)*. With radiolabeled tyrosine, the rate of DA synthesis in slices of rat striatum in vitro is estimated at 18.2 nmol/g/hr *(63)*. Remarkably this value is virtually identical to that calculated from earlier in vivo measurements of the rate of decay of DA levels in the caudate nucleus of the rabbit following inhibition of synthesis by α-methyl paratyrosine (α-MpT) (*136*). In the face of the vastly different functional states of DA neuron terminals under these two conditions, the absence of a difference between in vivo and in vitro results appears to preclude any correlation between synthesizing and neuronal activity; alternatively it casts doubt on the meaningfulness of such experiments.

Attempts to explain this discrepancy prompted reinvestigation of the disappearance rate of endogenous striatal DA levels following inhibition of its synthesis. Data obtained in the rat were interpreted as being indicative of a biexponential rather than monoexponential rate of decay reported in earlier studies (*136*). These conflicting interpretations have sparked a lively debate about the functional organization of mechanisms regulating steady-state DA levels within nigrostriatal neuron terminals. In supporting either a monoexponential or a biexponential decay rate, one conceives of DA synthesis, storage, and release as occurring either in a single, open compartment (*135, 136, 485*) or in two functionally separate compartments (*62, 226*). According to estimations based on the one-compartment model (*136, 485*), the respective turnover rates in the rabbit or guinea pig striatum, which under steady-state conditions can be assumed to equal the rate of synthesis, amount to 18.6 and 32 nmol/g/hr, with utilization rate constants of 0.32 and 0.44/hr.

According to the two-compartment hypothesis (*288*), the "functional" compartment comprises about 23% of the endogenous striatal DA in the rat, and the "main storage" compartment 77%. The respective fractional rate constants are 4.6/hr and 0.34/hr. With an average value of 65 nmol/g representing the endogenous DA in the rat striatum, the synthesis rate will be 70 nmol/g/hr in the functional compartment and 17 nmol/g/hr in the main storage compartment, for a total of 87 nmol/g/hr. This rate is five-fold greater than that obtained in vitro in the same laboratory, a difference that may well reflect the synthetic activity in the functional compartment. It is important to note that in such calculations one does not question the assumptions underlying the interpretation of disappearance rates of endogenous DA following inhibition of synthesis, assumptions that provide the rationale for equating disappearance rates with those of utilization and synthesis. Specifically 1) the synthesis inhibitor α-MpT is assumed *not* to alter the functional state (firing activity) of DA neurons; 2) the depletion is assumed *not* to trigger a compensatory mechanism; and 3) striatal DA terminals are presumed to represent a functional unit. More recent observations from both histofluorescence work (*41*) and studies on the mechanisms involved in the regulation of DA synthesis and release (*32,*

448, 449) seem to challenge the validity of these assumptions. Notwithstanding these reservations, the "two-compartment" model appears to accommodate a number of observations on the effects of various psychotropic agents on DA synthesis and release. (For further discussion see section on "Neuropharmacology" in this chapter.) Clearly the present controversy draws attention to the desirability of methods that would afford accurate in vivo determination of actual DA synthesis rates in the intact brain.

Release and Uptake

To qualify as a transmitter agent, a substance occurring intraneuronally must exhibit releasability (collectibility [*553*]), i.e., be capable of being released into the extraneuronal space in response to a biological stimulus, such as an action potential. A large number of pharmacological and electrophysiological manipulations have been effective in vivo and in vitro in causing the release of DA from brain tissues known to contain DA neuron terminals. For the purpose of this discussion, attempts to pinpoint the in situ source from which the released material originates are of primary interest. Considering that it is an exceedingly difficult—if not technically impossible—task to demonstrate that a substance is released from a specific type of nerve ending in the intact CNS, presently available techniques for collecting DA or its metabolites from interstitial or cerebrospinal fluid or from superfusates of cerebroventricular border structures seem inadequate to achieve this goal. Moreover, interpretation of experimental data obtained with such methods must acknowledge that the latter do not measure the release process per se but rather the resultant of at least four interacting variables, namely synthesis, catabolism, release, and reuptake. Nevertheless, experimental data from a number of studies make it likely that the release of DA in response to various pharmacological or neural stimuli originates from axon terminals of DA neurons.

For methodological reasons, in vivo studies of release have focused on the most accessible of DA-containing brain structures, namely, the corpus striatum. Early investigators employed Gaddum's push-pull cannula and while they were successful in recovering catechol compounds from perfusates of the feline caudate nucleus, both under resting conditions and following electrical stimulation of certain areas of the brain, they failed to establish the chemical identity of the extracted materials (*366*). Hence estimations of DA output in these experiments were too high since they probably included DA metabolites. In subsequent work the basal rate of release of striatal DA into lateroventricular (*427*) or push-pull intrastriatal (*361*) perfusates in the anesthetized cat was estimated to be in the range of

20 to 66 pg/min. These values approach the lower limits of sensitivity of the fluorimetric assay procedure (3 to 5 ng of DA) and thus may lack accuracy. More recently, radioisotope labeling techniques that permit the detection of DA in amounts of 5 to 10 pg confirmed these low resting output rates (437). Under similar experimental conditions, output rates of DOPAC and HVA amounted to 0.42 ng/min (361) and 2 to 8 ng/min (427), respectively. In view of the high endogenous DA concentration in striatal tissue, a possible interpretation of these findings is that DA released from nerve terminals is catabolized before it can reach the collecting cannula. Since pretreatment with inhibitors of MAO or COMT generally fails to enhance the basal efflux rate of DA, an even more plausible explanation is that a major fraction of released DA is recaptured through the highly active reuptake process which is known to operate in catecholamine fiber terminals.

Ever since the nigral origin of DA fiber terminals in the striatum was recognized, attempts have been made to induce the release of striatal DA by electrical stimulation of this pathway. The initial failure of early investigators to obtain consistent data may have been attributable to: 1) insufficient knowledge of the topography of nigrostriatal projections; 2) differences in experimental methods; and 3) the low sensitivity of fluorimetric assay procedures. As already mentioned, the third difficulty has been overcome by the development of radioisotope-labeling methods which, combined with various perfusion procedures, are now the principal techniques employed. Among them are 1) continuous administration of radiolabeled DA precursors, L-tyrosine, or L-DOPA, through intrastriatal or intraventricular push-pull cannulae (437) or through superfusion cups placed on the exposed ventricular surface of the caudate (61); and 2) perfusion of the forebrain ventricle in one hemisphere following a single intraventricular injection of radiolabeled L-tyrosine or DA (540–543).

During continuous push-pull perfusion with artificial CSF containing either ^{14}C-labeled L-tyrosine or L-DOPA, the efflux of ^{14}C-DA increases in the: 1) caudate nucleus when the nigrostriatal fiber tract is electrically stimulated at the level of the optic chiasm or at the level of the rostral (but not the caudal) SN (542, 543), and 2) putamen when stimulation is applied to the posterior portion of the SN (437). Dopamine efflux from the caudate nucleus probably cannot be altered by stimulation of other regions synaptically linked to the corpus striatum, such as the entopeduncular and subthalamic nuclei or the centromedian. Although such results are obtained with prolonged and intense stimulation procedures, they are in perfect agreement with available neuroanatomical and histochemical data concerning the topographical distribution of nigral dopamine fiber projections (*51, 53, 114*), and they also confirm results from earlier release studies. The rostromedial nigral pars compacta cells project to the medial aspect of the head of the caudate nucleus, whereas cells in its lateral and caudal portion supply the putamen. Moreover, fiber bundles of the nigrostriatal tract are most densely packed at the level at which stimulation most effectively releases DA (*121, 543*), and a small lesion placed in this area is known to cause a selective loss of striatal DA but not of 5-HT (*412, 541, 543*).

Whereas electrical stimulation of the nigrostriatal tract evokes the release of DA during continuous labeling of intracaudate nerve terminals with either radiolabeled L-tyrosine or radiolabeled L-DOPA, the possibility of artifacts must be kept in mind when the latter precursor substance is used (*437*). Although lesion studies make it likely that striatal tyrosine hydroxylase and AAAD are localized to DA fiber terminals (*240, 373*), the latter enzyme also occurs in 5-HT fibers (*28, 81, 262, 421*) and in the walls of brain capillaries (*56*). Dopamine may therefore be formed within or taken up into nondopaminergic fibers and subsequently released upon electrical stimulation. Indirect evidence favoring this possibility comes from the observation that nigrostriatal tract stimulation does not affect the striatal release of acidic DA catabolites when L-tyrosine is employed as a labeling agent; however, output is either increased or decreased when L-DOPA is used (*437*).

From the foregoing it is evident that the method of labeling striatal DA fiber terminals with exogenous DA entails a considerable risk of artifacts, owing to the apparent nonspecificity of membranal amine uptake systems in monoamine fibers (*312*). Thus, when administered into the lateral ventricle, DA is likely to enter monoamine fiber terminals in various periventricular structures, e.g., intracaudate DA and 5-HT axons; septal NE and 5-HT axons; as well as intrathalamic NE and 5-HT axon terminals. Nonetheless, since the bulk of radiolabeled NE, intraventricularly administered, accumulates in the caudate nucleus and hypothalamus (*115*), there is a good chance that this direct labeling technique, if combined with a careful selection of stimulation sites, may provide useful information.

It has thus been reported that short-term electrical stimulation of the pars compacta of the SN, the nigrostriatal tract, or the caudate nucleus, subsequent to intraventricular injection (and washout) of tritiated DA, causes a frequency- and intensity-dependent release of ^{3}H-DA (*542, 543*). It is interesting, however, that the resting release of ^{3}H-DA from brains with chronic nigrostriatal tract lesions is only slightly lower than that in control preparations, indicating that the background efflux of radioactive amine into the lateral ventricle does not originate primarily from striatal

DA axon terminals (*541*). In more recent work from the same laboratory (*121*), periventricular amine stores were labeled successively with ^{14}C-DA synthesized intraneuronally from ^{14}C-tyrosine and with exogenously administered ^{3}H-DA. When continuous electrical stimulation was applied to the lateral hypothalamic area, at the level of the fornix column, the output of endogenous- and exogenous-labeled amines rose by a factor of 2 and 3, respectively, and returned to control levels upon cessation of stimulation. According to these investigators (*121*), both DA synthesized endogenously in fiber terminals of nigrostriatal DA neurons and exogenous DA previously taken up into them are released on electrical activation of the corresponding preterminal axons. Since ^{14}C-tyrosine was administered intraventricularly, there is the distinct possibility of a simultaneous labeling of NE fiber terminals in other periventricular structures; therefore the contribution of these fibers to the observed evoked release of ^{14}C-catecholamines remains unknown.

Many of the pitfalls that await the unwary investigator who employs ventricular and push-pull perfusion procedures are perhaps avoided by use of the topical superfusion technique described by Glowinski and his colleagues (*61, 226*). This approach affords two major advantages: 1) DA caudate terminals can be selectively labeled in vivo; and 2) a larger number of fiber terminals can be sampled than is possible with local push-pull perfusion. Thus, when the caudate surface is continuously superfused with radiolabeled tyrosine, there is a steady release of labeled DA into the superfusate. This release is evidently related to impulse activity in DA fiber terminals, since it increases in response to local electrical stimulation or application of potassium, whereas local application of tetrodotoxin, an agent that prevents the generation of action potentials, causes a marked drop in amine output. Although electrical stimulation of the SN significantly raises DA output into the superfusate, this effect does not become manifest during the stimulation period itself but only after a certain latency. The reason for this delay is uncertain. Since stimulation of the nigrostriatal tract, when compared with nigral stimulation, is reported to be ten times more effective in releasing striatal DA (*543*), the possibility exists that stimulation in the ventromedial tegmentum may excite inhibitory inputs to nigral cell bodies or their striatal axon terminals, thus holding stimulation-induced release in abeyance.

In summary, the aforementioned data leave little doubt that the rates at which DA or its catabolites appear in perfusates of the striatum or the lateral ventricle reflect the functional state of DA substantia nigra neurons projecting to the forebrain. As all studies referred to have been performed on anesthetized animals, it appears that nigral DA neurons remain active even during surgical anesthesia, a conclusion corroborated in recent electrophysiological studies (*87, 89, 448*). In this context it is noteworthy that stimulation of the mesencephalic reticular formation adjacent to the SN produces the characteristic EEG arousal without influencing striatal output of HVA (*427*).

The sequence of subcellular events occurring between depolarization of the nerve terminal membrane and the extrusion of DA from intraneuronal vesicles into the synaptic gap is not yet completely understood. As with certain hormones and other transmitter candidates, Ca^{2+} seems to be required for the release of DA from nigrostriatal nerve terminals. For example, electrical field stimulation and potassium depolarization of striatal slices initiate a release of both newly synthesized and previously taken-up DA which is highly dependent, on the one hand, on Ca^{2+} in the incubating medium (*47, 94, 191*). On the other hand, the addition of Ca^{2+} to in vitro suspensions of striatal DA-storing vesicles enhances the release of DA in a concentration-related manner (*415*). This release is independent of temperature and insensitive to acetylcholine. Thus in DA neuron terminals, two sequential steps have been identified that form part of the postulated reaction scheme underlying stimulus-secretion coupling in other biological systems.

The rate at which DA passes into striatal interstitial or ventricular fluid as determined experimentally does not provide a direct estimate of the rate of the release process per se, but rather an estimate of the rate of overflow from the tissue. This well-recognized distinction takes into account a common property of catecholamine-secreting nerve terminals, viz., their capacity to retrieve released amines through a specific and highly active transport system. Much of our present knowledge about the existence and mode of operation of the neuronal DA uptake mechanism (the Uptake$_1$ of Iversen [*280–282*]) derives from in vitro studies on slices of striatal tissue and striatal synaptosomes, the latter consisting of pinched-off nerve endings isolated from striatal homogenates.

The striatal DA reuptake mechanism shares many features with that of NE in peripheral and CNS tissues. It is dependent on energy-yielding processes, since it occurs neither in the absence of oxygen and glucose nor at low temperatures nor in the presence of inhibitors of oxidative phosphorylation, such as dinitrophenol and cyanide (*243*). Its sensitivity to changes in extracellular concentrations of sodium and potassium closely resembles that of Na^+, K^+-activated ATPase of cell membranes and, like the latter, it is susceptible to the inhibitory actions of cardiac glycosides (*469*). Uptake of DA displays higher affinity for its natural substrate ($K = 3$ to 4×10^{-7} M [*144, 479*]) than for NE. Kinetic studies have established, on the basis of identity of K_i values for mutual uptake inhibition, that both amines use the same transport system *(144)*. When compared with NE neurons, uptake in DA neurons reveals remarkable differences with respect to substrate specificity and sensitivity to the action of inhibitory agents.

Thus, unlike cortical synaptosomal preparations containing NE fiber terminals, preparations from dopaminergically innervated tissues, such as the striatum and retina, do not exhibit stereoselectivity toward enantiomers of NE *(144, 254)*. In an analogous manner, while optical isomers of both amphetamine and ephedrine may differ by 10- to 100-fold, respectively, in their potency as inhibitors of NE uptake into NE neuron terminals, no—or much smaller—differences are found with DA (or NE) uptake into striatal synaptosomes *(255)*. Finally, desipramine, one of the most potent inhibitors of NE Uptake$_1$, exhibits weak activity on intrastriatal DA neurons *(100, 218, 243, 480)*, its inhibitory strength on the latter being approximately 1/1000 of that on hypothalamic neurons *(268)*, confirming earlier studies of DA uptake into striatal tissue slices.

Teleologically, the reuptake mechanism (or membrane amine pump) may serve the economic maintenance of neuron amine stores. For synaptic transmission there is now general consensus that it functions to terminate the action of released transmitter; in other words, Uptake$_1$ critically determines the duration of exposure of the postsynaptic membrane to effective transmitter concentrations *(43, 281, 282)*. Since precise estimates of the amount of brain DA that is thus recaptured do not appear to have been made, it remains to be seen how they might compare to values reported for peripheral sympathetically innervated structures, e.g., the rabbit *(276)* or rat *(326)* vas deferens, rabbit portal vein *(276)*, and the cat nictitating membrane *(241, 326)*. According to calculations from in vivo perfusion studies, the basal striatal output rate of DA is on the order of magnitude of 0.3 pmol/min, less than 0.2% of endogenous DA being present in a freely diffusible form *(361)*. Conversely both DOPAC, a completely diffusible catabolite, and HVA are continuously secreted in amounts 10- and 100-fold larger, respectively, their rates of formation being dependent on firing activity of nigral neurons. These differences in output rates are readily interpreted as reflecting the activity of the reuptake process which effectively prevents escape of the released nonmetabolized amine into the perfusate, notwithstanding the continuous impulse flow in these neurons (Figure 2).

As already indicated, amphetamine compounds effectively inhibit the uptake of catecholamines into synaptosomes isolated from a number of adrenergically innervated tissues. Such findings provide an important clue for the interpretation of the effects of these substances on striatal DA release. Push-pull perfusion of the feline caudate nucleus with (+)-amphetamine (0.5 to 1 mg/ml) causes a sustained and manifold increase of DA output (not of DOPAC) *(361)*, a sharp, albeit transient, rise in ^{14}C-DA output during continuous perfusion with ^{14}C-tyrosine or ^{14}C-L-DOPA (not of acidic ^{14}C-DA catabolites) *(437)*. In caudate superfusion experiments, (+)- and (−)-amphetamine administered either topically or systemically enhance the output of DA either newly synthesized or previously taken up *(61)*. Analogous dose-related effects with (+)-amphetamine are obtained in ventricular perfusion experiments that utilize direct labeling of the striatum with ^{3}H-DA *(541)*. These effects, however, are greatly diminished in brains in which acute or chronic lesions of the nigrostriatal bundle have been made; moreover, a supraadditive effect on release evoked by low-frequency stimulation of the latter is observed in intact brains. This dependency on impulse traffic in nigrostriatal neurons strongly supports the hypothesis that the DA releasing effect of (+)-amphetamine is attributable, at least in part, to blockade of reuptake.

By contrast, the output-enhancing effect of tyramine is unaltered by acute nigrostriatal tract lesions and may therefore represent a pharmacological release phenomenon, indistinguishable from the Ca^{2+}-independent effect of this drug on peripheral noradrenergic tissues *(340, 515)*. However, none of the preceding observations precludes a direct facilitatory action of drugs, such as the amphetamines, on the release process per se. At present, little information is available to indicate how this process can be modulated in brain monoamine neurons. Several reports favor the existence of α-adrenergic receptors on peripheral presynaptic terminals with which NE might interact to inhibit its own release *(174, 192, 309, 488)*; and as suggested by recent observations on striatal slices, this autoinhibition hypothesis may also hold for central DA neurons *(191)*.

In vitro studies on striatal slices from animals pretreated (in vivo) with the purported DA receptor blocker, thioproperazine, demonstrate increased quantities of newly synthesized DA in the incubating medium *(119)*; however, it is doubtful if this can be attributed to an enhanced release since the tissue:medium ratios for DA in control and experimental animals did not differ. Additional studies are needed to permit a differentiation between enhanced overflow owing to increased synthesis, on the one hand, and release, on the other. Furthermore, release of DA from tissue slices deprived of their continuous impulse input may not be comparable to that occurring in vivo. Thus potassium depolarization is shown to result in a preferential release of "reuptake DA" and not of the newly synthesized amine *(554)*, in contrast to the situation assumed to exist in vivo *(62, 226)*. Results from other in vivo experiments with presumptive DA blockers are no less controversial. Thioproperazine administered systematically to cats, e.g., induces small but persistent rises in ^{3}H-DA output as revealed in superfusion studies (226); however, neither systemic nor topical administration of haloperidol has any effect on the output of previously stored DA into cerebral ventricular perfusates (540). Furthermore, other observations in push-pull perfusion experiments do not entirely support a facilitation of DA release following pretreatment with chlorpromazine (360).

In addition to autoinhibition, which may represent a special case of presynaptic inhibition, other receptor-mediated presynaptic mechanisms may operate in the control of DA release and uptake into nigrostriatal fiber terminals. Certain aspects pertinent to the topic of presynaptic inhibition have been discussed in Ch. 1. As first shown by Muscholl and his colleagues, there is strong evidence for muscarinic and nicotinic receptors on cardiac NE fibers, which mediate inhibition and facilitation of amine release, respectively (339). As yet we do not know whether or not an analogous cholinergic mechanism operates on nigrostriatal fiber terminals, although a wealth of pharmacological and pathophysiological data draw attention to the close functional interrelationship between cholinergic and dopaminergic transmission systems at the level of the corpus striatum.

When added to striatal tissue slices, acetylcholine increases DA output (62); however, this effect appears to be negligible in caudate superfusion experiments in vivo (226). Besides, anticholinergic agents such as benztropine (96) and atropine (360) increase striatal DA output in vivo, an effect that most probably results from blockade of DA reuptake, because some of these compounds are potent inhibitors of synaptosomal catecholamine uptake in vitro (268). However, since there is no consistent relationship between anticholinergic potency and the ability to block in vitro (268) or in vivo (214) uptake, one must be cautious about attributing these effects to a blockade of presynaptic receptor-mediated control mechanism. Despite this note of caution, the hypothesis still remains attractive since there is substantial, although indirect, evidence in its favor, such as the accelerating effects of centrally-acting cholinomimetic agents on striatal DA catabolism (129, 333). (For further discussion see section on "Acetylcholine-Dopamine Balance" in this chapter.)

Serotonin, another putative transmitter at striatal synapses, also enhances DA output from striatal slices and the superfused caudate nucleus (62, 226). The underlying mechanism for this action appears to be one of heteroneuronal uptake and displacement of autochthonous transmitter (469, 508), reflecting the nonspecificity of axolemmal and vesicular amine transport and storage mechanisms.

At present, information concerning the influence of striatal amino acid transmission systems on nigral DA neurons is relatively sparse. On the basis of electrophysiological (*429*) and neurochemical (*223, 308, 403*) findings, the proposal has been made that the striatonigral projection utilizes GABA as its transmitter. By this pathway, striatal neurons could exert inhibitory influences on pars reticulata cells and, to a certain extent, a direct inhibition on pars compacta cells (*237, 386, 438*). (For additional references, see [*237*].)

Until more data on the synaptic organization of this nigropetal system become available, particularly as regards the majority of pars compacta DA cells, one can only speculate on its relationship to the postulated transneuronal feedback mechanism controlling striatal DA turnover and the maintenance of a functional balance between cholinergic and dopaminergic striatal receptor activity. In this context, several other findings deserve mention; γ-hydroxybutyric acid (GBH) and its precursor, γ-butyrolactone (GBL) are capable of depressing the firing of nigral cells (*448*), and of producing the entire spectrum of neurochemical changes associated with transection of nigrostriatal DA fibers (see "Feedback Regulation" in this section). In addition to elevating striatal DA levels (*222, 223, 447*) and activating tyrosine hydroxylation in vivo (*448, 545*), these compounds exert a dual effect on striatal DA catabolites, DOPAC levels being slightly lowered acutely (*448, 545*) but markedly increased together with those of HVA after longer intervals (*278, 545*).

All of these effects appear to reflect cessation of impulse traffic in nigrostriatal neurons; furthermore, they are consistent with receptor feedback control of DA synthesis, as already discussed earlier in this chapter. It appears, however, that GBH can affect DA release at the level of nigrostriatal fiber terminal itself as indicated by its blocking effect on potassium-induced (Ca^{2+}-dependent) release of newly synthesized but not previously stored DA from striatal tissue slices (*94*). HA966, which resembles these agents in its effects on striatal DA metabolism (*53*), may act by a similar mechanism (see also Ch. 3, "Excitatory Amino Acids"). Furthermore, (+)-tubocurarine, unrecognized as a potent GABA antagonist until recently (*152, 260*), not only enhances in vivo efflux of HVA (*426*) (and probably also of DA from the caudate nucleus), but also facilitates caudatal release of newly synthesized DA in response to local electrical stimulation (*360*).

NEUROPHARMACOLOGY

Dopamine Receptor Agonists

The pharmacological work that led to the recognition of DA as a transmitter in the brain has greatly stimulated interest in drugs that influence basal ganglia function. Of the large number of such compounds, few exert a selective action on dopaminergic transmission. Agents that directly activate the DA receptor are of particular concern for pharmacological investigations of the role of brain DA, since this amine penetrates only poorly from the blood into brain tissue (56). At the outset it was necessary to rely on L-DOPA, the precursor of DA, in attempts to activate DA-sensitive neural substrates. The L-amino acid readily crosses the blood-brain barrier and undergoes rapid decarboxylation in the cerebral parenchyma (56, 58, 59, 109, 153, 347, 422).

Since not only L-DOPA, but also metabolites (3-O-methyl-L-DOPA and HVA) become detectable in brain regions normally devoid of these substances (161), one may conclude that initially there is a widespread formation of DA throughout the brain, owing to the ubiquitous occurrence of AAAD in other monoamine neurons and vasculature; however, the highest amounts are formed and retained in areas known to be rich in DA fiber terminals, i.e., the striatum. Enhanced accumulation can be achieved by: 1) the administration of AAAD inhibitors (Ro 4-4602 and MK486) in doses that preferentially inhibit the extracerebral and cerebral vascular enzyme (50, 425) or 2) monoamine oxidase (MAO) inhibitors (99, 186, 417). To the extent that these procedures enhance the uptake and decarboxylation of L-DOPA by other AAAD-containing neurons, the probability is increased that DA thus formed displaces the autochthonous amines (NE and 5-HT) and acts as a false transmitter (49, 95, 184, 394).

The heuristic potential of L-DOPA as a dopaminomimetic agent is also compromised by the fact that with dosages commonly employed, a general flooding of postsynaptic DA-sensitive substrates occurs by a process of synthesis and release which does not require the functional integrity of the presynaptic neuron (102). While such features underline the shortcomings of L-DOPA as a pharmacological tool, they do have quite important implications for its therapeutic potential in the treatment of cerebral DA deficiency syndromes, such as Parkinson's disease (229, 341, 394).

So far the only two substances to have gained acceptance as DA receptor agonists are apomorphine and piribedil (ET 495). Long established as an emetic (220), apomorphine attracted the attention of pharmacologists more than 100 years ago on account of its unique behavioral effects (8, 246); however, it was not recognized as a DA mimetic agent until the 1960s (34, 176, 178, 180). In addition to catalyzing psychopharmacological research in stereotyped and compulsive animal behavior, this substance was soon to become an indispensable tool for the dissection of DA receptor mechanisms. Piribedil, first described as a cardiovascular agent (329), owes its identification as a DA mimetic agent to systematized screening (128).

In assessing the DA-like properties of these two substances, it should be remembered that few direct criteria of agonist intrinsic activity are available most of which relate to extrapyramidal motor and behavioral effects of DA (and L-DOPA), whereas the larger number of indirect indices derives from current basic concepts of monoamine neurochemistry. As demonstrated by a host of pharmacological paradigms to be discussed in detail in a subsequent section, apomorphine duplicates virtually all of those CNS effects of L-DOPA that are believed to be mediated by DA. At this point only the most salient features of this pharmacodynamic parallelism will be emphasized.

Like L-DOPA (*53, 68, 142*), apomorphine displays anti-Parkinson activity (*140, 450, 539*) the clinical significance of which is overshadowed by the very powerful emetic action of the drug. Relatedly, the alkaloid stimulates locomotor activity (*8*), induces behavioral stereotypes (*8, 176, 180*), and reverses hypokinetic-rigid (cataleptic) states attendant on pharmacological and nonpharmacological procedures that bring about a presynaptic blockade of dopaminergic transmission (*82, 176, 324, 351*); e.g., brain DA depletion induced by blockade of synthesis, inactivation of amine storage mechanisms, or physical destruction of DA fiber projections. Moreover, the effects of apomorphine are selectively blocked by agents assumed to act as DA receptor antagonists (*15, 67, 284, 528*).

Inasmuch as the various neurochemical indices of the receptor stimulant action of apomorphine and piribedil relate to the deceleration of telencephalic DA metabolism and turnover, they probably reflect the ability of both agents to trigger receptor-mediated feedback inhibition of DA turnover. Thus apomorphine 1) reduces DA formation from tyrosine in vivo (*401*); 2) retards the disappearance of endogenous amine (*13, 34*) or specific fluorescence (*15*) from striatal and mesolimbic nerve terminals following arrest of catecholamine synthesis; 3) lowers HVA production (*230, 324, 442*); 4) counteracts postaxotomy rises in tyrosine hydroxylase activity (*297*) and telencephalic DA levels (*32*); and 5) reduces the overflow of DA from field-stimulated striatal slices (*191*). Piribedil (*128*) behaves similarly with respect to items 1 and 2; however, unlike apomorphine, it does accelerate DA turnover in median eminence neurons, fails to alter DA release from striatal slices, and is not susceptible to atropine with respect to its effect on tyrosine-DA conversion (*401*).

Other relevant pharmacological properties of apomorphine include its substrate relationship to COMT, which is apparently concerned with its catabolism (*362*). Dimethylation of the apomorphine molecule in its 10,11-hydroxy positions is known to abolish DA-like activity (*324*). High doses of the drug also inhibit tyrosine hydroxylase (*230*) and cause a release of endogenous NE (*32*). With regard to the "dualist" properties of apomorphine and of α-methyl-dopamine, refer to sections on DA synthesis inhibitors and DA receptors.

Indirect Dopamine Mimetic Substances

Various drugs, considered under the rubric of "indirect DA mimetics," exert an action at central synapses analogous to that of neurosympathomimetic and tyramine-like substances that simulate neurally evoked responses of sympathetically innervated peripheral organs. Characteristic of most of these substances are their alerting and motor-stimulant effects owing to which they are subsumed under the label of either psychostimulant or psychoanaleptic agents. Relevant compounds include 1) ephedrine, amphetamine, and its various congeners, viz., methamphetamine and benzphetamine, phentermine and phenmetrazine; 2) pipradol, methylphenidate, and NCA; 3) the metatyramines with or without a ring methoxy or side chain α-methyl group; and 4) miscellaneous compounds, such as amantadine and cocaine.

The essential feature of indirect mimetic drug action is its dependence on presynaptic factors, such as saturation of amine stores, ongoing synthesis, and nerve impulse flow. Given the complex morphological and functional organization of the nerve terminal, it is hardly surprising that the aforementioned compounds produce their DA-like effects by more than a single mechanism of action. While it has become obvious that dependence on the integrity of presynaptic nerve terminals is the common and identifying feature of such agents, the biochemical concept of intraneuronal compartmentalized amine pools has helped to refine the taxonomy of indirect DA mimetics and to resolve a number of controversial issues. It now appears justifiable to distinguish between agents that depend on the functional DA pool and those that act on the main storage compartment (Figure 2). Both types can be differentiated pharmacologically and biochemically by virtue of their interactions with 1) drugs that selectively inactivate the vesicular amine storage organelle, e.g., the Rauwolfia alkaloids and benzoquinolizines; and 2) drugs that selectively inhibit catecholamine synthesis. On this basis a fairly coherent picture has been established, fully consistent with the two-compartment hypothesis of intraneuronal DA storage and turnover.

On the one hand, there are the reserpine-resistant, α-MpT-sensitive agents which include ephedrine, amphetamine, methamphetamine, phentermine and phenmetrazine (*217, 372, 431, 433, 434, 453, 551, 556*). It seems clear that the ability of these drugs to reverse reserpine-catalepsy and to produce stereotyped behaviors in reserpinized animals when cerebral endogenous DA levels are almost undetectable can no longer be taken as evidence for a direct receptor stimulant action (*532*). On the contrary, their refractoriness to reserpine-like drugs, coupled with their marked susceptibility to α-MpT, is fully compatible with the view that they act specifically to release a reserpine-resistant extragranular DA pool (*100, 243*). The latter is presumed to contain newly synthesized amine and to contribute largely to that fraction which is available for release by nerve action potentials (*227*).

On the other hand, there are those compounds the DA-like action of which, although readily abolished by Rauwolfia alkaloids, is relatively resistant to α-MpT unless the dosage regimen is such as to cause depletion of the reserpine-sensitive pool (*453*). This group comprises benzphetamine, pipradol, methylphenidate, NCA, cocaine, and probably α-methyltyramine (*1, 372, 453, 532, 556*). There is reason to believe that these agents preferentially interact with DA stored in the main storage compartment.

A suggestive biochemical corollary to these two different modes of action is provided by the finding that reserpine-resistant, but not reserpine-susceptible, agents are able to stimulate DA catabolism (3-methoxytyramine formation) in the presence of reserpine and MAO inhibitors (*453*). This observation is also consistent with a shift of DA catabolism from intraneuronal to extraneuronal sites (*441*).

In addition to acting on other monoamine neurons, all of these "indirect-acting" compounds lack selectivity since they exhibit multiple actions that may either mask or reinforce the effect on DA release, e.g., inhibition of MAO (*44, 64, 76*), inhibition of SN cell firing (*87, 89*), and triggering of intraneuronal feedback inhibition of synthesis as evidenced by their inhibitory effect on DA synthesis (*63, 227*). Moreover, in the light of data from both in vivo and in vitro studies on striatal catecholamine uptake and release, Uptake$_1$ inhibition appears to be a significant factor not only with amphetamine (*121, 144, 268, 361, 437, 511, 541*), but also with (−)-ephedrine (*255*), (+)-threo-methylphenidate (*255*), amantadine (*252, 541*), and related psychomotor stimulants. Mimetic agents with such a mechanism of action appear to depend on impulse flow in DA fibers (*541*).

Of the numerous other drugs that inhibit DA Uptake$_1$, few if any are sufficiently selective to qualify as experimental tools. Cocaine has played a minor role in studies of dopaminergic activity of stimulant drugs. Its stimulant effects in rodents, unlike those of (+)-amphetamine, are completely abolished by reserpine but reinstated by both L-DOPA and metatyrosine (*532*). This pattern is compatible with Uptake$_1$ blockade (*191, 243*) which is generally held to constitute the principal action of cocaine at peripheral noradrenergic neuroeffector junctions (*276, 326, 327, 518*). In view of the well-known antagonism exerted by cocaine against tyramine (*92, 204, 257*), which has been imputed to possess cocaine-like properties of its own (*257*), it is pertinent to note that tyramine-induced amine release may originate from a cytoplasmic amine pool (*211*), differing from the functional and reserpine-sensitive granular pools (*255*). Cocaine does not seem to interfere with the "releasing" action of agents that interact primarily with the latter two DA pools.

Some of the most potent uptake blockers are found among anticholinergic and antihistaminic drugs, e.g., benztropine, diphenpyraline, and chlorpheniramine (*214, 268*). There is good reason to believe that this property contributes significantly to the therapeutic value of benztropine in parkinsonian disorders of various etiologies (*145, 481*).

Dopamine Receptor Antagonists

Neuroleptics The concept of DA receptor antagonism, originally inferred by transitivity from adrenergic receptor pharmacology, was indeed highly speculative at the time, but it has since withstood numerous tests. The first two compounds to be cast in the role of DA receptor antagonists were the neuroleptics chlorpromazine and haloperidol (*105, 273, 531*) and, ever since, an impressive number of congeners has been added to this category. These substances abolish peripheral (*271*) and CNS actions of both "direct" and "indirect" DA mimetics (*15, 23, 34, 65, 127, 284, 286, 375, 528*) and without causing significant alterations in endogenous DA levels (*22, 333, 404*) they induce behavioral and extrapyramidal motor deficits indistinguishable from those produced by Rauwolfia alkaloids that deplete the brain of DA (*273*). Owing to local anesthetic effects of these compounds, microiontophoretic experiments have been hampered in attempts to demonstrate DA receptor blockade (*124, 564*) at the cellular level. Some evidence suggests, however, that chlorpromazine may satisfy the rather crucial criterion of specific antagonism to the synaptic effects of DA (*565*).

An outstanding feature of neuroleptic substances is their ability to accelerate DA turnover in telencephalic structures. This is evidenced by a large number of biochemical indices: first, increased DA synthesis as revealed by 1) an augmented rate of DOPA accumulation in the forebrain in the presence of AAAD inhibition (*117, 297*), 2) an enhanced rate of striatal DA formation following systemic (*224, 399–401*), intracerebral (*90, 289*), or in vitro (*63*) administration of radiolabeled tyrosine, and 3) an enhanced DA fluorescence in nigrostriatal and mesolimbic DA neurons (*22, 130*); second, increased rate of DA utilization as shown by 1) an acceleration of α-MpT-induced disappearance of endogenous amine (*390, 404, 468*) (as determined chemically) or of specific histofluorescence (*21, 130*), and 2) a faster disappearance of newly synthesized DA from the brain (*224, 398, 398, 401*); and third, an increased DA breakdown as indicated by an augmented striatal output of MT (*105, 507*), DOPAC (*33, 448*) or HVA (*31, 160, 332, 404, 468*).

Considering that these metabolic changes: 1) are not unique to neuroleptics, 2) exhibit marked interspecies variations, and 3) often require dosages in excess of clinically effective levels, the question to be asked is whether they in fact relate to a single physicochemical action on DA neurons in the CNS. Indeed, more than one hypothesis is currently entertained. According to the widely favored receptor feedback theory of Carlsson and Lindqvist (*105*), the neuroleptic-induced acceleration of DA turnover represents a counter-regulatory increase in presynaptic impulse flow operating to restore DA transmission in the presence of postjunctional receptor blockade. The enhanced presynaptic activity is evidenced by increased tyrosine hydroxylation and DA release (the corollary process of receptor-mediated changes in the opposite direction by DA and related agonists has already been considered).

As shown in numerous in vivo studies of striatal DA metabolism, an anatomically and functionally intact nigrostriatal pathway is a prerequisite for this process, since temporary or permanent elimination of DA cell activity by drugs or by discrete lesions markedly attenuates the enhanced accumulation of DA from radiolabeled tyrosine in vivo (*398*), the accelerated loss of DA fluorescence following α-MpT treatment (*21*), and the enhanced production of HVA or DOPAC (*319, 446, 448*). Nonetheless, indirect evidence from axotomy experiments (*297*) suggests that receptor-mediated regulation of tyrosine hydroxylation remains susceptible to the influence of neuroleptic agents even in transected DA neurons. Furthermore, chlorpromazine and pimozide augment field-stimulated efflux of DA from striatal slices in vitro (*191*). The concept of neuroleptic postjunctional DA receptor blockade, hence the receptor feedback theory, is most convincingly supported by pharmacological observations demonstrating the remarkable specificity of some of these substances. In particular the rank-order of stimulatory action parallels that of antagonistic potency, as determined by various pharmacological paradigms (*15*). *Mutatis mutandis*, the ability of a given neuroleptic to accelerate NE turnover, is correlated with its degree of antagonism at NE receptors. Moreover, some of the most powerful "dopaminolytic" diphenylbutylamines, which are virtually devoid of NE antagonism, markedly stimulate striatal DA turnover but fail to alter NE synthesis rates (*397*). Nonetheless certain of these drugs (fluspirilene and pimozide) accelerate disappearance of NE from the brain when its synthesis has been arrested at the step either of tyrosine hydroxylation or of DA β-hydroxylation (*15, 19*) (but see Nybäck et al. [*397*]). Conceivably other factors may become critical under these circumstances, such as impaired permeability of axolemmal or storage granular membranes, or both.

As to the acceleration of neuroleptic-induced DA turnover, the receptor feedback hypothesis remains vulnerable until one of its principal predictions is directly verified, i.e., the increased release of DA into the extraneuronal space. So far, only two in vivo studies (*360, 540*) are available, neither of which has borne out this prediction, and most recent in vitro observations on striatal slices show that chlorpromazine and haloperidol, in lower concentrations than those used in the study already referred to, diminish electrically induced release (*462*).

Pending additional corroboration of the receptor feedback theory, alternative explanations deserve attention. Moderate-to-slight decreases in whole brain or striatal DA levels of different species have been reported with many neuroleptic drugs tested, although usually after amounts that are considerably in excess of behaviorally active levels: chlorpromazine and haloperidol (mouse [*404*], rat [*190, 390*], rabbit [*404, 507*], and cat [*404*]), thioproperazine, clothiapine, fluspirilene, pimozide, and spiroperidol (rat [*15, 119*]). Since the amount of DA lost closely agrees with that calculated to be in the functional compartment, the latter has been invoked as its likely source (*226*). Accordingly neuroleptic-induced acceleration of DA synthesis is thought to be contingent on the depletion of this functional pool, the trigger mechanism being the removal of end-product inhibition. Obviously this scheme implies an actual release of DA or, alternatively, blockade of its reuptake.

Another hypothesis attributes the metabolic actions of neuroleptic DA receptor antagonists to an increased *intraneuronal* destruction of DA rather than to an "incontinence" of the presynaptic terminal. This concept draws on certain analogies between the effects of chlorpromazine and reserpine on striatal DA catabolism in the rabbit (*507*). Both drugs elicit a decrease in striatal levels of methoxytyramine, the principal extraneuronal DA catabolite, i.e., they decrease extraneuronal catabolism. If intraneuronal DA metabolism is blocked by a MAO inhibitor, there is an increase in methoxytyramine. This may indicate that chlorpromazine, like reserpine, attacks the storage organelle itself, differing from the former only in that it impairs the permeability of the vesicular membrane without destroying the binding capacity of the intravesicular matrix. By virtue of such an action, chlorpromazine could divert the flow of freshly synthesized DA from storage toward catabolic sites (mitochondrial and microsomal MAO [*468*]). The failure of haloperidol to affect the efflux rate of tritiated DA from the feline caudate into cerebroventricular perfusates would be consistent with this hypothesis (*540*).

Other Compounds Bulbocapnine, a Corydalis alkaloid bearing close structural similarity to apomorphine, exhibits numerous actions suggestive of DA-receptor blockade. The alkaloid produces a cataleptic syndrome (*165, 168, 180, 494*) that resembles the one caused by neuroleptics, raises striatal HVA concentrations (*468, 495*), blocks synaptic transmission in the nigrostriatal pathway (*341*) (probably by a postsynaptic mechanism of action), and is reported to block vascular DA receptors (*519*). It is significant that bulbocapnine and other naturally occurring alkaloids that possess cataleptogenic activity contain within their molecular structure a para-methoxy-phenylethylamine moiety. In all likelihood the latter confers cataleptogenic (DA-antagonistic) activity on these compounds (*179*), since para-methoxylated DA derivatives, e.g., homoveratrylamine (DMPEA) (*495*), para-methoxy-metahydroxy-phenylethylamine (*177, 495*), and mescaline (*177*), are known to induce the "hypokinetic-rigid syndrome" and antagonize postsynaptic actions of DA (*178, 232*).

Dopamine-depleting Drugs

Reserpine, albeit a pharmacological blunderbuss, proved to be indispensable in pioneering investigations of brain DA. The drug depletes the brain of catechol and indoleamines alike (*109, 418, 472*), and most of its acute CNS effects, such as catalepsy and parkinsonian signs, appear attributable to a loss of DA since they are readily reversed by L-DOPA (*78, 108*), but neither by 5-HTP (*108*) nor by threo-DOPS (*6, 147, 455*), which gives rise to NE, by-passing the intermediary formation of DA. The depletory effect is generally believed to result from impairment of the amine-binding capacity of the storage organelle (*181, 183*) (the granular or dense-core vesicle). Neither synthesis (*33, 239, 290, 414, 455*) nor neuronal DA Uptake$_1$ (*243*) is inhibited, the former probably even being stimulated (*239*) (but see Besson et al. [*63*]). The molecular mechanism by which reserpine alters the vesicular uptake–storage process is not completely understood. Amine transport across the vesicular membrane is dependent on ATP, Mg^{2+}, and temperature, is stereoselective (*497, 509*), and is inhibited in vitro by nanomolar concentrations of reserpine (*498*). Although the impairment of vesicular amine binding seems to be irreversible, in vitro studies on uptake of NE and DA into splenic nerve storage granules (*183a, 498*) and striatal slices (*263*), respectively, suggest that the loss of binding capacity is not complete; that is to say, despite the presence of reserpine, amines can still be taken up if offered in sufficiently high concentrations.

Furthermore, in mice, it is possible to protect the storage organelle for catecholamines by the repeated administration of metatyrosine shortly before and after reserpine (*103*). In this manner, flooding the brain with metatyrosine will to a certain degree protect brain catecholamines, but not 5-HT stores, against the reserpine depletory effect, at the same time affording a partial prophylaxis against its cataleptogenic effect. If, in addition, animals are treated with protriptyline, the protective action of metatyrosine on brain NE is absent, but that on brain DA and behavior remains.

Numerous other substances, such as tetrabenazine, oxypertine, prenylamine, and benzquinamide share the neuroleptic properties of reserpine as well as its ability to interfere with the $ATP-Mg^{2+}$-dependent vesicular uptake–storage mechanism (*104, 294, 404, 416, 550, 552*). The last two compounds apparently also block catecholamine receptors (*27*).

Dopamine Synthesis Inhibitors

Together with its methyl ester H44/68, α-MpT is the only DA synthesis inhibitor to have achieved prominence as an investigative tool. Both compounds interfere with catecholamine biosynthesis at the tyrosine hydroxylation step, inhibition of that enzyme being competitive with the natural substrate tyrosine but noncompetitive with the synthetic cofactor $DMPH_4$ (*379*). As a result, DA as well as NE levels are lowered in both peripheral adrenergic tissues and the CNS, while 5-HT synthesis is unaffected (*11, 18, 244*). Results from fluorescence, microscopical, and biochemical studies, in rats pretreated with H44/68, suggest that the uptake and storage mechanisms of catecholamine neurons remain intact (*131, 215*) since L-DOPA injection restores intraneuronal catecholamine levels to normal values in rats pretreated with H44/68; however, electron microscopic studies on vesicular amine uptake in striatal slices suggest that H44/68 reduces accumulation of α-methyl-noradrenaline in small granular vesicles (*263*).

Unlike that of reserpine, the depletory effects of α-MpT are singularly dependent on neuronal impulse activity (*134*). This feature, in conjunction with its relative selectivity of action, may explain why the substance is extensively used in catecholamine turnover studies, although there appears to be no hard neurophysiological evidence to preclude an alteration of the functional state of catecholaminergic neurons by the drug itself. Such an effect may account for the inability of α-MpT to bring about a complete loss of striatal DA in certain species, e.g., the mouse (*468*). As expected, cessation of impulse activity in nigrostriatal DA fibers following the administration of γ-hydroxybutyrate or axotomy markedly slows down α-MpT-induced disappearance of telencephalic DA (*499*). Most of its CNS depressant effects in animals, such as depression of locomotor activity (*133, 487*), catalepsy (*52*), disruption of conditioned behavior (*133, 244*), and amphetamine antagonism (*431, 453, 551*), may be causally related to DA depletion, since they are readily reversed by L-DOPA; however, one must not overlook their relationship to NE depletion, as will be discussed in the section "Locomotor Mechanisms and Hyperkinesias" in this chapter. In man, the drug has no psychotolytic effect, but it significantly augments the action of neuroleptic phenothiazines and butryophenones (*110*). In addition it exacerbates the symptoms of parkinsonism (*484*).

Attempts to inhibit DA synthesis at the level of L-DOPA decarboxylation may have met with the expectations of biochemists but not with those of clinical pharmacologists. The α-methyl analogues of DOPA and metatyrosine, both competitive inhibitors of AAAD in vivo and in vitro, lower brain levels of catechol- and indoleamines if given in sufficiently high amounts[1] (*259, 417, 484*); even so, attending behavioral and neurological deficits are slight (*273*). Two reasons may account for this discrepancy: 1) the high reaction velocity of AAAD owing to which a residual enzyme activity below 10% may be sufficient to sustain the functional pool (*341, 370*); and 2) the formation of biologically active metabolites from α-methyl-DOPA and α-methyl-metatyrosine (α-MMT) (*106, 148, 170*). The resulting α-methylated phenylethylamines, α-methyl-DA, α-methyl-NE, α-methyl-*m*-tyramine, and metaraminol, not only are resistant to MAO but also act as substrates of both axolemmal and vesicular uptake systems (*243, 263, 515*).

There is reason to believe that α-methyl-metatyramine, similar to α-methyl-NE and metaraminol in peripheral adrenergic neurons, behaves as a false transmitter in brain DA neurons (*169*). Since both α-methyl-DA and α-methyl-metatyramine appear to possess (partial) intrinsic activity at DA receptors, their formation and release from DA fiber terminals could be expected to buffer the functional sequels of DA depletion (*106, 124, 532*). The reader interested in other AAAD inhibitors of the hydrazine and oxyamine types and their potential clinical uses is referred to several excellent sources (*50, 370, 417, 483*).

---

[1] Results from a recent in vitro study indicate that the inhibitory effect of α-methyl-DOPA on striatal DA synthesis can be fully accounted for by inhibition of tyrosine hydroxylase (*528a*).

Surgical and Drug-induced Lesions

Depigmentation and neuronal fallout in the pars compacta of the SN are the common and probably the most specific pathomorphological features of parkinsonian syndromes of various etiologies (*53, 171, 248, 409*). This correlation has prompted investigators to attempt to reproduce experimentally the neurological signs of this syndrome in laboratory animals either by the placement of discrete lesions in this nucleus or by severing its fiber connections with the striatum (*366, 419, 493*). In general surgical methods have met with limited success because they cannot avoid damaging adjoining fiber systems in the cerebral peduncle and ventral tegmentum, as well as transnigral fibers. Nonetheless, this approach has great heuristic value since it induces biochemical alterations in the animal brain that closely resemble the pathoneurochemical correlates of human parkinsonism. It is now well established that stereotaxically placed electrolytic or electrothermic lesions in or rostral to the SN or at various levels of the medial forebrain bundle in monkeys, cats, and rats cause a loss of striatal DA (*16, 26, 194, 229, 262, 373, 412, 421, 424*) and its catabolites (*411, 448*), as well as a significant reduction in the activity of biosynthesizing enzymes—tyrosine hydroxylase (*229, 373, 423*) and AAAD (*26, 229, 262, 341, 373*). Moreover, following chronic ventromedial, tegmental, and lateral hypothalamic lesions, there is a reduced capacity of the striatum to take up and accumulate exogenous DA (*229, 541*) and release it in response to the administration of indirect DA mimetic agents (*541*). With conventional histological methods (*51, 423*), histofluorescence (*26*), and silver impregnation techniques (*113, 114, 373, 471*), similar lesions have been shown to cause retrograde and anterograde degeneration, respectively, of pars compacta cells and their fiber tracts ascending to the corpus striatum. These data, coupled with biochemical findings, most decisively support the existence of a nigrostriatal DA fiber projection and, furthermore, they provide a well-founded rationale for the use of animals in studies of the role of DA in the pathogenesis of Parkinson's disease.

Chemical Lesions While 5-hydroxydopamine (5-OHDA) is recognized as a false transmitter, albeit of minor importance, in central and peripheral monoaminergic structures (*333, 513*), its neurotoxic analogue 6-OHDA has gained widespread use since it is claimed to produce a selective depletion of brain catecholamines with less damage to noncatecholaminergic elements than that caused by surgical or electrocoagulation techniques. In a sense 6-OHDA represents a special case of a false transmitter, a "malignant" one, as it were. Monoamine neurons, catechol more than indole, fail to distinguish between this compound and other monoamines that fulfill the structural and stereochemical requirements for

Uptake$_1$. As a result, catecholamine neurons actively accumulate 6-OHDA into their cytoplasm and storage granules at lower extracellular concentrations than do other neurons, the uptake being susceptible to inhibition by tricyclic antidepressants (*186, 346, 500*). Beyond this step the ensuing catabiotic events appear to be essentially nonspecific in that they apply to neural as well as to nonneural cell structures.

The destructive action of 6-OHDA requires the buildup of a critical cytoplasmic level (in this sense it is virtually an all-or-none phenomenon) and may be attributable to 1) either the formation of electrophilic p-quinones or indole derivatives, which bind covalently to functionally important macromolecules (*513*); or 2) the formation of hydrogen peroxide (*250*). The amount required to induce irreversible structural damage probably also depends on variables, such as regional angiocytoarchitectonics. Since at sufficiently high extracellular concentrations any nerve cell or fiber will accumulate the "lethal" amount, selectivity depends to a high degree on the dose administered. Susceptibility to the neurotoxic action of the drug is higher in nigral than in mesolimbic neurons and least in tuberoinfundibular neurons (*522*). Owing to this differential sensitivity, it is apparently possible to destroy striatal DA terminals in rats without affecting serotoninergic elements (*91, 393, 529*). In the cat the latter elements are sufficiently susceptible to 6-OHDA so as to invalidate its use as a selective depletor (*322*). In addition it is possible to protect NE neurons with protriptyline, a selective uptake inhibitor at NE but not at DA nerve terminals (*186, 522*). Reserpine does not afford any protective effect (*513*). A more localized, although not necessarily a more selective, action of 6-OHDA may be achieved by intracerebral stereotaxic microinjection (*521*). In the rat, systematic application of this technique has not only facilitated hodological studies of catecholamine neuron projection systems, but also contributed extensive information on functional and pharmacological aspects of striatal DA receptor "denervation." (For a comprehensive review of the literature and discussion of morphological and histochemical effects, the interested reader may consult two excellent sources ([*513, 522*].)

In the rat, bilateral microinjections of 6-OHDA into the SN produce severe sensorimotor and behavioral disabilities. In several respects these abnormalities resemble those observed in animals subjected to reserpine treatment or brain lesions that destroy the ventral midbrain tegmentum (*29, 291, 420*) or the posterior or lateral hypothalamus (*46, 348, 408*), that is to say, areas containing DA cell bodies or their projections. In addition to pronounced deficits in locomotor and exploratory activity, catatoniform manifestations, loss of postural and exteroceptive reflexes, and brady-kinesia, severe aphagia and adipsia are regular features, the latter condition being fatal if the animal is not nursed. Unilateral injections of 6-OHDA into the SN of rats (526) and cats (208) cause marked asymmetries of posture and locomotion, such as torticollis, compulsive circling, and sensorimotor disturbances, e.g., ataxic gait and inability to step or jump downward. Whereas the direction of circling follows a predictable pattern in rats (see section on "Locomotor Mechanisms and Hyperkinesias" in this chapter), no such relationship obtains in the cat (420).

Other Drugs
The recognition of relationships between psychotropic drug activity and monoaminergic transmission prompted widespread screening of such agents for potential actions on brain DA metabolism. As this strategy has now become routine, one cannot help but wonder if any such compound has escaped this type of scrutiny. In view of the literature glut in recent years, even a cursory discussion of this topic is beyond the scope of the present chapter.

Narcotic Analgesics Narcotic analgesics have recently received a great deal of attention because of their interaction with central dopaminergic mechanisms. Morphine, methadone, and related substances alter brain DA metabolism in a manner similar to that of the neuroleptics, and in rats they share the cataleptogenic action of the latter (353, 493, 546). Accordingly these drugs augment striatal DOPAC (210) and HVA (210, 321, 452, 468, 495) production and tyrosine-dopamine conversion (219, 444, 452) and they accelerate α-MpT–induced disappearance of endogenous DA (240, 468). Effects observed in histofluorescence studies are consistent with enhanced DA turnover (253). On the basis of such analogies some investigators have imputed to these compounds a blocking action on DA receptors (217, 452).

Nonetheless a more careful comparison between morphine and chlorpromazine catalepsy reveals remarkable differences: 1) The catalepsy and rise in HVA induced by morphine, unlike those caused by chlorpromazine, are abolished by naloxone, a specific morphine antagonist, but are unaffected by atropine. 2) L-DOPA and apomorphine overcome the cataleptogenic effects of morphine more readily than those of chlorpromazine (321). 3) Although apomorphine is found to reverse catalepsy caused by morphine or methadone (452), its stereotypy-inducing effect is apparently enhanced by morphine pretreatment (533).

Such findings militate against a common mode of action at central DA synapses. If one assumes that morphine catalepsy is unrelated to cholinergic synaptic activity, and that neuroleptic catalepsy is indeed attributable to a postsynaptic DA receptor blockade, one can envisage a presynaptic effect of morphine, in keeping with the "diversion" hypothesis proposed by Sharman (*321, 468*). Although morphine produces these effects in analgetic doses, evidence so far available from studies on pain is equivocal as to whether the antinociceptive action of opioids is attributable to altered DA transmission and, if so, either to impaired or to enhanced transmission at telencephalic DA synapses. Injected intraventricularly in low doses, DA enhances antinociceptive effects of morphine in the mouse (*96*), whereas in high doses its action is antagonistic (*97*). Relatedly intraventricular administration of ouabain, which produces an elevation in telencephalic DA levels (*9*), exhibits a synergism with morphine (*96*). Consequently it is argued that the morphine antagonism of high DA doses may result from its conversion to NE, itself a morphine antagonist (*97*).

It remains unclear if the synergistic effect of ouabain relates to blockade of DA Uptake$_1$ or is independent of the reported elevation of DA levels, inasmuch as the latter action may reflect the characteristic response of DA axon terminals to cessation of impulse flow, an effect compatible with the inhibitory action of the glycoside on excitable membranes. Furthermore, ouabain may alter transmembrane fluxes or intraaxoplasmic concentrations of Ca^{2+}, and this could conceivably influence Ca^{2+}-dependent excitation-secretion coupling at DA axon terminals. It also ought to be pointed out that rodents become rapidly tolerant to opiate-induced alterations of DA metabolism (*240, 477*). Whether this is attributable to the development of supersensitivity of DA receptors or to formation of new receptors, and whether or not withdrawal and craving phenomena result from a sudden "unmasking" of postjunctional supersensitivity on restoration of normal presynaptic function, remain to be seen.

Concerning the relationship between dopaminergic transmission and opiate addiction, it may be pertinent to point out the existence of DA fiber innervation of limbic cortical structures which are believed to contain neural substrates concerned with telencephalic integration of nociception, i.e., the affective-emotive processing of pain. For additional discussion of the relationship between monoamines and morphine-like drug action, see the section on "Morphine Analgesia" in Ch. 7 and also recent brief reviews (*247, 546*).

NEUROPHYSIOLOGY

A wealth of data from histochemical, neurochemical, and pharmacological investigations attests to the significance of DA in CNS functions; however, a complete understanding of the synaptic function of central DA neurons hinges upon neurophysiological information about the intrinsic organization of regions of the brain innervated by these neurons. An excellent account of the microscopic anatomy of the caudate nucleus is presented by Kemp and Powell (*299–302*) and may be consulted by the interested reader. For the nigroneostriatal projection, the task of assigning recorded synaptic events to specific striatal cell types has not yet been performed, owing perhaps to the small size of striatal cells as well as the cytoarchitectonic monotony of the striatal neuron network. Such handicaps, however, have not discouraged attempts to examine the responsiveness of striatal units to microiontophoretically applied DA and to elucidate their possible relationships to synaptic inputs from the SN.

Caudate Nucleus

Curtis and Davis (*151*) were the first investigators to demonstrate that DA altered neuronal activity in the mammalian brain. They reported that activation of lateral geniculate cells by optic nerve stimulation was suppressed by microiontophoretic application of DA to these units. Subsequent to the neurochemical and histofluorimetric characterization of the nigroneostriatal DA pathway, it was but a logical step to extend the microiontophoretic studies to this neuron system. While a modest body of pertinent data has accumulated, it is generally difficult to make meaningful comparisons between results from different laboratories, especially since the cells under study have not been precisely identified.

Table 1 represents an attempt at categorizing the unit data from experiments on the caudate nucleus on the basis of 1) firing characteristics; and 2) whether or not stimulation of the SN was used to ascertain the presence and mode of nigral synaptic input. At this point it may be appropriate to recall that the spontaneous activity levels of caudate units appear to vary with imposed experimental conditions (*2, 80, 124, 330, 364, 461*). The overall impression is that these cells are not very reactive when tested with central and peripheral stimuli (*80, 277, 330, 430*). (For a detailed discussion on spontaneous activity patterns in caudate units, see [*198*].)

Results obtained from unit types with an unspecified or unknown input (*80, 124, 364, 564*) add little substance to the suggestion that DA may be acting as an inhibitory transmitter in the caudate nucleus. In order to gain more direct information about the synaptic effects of DA on striatal neurons, it becomes mandatory to activate the nigral fiber input, a venture that, for various reasons, is not without technical problems: 1) The topographical nature of nigrostriatal projections dictates that stimulation

Table 1. Caudate unit responses to microiontophoretically applied dopamine and SN stimulation

Unit	Response to DA	Treatment	Ref.
Normally silent; synaptically driven from SN	I:X E:X	D	364
	None	D	197
	None	A[a]	201
	I:8%	6-OHDA	201
Normally silent; microiontophoretically driven excitatory SN input	I:20% E:47%	+	123 123a
	I:43% E:X	A[a]	201
	I:100%	6-OHDA	201
Normally silent; microiontophoretically driven; excitatory intralaminar input	I:X E:X	U	258
Normally silent; microiontophoretically driven; inhibitory SN input	I:77% E:16%	D	123 123a
Normally silent; microiontophoretically driven; input unknown	I:63% E:10%	D	565
	I:60% E:9%	D	364
	I:62% E:2%	D	123 123a
Spontaneously active; inhibitory SN input	I:90% E:X	A[b]	231
Spontaneously active; input unknown	I:50% E:14%	D	80
	I:60% E:2.5%	A[c]	80

[a]Dial-urethane
[b]Urethane
[c]Barbiturate

Legend: I, inhibited; E, excited; X, not reported; D, decerebrate; A, anesthetized; U, unanesthetized: gallamine.

sites match the recording sites. 2) Microelectrode sampling bias favors large exit neurons which constitute only 5% of the total neuron population but are most liable to be invaded antidromically. 3) Owing to unavoidable activation of transnigral, ascending or descending fibers, allowance must be made for unit responses secondary to corticothalamic excitation.

In retrospect it appears that the early studies (*364, 365*) were plagued by these handicaps. As they failed to find any correlation between nigrally evoked responses (mainly excitatory) and microiontophoretically induced DA effects in amino-acid driven cells (mainly inhibitory), York and McLennan (*363, 364*) postulated a dopaminergic interneuron lying entirely within the caudate nucleus and receiving a major input from the centromedian (CM) of the thalamus. Work from the same laboratory demonstrated a DA release from the caudate to CM stimulation (*363*). (See section on "Release and Uptake" under "Neurochemical Properties" in this chapter.) These results are not compatible with existing neuroanatomical evidence which fails to support a dopaminergic CM projection to the caudate nucleus, apart from the fact that this region of the intralaminar group appears to project preferentially to the putamen. Above all they do not meet the criterion of identity of synaptic action (*553*). This goal appears to have been achieved in more recent studies (*124, 124a*) with midpontine decerebrate cats in which the firing of "silent" caudate neurons was maintained by microiontophoretic application of DL-homocysteic acid or L-glutamic acid. Under these conditions 31 of 40 neurons depressed by electrical stimulation of the SN were also depressed by microiontophoretic application of DA, whereas only three of 15 units exhibiting excitatory responses to nigral stimulation were depressed by application of DA. Thirty-four of 55 units that failed to respond to nigral stimuli displayed inhibitory responses to DA. In this latter population only six units were facilitated by the amine. Data from another investigation with anesthetized rats (*231*) show that electrical stimulation of the SN inhibits spontaneously active units in the caudate and that most of these (34 of 40) are similarly affected by microiontophoretic application of DA. To date these findings offer the most conclusive evidence that DA acts as the transmitter of the nigrocaudatal pathway. In conjunction with other data listed in Table 1, they argue strongly in favor of an inhibitory function for this projection.

Comparisons between different studies make apparent the considerable variability of unitary responses evoked in the caudate nucleus by SN stimulation. To a certain degree, synaptically evoked responses appear to vary with the level of resting activity. Most investigators report that responses, whether excitatory or inhibitory, are detected in all portions of the caudate nucleus, observations perhaps not difficult to reconcile with the neuroanatomical monotony of this structure and the widespread terminal arborization of nigroneostriatal DA fibers.

According to a recent study (*198*) in which the caudate nucleus was extensively explored during electrical stimulation of the SN, two populations of units are discernible on the basis of their localization and responsiveness. One, located in the medial two-thirds of the head of the caudate nucleus, can be activated by nigral stimulation and shows no spontaneous activity (NCE population). The second is scattered throughout the caudate nucleus and consists of spontaneously active elements that are inhibited by nigral stimulation (NCI population). When cells of this type are depressed by anesthetics, they can be readily activated by glutamate or homocysteic acid. Units comprising the first population exhibit latencies of approximately 18.5 msec when activated by electrical stimulation of the SN, confirming earlier results (*209*). In the opinion of these investigators, this "nigrostriatal" pathway differs from the DA projection in that it might function to activate inhibitory interneurons (connected with NCI?) within the caudate nucleus. In this context it is noteworthy that other workers (*428*) recently reported monosynaptic inhibition of feline nigral neurons on stimulation of the caudate nucleus: the pathway mediating the IPSP is slow conducting and response latencies average 18.2 msec, a description that closely resembles the proposed non-DA nigrostriatal projection. The caudate-evoked inhibition of nigral neurons can be blocked by systemic application of picrotoxin, suggesting that the response is mediated by GABA (*429*).

Evidence from neurochemical, pharmacological, and morphological studies demonstrates that 6-OHDA destroys DA fibers. This information has attracted the attention of neurophysiologists since it suggested a method that would permit selective destruction of DA projections, and the implications of such a technique were fraught with promise. Contrary to expectations, destruction of intracaudate DA-containing nerve terminals following the administration of 6-OHDA to cats appears to have no effect on caudate neuronal responses to nigral stimulation (*200*). However, unit responses to microiontophoretically applied DA are markedly altered. Thus in untreated animals, microiontophoretically applied DA appears to have no effect on synaptically driven excitation in the absence of excitatory amino acids; however, when similarly identified neurons are driven with amino acids, application of DA inhibits 43% of the neurons so treated. Following repeated intraventricular injections of 6-OHDA, amino acid-driven caudate units—identified either by short-latency antidromic or long-latency orthodromic activation—respond with inhibition to DA in virtually all cases tested (196 identified as orthodromic and 47 as antidromic) (Table 1). These findings were interpreted as supporting the notion that "... DA operates in the entire caudate by adjusting the background membrane properties of neurons excited and inhibited by other possible neurotransmitters." Considering the intricate synaptic organization of the caudate nucleus as revealed by electron microscopic experiments (*300*), this vague proposal adds little to our understanding of the synaptic action of DA.

Putamen

A sizeable portion of the nigral pathway arising from caudal segments of the pars compacta of the SN supplies the lateroventral aspect of the corpus striatum, namely, the putamen. When compared with the caudate nucleus, the putamen generally exhibits greater spontaneous unit activity and a greater responsiveness to central and peripheral inputs (*430, 464*). As revealed in a study of 120 amino acid-driven putaminal units (*564*), a large proportion (77%) of the 53 units showing excitatory responses to DA responded in a similar manner to nigral stimulation. The inhibitory effect of DA was preferentially antagonized by β-adrenergic receptor blocking drugs, such as DCI, D-INPEA, and MJ 1999, whereas the excitatory effects were generally more susceptible to inhibition by α-adrenergic receptor blockers, chlorpromazine, phentolamine, and dibenzyline. On the basis of these findings, it was suggested that two receptors may be involved in mediating the action of DA. It seems premature to label the receptor mediating excitation as "β-like" and the receptor mediating inhibition as "α-like" since MJ 1999, generally recognized as a more selective β-receptor antagonist than the other agents employed in this study, blocked inhibitory as well as excitatory DA effects.

Two other arguments, cited in support of the dual receptor scheme, are based on rather indirect evidence: 1) Unlike the inhibitory, the late excitatory effect of DA exhibits tachyphylaxis. 2) Following application of MAO inhibitors, excitatory but not inhibitory responses are enhanced. In the absence of more conclusive data, the excitatory DA effects could be equally well attributed to an indirect, i.e., releasing, action on other monoamine terminals (see section on "Release and Uptake" and "Dopamine Receptor Agonists" in this chapter). In a subsequent study in the squirrel monkey (*562*), putaminal units appear to display mainly an inhibitory response or a mixed inhibitory-excitatory response to microiontophoretically applied DA. It would seem that these inconsistencies reflect not so much species differences as they do methodological problems related to the use of excitatory amino acids to establish base-line patterns of unit-firing activity. The question of the receptor mechanism underlying synaptic excitatory vs depressant effects of DA cannot be answered in the absence of additional data, and lesion experiments have failed to clarify this issue.

In encéphale isolé cats with chronic unilateral electrolytic lesions interrupting the nigroneostriatal pathway, unit firing was reported to be greater in the putamen on the lesioned side when compared with the contralateral side, or in the putamen of one control animal (*402*). Medium-to-high doses of L-DOPA administered systemically further increased this abnormally enhanced activation without affecting unit firing on the side opposite the lesion. The relationship of these pathological changes to the therapeutic action of L-DOPA in DA-neuron degeneration syndromes in man remains unclear. At any rate the data presented are insufficient to support a causal relationship to DA-receptor supersensitivity of the denervation type.

Pallidum

Experimental evidence for a transmitter role for DA in the globus pallidus is sparse. In one study already referred to (*563*), it was shown that of 40 pallidal units identified in the cat, microiontophoretic application of DA depressed 35% and excited 37%. A correlative examination of the effects of nigral stimulation on such units would be of interest, since a direct dopaminergic nigropallidal pathway has been proposed on the basis of histofluorescent, microscopical, and neurochemical observations in man (*272, 395*). These data afford but a glimpse into an area that obviously requires extensive investigation.

Drug Effects on Striatal Neurons

Amphetamine In the section on "Indirect Dopamine Mimetic Substances" in this chapter, attention was drawn to important features of "indirect" DA-mimetic drug action, namely, its dependence on presynaptic parameters, such as saturation of amine stores, ongoing synthesis, and nerve impulse flow. Amphetamine compounds, in particular, are believed to mimic the action of catecholamines by promoting release or blocking reuptake, or both, at presynaptic terminals.

Intravenous administration of (+)-amphetamine to both anesthetized and unanesthetized rats inhibits the firing of DA cells in the SN and markedly decreases striatal levels of DOPAC (*89*). This inhibitory action on DA neurons is antagonized by numerous antipsychotic compounds, including chlorpromazine, thioridazine, fluphenazine, perphenazine, trifluoperazine, and haloperidol. A follow-up study (*86*) showed that nigral DA cells in the rat are relatively unresponsive to (−)-amphetamine. The depressant effects of (+)-amphetamine are abolished by a transection between the striatum and the SN. This may be suggestive of an indirect mechanism of action, in accordance with the proposal of *transneuronal* feedback (see the section on "Feedback Regulation" under "Regulation of Dopamine Synthesis" in this chapter).

A divergent picture of the synaptic action of amphetamine emerges from studies on cats pretreated with intraventricular administration of 6-OHDA *(199, 201)*. These experiments revealed that microiontophoretic application of both (+)- and (−)-amphetamine produces identical depressant effects on 78 caudate units that were activated either by amino acids or by nigral stimulation. These neurons are localized in ". . . the main zone of the caudate nucleus where the green fluorescence of the dopamine-containing neurons was always shown to have totally disappeared." In control animals both isomers of amphetamine invariably depress microiontophoretically driven units; however, they fail to affect synaptically driven elements in more than 50% of the cases tested. In addition, after depletion by reserpine and blockade of synthesis by α-MpT, there were no differences between the depressant actions of (+)- and (−)-amphetamine on caudate neurons. One wonders if the alleged postsynaptic effects of both isomers of amphetamine in barbiturate-anesthetized preparations might be related to the nicotine-like action of amphetamine described in autonomic ganglia *(435)*. This possibility is also consonant with the observation that striatal cholinoceptive elements, which in the unanesthetized preparations are predominantly facilitated by acetylcholine (ACh), are predominantly inhibited by ACh in the presence of barbiturates *(81)*. The burden of proof that these effects can indeed be attributed to a specific postsynaptic amphetamine action on DA receptors thus remains with the investigators *(199, 201)*.

Neuroleptics A number of attempts have been made to demonstrate DA receptor blockade at striatal synapses with microiontophoretic application of chlorpromazine and haloperidol *(124, 196, 562, 564)*. Owing to the local anesthetic properties of these compounds, the outcome of such experiments has been disappointing, and interpretation of the data most difficult. According to a recent report *(562)*, it is possible to dissociate the local anesthetic effects of chlorpromazine from a *competitive* postsynaptic DA antagonism. The latter notion, however, receives little support from indirect pharmacological and clinical data *(13, 132, 166, 273)*. (See also the section on "Dopamine Receptor Antagonists" and "Locomotor Mechanisms and Hyperkinesias".)

Surprisingly there is little information on the effects of systemic administration of presumptive DA blockers. In the presence of chlorpromazine, the excitability of ventral caudate and medial globus pallidus units appears to be enhanced as evidenced by an augmentation of stimulus-locked rhythmic bursts induced by paired stimulus trains to the amygdala and sciatic nerve (2). Remarkably, spontaneous firing rates in the globus pallidus are slowed by chlorpromazine, and the slowing effect of sciatic stimulation on unit firing is changed into an excitatory response. As mentioned in an earlier section, findings in the rat (89) indicate that antipsychotic compounds known for their extrapyramidal side effects, e.g., chlorpromazine and haloperidol, increase the rate of firing of SN neurons when administered intravenously. It is noteworthy that drugs lacking antipsychotic activity or ability to cause motor disturbances appear to have no effect on SN unit activity. Such studies illustrate the importance of comparing different methods of administration in relation to the potential effects of these compounds on dopaminergic or dopaminoceptive cells, even though some fall short of the objective of pinpointing synaptic sites of action. Nevertheless they do provide us with data pertinent to an understanding of the clinical effects of these drugs.

THE DOPAMINE RECEPTOR

The peripheral nonneural β-adrenoceptor has been equated with the regulatory subunit of adenylcyclase (EC 4.6.1.1) which, oriented toward the cell membrane surface, communicates allosterically with the catalytic site of the enzyme oriented toward the interior of the cell (440). Recent biochemical and electrophysiological studies, designed to test the validity of this concept as it may apply to neural catecholaminergic transmission systems, furnish data that link the functional state of DA and NE synapses at sympathetic ganglia (296, 354) and cerebellar Purkinje cells (see section on "Cyclic Adenosine 3',5'-Monophosphate . . ." in Ch. 6) to adenylcyclase activity in these tissues. Evidence for a similar relationship at DA synapses in the brain is derived from biochemicopharmacological in vitro studies of the enzyme in cell-free homogenates of the rat striatum (295). Under such conditions DA elicits a two-fold increase in enzyme activity, the equieffective concentrations of NE being four to five times higher than those of DA, whereas the pure β-agonist, isoproterenol, is virtually inactive. Apomorphine displays a bell-shaped, dose-response relationship with an apparent peak effect one-third that of DA. In combination with DA or NE the drug produces infraadditive effects, and higher concentrations are inhibitory. These findings are in agreement with neuropharmacological evidence supporting the proposal that apomorphine is a partial agonist at DA receptors (68). Dopamine-stimulated activity can be antagonized by low concentrations of the neuroleptic agents haloperidol and chlorpromazine, as well as by higher concentrations of conventional α-adrenergic antagonists, such as phentolamine and phenoxybenzamine; β-receptor blockers are ineffective.

At present the regulatory mechanisms controlling cyclic adenosine monophosphate (AMP) levels in neural tissue slices, as well as the parameters controlling the activity of adenylcyclase in tissue homogenates, are only partially understood. This is particularly true of in vitro effects as they relate to the unique interactions between biogenic amines and adenosine at this membrane-bound enzyme, as well as to the potential significance of the latter for neuronal functioning in the intact brain (*158*). Nonetheless the striking parallels to data obtained with more indirect pharmacological assay methods have greatly encouraged speculation about the identity of the neural DA receptor and the presumptive binding site on adenylcyclase of mammalian neural tissues with a dopaminergic innervation, e.g., sympathetic ganglia and the corpus striatum. In other words, the synaptic effects of DA on neurons in these structures are postulated to be mediated via the intracellular formation of cyclic AMP within the postsynaptic neural elements (*296, 547*). The cyclic nucleotide could then activate a protein kinase, resulting in phosphorylation of cell-membrane proteins that control electrogenic properties of the membrane (*369*).

While this hypothesis still awaits rigorous testing, it has prompted the suggestion that ganglionic adenylcyclase may mediate the slow dopaminergic IPSP in sympathetic ganglia (*336*). It would be of interest to know whether or not this system also plays a role in the generation of long-lasting, DA-mediated facilitation of ganglionic muscarinic slow EPSPs (*336*). As shown in amphibian sympathetic ganglia, the electrogenic mechanism underlying *slow* PSPs, inhibitory as well as excitatory, consists in a lowering of ionic conductance of the neuronal membrane (*547, 548*). Thus the facilitation of the muscarinic slow EPSP by DA may reflect the effect of increased membrane resistance on the voltage-current relationships of the postsynaptic membrane, i.e., for a given membrane current, synaptic potentials will be larger, the higher the membrane resistance. One may ask if 1) this mode of synaptic action constitutes monoaminergic transmission as opposed to amino acids and ACh that, with the exceptions discussed in Chs. 1 and 4, produce synaptic potentials characterized by an *increase* in ionic membrane conductance; and 2) heterosynaptic facilitation of this mode also occurs in dopaminergically innervated brain structures, specifically at DA synapses within the striatal neuropil.

Perhaps owing to the unavailability of a simple preparation, few systematic attempts have been made to characterize the mammalian neural

DA receptor with classic pharmacological methods of receptor taxonomy, i.e., quantitative structure—activity studies of compounds that interact with a presumptive receptor.

Microiontophoretic work on striatal neurons has so far failed to identify a specific DA receptor qualitatively distinguishable from other catecholamine receptors (see section on "Putamen" in this chapter). Available data are equivocal in that one cannot ascertain whether the presumptive DA receptor bears more similarity to α- than to β-adrenergic receptors as found at visceromotor synapses. At present there is still uncertainty about the structural requirements essential for DA-like activity; however, much has been learned from comparative studies of DA and apomorphine at the submolecular, physicochemical level. NMR (nuclear magnetic resonance) spectroscopic measurements and HMO (molecular orbital) determinations (*93*) have made it likely that the active portion of the apomorphine molecule resides in its dihydroxy-tetrahydroamino-naphthalene moiety which features an N-OH distance practically identical with that existing in the *anti*-, and perhaps also in the *gauche*-, conformers of DA (*436*).

A recent attempt was made to explain the DA antagonism of neuroleptic phenothiazines in terms of congruencies between preferred conformations of these substances and the *anti*-conformer of DA as revealed by X-ray crystallography. The proposed model (*269*) would seem to account reasonably well for essential structural requirements of DA antagonists of this general chemical configuration, such as electron-dense ring substituents at the two-position and three-carbon side chain. It remains to be shown, however, whether or not this model can accommodate other presumptive neuroleptic DA receptor antagonists, presynaptic actions having been imputed to many of them (see section on "Dopamine Receptor Antagonists" in this chapter).

Since rather detailed studies of structure-activity relationships of DA-like substances have been performed on either nonneural or nonmammalian substrates (*544, 557*), such work merits more than a passing reference, allowing for obvious differences between these and vertebrate central DA synapses. Extensive morphological, biochemical, and electrophysiological data indicate that neurons in the CNS of the land snail, *Helix aspersa*, engage in synaptic relations with DA-containing cells (*304*). In identified cells of the parietal ganglion, DA induces an IPSP (the ILD response), probably by increasing membrane K^+ permeability (*305*). While the postsynaptic receptor mediating this response bears a superficial resemblance to the α-adrenergic type, it differs in its structural requirements vis-à-vis agonistic and antagonistic substances. Most notable are a relatively low sensitivity to noradrenaline and adrenaline and a unique affinity to ergometrine, the most potent antagonist at this receptor so far discovered (*544, 577*).

The possible use of the *Helix* neuronal DA receptor for general bioassay purposes has been hampered by its apparent insensitivity to other substances that are believed to act as agonists or antagonists at *vertebrate* DA receptors, such as apomorphine and haloperidol. However, the most recent discovery of ergometrine-resistant excitatory DA receptors in other neural tissues of *Helix* may help to clarify the pharmacological picture, inasmuch as the aforementioned compounds do indeed produce their expected effects on this second type of receptor (*501*). At present, many unresolved questions remain to be answered before one can attempt to assess the implications of such findings for mammalian DA neuron systems. *Inter alia*, it would be desirable to know whether or not: 1) the ionic mechanism effecting the excitatory synaptic action of DA differs from that operating in the ILD response, and 2) both response types can be linked to the adenylcyclase–cyclic AMP system. With regard to the latter point, it is of interest to note that in cockroach ganglia where DA is known to enhance firing activity (*306*), stimulation of adenylcyclase has been achieved with various catecholamines, including DA (*385*).

The demonstration of excitatory DA synapses in invertebrate neuron systems, coupled with the antagonism exhibited by neuroleptics at such sites, has prompted speculation (*125*) that excitatory rather than inhibitory DA synapses are the neural substrate of certain pharmacogenic behavioral stereotypes. To date single unit work on striatal tissue in various mammals has provided little support for this notion, and it remains for future studies of extrastriatal DA synapses to furnish more cogent evidence.

FUNCTIONAL CONSIDERATIONS

Empirical models for the pharmacological evaluation and quantitation of DA receptor activity address themselves primarily to motor aspects of basal ganglia function and, beyond this level, to their sensorimotor and behavioral correlates. Existing experimental evidence in fact implicates brain DA neurons in a broad array of behaviors traditionally subsumed under the rubric "survival of the individual and of the species." The basal ganglia, specifically the corpus striatum, are believed to be concerned with programming and enacting tactics of behavior, and in this capacity they appear to control functions from the most elementary to the most complex, enabling the mammalian organism to structure patterns of adaptive behavior (*293, 317, 330, 368*). It is perhaps with this perspective that one has to view the kaleidoscopic pharmacology of brain dopaminergic mechanisms.

Sensorimotor Functions

Disturbances of bodily motility associated with either a depletion or an overabundance of DA at striatal synapses illustrate the aptness of terming DA a kinetic (movement-producing) substance. Both in laboratory mammals and in man, loss of striatal DA leads on the one hand to 1) a cessation of spontaneous motor activity; 2) an impaired ability to initiate purposive movement; or 3) the inability to make transitions from one movement pattern to another or to terminate activity once it has been initiated. On the other hand, an overabundance of DA at striatal synapses expresses itself in excessive motor activity which is manifested through: 1) involuntary quasipurposive choreiform head and limb movements; 2) automatisms of muscle synergies of face, mouth, tongue, and pharynx; and 3) a continuum of more complex patterns of compulsive activities (stereotypies).

Akinesia and Catalepsy If in animals transmission at telencephalic (striatal) DA synapses is blocked either by pharmacological (*273*) or surgical (*29, 423, 476, 492, 552*) manipulations a "motor inactivation" syndrome develops that, in many respects, qualifies as a homologue of human parkinsonian brady- and akinesia. As in man (*460*) this state of immobilization cannot be attributed to rigidity of skeletal musculature subserving locomotion. It appears significant that akinetic states attending dopaminergic hypofunction bear a close relationship to catalepsy. As a distinct pharmacogenic sensorimotor syndrome, catalepsy was first associated with the Corydalis alkaloid, bulbocapnine (*165, 168*), and later on with centrally-acting cholinergic substances, such as eserine and arecoline (*512, 567*). (For other cataleptogens, see section on "Other Compounds" under "Dopamine Receptor Antagonists" in this chapter.) Following neuropharmacological convention, catalepsy[2] can be defined as the tendency of an experimental subject to retain awkward, unnatural, externally enforced body postures. An essential feature of this syndrome is a peculiar change in skeletal muscle tone commonly referred to as "waxy flexibility" or "lead-pipe rigidity." In contradistinction to states induced by sedative-hypnotic substances, there is no impairment in arousability by sensory stimuli (if sufficiently intense), in nociception, in muscle power and

---

[2] In the authors' opinion the term *catatonia*, although often used synonymously with catalepsy, should be reserved for states presenting mixtures or alternations of motor excitement and stupor.

coordination, and in the righting reflex (*315*). When compared across species, neuroleptic-induced catalepsy appears to correlate more closely with altered striatal DA metabolism than do other "dopaminoprivic" motor disabilities (*404*). Although this correlation is not absolute, and although the cataleptic syndrome as such is not a specific indicator of DA receptor antagonism or DA neuron blockade, its reliability in predicting parkinsonogenic potency of neuroleptic agents is unquestioned (*123, 286*).

Furthermore, several observations point to the pathogenetic relationship between dopaminolytic catalepsy and disordered striatal neuron activity: first, induction of a spreading depression in the striatum reverses reserpine catalepsy (*496*); and second, discrete chronic lesions of the corpus striatum (caudate, putamen, or globus pallidus) (*205, 388*) or combined ablation of the cortex and basal ganglia (thalamic preparation [*84*]) attenuate catalepsy induced either by phenothiazines or by a butyrophenone. Thus neuroleptic catalepsy, unlike that induced by cholinergic drugs, is dependent on the functional integrity of the corpus striatum. Since catalepsy produced by blockade of striatal DA synapses is antagonized by anticholinergic agents (*132, 137, 328*), rebound hyperactivity of cholinergic striopallidofugal elements could be its principal pathoneurophysiological correlate. Obviously catalepsy can only manifest itself if akinesia is present. It is therefore noteworthy that anticholinergics are relatively little or not effective in relieving either human parkinsonian akinesia or neuroleptic immobilization in the laboratory animal. Hence akinesia and catalepsy may represent separate, although interdependent, symptoms of deficient transmission at striatal DA synapses.

Akinesia caused by DA deficiency can perhaps be viewed as a breakdown in striatal processing of *parietal* cortical action programs insofar as they concern teleokinetic movement, catalepsy being a corollary symptom, insofar as it results from the concomitant suspension of postural regulatory mechanisms that constitute supportive motility. Inasmuch as akinesia and loss of postural regulation, in contradistinction to rigidity and tremor, are customarily classified as "negative" signs, i.e., reflections of neuronal hypofunction, the preferential responsiveness of akinetic symptoms to L-DOPA merits particular attention. *Prima facie*, such a classification is difficult to reconcile with an inhibitory synaptic action of DA, unless one assumes that the neuronal hypofunction relates not to a state of inactivity of dopaminoceptive cells but rather to a tonic inhibition of intrastriatal neurons synaptically linked with them (see section on "Concluding Remarks" in this chapter).

At this point it is perhaps pertinent to comment on the seeming absence of cataleptic signs in the symptomatology of *chronic* striatal DA deficiency in man, viz., in organic parkinsonism. From a semiotic point of view, catalepsy would be more difficult to detect when tremor and rigidity of the "cogwheel" type are simultaneously present. Nonetheless true catalepsy may occur in man during the acute phase of nigral neuron infection in virally induced encephalitis lethargica (*172*), a condition whose etiological relationship to parkinsonism is now established. (For a detailed account, including references, see Hassler [*248*] and Bernheimer et al. [*53*].)

Rigidity and Tremor As already pointed out, pharmacological blockade of central DA synapses can induce a parkinsonian syndrome in man that is clinically indistinguishable from that of organic degenerative origin. In addition to the impairment in teleokinetic motility, the disturbance of supportive motility manifests itself in muscle rigor, postural tremor, and loss of postural reflexes. The pathogenetic relationship of these motor disabilities to deficient nigrostriatal DA transmission is supported by a substantial corpus of clinical, neurochemical, and histopathological evidence (*53, 270, 271, 273, 275*). Despite much work aimed at elucidating the striking parallels between the organic and drug-induced syndrome, the pathoneurophysiological and neuroanatomical substrates of dopaminoprivic rigor and tremor are still far from being understood. Repeated attempts have been made to reproduce parkinsonian symptomatology in laboratory animals (see section on "Surgical and Drug-induced Lesions" in this chapter). The inconclusiveness of data generated by such experiments may mirror differences in motor organization between primates and nonprimates, or bipeds and quadrupeds, as much as inconsistent use of nomenclature or inappropriate neurological classification. The difficulties in reproducing Parkinson-like deficits in higher mammals by neurosurgical destruction of DA neurons have already been mentioned.

Similar obstacles are encountered in pharmacological manipulations of dopaminergic synaptic transmission: compared across species the various biochemical indicators of impaired dopaminergic transmission (with the exception of DA depletion) have not proved to be absolutely reliable as predictors of parkinsonogenic drug potency in man (*404*). Thus in certain species significant increases in DA turnover can be induced by neuroleptics in the absence of Parkinson-like motor deficits. Conversely, if motor disabilities occur during neuroleptic drug treatment, they are either transient, dissimilar from human parkinsonism, or quantitatively unrelated to the magnitude of biochemical changes. Notwithstanding these notes of caution, it is perhaps safe to say that the drugs that induce a parkinsonian syndrome in man, as a rule, modify cerebral DA metabolism in animals, irrespective of their motor effects on the latter.

Rigidity In keeping with the well-documented relationship between rigidity and proprioceptive motor innervation *(293)*, neurophysiological analyses of spinal reflexes reveal that the rigidity of tail and hind limb muscles induced by reserpine, chlorpromazine, or haloperidol is characterized by α-motoneuron hyperactivity and γ-motoneuron hypoactivity *(31, 42, 490)*. Administration of L-DOPA restores to normal the ratio between α and γ unit discharges. Since descending tegmentospinal NE fiber systems are known to play a part in the suprasegmental control of γ-motoneurons *(236)*, the latter finding by itself does not necessarily implicate striatal dopaminergic transmission failure as a primary cause of drug-induced rigidity. Evidently the NE system can be expected to respond to such pharmacological manipulations in much the same fashion.

Despite these uncertainties, the pathogenetic contribution of striatal neurons to rigidity is evinced by systematic ablation studies in rats, demonstrating the disappearance of electromyographic signs of rigidity only on bilateral extirpation of the corpora striata *(41)*, but not on decortication, pyramidectomy, cerebellectomy, or coagulation of the red nucleus. Consistent with these observations is the enhanced excitability of rat striatal neurons as tested with iontophoretically applied glutamate following systemic, rigidity-producing doses of reserpine and its reversal by L-DOPA *(489)*. In summary these observations explain why parkinsonian rigidity is effectively relieved in man by L-DOPA and other dopaminomimetics *(53, 141, 142, 411, 560)* (for additional references, see Hornykiewicz [*271*]).

Tremor Whether resulting from cerebral DA depletion or antidopaminergic drug treatment, parkinsonian tremor in man and its equivalent in laboratory animals are still the least understood of DA deficiency symptoms. In the human subject tremor is not infrequent during chronic neuroleptic drug treatment *(273)*, whereas in monkeys *(423)* it is rarely or not at all observed after such treatment or from lesions confined to the nigrostriatal DA pathway. However, if damage is inflicted solely on the rubroolivocerebellar connections, blockade of DA transmission by depleting or receptor blocking agents will cause the appearance of parkinsonian postural tremor in limbs contralateral to the lesion. Small doses of L-DOPA temporarily inhibit this symptom *(52)*.

Parkinsonian tremor in man, irrespective of its etiology, is also alleviated by various pharmacological measures aimed at restoring striatal DA-receptor activity. With the usual regimen of oral administration of L-DOPA, a temporary exacerbation may be seen before the gradual establishment of the therapeutic response which takes from several weeks to months. In the presence of peripheral AAAD inhibition, a much prompter onset of antitremor effect is obtained *(559)*. Concerning the role of GABA in tremorigenesis, see sections on "Release and Uptake" and "Acetylcholine-DA Balance" in this chapter.

Locomotor Mechanisms and Hyperkinesias In rodents the kinetic effects of DA and dopaminomimetics as well as the extrapyramidal effects of antidopaminergic substances have been advantageously employed as experimental variables in both qualitative and quantitative studies of striatal DA-receptor mechanisms. Unilateral activation (or inhibition) of the uncrossed nigrostriatal DA pathway leads to asymmetries in movement and posture that manifest themselves in rotational locomotion or contraversion *(12)*. Figure 3 reflects the experimental designs employed to achieve an interhemispheric asymmetry in telencephalic DA receptor activity. With both corpora striata intact, the direction of circling is away from the side the dopaminergic activity of which is enhanced over that of the other. Similarly when the corpus striatum of one hemisphere is removed, the direction of circling will be away from the intact side, if its DA synapses are stimulated, but toward the same side if they are blocked either pre- or postsynaptically. Since asymmetries following unilateral striatectomy or DA tract lesion do not persist, one must conclude that other neural mechanisms compensate for the interhemispheric imbalance in DA synaptic activity. The cholinergic nature of these compensatory readjustments may be inferred from the observation that anticholinergic agents induce ipsiversive circling in rats after nigrostriatal DA tract lesions *(132)*, although not after acute unilateral striatectomy *(13)*. This may be suggestive of a deactivation of the remaining (contralateral) DA fiber projection in conjunction with a corollary increase in cholinergic activity. Since relatively high doses of anticholinergic or anti-Parkinson substances are required to produce such effects, the question arises as to the contribution of other mechanisms, such as blockade of neuronal uptake of DA in the striatum, that might initiate indirect dopaminomimetic effects. It is unlikely, however, that such an action could account for the elimination of neuroleptic-induced asymmetries in the unilaterally lesioned animals since 1) there is no evidence to indicate an actual reversal of DA receptor blockade under such conditions *(13, 15, 23, 132)*; and 2) anti-Parkinson agents, which are potent Uptake$_1$ inhibitors, are not significantly more effective than anticholinergics which are poorly active in this respect *(214)*.

During the acute phase of DA fiber injury (caused by intranigral administration of 6-OHDA on one side), (+)-amphetamine-evoked rotational locomotion displays complex changes in direction *(525)*: until three hr following intranigral application of 6-OHDA, there is ipsiversion only. Subsequently contraversion precedes ipsiversion, the former reaching peak intensity and duration about one day postoperatively. After two days

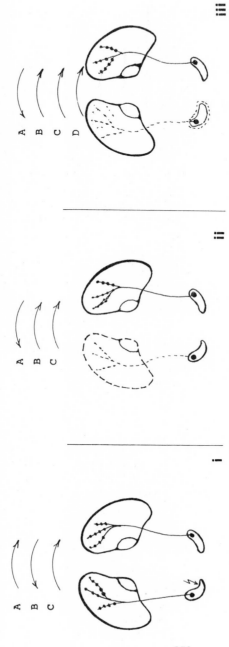

Figure 3. Rotational response as an indicator of DA receptor activity. Diagram depicts uncrossed, paired nigroneostriatal projections of the rat brain, as symbolized by two individual DA neurons. Lettered arrows indicate direction of postural and/or locomotor deviation with regard to side of experimental manipulation. i, unilateral alteration of dopaminergic input (on left) by (A) intrastriatal DA mimetics (*360, 528*), (B) intrastriatal DA antagonists (*360, 358*), and (C) electrical stimulation (*4*) of substantia nigra. ii, unilateral striatectomy combined with (A) systemically administered direct or indirect DA mimetics or releasing drugs (*13, 23, 34*), (B) DA receptor antagonists (*13, 15, 23*), and (C) DA depletors (*23*). iii, unilateral presynaptic lesion combined with (A) systemically administered indirect DA mimetics or releasing drugs (*8, 23, 34, 122, 149, 527*), (B) direct DA mimetics (*23, 38, 128, 525–527*), (C) DA receptor antagonists (*15, 23, 132*) and (D) DA depletors (*23, 31*).

272

ipsiversion becomes established as the response typical of the chronic stage of unilateral DA fiber loss. Given the impulse-flow dependence of (+)-amphetamine effects (see section on "Release and Uptake" in this chapter), it is conceivable that the contraversive episode (during which dopaminergic activity prevails on the lesioned side) reflects not only spontaneous amine leakage from disintegrating axon terminals, but also a pathologically enhanced, impulse-dependent release attendant on increases in axolemmal Ca^{2+} permeability (*513*).

In the chronic stage of unilateral "chemonigrectomy" systemic administration of dopaminomimetics induces rotatory locomotion, the direction of which depends on the synaptic locus of action. Thus (+)-amphetamine and other indirect dopaminomimetics cause ipsiversive circling, since they apparently can only activate intact DA synapses via released DA from the presynaptic terminal. In contrast, direct dopaminomimetics, L-DOPA, apomorphine, and piribedil, in doses that are ineffective in unilaterally striatectomized animals, cause contraversive circling, i.e., away from the lesioned side. Hence they appear to act primarily on the lesioned side, i.e., on DA receptors deprived of their presynaptic connections. Similarly a lowered threshold to apomorphine is detected in rats with a unilateral striatal lesion following reserpine-induced presynaptic blockade of striatal DA transmission.

In an attempt to explain these *supersensitivity* phenomena (*524*), the proposal has been made that in all these instances the DA receptor develops a denervation supersensitivity analogous to adrenoreceptors in peripheral tissues upon destruction of their noradrenergic nerve supply (*327*). This attractive explanation cannot be accepted without certain reservations until more definitive evidence becomes available, e.g., from quantitative electrophysiological measurements of sensitivity shifts of "denervated" dopaminoceptive striatal neurons. In a recent study with cats pretreated with repeated intraventricular injections of 6-OHDA, an increased percentage of striatal neurons was found to be sensitive to inhibition by microiontophoretic application of DA (*201*). The significance of these changes is not entirely clear since the glutamate-driven neurons apparently not only possessed an excitatory input from the SN but also showed signs of generalized hyperexcitability. Furthermore the dopaminomimetic (allegedly postsynaptic) effects of (+)-amphetamine, as well as inhibitory effects of nigral stimulation, were unaffected by 6-OHDA treatment.

Initially, the results of "repletion" experiments with L-DOPA in reserpinized rodents were interpreted as favoring a predominant role for DA, with no role or only a minor one for NE, in the regulation and maintenance of locomotor activity (*272*). There is now general agreement that such experiments suffer from at least two serious flaws: 1) the heteroneuronal uptake and formation of precursor amino acids and catecholamines, respectively; and 2) the fact that amine release will obviously occur onto nonphysiological postsynaptic sites. Besides, this "release" appears unrelated to physiological nerve activity as evidenced by the persistence of antireserpine effects of L-DOPA after chronic transection of DA fiber projections (*23*). Primarily on the basis of studies with amphetamines, other workers have held NE to be the cardinal mediator of drug-induced hypermotility (*146, 271, 455, 511*).

Recently, attention has been directed toward a functional interdependence between NE and DA in relation to rodent locomotor activity (*502, 503*). Specifically if care is taken to stimulate DA receptors while concurrent activation of NE synapses is precluded, maximal locomotor activation clearly appears reduced when compared with combined stimulation of DA and NE synapses (*375, 455*). Other examples are: 1) the diminution of apparent "intrinsic activity" of L-DOPA as a locomotor stimulant in the presence of DA-β-hydroxylase inhibitors and its restoration by the NE receptor agonist clonidine (*320*); and 2) the supraadditive stimulation obtained with combinations of apomorphine and clonidine in rodents pretreated with reserpine, α-MpT (H44/68), or FLA-63, a DA-β-hydroxylase inhibitor (*20, 345*). In the latter interaction a contributory factor may be the restraining influence of NE receptor stimulation on DA-mediated "stereotyped" activity, and thus the elimination of interference with running activity (*5*). Although these findings validate the concept of a functional synergism between dopaminergic and noradrenergic locomotory mechanisms, the question of the hierarchical organization of the two systems bears further investigation.

Clinical and pathoneurochemical evidence gleaned from parkinsonian patients favors *striatal* DA synaptic activity as playing a prepotent role in the initiation of locomotion (*53*). Ascending NE-containing ceruleocortical projections probably form an integral component of the "reticular activating system," and hence may be directly concerned with the control of corticostriate input and with EEG arousal. Bilateral destruction of the locus ceruleus in the cat largely diminishes ECoG activation but does not eliminate the animal's ability either to orient toward exteroceptive stimuli or to respond with behavioral arousal (*291*).

By contrast, bilateral destruction of the feline DA pathway is followed by a persistent comatose state during which spontaneous EEG alternations between sleep and waking patterns are present. Although the EEG can be desynchronized by sensory stimuli, the animal can no longer be aroused behaviorally. While these observations once again draw attention to the well-known dissociability between electrocorticographical and behavioral arousal, they also strengthen the view that the functional organization of neural processes as complex as the structuring of locomotor patterns depends on the interaction between NE- and DA-operated projections to the individual components of the corticostriopallidothalamocortical circuitry. Parenthetically the elimination of the stimulant locomotory effect of apomorphine by decortication, described more than 50 years ago, merits attention (*8*).

Recent neurophysiological work strongly supports the participation of tegmentospinal NE fiber projections in supraspinal fusimotor control (*236*). The differential modulation exerted by this projection on static and dynamic γ-motoneurons innervating flexor and extensor muscle groups may generate appropriate motoneuronal background discharges both in preparation for and during locomotion. Thus NE neurons may well be involved in locomotor regulation at both the telencephalic and segmental levels.

In attempts to delineate a neural substrate of the stimulant (loco-) motor effects of DA, intrastriatal microinjection experiments have been performed in freely moving cats and rats. While this approach can be criticized on methodological grounds, one is struck by the uniformity of motor and behavioral effects induced by topical application of DA (*127, 360, 528*), L-DOPA, (*127*), KCl (spreading depression) (*499*), or local electrical stimulation of the corpus striatum (*167, 293, 330*). Applied unilaterally, all of these manipulations lead to contraversive head and contralateral limb movements. The latter are considered by some workers to result from activation of internal capsular fibers (*7, 330*) and by others to originate from the caudate (*207, 293, 388*). It is remarkable that in cats, contrary to rats, intrastriatal injection of L-DOPA or DA causes no increase in locomotor activity (*127*). On the contrary, there appears to be a decrease similar to that obtained with low-frequency electrical stimulation of the caudate or putamen (*167*).

Also of interest is the change in direction of compulsive head movement from contraversion with lower doses to ipsiversion with higher doses. In later work (*126*), compulsive fore and hind limb movements following intracaudatal application of L-DOPA, DA, (+)-amphetamine, and methoxytyramine were reinterpreted as being equivalent to the persistent choreiform limb hyperkinesias observed in cats after electrolytic coagulation of the anteroventral portion of the head of the caudate nucleus (*338*). This region is reported to exert inhibitory influences on cortically evoked tetanic limb muscle contractions (*337*). It adjoins a facilitatory area that in turn overlaps with those sites (the rostromedial head of the caudate) that are responsive to DA agonists (*126*). Consequently it is argued that 1) DA activates inhibitory, intrinsic caudate neurons projecting to the aforementioned inhibitory region; and 2) in pathogenetic terms, the observed involuntary limb movements are equivalent to choreoathetoid hyperkinesias occurring as a result of L-DOPA medication in patients with parkinsonism or Huntington's chorea. Other workers, however, believe that L-DOPA hyperkinesias develop when DA is formed and released at synapses where it does not occur physiologically, e.g., 5-HT terminals (*410*). At present it would be premature to preclude an involvement of extrastriatal, nondopaminergic sites in the hyperkinetic effects of L-DOPA.

Neuropharmacological analyses of L-DOPA hyperkinesias in rats have provided additional data implicating the nigrostriopallidum in choreiform dyskinesias of the head-mouth-neck region, a particularly troublesome adverse effect in parkinsonian patients under L-DOPA medication (*69, 141, 411, 560*). A striking parallel to the human buccolinguofacial syndrome are the hyperkinesias of the branchiomeric oropharyngeal musculature regularly exhibited by rats and cats when treated acutely with high doses of L-DOPA (*66, 67*). Specifically an obligatory component of the hyperkinesias, probably related to swallowing, has proved amenable to quantitative evaluation of dopaminomimetic or antagonist drug activity. In the rat substantial neuropharmacological evidence supports the contention that the rate of repetitive swallowing as evoked by L-DOPA and various dopaminomimetics provides a functional index of the degree of central DA receptor activity. Data obtained from microinjection and lesion experiments suggest that this response involves thalamus-bound pathways originating in the striopallidum and basal forebrain (*65, 67*); however, more work is required to delineate the precise topography of the latter and connections with diencephalic and bulbar serotoninergic structures that appear to play an equally important part in this bulbar motor synergy.

In summary, excessive activity at striatal DA synapses may well represent a central pathogenetic mechanism in choreatic hyperkinesias of organic degenerative and pharmacogenic origin. Undoubtedly additional factors contribute to their etiology in neuropathological conditions in the human subject. In the parkinsonian syndrome, degeneration of nigro-neostriatal DA neurons would entail a loss of DA storage capacity and possibly a receptor supersensitivity of the "denervation" type (*133*). Furthermore an altered interplay between caudatal and putaminal neuron populations, in conjunction with dopaminergic-cholinergic imbalance in either, may also be relevant, the putamen usually being more severely affected *(133)*. In Huntingtonian chorea there may exist a relative oversupply of DA at striatal dopaminoceptive sites owing to near-normal and slightly subnormal DA levels in putamen and caudate, respectively. In addition strionigral feedback regulation of DA cell activity may be defective as a result of degeneration of the fasciculus strionigralis *(535)*. Accordingly small doses of L-DOPA may exacerbate choreatic movements in subclinical cases and thus afford a diagnostic aid in the early detection of this condition *(116, 310)*.

Behavior

Stereotypies (Compulsive Behavior) In addition to nonpatterned hyperkinetic phenomena, pharmacological stimulation of telencephalic (striatal, mesolimbic) DA synapses also elicits diverse integrated motor activities. They manifest themselves through a spectrum of behaviors that incorporate basic components of species-specific motor repertoires. To a certain degree the incidence of these activities is correlated with the intensity of DA receptor stimulation, as evidenced in rodents by the effects of increasing doses of DA mimetics, such as apomorphine and (+)-amphetamine *(8, 246, 375)* (unpublished observations, D.B.). In ascending order one commonly observes freezing, tenseness, exaggerated startle reactions, predatory activities, and threatening or fighting behavior. Eventually oral activities such as sniffing, licking, gnawing, and biting become predominant, whereas locomotion and exploration subside. Likewise, both birds and larger mammals regularly display generalized motor excitement and compulsive oral behavior that, in the latter, may include excessive grooming to the point of self-mutilation *(238)*.

The salient feature of these behavioral patterns is their nonadaptive, nonpurposive, perseverative, and automatic nature. In other words, *stereotypy* appears to characterize not only nonlocomotory components (scanning head and neck movements and masticatory and oropharyngeal activities) but also those activities that relate either to locomotion (circus movements, running) or motor concomitants of aggression, described under the misleading label of "bizarre social behavior" *(325)*. Many investigators have somewhat arbitrarily excluded the latter activities from the category of stereotyped behavior. On the whole, with increasing intensity of dopaminergic stimulation, there seems to develop a progressive disintegration of complexly patterned behavior, a loss in ability to respond to prevailing environmental contingencies and a general shift toward oral approach behaviors. The common denominator of such deficits may be the animal's failure to attend to stimuli other than those arising from its immediate surroundings (or perhaps its visceral sphere).

A brief remark may be pertinent here on the relationship between the stereotyped patterns of the psychopharmacologist and the stimulus-bound activities of the behavioral physiologist. Stimulus-bound activities can be altered by modifying the environmental contingencies that elicit a particular behavior (*530*). By contrast, stereotyped activity of a seemingly goal-directed nature may occur in the absence of appropriate goal objects. It seems that neuropharmacologists, in studying stereotyped activity related to the manipulation of striatal DA mechanisms, have largely ignored the ethological aspects of drug-induced behavior patterns. As a result it is not always clear if the behavior under study is stereotyped or indeed represents a "fixed action pattern" (stimulus-bound behavior such as that occurring to electrical stimulation of the hypothalamus).

Many attempts have been made to identify the telencephalic target structures of DA fiber projections as the neural substrates of stereotyped behavior, particularly apomorphine- and (+)-amphetamine-induced stereotypies (*375*). As revealed by a recent topological study (*84*), stereotyped behavior patterns not only occur spontaneously in rats after chronic, total ablation of telencephalic dopaminergic structures, but can even be further enhanced by (+)-amphetamine. Significantly, however, the depressant effects of chlorpromazine are much reduced. Thus one may have to conclude that 1) the presence and integrity of dopaminergic forebrain areas, at least in the rat, are not an absolute prerequisite for the elaboration of stereotyped hyperactivity by dopaminomimetics; 2) patterns for stereotyped hyperactivity are likely to be integrated at hypothalamo-mesencephalic levels; and 3) neuronal circuits in telencephalic dopaminergic areas (corpus striatum, mesolimbic forebrain, and dopaminergic mesocortex) may exert inhibitory control over brain stem "analyzer-integrator" systems (*387*), while being themselves subject to inhibitory control through DA synapses. Although these considerations derive chiefly from work on amphetamine-induced stereotypies, they can equally well be reconciled with observations on the effects of apomorphine on animals with striatal lesions.

As first observed by Amsler in his pioneering studies (*8*), compulsive gnawing induced by apomorphine in guinea pigs and rats is abolished by acute bilateral extirpation of the caudate-putamen in previously decorticated animals. Such animals, however, also exhibit a lowered threshold to the "paralyzing" action of larger doses of apomorphine, a finding that Amsler interpreted as blockade of striatothalamic pathways. Our own observations favor an inhibitory action for apomorphine on brainstem structures below the thalamus. Fairly small amounts of apomorphine, if injected into the vertebral artery, exert a powerful inhibition of spontaneous swallowing or mastication evoked by close-arterial administration of the cholinomimetics, eserine and ethylarecaidine, in rats anesthetized with urethane (*67*) (unpublished observations, D.B.). By drawing attention to extrastriatal sites of action of apomorphine, these observations caution against attributing any CNS effect of the drug to an action on striatal DA receptors.

Schizophrenia The wealth of neuropharmacological data implicating the ascending DA pathway in drug-induced behavioral stereotypies in laboratory animals has left a strong impact on psychiatric research into the etiology of human psychoses. In attempts to bridge the gap between neuropharmacological animal studies and clinical investigations, work on amphetamine compounds has proved to be of eminent importance. Thus neuroleptic receptor antagonists display a fairly selective antiamphetamine (antistereotype or antihyperkinetic) activity in freely moving animals. In turn, antagonism to amphetamines (and also cataleptogenicity) shows a highly positive correlation with clinical antischizophrenic efficacy (*278, 375*). Yet the most suggestive lead is the striking phenomenological parallelism between amphetamine-induced psychosis, a schizoid paranoic delusional syndrome, and genuine paranoid schizophrenia (*36, 234, 235, 256*). Accordingly, it is hypothesized that the stereotypy-inducing action of amphetamines and its inhibition by antipsychotic neuroleptic agents in animals directly involve dopaminergic substrates in the basal ganglia, specifically, the corpus striatum (*206*); and that, *mutatis mutandis*, in human subjects the same neuronal system plays a central role in the psychotogenic and psychotolytic actions of amphetamines and neuroleptic agents, respectively (*432, 478*). Although ample clinicopharmacological evidence attests to the heuristic value of this hypothesis—essentially a restatement of Mettler's original thesis (*368*)—there is still insufficient neurochemical evidence to indicate a pathological alteration of brain DA neuron function in any of the subvarieties of schizophrenia (*406*). Nonetheless several clinical findings appear to provide additional leads. Of these we may mention 1) the potentiation of antipsychotic drug effects by agents that interfere with brain catecholamine synthesis, e.g., α-MpT and α-methyl-DOPA (*110*); 2) the activation of schizophrenic symptomatology by pharmacological procedures that are likely to enhance activity at central DA synapses, e.g., treatment with L-DOPA (*37, 195*), MAO inhibitors (*195*), amphetamine (*162*), methylphenidate (*60, 283*), and the DA β-hydroxylase inhibitor, disulfiram (*249*); and 3) the effective suppression of (+)-amphetamine-induced psychosis by antischizophrenic phenothiazines (*478*), especially chlorpromazine (refer also to preceding section on neuroleptic DA antagonists and striatal DA metabolism).

It is commonly held that anticholinergic agents alleviate most neuroleptic-induced extrapyramidal (parkinsonian) signs of striatal DA deficiency without significantly altering the antipsychotic activity of these compounds (note, however, [*611*]). Since 1) this functional antagonism is paralleled by a normalization of enhanced DA turnover within the corpus striatum (*10*), but not within the mesolimbic dopaminergic regions (nucleus accumbens); and 2) presumptive DA receptor blockers may augment ACh release from the striatum but not from the nucleus accumbens (*343*), some workers emphasize the possibility that mesolimbic regions may function as a major substrate of antipsychotic drug action. Although this argument ignores the poor therapeutic effect of anticholinergic drugs on parkinsonian akinesia, it deserves mention, especially since it has been shown that lesions of the rostromedial striatum involving the nucleus accumbens abolish the locomotor stimulation as well as the agitated hyperactivity induced by L-DOPA in cats, without interfering with the production of stereotyped movements (*245*).

Affective Disorders The argument favoring primary involvement of *extrastriatal,* nonnigral DA pathways in the pathogenesis of schizophrenia is compatible with the absence of an epidemiological correlation between parkinsonism and schizophrenia (note, however, Mettler [*368*]). There is growing interest in projections to limbic system regions, e.g., the "paleocortical" gray of the forebrain, the fundus striati, the interstitial nucleus of the diagonal band, the nucleus centralis of the amygdala, and the cingulate cortex—structures believed by some workers to contain neural circuits subserving telencephalic integration of visceroendocrine and reproductive activities (*225, 561*). Thus investigations into the pathobiology of brain DA neurons now form an important part of research strategies aimed at establishing a neurochemically based, unifying nosology of mood disorders. The desirability of such attempts is all the more apparent if one considers the large diagnostic vacillations between the so-called undifferentiated subcategories of schizophrenia and manic-depressive syndromes.

Inasmuch as the nosological boundaries between schizophrenia and affective disorders are vague, it is presently difficult to judge the merits of any hypothesis that implicates faulty DA metabolism as a pathogenetic factor in the latter. Notwithstanding a large body of neuropsychiatric and biochemical data (*88, 233, 376, 439, 457, 458*), the precise relationship between manic-depressive states and possible alterations in DA neuron function is far from being understood. To compound the problem, L-DOPA is reported to produce both antidepressant and depressant effects in patients with mood or movement disorders, or both (*120, 159, 351, 352*). It remains to be established 1) whether or not the antidepressant effect is primarily attributable to an increased availability of DA or NE, or both, at central synapses; and 2) to what degree the displacement of brain NE or serotonin by DA is responsible for the depressant action, particularly during long-term medication with large doses of DOPA.

In summary, it appears that neurochemicoclinical correlations in psychotic patients, in conjunction with neuropharmacological animal studies of drug-induced stereotypies and catalepsy, have drawn attention to the role of DA in psychotic disorders. More knowledge about dopaminergic synaptic functions in mesocortical and limbic areas may well prove to be of significance in the elucidation of schizophrenia. Certainly the information so far gained has contributed much to a better understanding of neuroleptic drug action in man, as well as to the development of more selective and powerful compounds. However, one does well to recognize that these chemical agents afford little more than a palliation of psychotic symptomatology, perhaps by inactivating DA-operated neural substrates concerned with the emotive or ideational elaboration of the psychotic process.

To date, few other experimental animal models have played a significant role in attempts to elucidate the psychobiological aspects of central DA neuron malfunction. During the past two decades extensive interest has been lavished on various paradigms of operant conditioning, including intracranial self-stimulation. It is well established that neuroleptic agents selectively depress schedule-controlled behaviors under aversive or food-reinforcement contingencies (*285, 391, 392*) as well as intracranial self-stimulation (*491*).

In cats (*307*), rats, and mice (*465*), L-DOPA partially reverses reserpine-induced suppression of nondiscriminate conditioned avoidance responding, restoration of the latter in rodents being dependent on the formation of DA and NE in the brain (*466, 467*). In self-stimulation experiments, data obtained with DA-β-hydroxylase inhibitors favor a central role for NE. Nonetheless the observation that positive self-stimulation sites are also located in ventral tegmental regions containing DA perikarya or their fiber projections points to the participation of dopaminergic mechanisms.

While much evidence gathered from ablation and stimulation studies in rats, cats, and monkeys indicates that striopallidal structures are involved in conditioned behaviors, such as delayed responding (*163*), delayed alternation (*449*), and conditioned avoidance (*330*), the perceived relationships between these findings and more recently accumulated knowledge about nigrostriopallidal and mesolimbic dopaminergic transmission systems are currently the subject of intensive investigations in many laboratories.

Acetylcholine-Dopamine Balance

Like DA receptor antagonists, cholinomimetic substances such as oxotremorine and eserine activate striatal DA turnover (*129, 333*). In either case cholinolytic agents can prevent the acceleration of telencephalic DA turnover without markedly altering DA steady-state levels, control turnover rates being either slightly decelerated or unaffected (*13, 132, 404*). Neuroleptics appear less susceptible than cholinomimetic agents to this antagonism which is not detectable in mesolimbic and tuberoinfundibular DA neurons (*10, 132*). Conversely either type of agent can give rise to Parkinson-like disabilities in animals (*137, 138, 303, 328, 360, 512, 567*) or aggravate human parkinsonism (*171a, 273*). Nearly all of these drug-induced extrapyramidal symptoms can be either partially or fully abolished by anticholinergics (*13, 118, 137, 328*). For a review of earlier literature, see Hornykiewicz (*273*). In contradistinction to metabolic alterations, certain of these neurological effects appear to be reversed more readily by anticholinergic agents when they are induced by neuroleptics rather than by cholinomimetics. Taken together, the foregoing data suggest that the proper functioning of the extrapyramidal motor system requires the maintenance of an equilibrium between dopaminergic and cholinergic synaptic activities.

A second line of evidence consistent with such a dopamine-acetylcholine balance evolved from studies demonstrating modulatory influences of central cholinergic mechanisms on animal behavior abnormalities (*40, 206, 214, 375, 454, 456*) or human neurological syndromes (*37a*) arising from DA receptor hyperactivity (see sections on "Locomotor Mechanisms and Hyperkinesias" and "Stereotypies [Compulsive Behavior]" in this chapter).

Current pharmacological models of the DA-ACh interdependence relate to three main concepts: 1) the functional antagonism between central dopaminergic and cholinergic neuron circuits; its corollary 2) a neural tuning mechanism operating to maintain a balance between both transmission lines; and 3) the cholinergic nature of feedback loops that effectuate compensatory activation of DA synthesis and turnover in response to blockade of postsynaptic receptors.

Despite a wealth of pertinent biochemical and pharmacological data, the synaptic makeup of and control modes governing the DA-ACh equilibrium are far from being understood. Although it is reported that 1) presumptive DA receptor antagonists augment the release of striatal ACh

(*343*), and 2) anticholinergic agents increase intrastriatal DA release (*360*), the existence of an axoaxonal synaptic linkage between ACh and DA fibers has been questioned (*132*), and hence cholinergic modulation of DA release at the level of the axon terminal. In view of the well-documented cholinergic striatonigral projection (*405*), emphasis has been placed on modulation of nigrostriatal DA input via cholinergic influences impinging on nigral pars compacta cells. If such be the case, the error signal fed back to nigral DA perikarya may represent a derivative of striatal dopaminergic activity.

Pharmacological observations are indeed compatible with the operation of excitatory cholinoceptive synapses at the level of the SN. Results from a study in which eserine crystals were implanted into the SN of rats suggest that pars compacta DA neurons may be activated by excitatory cholinergic synapses (*476*). However, in recent microiontophoretic studies of SN neurons, only cells in the *pars reticulata* exhibited a high sensitivity to ACh (*4*). Conversely, since depressed nigrostriatal synaptic activity enhances intrastriatal ACh release, the assumption seems viable that striopallidofugal efferent fibers (when released from DA inhibition) act to increase the activity in cholinergic fiber projections from brain stem to striatum. That is to say, striatofugal feedback loops, the transmitters of which are unknown (GABA?), may modulate not only the dopaminergic but also the cholinergic input into the striatum.

As already pointed out, some accompaniments of deficient dopaminergic activity, e.g., rigidity, tremor, and catalepsy, can be antagonized by anticholinergic agents; however, akinesia is either little affected (*132, 328*) or even exacerbated (*460*). Indeed, the antagonism between anticholinergics and neuroleptics raises certain questions about the proposed feedback role of cholinergic striatonigral fibers. If the *sole* function of this tract were to regulate nigral DA fiber input into striatal command neurons, say in a proportional control mode, one would expect anticholinergic agents to exert counterproductive effects in relation to those extrapyramidal manifestations that are caused solely by a failure in striatal dopaminergic transmission. For that reason neuroleptic *rigidity, tremor,* and *catalepsy* are probably a direct reflection of unrestrained hyperactivity of cholinoceptive rather than of dopaminoceptive effector neurons located either in the striopallidum or the pars reticulata of the SN. In keeping with this notion, neuroleptic catalepsy is, on the one hand, largely abolished in rats by neostriatal, pallidal, and SN lesions. By contrast, cholinergic catalepsy, which is enhanced by neostriatal lesions, is unaffected by pallidal lesions but is abolished by nigral destruction (*137, 138*). On the other hand, striatal lesions eliminate rigidity induced either by neuroleptics or by cholinomimetics *(23, 41, 137)*. The situation is not so clear with regard to tremor *(328, 423)*: the hyperactive cholinergic synapse may be located within the striopallidal complex *(423)*. It is noteworthy that GABA derivatives (3,4,5-trimethoxybenzoyl-1,4-aminobutyrate *(48)* and 1-hydroxy-3-amino-pyrrolidone-2 *[83]*) antagonize tremor produced by oxotremorine, and that the former effectively relieves parkinsonian tremor in man without altering akinesia and rigidity *(150)*. Such findings draw attention to the possible role of GABA-operated striopallidal fiber projections in the genesis of tremor.

Akinesia and probably certain noncholinergic forms of catalepsy may relate more directly to impaired dopaminergic transmission and hence to unbalanced activity of noncholinergic striopallidal efferent projections, perhaps the striatofugal GABA fiber system (see also section on "Release and Uptake" in this chapter). The latter may participate in a feedback circuit not only governed by a different control mode but also primarily concerned with the regulation of dopaminoceptive effector neurons at the level of the striatum. It appears that this pathway undergoes a total degeneration in Huntington's chorea *(535)*.

Directly related to the foregoing considerations is the question of the role of the ascending striatopetal cholinergic pathway. As revealed in studies with cholinesterase staining, this fairly substantial fiber system of unknown origin arises in the mesencephalon and travels upward in close proximity to the SN *(473)*, which finding would explain the release of striatal ACh upon electrical stimulation of the SN *(427)*. One may conjecture that cholinergic influences on DA-sensitive striatal neurons are brought to bear via this pathway through axodendritic synapses. Although pertinent data from electron microscopic and pharmacological studies are scant, one should perhaps consider more seriously than heretofore the possibility of a heterosynaptic interaction between DA and ACh at the same postsynaptic element, such as proposed by Libet and Tosaka *(336)* for peripheral sympathetic ganglia. In these, DA has been implicated as the ultimate mediator of the muscarinic slow inhibitory postsynaptic potential and, in addition, as a selective facilitator of the muscarinic slow excitatory postsynaptic potential. Dopaminergic modulation of the latter, once established, is indeed remarkable for both its extraordinary persistence and its complete insensitivity to subsequent exposure to receptor blocking agents.

According to Libet and Tosaka, this process may be interpreted as the result of "long-lasting metabolic and/or structural change in the postsynaptic neuron." A similar heterosynaptic interaction may operate at the level of the dendritic spines of striatal neurons upon which different synaptic inputs converge, in particular cholinergic and dopaminergic ones. Certain observations from intracellular experiments on caudate neurons, such as the nigrally evoked slow depolarizing membrane potential shifts and their concomitant modulations of spike discharges, are compatible with this view (277).

The model of Libet and Tosaka affords a conceptual departure from unitary schemes of ACh-DA antagonism within the nigrostriopallidal circuitry. Thus one can envisage a facilitatory effect of DA on ACh synapses at dendritic shafts and spines that could effectively modulate the degree of convergence between various inputs and that may be crucial for the integrating properties of the entire network.

Admittedly it remains for future research to determine how neuroleptic actions may fit into this conceptual frame. The neuroleptic blockade of dopaminergic transmission may indeed be infinitely more complex than the commonly presumed occlusion of the DA receptor. *Inter alia*, an excessive release of DA in the presence of such agents may bring about a spillover of the amine into the wider synaptic surroundings and reinforce such heterosynaptic interaction, irrespective of an existing DA receptor block.

Although the foregoing conclusions are largely of an inferential nature, they appear to be pertinent to any attempt to relate biochemically based models of DA turnover regulation to a rational pathoneurophysiology of "striatal DA deficiency syndromes" (270, 271).

CONCLUDING REMARKS

Pathophysiology of the Basal Ganglia

The nigrostriopallidum, a major component of the extrapyramidal system, has traditionally been regarded as forming an integral part of the supraspinal system that regulates voluntary movement and posture (165a, 293, 349). This concept evolved from clinical and neuropathological observations in the human, in conjunction with ablation and electrical stimulation studies in animals (330). With the advent of more sophisticated biochemical and histochemical techniques, research in this field has advanced beyond the merely descriptive juxtaposition of pathomorphological and neurological data, and functional relationships between components of the nigrostriopallidal circuit are now being defined in neurochemical terms. This rapid expansion of pharmacological and neurochemical knowledge has ushered in a new era in neurophysiological and clinical investigation. And an elucidation of the synaptic and cellular mechanisms underlying the physiological functions of the nigrostriopallidal system may no longer be an elusive goal.

The discovery that striatal DA deficiency is the common pathogenetic denominator of parkinsonian disorders of various etiologies (*53, 270*) now stands as a major scientific breakthrough, thrusting the nigrostriopallidum squarely onto the center stage of experimental endeavor. The realization that striatal DA deficiency, per se, appears to be a sufficient and necessary etiopathogenetic factor in parkinsonian dysfunctions, such as akinesia and rigidity, gave birth to the postulate that DA is crucial in the interneuronal transmission process at the striopallidal level, and it prompted a reevaluation of traditional concepts of the role of the nigrostriopallidum in telencephalic motor control.

According to the working hypothesis advanced by Hornykiewicz (*272*), the effects of DA on motor activity can be considered to reflect its inhibitory transmitter action on neurons of the striatum, a structure traditionally believed to be concerned with the "... elaboration of inhibitory impulses that modify the primitive motor patterns of the pallidum in such a way as to permit the performance of purposeful motor activity of the cortex." Viewed from this perspective, the stimulatory effect of DA on motor activity essentially appears to be one of disinhibition, resulting from a release of the pallidum from striatal or cortical inhibitory influences, or both. Attention has to be given, however, to the possibility of a direct facilitatory synaptic action to account for the apparent excitatory effect produced by DA on neurons of the putamen.

The question thus arises as to how the perceived sequelae of deficient dopaminergic transmission at striatal synapses fit into conventional neurological concepts of parkinsonian symptomatology. If the sole synaptic action of DA were one of inhibition, then rigidity, tremor, and akinesia should correspond to positive signs, i.e., release phenomena. If, alternatively, one imputes an additional direct facilitatory action to DA, akinesia and loss of postural reflexes may represent negative symptoms as customarily assumed. In either case tremor and rigidity would still be regarded as positive signs because they are relieved by pallido- or thalamotomies, which often aggravate akinesia. In the absence of more conclusive experimental evidence, and in view of the neurophysiological oversimplifications inherent in such interpretations, it may be gratuitous to judge the relative merits of the preceding concepts, as well as to look for still other schemata. Nonetheless it is tempting to conceptualize the relationship between dopaminergic synaptic transmission and either positive or negative signs on the basis of a *single*, viz., inhibitory, DA action:

So-called positive symptoms may be viewed as reflecting rebound *pallidal overactivity* owing to the loss of inhibitory input via nigroneostriatal DA fibers. The latter may result in disinhibition of intrinsic inhibitory striatal elements that impinge on the long-axon neurons projecting to the pallidum, including the entopeduncular nucleus of subprimates (*430, 505, 506, 538, 565*).

These striopallidal fibers may be identical with collaterals of striatonigral GABA fibers (*428, 429*). On the other hand, so-called negative symptoms may reflect either hypoactivity of certain pallidal elements owing to the overactivity of inhibitory striatofugal (GABA) neurons normally under direct control by nigral DA fibers, or an altered balance between caudate/putamen input to pallidal neurons. The basic assumption necessary is that DA should be able to inhibit intrinsic, inhibitory, striatal neurons as well as inhibitory long-axon (GABA) *exit* neurons projecting to the globus pallidus and tegmentum. Loss of DA inhibition would therefore increase the activity of exit neurons if they are *not* predominantly controlled by intrinsic inhibitory elements, or decrease the activity of exit neurons if they are primarily under intrinsic neuronal control. One must also consider the possibility that pallidal neurons may be inhibited directly by midbrain catecholamine fibers. At present, however, there is only sparse information on DA sensitivity of pallidal cells and the mode of termination of nigropallidal DA fibers.

Kornhuber (*313, 314*) recently postulated that one aspect of nigrostriopallidal motor function may be operationally defined as ramp generation, i.e., the elaboration of thalamocortical activation patterns which underlie "voluntary-speed smooth movement." The salient feature of this hypothesis is that the motor cortex—in its capacity as a somatosensory and vestibular association area—is primarily concerned with the continuous monitoring and feedback regulation of striopallidothalamic output patterns. Concerning information flow, the motor cortex would thus follow rather than lead the striopallidum as was originally assumed. In support of his own theory, Kornhuber cited data by M. R. De Long reported at an NRP conference held at Boulder, Colorado in 1972 (*313*).

At this point our rudimentary knowledge about input-output relations governing the mode of operation of the striatal neuron network forestalls any attempt to define the nature of striatal ramp discharge patterns in precise neurophysiological terms. The question of how DA may modulate this process can only be answered in qualitative, descriptive terms. Evidently the presence of DA at striatal synapses is a prerequisite for ramp generation. The lack or absence of DA would thus lead to deficient ramp production as reflected by the symptoms of brady-, hypo-, or akinesia, whereas its overabundance would entail released ramp production as manifested by the signs of choreoathetoid hyperkinetic disturbances.

Transmitter Action of Dopamine

Brain DA satisfies both ancillary and cardinal criteria of transmitter function: First, despite gaps in our knowledge about the ultrastructure of central DA synapses, the *criterion of synaptic localization* has been established. In addition the functional dynamics of neuronal DA *storage mechanisms* have been delineated. Second, the topochemistry and kinetics of DA *synthesizing and catabolizing enzymes* have been elucidated in great detail and, in conjunction with advances in immunohistochemical methodology, afford a fairly coherent picture of DA metabolism at intraneuronal and extraneuronal sites. Third, the experimental validation of the *criterion of collectibility* comes from repeated demonstrations of DA release by neural stimuli, Fourth, the functional counterpart to release—neuronal reuptake of liberated amine ($Uptake_1$)—has been identified as the major *mechanism of termination of synaptic activity*. The pharmacological effects of a number of centrally acting drugs, e.g., the neuroleptics and psychoanaleptics, can be related to their actions on release and reuptake of brain DA.

Attempts to ascertain the *identity of synaptic action* are severely constrained by the present inability to identify, either in neurophysiological or morphological terms, or both, the neuronal substrates possessing DA synapses. As shown in the striatum, there is a high degree of concordance between extracellular unit responses evoked by manipulations known to activate DA release, on the one hand, and microiontophoretic application of this amine, on the other. However, the identity of the active chemical principle released by presynaptic electrical stimulation with endogenous DA has merely been inferred by analogy from perfusion experiments.

At this point a great deal of work remains to be done before one can hope to particularize in physicochemical terms the transductive process underlying receptor-triggered alterations in neuronal activity. The hypothesized link with adenylcyclase appears to hold promise; its corroboration calls for more rigorous testing of its relationship to dopaminergic synaptic activity in the intact brain. This may entail an examination of the time courses of ion conductance changes at postsynaptic membranes and of intracellular biochemical parameters, say, protein kinase activity. Although intracellular recordings are indispensable to the elucidation of the synaptic transmission process at the cell membrane level, application of this technique to small neuron networks, such as the striatal neuropil, poses a rigorous challenge. This stricture applies even more to microiontophoretic methodology, and one cannot help but wonder how such "diluvian" techniques can come to grips with the forbidding complexity of the microstructure of dendritic synaptic mosaics encountered in the striatal neuropil. In such a synaptic microcosm the function of the individual neuron transcends the mere relaying of incoming information, making difficult any attempt to derive conclusions on the functional properties of the network from changes in single unit activity.

To end this chapter on a gloomy note would be unjustified. Certainly the discovery of brain DA deficiency as the underlying pathogenetic correlate of the parkinsonian syndrome stands as a milestone since it has established, for the first time, disordered metabolism of a natural constituent of the brain as the causal factor of a CNS disease. Thus Thudichum's venerable theory (516) was finally brought to fruition.

LITERATURE CITED

1. Aceto, M. D., Harris, L. S., Lescher, G. Y., et al.: Pharmacological studies with 7-benzyl-1-ethyl-1,4-dihydro-4-oxo-1,8-napthyridine-3-carboxylic acid. *J. Pharmacol. Exp. Ther.* 158:286–293, 1967.
2. Adey, W. R. and Dunlop, C. W.: Amygdaloid and peripheral influences on caudate and pallidal units in the cat and effects of chlorpromazine. *Exp. Neurol.* 2: 348–363, 1960.
3. Aghajanian, G. K. and Bunney, B. S.: Central dopaminergic neurons: Neurophysiological identification and responses to drugs. In Usdin, E., and Snyder, S. J. (Eds.): *Frontiers in Catecholamine Research.* Oxford: Pergamon Press, 1973, pp. 643–648.
4. Aghajanian, G. K. and Roth, R. H.: γ-Hydroxybutyrate-induced increase in brain dopamine: Localization by fluorescence microscopy. *J. Pharmacol. Exp. Ther.* 175: 131–138, 1970.
5. Ahlenius, S. and Engel, J.: Behavioral and biochemical effects of L-DOPA after inhibition of dopamine-β-hydroxylase in reserpine pretreated rats. *Naunyn Schmiedebergs Arch. Pharmacol.* 270: 349–360, 1971.
6. Aigner, A., Hornykiewicz, O., Lisch, H.-J., et al.: Beeinflussung der Gehirn-Katecholamine, der Spontanaktivität und der L-DOPA-Hyperaktivität durch Diäthyldithiocarbamat. *Med. Pharmacol. Exp.* 7: 576–585, 1967.
7. Akert, K. and Anderson, B.: Experimenteller Beitrag zur Physiologie des Nucleus Caudatus. *Acta Physiol. Scand.* 22: 281–298, 1951.
8. Amsler, C.: Beiträge zur Pharmakologie des Gehirns. 2. Über einige Wirkungen des Apomorphins. *Naunyn Schmiedebergs Arch. Exp. Pathol. Pharmacol.* 97: 1–14, 1923.
9. Anagnoste, B. and Goldstein, M.: The effects of ouabain on catecholamine biosynthesis in different areas of rats' brains. *Pharmacologist 9:* 210, 1967.
10. Andén, N.-E.: Dopamine turnover in the corpus striatum and the

limbic system after treatment with neuroleptic and anti-acetylcholine drugs. *J. Pharm. Pharmacol. 24:* 905–906, 1972.
11. Andén, N.-E.: Effects of reserpine and a tyrosine hydroxylase inhibitor on the monoamine levels in different regions in the rat central nervous system. *Eur. J. Pharmacol. 1:* 1–5, 1967.
12. Andén, N.-E.: On the function of the nigrostriatal dopamine pathway. In von Euler, U. S., Rosell, S., and Uvnas, B. (Eds.): *Mechanisms of Release of Biogenic Amines.* Oxford: Pergamon Press, 1966, pp. 357–359.
13. Andén, N.-E. and Bédard, P.: Influences of cholinergic mechanisms on the function and turnover of brain dopamine. *J. Pharm. Pharmacol. 23:* 460–462, 1971.
14. Andén, N.-E., Bédard, P., Fuxe, K., et al.: Early and selective increase in brain dopamine levels after axotomy. *Experientia 28:* 300–301, 1972.
15. Andén, N.-E., Butcher, S. G., Corrodi, H., et al.: Receptor activity and turnover of dopamine and noradrenaline after neuroleptics. *Eur. J. Pharmacol. 11:* 303–314, 1970.
16. Andén, N.-E., Carlsson, A., Dahlström, A., et al.: Demonstration and mapping out of nigro-neostriatal dopamine neurons. *Life Sci. 3:* 523–530, 1964.
17. Andén, N.-E., Carlsson, A., and Häggendal, J.: Adrenergic mechanisms. *Ann. Rev. Pharmacol. 9:* 119–134, 1969.
18. Andén, N.-E., Corrodi, H., Dahlström, A., et al.: Effects of tyrosine hydroxylase inhibition on the amine levels of central monoamine neurons. *Life Sci. 5:* 561–568, 1966.
19. Andén, N.-E., Corrodi, H., and Fuxe, K.: Effects of neuroleptic drugs on central catecholamine turnover assessed using tyrosine- and dopamine-β-hydroxylase inhibitors. *J. Pharm. Pharmacol. 24:* 177–182, 1972.
20. Andén, N.-E., Corrodi, H., Fuxe, K., et al.: Evidence for a central noradrenergic stimulation by clonidine. *Life Sci. 9:* 513–523, 1970.
21. Andén, N.-E., Corrodi, H., Fuxe, K., et al.: Importance of nervous impulse flow for the neuroleptic induced increase in amine turnover in central dopamine neurons. *Eur. J. Pharmacol. 15:* 193–199, 1971.
22. Andén, N.-E., Dahlström, F., Fuxe, K., et al.: The effect of haloperidol and chlorpromazine on the amine levels of central monoamine neurons. *Acta Physiol. Scand. 68:* 419–420, 1966.
23. Andén, N.-E., Dahlström, A., Fuxe, K., et al.: Functional role of the nigro-neostriatal dopamine neurons. *Acta Pharmacol. Toxicol. 24:* 263–274, 1966.
24. Andén, N.-E., Dahlström, A., Fuxe, K., et al.: Further evidence for the presence of nigro-neostriatal dopamine neurons in the rat. *Am. J. Anat. 116:* 329–333, 1965.
25. Andén, N.-E., Dahlström, A., Fuxe, K.: et al.: Mapping out of catecholamine and 5-hydroxytryptamine neurons innervating the telencephalon and diencephalon. *Life Sci. 4:* 1275–1279, 1965.
26. Andén, N.-E., Dahlström, A., Fuxe, K., et al.: Ascending monoamine

neurons to the telencephalon and diencephalon. *Acta Physiol. Scand. 67:* 313–326, 1966.
27. Andén, N.-E., and Fuxe, K.: The influence of benzquinamide, oxypertine and prenylamine on monoamine levels and on monoamine effects in the spinal cord. *Acta Pharmacol. Toxicol. 30:* 225–237, 1971.
28. Andén, N.-E., Fuxe, K., and Hökfelt, T.: Effect of some drugs on central monoamine terminals lacking nerve impulse flow. *Eur. J. Pharmacol. 1:* 226–232, 1967.
29. Andén, N.-E., Fuxe, K., and Larsson, K.: Effects of large mesencephalic and diencephalic lesions on the noradrenaline, dopamine and serotonin neurons of the central nervous system. *Experientia 22:* 842–843, 1966.
30. Andén, N.-E., Fuxe, K., Hamberger, B., et al.: A quantitative study on the nigro-neostriatal dopamine neuron system in the rat. *Acta Physiol. Scand. 67:* 306–312, 1966.
31. Andén, N.-E., Larsson, K., and Steg, G.: The influence of the nigro-neostriatal dopamine pathway on spinal motoneuron activity. *Acta Physiol. Scand. 82:* 268–271, 1971.
32. Andén, N.-E., Magnusson, T., and Stock, G.: Effects of drugs influencing monoamine mechanisms on the increase in brain dopamine produced by axotomy or treatment with gammahydroxybutyric acid. *Naunyn Schmiedebergs Arch. Pharmacol. 278:* 363–372, 1973.
33. Andén, N.-E., Roos, B.-E., and Werdinius, B.: On the occurrence of homovanillic acid in brain and cerebrospinal fluid and its determination by a fluorimetric method. *Life Sci. 7:* 448–458, 1963.
34. Andén, N.-E., Rubensson, A., Fuxe, K., et al.: Evidence for dopamine receptor stimulation by apomorphine. *J. Pharm. Pharmac. 19:* 627–629, 1967.
35. Angelakos, E. T.: Regional distribution of catecholamines in the dog heart. *Circ. Res. 16:* 39–44, 1965.
36. Angrist, B. M., and Gershon, S.: Amphetamine induced schizophreniform psychosis. In Siva Sankar, D. V. (Ed.): *Schizophrenia. Current Concepts and Research.* New York: PJD Publications, 1969, pp. 508–524.
37. Angrist, B. M., Sathananthan, G., Wilk, S., et al.: Behavioral and biochemical effects of L-DOPA in psychiatric patients. In Usdin, E. and Snyder, S. H. (Eds.): *Frontiers in Catecholamine Research.* Oxford: Pergamon Press, 1973, pp. 991–993.
37a. Aquilonius, S. M. and Sjöström, R.: Cholinergic and dopaminergic mechanisms in Huntington's chorea. *Life Sci. 10*:405–414, 1971.
38. Arbuthnott, G. W. and Crow, T. J.: The relationship between turning behavior and unilateral release of dopamine in the rat. *Exp. Neurol. 30:* 484–491, 1971.
39. Arbuthnott, G. W., Crow, T. J., Fuxe, K., et al.: Behavioral effects of stimulation in the region of the substantia nigra. *J. Physiol.* (Lond.) *210:* 61–62P, 1970.
40. Arnfred, T. and Randrup, A.: Cholinergic mechanisms in brain inhib-

iting amphetamine-induced stereotyped behavior. *Acta Pharmacol. Toxicol. 26*:384–394, 1968.
41. Arvidsson, J., Jurna, I., and Steg, G.: Striatal and spinal lesions eliminating reserpine and physostigmine rigidity. *Life Sci. 6:* 2017–2020, 1967.
42. Arvidsson, J., Roos, B.-E., and Steg, G.: Reciprocal effects on α and γ motoneurons of drugs influencing monoaminergic and cholinergic transmission. *Acta Physiol. Scand. 67:* 398–404, 1966.
43. Axelrod, J.: The metabolism storage and release of catecholamines. *Recent Prog. Horm. Res. 21:* 597–622, 1965.
44. Axelrod, J.: Metabolism of epinephrine and of the sympathomimetic amines. *Physiol. Rev. 39:* 751–776, 1959.
45. Bagchi, S. P. and McGeer, P. L.: Some properties of tyrosine hydroxylase from the caudate nucleus. *Life Sci. 3:* 1195–1200, 1964.
46. Balagura, S., Wilcox, R. H., and Coscina, D. V.: The effects of diencephalic lesions on food intake and motor activity. *Physiol. Behav. 4:* 629–633, 1969.
47. Baldessarini, R. J. and Kopin, I. J.: The effect of drugs on the release of norepinephrine-H^3 from central nervous system tissues by electrical stimulation *in vitro. J. Pharmacol. Exp. Ther. 156:* 31–38, 1967.
48. Baraldi, M., Bertolini, A., Baggio, G., et al.: Selective inhibition of oxotremorine tremor with sodium 3,4,5-trimethoxybenzoyl-1,4-amino butyrate. *Riv. Farmacol. Ter. 3:* 179–193, 1972.
49. Bartholini, G., Da Prada, M., and Pletscher, A.: Decrease of cerebral 5-hydroxytryptamine by 3,4-dihydroxyphenylalanine after inhibition of extracerebral decarboxylase. *J. Pharm. Pharmacol. 20:* 228–229, 1968.
50. Bartholini, G. and Pletscher, A.: Cerebral accumulation and metabolism of C^{14}-dopa after selective inhibition of peripheral decarboxylase. *J. Pharmacol. Exp. Ther. 161:* 14–20, 1968.
51. Bédard, P., Larochelle, L., Parent, A., et al.: The nigrostriatal pathway: A correlative study based on neuroanatomical and neurochemical criteria in the cat and monkey. *Exp. Neurol. 25:* 365–377, 1969.
52. Bédard, P., Larochelle, L., Poirier, L. J., et al.: Reversible effect of L-DOPA on tremor and catatonia induced by α-methyl-*p*-tyrosine. *Can. J. Physiol. Pharmacol. 48:* 82–84, 1970.
53. Bernheimer, H., Birkmayer, H., Hornykiewicz, O., et al.: Brain dopamine and the syndromes of Parkinsonism and Huntington: Clinical, morphological and neurochemical correlations. *J. Neurol. Sci. 20:* 415–455, 1973.
54. Bertler, Å.: Occurrence and localization of catecholamines in human brain. *Acta Physiol. Scand. 51:* 97–107, 1961.
55. Bertler, Å., Falck, B., Gottfries, C. G., et al.: Some observations on adrenergic connections between mesencephalon and cerebral hemispheres. *Acta Pharmacol. Toxicol.* (Kbh.) *21:* 283–289, 1964.
56. Bertler, Å., Falck, B., Owman, Ch., et al.: The localization of mono-

aminergic blood-brain barrier mechanisms. *Pharmacol. Rev. 18:* 369–385, 1966.
57. Bertler, Å, Hillarp, N.-Å., and Rosengren, E.: "Bound" and "free" catecholamines in the brain. *Acta Physiol. Scand. 50:* 113–118, 1960.
58. Bertler, Å. and Rosengren, E.: On the distribution of monoamines and of enzymes responsible for their formation. *Experientia 15:* 382–384, 1959.
59. Bertler, Å. and Rosengren, E.: Occurrence and distribution of catecholamines in brain. *Acta Physiol. Scand. 47:* 350–361, 1959.
60. Bertler, Å. and Rosengren, E.: Occurrence and distribution of dopamine in brain and other tissues. *Experientia 15:* 10–11, 1959.
61. Besson, M. J., Cheramy, A., Feltz, P., et al.: Dopamine: Spontaneous and drug induced release from the caudate nucleus in the rat. *Brain Res. 32:* 407–424, 1971.
62. Besson, M. J., Cheramy, A., Feltz, P., et al.: Release of newly-synthesized dopamine from dopamine-containing terminals in the striatum of the rat. *Proc. Nat. Acad. Sci. U.S.A. 62:* 741–748, 1969.
63. Besson, M. J., Cheramy, A., and Glowinski, J.: Effects of some psychotropic drugs on dopamine synthesis in the rat striatum. *J. Pharmacol. Exp. Ther. 177:* 196–205, 1971.
64. Beyer, K. H.: Sympathomimetic amines: The relation of structure to their action and inactivation. *Physiol. Rev. 26:* 169–197, 1946.
65. Bieger, D.: Influence of striatal dopamine receptor blockade on a bulbar motor reaction. *Neuropharmacol. 13:* 1141–1152, 1974.
66. Bieger, D. and Hockman, C. H.: Unpublished observations.
67. Bieger, D., Larochelle, L., and Hornykiewicz, O.: A model for the quantitative study of central dopaminergic and serotoninergic activity. *Eur. J. Pharmacol. 18:* 128–136, 1972.
68. Birkmayer, W. and Hornykiewicz, O.: Der L-3,4-Dioxyphenylalanin (=DOPA)-Effekt bei der Parkinson-Akinese. *Wien Klin. Wochenschr. 73:* 787–788, 1961.
69. Birkmayer, W. and Mentasti, M.: Weitere experimentelle Untersuchungen über den Catecholaminstoffwechsel bei extrapyramidalen Erkrankungen (Parkinson-und Chorea-Syndrom). *Arch. Psychiatr. Nervenkr. 210:* 29–35, 1967.
70. Björklund, A., Ehinger, B., and Falck, B.: A method for differentiating dopamine from noradrenaline in tissue sections by microspectrofluorometry. *J. Histochem. Cytochem. 16:* 263–270, 1968.
71. Björklund, A. and Nobin, A.: Fluorescence histochemical and microspectrofluorimetric mapping of dopamine and noradrenaline cell groups in rat diencephalon. *Brain Res. 51:* 193–205, 1973.
72. Blanc, G., Glowinski, J., Stinus, L., et al.: Is cortical dopamine only the precursor of noradrenaline? *Br. J. Pharmacol. 47:* 648P, 1973.
73. Blaschko, H.: Catecholamine biosynthesis. *Br. Med. Bull. 29:* 105–109, 1973.
74. Blaschko, H.: Metabolism and storage of biogenic amines. *Experientia 13:* 9–12, 1957.

75. Blaschko, H.: Formation of catecholamines in the animal body. *Br. Med. Bull. 13:* 162–165, 1957.
76. Blaschko, H.: Amine oxidase and amine metabolism. *Pharmacol. Rev. 4:* 415–458, 1952.
77. Blaschko, H.: The specific action of L-DOPA decarboxylase. *J. Physiol.* (Lond.) *96:* 50–51P, 1939.
78. Blaschko, H. and Chrusciel, T. L.: The decarboxylation of amino acids related to tyrosine and their awakening action in reserpine-treated mice. *J. Physiol.* (Lond.) *151:* 272–284, 1960.
79. Blaschko, H. and Muscholl, E.: Catecholamines. *Handbuch der experimentellen Pharmakologie.* Berlin: Springer, 1972.
80. Bloom, F. E., Costa, E., and Salmoiraghi, G. C.: Anesthesia and the responsiveness of individual neurons of the caudate nucleus of the cat to acetylcholine, norepinephrine and dopamine administered by microiontophoresis. *J. Pharmacol. Exp. Ther. 150:*244–252, 1965.
81. Bogdanski, D. F., Weissbach, H., and Udenfriend, S.: The distribution of serotonin, 5-hydroxytryptophan decarboxylase and monoamine oxidase in brain. *J. Neurochem. 1:* 272–278, 1957.
82. Boissier, J. R., Etevenon, P., Piarroux, M. C., et al.: Syndrome cataleptique après lésion du faisceau médian du telencéphale chez le rat. *C. R. Acad. Sci.* (Paris) *269:* 785–787, 1969.
83. Bonta, I. L., de Vos, C. J., Grijsen, H., et al.: 1-Hydroxy-3-aminopyrrolidone-2 (HA-966), a new GABA-like compound, with potential use in extrapyramidal diseases. *Br. J. Pharmacol. 43:* 514–535, 1971.
84. Borbély, A. A., Huston, J. P., and Baumann, I. R.: Body temperature and behavior in chronic brain-lesioned rats after amphetamine, chlorpromazine, and γ-butyrolactone. In Schönbaum, E. and Lomax, P. (Eds.): *The Pharmacology of Thermoregulation.* Symposium, San Francisco, 1972. Basel: Karger, 1973, pp. 447–462.
85. Broch, O. J., Jr. and Marsden, C. A.: Regional distribution of monoamines in the corpus striatum of the rat. *Brain Res. 38:* 425–428, 1972.
86. Bunney, B. S. and Aghajanian, G. K.: Electrophysiological effects of amphetamines on dopaminergic neurons. In Usdin, E. and Snyder, S. H. (Eds.): *Frontiers in Catecholamine Research.* Oxford: Pergamon Press, 1973, pp. 957–962.
87. Bunney, B. S., Aghajanian, G. K., and Roth, R. H.: L-DOPA amphetamine and apomorphine: Comparison of effects on the firing rate of dopaminergic neurons. *Nature (New Biol.) 245:* 123–125, 1973.
88. Bunney, B. S., Walters, J. R., and Roth, R. H., et al.: Dopaminergic neurons: Effects of antipsychotic drugs and amphetamine on single cell activity. *J. Pharmacol. Exp. Ther. 185:* 560–571, 1973.
89. Bunney, W. E., Jr., Brodie, H. K. H., Murphy, D. L., et al.: Studies of alpha-methyl-para-tyrosine, L-DOPA, and L-tryptophan in depression and mania. *Am. J. Psychiatr. 127:* 872–881, 1971.
90. Burkard, W. P., Gey, K. F., and Pletscher, A.: Activation of tyrosine hydroxylation in rat brain *in vivo* by chlorpromazine. *Nature 213:* 732–733, 1967.

91. Burkard, W. P., Jalfre, M., and Blum J.: Effect of 6-hydroxydopamine on behavior and cerebral amine content in rats. *Experientia 25*:1295–1296, 1969.
92. Burn, J. H.: *The Autonomic Nervous System.* Oxford: Blackwell, 1971, pp. 72–81.
93. Bustard, T. M. and Egan, R. S.: The conformation of dopamine hydrochloride. *Tetrahedron 27:* 4457–4469, 1971.
94. Bustos, G. and Roth, R. H.: Release of monoamines from the striatum and hypothalamus: Effect of γ-hydroxybutyrate. *Br. J. Pharmacol. 46:* 101–115, 1972.
95. Butcher, L. L., Engel, J., and Fuxe, K.: L-DOPA induced changes in central monoamine neurons after peripheral decarboxylase inhibition. *J. Pharm. Pharmacol. 22:* 313–316, 1970.
96. Calcutt, C. R., Doggett, N. S., and Spencer, P. S. J.: Modification of the anti-nociceptive activity of morphine by centrally-administered ouabain and dopamine. *Psychopharmacologia 21:* 111–117, 1971.
97. Calcutt, C. R. and Spencer, P. S. J.: Activities of narcotic and non-narcotic analgesics following the intraventricular injection of various substances. *Br. J. Pharmacol. 41:* 401–402P, 1971.
98. Carlsson, A.: Morphologic and dynamic aspects of dopamine in the central nervous system. In Costa, E., Côté, L. J., and Yahr, M. D. (Eds.): *Biochemistry and Pharmacology of the Basal Ganglia.* New York: Raven Press, 1966, pp. 107–113.
99. Carlsson, A.: The occurrence, distribution and physiological role of catecholamines in the nervous system. *Pharmacol. Rev. 11:* 490–493, 1959.
100. Carlsson, A., Fuxe, K., Hamberger, B., et al.: Biochemical and histochemical studies on the effects of imipramine-like drugs and (+)-amphetamine on central and peripheral catecholamine neurons. *Acta Physiol. Scand. 76:* 481–497, 1966.
101. Carlsson, A., Bédard, P., Lindqvist, M., et al.: The influence of nerve impulse flow on the synthesis and metabolism of 5-hydroxytryptamine in the central nervous system. *Biochem. Soc. Symp. 36:* 17–32, 1972.
102. Carlsson, A., Kehr, W., Lindqvist, M., et al.: Regulation of monoamine metabolism in the central nervous system. *Pharmacol. Rev. 24:* 371–384, 1972.
103. Carlsson, A. and Lindqvist, M.: Metatyrosine as a tool for selective protection of catecholamine stores against reserpine. *Eur. J. Pharmacol. 2:* 187–192, 1967.
104. Carlsson, A. and Lindqvist, M.: The interference of tetrabenazine, benzquinamide, and prenylamine with the action of reserpine. *Acta Pharmacol. Toxicol. 24:* 112–120, 1966.
105. Carlsson, A. and Lindqvist, M.: Effect of chlorpromazine and haloperidol on formation of 3-methoxytyramine and normetanephrine in mouse brain. *Acta Pharmacol. Toxicol. 20:* 140–144, 1963.
106. Carlsson, A. and Lindqvist, M.: *In vivo* decarboxylation of α-methyl DOPA and α-methyl metatyrosine. *Acta Physiol. Scand. 54*:87–94, 1962.

107. Carlsson, A., Lindqvist, M., and Magnusson, T.: On the biochemistry and possible functions of dopamine and noradrenaline in brain. In Vane, J. R., Wolstenholme, G. E. W., and O'Connor, M. (Eds.): *Adrenergic Mechanisms.* London: Churchill, 1960, pp. 432–439.
108. Carlsson, A., Lindqvist, M., and Magnusson, T.: 3,4-Dihydroxyphenylalanine and 5-hydroxytryptophan as reserpine antagonists. *Nature 180:* 1200, 1957.
109. Carlsson, A., Lindqvist, M., Magnusson, T., et al.: On the presence of 3-hydroxytyramine in brain. *Science 127:* 471, 1958.
110. Carlsson, A., Persson, T., Roos, B.-E., et al.: Potentiation of phenothiazines by α-methyltyrosine in treatment of chronic schizophrenia. *J. Neural Trans. 33:* 83–90, 1972.
111. Carlsson, A. and Waldeck, B.: A method for the fluorimetric determination of 3-methoxytyramine in tissues and the occurrence of this amine in brain. *Scand. J. Clin. Lab. Invest. 16:* 133–138, 1964.
112. Carlsson, A. and Waldeck, B.: A fluorimetric method for the determination of dopamine (3-hydroxytyramine). *Acta Physiol. Scand. 44:* 293–298, 1958.
113. Carpenter, M. B. and McMasters, R. E.: Lesions of the substantia nigra in the Rhesus monkey. Efferent fiber degeneration and behavioral observations. *Am. J. Anat. 114:* 293–319, 1967.
114. Carpenter, M. B. and Peter, R.: Nigrostriatal and nigrothalamic fibers in the Rhesus monkey. *J. Comp. Neurol. 144:* 93–116, 1972.
115. Carr, L. A. and Moore, K. E.: Distribution and metabolism of norepinephrine after its administration into the cerebroventricular system of the cat. *Biochem. Pharmacol. 18:* 1907–1918, 1969.
116. Cawein, M. and Turney, F.: Test for incipient Huntington's chorea. *New Engl. J. Med. 284:* 504, 1971.
117. Cegrell, L., Nordgren, L., and Rosengren, A. M.: Effect of decarboxylase inhibition and neuroleptic drugs on the dopa level in rat brain. *Res. Comm. Chem. Pathol. Pharmacol. 1:* 479–484, 1970.
118. Chase, T. N.: Drug-induced extrapyramidal disorders. In: *Neurotransmitters. Res. Publ. Assoc. Nerv. Ment. Dis. 50:* 448–471, 1972.
119. Cheramy, A., Besson, M. J., and Glowinski, J.: Increased release of dopamine from striatal dopaminergic terminals in the rat after treatment with a neuroleptic: Thioproperazine. *Eur. J. Pharmacol. 10:* 206–214, 1970.
120. Cherington, M.: Parkinsonism, L-dopa and mental depression. *J. Am. Geriatr. Soc. 18:* 513–516, 1970.
121. Chiueh, C. C. and Moore, K. E.: Release of endogenously synthesized catechols from the caudate nucleus by stimulation of the nigrostriatal pathway and by the administration of D-amphetamine. *Brain Res. 50:* 221–225, 1973.
122. Christie, J. E. and Crow, T. J.: Turning behavior as an index of the action of amphetamines and ephedrine on central dopamine-containing neurons. *Br. J. Pharmacol.* 43:658–667, 1971.
123. Connor, J. D.: Caudate nucleus neurones: Correlation of the effects of substantia nigra stimulation with iontophoretic dopamine. *J. Physiol.* (Lond.) 208:691–703, 1970.

123a. Connor, J. D.: The nigro-neostriatal pathway: The effects produced by iontophoretic dopamine. In Kopin, I. J. (Ed.): *Neurotransmitters. Res. Publ. Assoc. Nerv. Ment. Dis. 50*:193–206, 1972.
124. Cole, J. O. and Edwards, R. E.: Prediction of clinical effects of psychotropic drugs from animal data. In Steinberg, H., de Reuck, A. V. S., and Knight, J. (Eds.): *Animal Behavior and Drug Action.* Ciba Foundation Symposium. London: Churchill, 1964, pp. 286–298.
125. Cools, A. R.: Chemical and electrical stimulation of the caudate nucleus in freely moving cats: The role of dopamine. *Brain Res. 58:* 437–451, 1973.
126. Cools, A. R.: Athetoid and choreiform hyperkinesias produced by caudate application of dopamine in cats. *Psychopharmacologia 25:* 229–237, 1972.
127. Cools, A. R. and van Rossum, J. M.: Caudal dopamine and stereotype behavior of cats. *Arch. Intern. Pharmacodyn. Ther. 187:* 163–173, 1970.
128. Corrodi, H., Farnebo, L.-O., Fuxe, K., et al.: ET 495 and brain catecholamine mechanisms: Evidence for stimulation of dopamine receptors. *Eur. J. Pharmacol. 20:* 195–204, 1972.
129. Corrodi, H., Fuxe, K., Hammer, W., et al.: Oxotremorine and central monoamine neurons. *Life Sci. 6:* 2557–2566, 1967.
130. Corrodi, H., Fuxe, K., and Hökfelt, T.: The effect of neuroleptics on the activity of central catecholamine neurons. *Life Sci. 6:* 767–774, 1967.
131. Corrodi, H., Fuxe, K., and Hökfelt, T.: Refillment of the catecholamine stores with 3,4-dihydroxyphenylalanine after depletion induced by inhibition of tyrosinehydroxylase. *Life Sci. 5:* 605–611, 1966.
132. Corrodi, H., Fuxe, K., and Lidbrink, P.: Interaction between cholinergic and catecholaminergic neurons in rat brain. *Brain Res. 43:* 397–416, 1972.
133. Corrodi, H. and Hanson, L. C. F.: Central effects of an inhibitor of tyrosine hydroxylation. *Psychopharmacologia 10:* 116–125, 1966.
134. Corrodi, H. and Malmfors, T.: The effect of nerve activity on the depletion of the adrenergic transmitter by inhibitors of noradrenaline synthesis. *Acta Physiol. Scand. 67:* 352–357, 1966.
135. Costa, E.: Simple neuronal models to estimate turnover rate of noradrenergic transmitters in vivo. *Adv. Biochem. Psychopharmacol. 2:* 169–204, 1970.
136. Costa, E. and Neff, N. H.: Isotopic and nonisotopic measurements of the rate of catecholamine biosynthesis. In Costa, E., Côte, L. J., and Yahr, M. D. (Eds.): *Biochemistry and Pharmacology of the Basal Ganglia.* New York: Raven Press, 1966, pp. 141–155.
137. Costall, B. and Olley, J. E.: Cholinergic and neuroleptic induced catalepsy: Modification by lesions in the caudate-putamen. *Neuropharmacology 10:* 297–306, 1971.
138. Costall, B. and Olley, J. E.: Cholinergic and neuroleptic induced

catalepsy: Modification by lesions in the globus pallidus and substantia nigra. *Neuropharmacology 10:* 581–594, 1971.
139. Côté, L. J. and Fahn, S.: Some aspects of the biochemistry of the substantia nigra of the Rhesus monkey. In Barbeau, A. and Brunette, J.-R. (Eds.): *Proc. 2nd Internat. Congr. on Neurogenetics and Neuroophthalmology.* Amsterdam: Excerpta Medica, 1968, pp. 311–317.
140. Cotzias, G. C., Papavasiliou, P. S., Fehling, G., et al.: Similarities between neurological effects of L-DOPA and apomorphine. *New Engl. J. Med. 282:* 31–33, 1970.
141. Cotzias, G. C., Papavasiliou, P. S., and Gellene, R.: Modification of Parkinsonism. Chronic treatment with L-DOPA. *New Engl. J. Med. 280:* 337–345, 1969.
142. Cotzias, G. C., Van Woert, M. H., and Schiffer, L. M.: Aromatic amino acids and modification of Parkinsonism. *New Engl. J. Med. 276:* 374–379, 1967.
143. Coyle, J. T.: The development of catecholaminergic neurons of the central nervous system. *Neurosci. Res. 5:* 35–52, 1973.
144. Coyle, J. T. and Snyder, S. H.: Catecholamine uptake by synaptosomes in homogenates of rat brain: Stereospecificity in different areas. *J. Pharmacol. Exp. Ther. 170:* 221–231, 1969.
145. Coyle, J. T. and Snyder, S. H.: Antiparkinsonian drugs: Inhibition of dopamine uptake in the corpus striatum as a possible mechanism of action. *Science 166:* 899–901, 1969.
146. Creese, I. and Iversen, S. D.: Amphetamine responses in rat after dopamine neuron destruction. *Nature (New Biol.) 238:* 247–248, 1972.
147. Creveling, C. R., Daly, J., Tokuyama, T., et al.: The combined use of alpha-methyltyrosine and threo-dihydroxy-phenylserine—selective reduction of dopamine levels in the central nervous system. *Biochem. Pharmacol. 17:* 65–70, 1968.
148. Creveling, C. R., Daly, J. W., Witkop, B., et al.: Substrates and inhibitors of dopamine-β-oxidase. *Biochim. Biophys. Acta 64*: 125–134, 1962.
149. Crow, T. J.: The relationship between lesion site, dopamine neurons, and turning behavior. *Exp. Neurol. 32:* 247–255, 1971.
150. Curci, P. and Prandi, J.: Inhibition of tremors in Parkinson's disease with sodium 3,4,5-trimethoxybenzoyl-γ-aminobutyrate. Preliminary clinical trial. *Riv. Farmacol. Ter. 111:* 197–203, 1972.
151. Curtis, D. R. and Davis, R.: A central action of 5-hydroxy-tryptamine and noradrenaline. *Nature 192:* 1083–1084, 1961.
152. Curtis, D. R., Duggan, A. W., Felix, D., et al.: Bicucullin, an antagonist of GABA and synaptic inhibition in the spinal cord of the cat. *Brain Res. 32:* 69–96, 1971.
153. Dagirmanjian, R., Laverty, R., Mantegazzini, P., et al.: Chemical and physiological changes produced by arterial infusion of dihydroxyphenylalanine into one cerebral hemisphere of the cat. *J. Neurochem. 10:* 177–182, 1963.

154. Dahlström, A.: Axoplasmic transport (with particular respect to adrenergic neurons). *Philos. Trans. R. Soc. Lond. (Biol.) 261:* 325–358, 1971.
155. Dahlström, A. and Häggendal, J.: Some quantitative studies on the noradrenaline content of the cell bodies and terminals of a sympathetic adrenergic neuron system. *Acta Physiol. Scand. 67:* 271–277, 1966.
156. Dahlström, A., Häggendal, J., and Hökfelt, T.: The noradrenaline content of the terminal varicosities of sympathetic adrenergic neurons in the rat. *Acta Physiol. Scand. 67:* 289–294, 1966.
157. Dairman, W., Christenson, J., and Udenfriend, S.: Characterization of dopa decarboxylase. In Usdin, E. and Snyder, S. H. (Eds.): *Frontiers in Catecholamine Research.* Oxford: Pergamon Press, 1973, pp. 61–67.
158. Daly, J. W.: Regulation of cyclic AMP levels in brain. In Usdin, E. and Snyder, S. H. (Eds.): *Frontiers in Catecholamine Research.* Oxford: Pergamon Press, 1973, pp. 301–305.
159. Damasio, A. R., Antunes, J. L., and Macedo, C.: L-DOPA Parkinsonism and depression. *Lancet II:* 611–612, 1970.
160. Da Prada, M. and Pletscher, A.: Acceleration of the cerebral dopamine turnover by chlorpromazine. *Experientia 22:* 465–466, 1966.
161. Davidson, L., Lloyd, K. G., Dankova, J., et al.: L-DOPA treatment in Parkinson's disease: Effect on dopamine and related substances in discrete brain regions. *Experientia 27:* 1048–1049, 1971.
162. Davis, J. M. and Janowsky, D. S.: Amphetamine and methylphenidate psychosis. In Usdin, E. and Snyder, S. H. (Eds.): *Frontiers in Catecholamine Research.* Oxford: Pergamon Press, 1973, pp. 977–981.
163. Dean, W. H. and Davis, G. D.: Behavior changes following caudate lesions in Rhesus monkeys. *J. Neurophysiol. 22:* 524–537, 1959.
164. Dearnally, D. P., Fillenz, M., and Woods, R. I.: The identification of dopamine in the rabbit's carotid body. *Proc. R. Soc. Lond. 170:* 195–203, 1968.
165. De Jong, H.: *Experimental Catatonia.* Baltimore: Williams and Wilkins, 1945.
165a. Denny-Brown, D.: *The Basal Ganglia and Their Relation to Disorders of Movement.* London: Oxford University Press, 1962.
166. Derkach, P., Larochelle, L., Bieger, D., et al.: L-DOPA-chlorpromazine antagonism on running activity in mice. *Can. J. Physiol. Pharmacol. 52:* 114–118, 1974.
167. Dieckmann, G. and Hassler, R.: Reizexperimente zur Funktion des Putamen. *J. Hirnforsch. 10:* 187–225, 1968.
168. Divry, P. and Evrard, E.: Recherches sur certaines substances antagonistes de la bulbocapnine. *J. Belg. Neurol. Psychiatr. 34:* 506–523, 1934.
169. Dorris, R. L. and Shore, P. A.: Evidence for α-methyl-*m*-tyramine as a false dopamine-like transmitter. *J. Pharm. Pharmacol. 24:* 581–582, 1972.

170. Dorris, R. L. and Shore, P. A.: Localization and persistence of metraraminol and α-methyl-*m*-tyramine in rat and rabbit brain. *J. Pharmacol. Exp. Ther. 179:* 10–14, 1971.
171. Duffy, P. E. and Tennyson, V. M.: Phase and electron microscopic observations of Lewy bodies and melanin granules in substantia nigra and locus coeruleus in Parkinson's disease. *Neuropath. Exper. Neurol. 24:* 398–414, 1965.
171a. Duvoisin, R. C.: Cholinergic-anticholinergic antagonism in Parkinsonism. *Arch. Neurol. 17:* 124–136, 1967.
172. Economo, C. von: *Die Encephalitis lethargica.* 2. Aufl. Berlin, 1929.
173. Ehringer, H. and Hornykiewicz, O.: Verteilung von Noradrenalin und Dopamin (3-Hydroxytyramin) im Gehirn des Menschen und ihr Verhalten bei Erkrankungen des extrapyramidalen Systems. *Klin. Wochenschr. 38:* 1236–1239, 1960.
174. Enero, M. A., Langer, S. Z., Rothlin, R. P., et al.: Role of the α-adrenoceptor in regulating noradrenaline overflow by nerve stimulation. *Br. J. Pharmacol. 44:* 672–688, 1972.
175. Eränkö, O. and Eränkö, L.: Differentiation of dopamine from noradrenaline in tissue sections by microspectrofluorimetry. *J. Histochem. Cytochem. 19:* 131–132, 1971.
176. Ernst, A. M.: Mode of action of apomorphine and dexamphetamine in gnawing compulsion in rats. *Psychopharmacologia 10:* 316–323, 1967.
177. Ernst, A. M.: Relation between the structure of certain methoxyphenylethylamine derivatives and the occurrence of a hypokinetic rigid syndrome. *Psychopharmacologia 7:* 383–390, 1965.
178. Ernst, A. M.: Relation between the action of dopamine and apomorphine and their O-methylated derivatives upon the CNS. *Psychopharmacologia 7:*391–399, 1965.
179. Ernst, A. M.: Phenomena of the hypokinetic-rigid type caused by O-methylation of dopamine in the para position. *Nature 193:* 178–179, 1962.
180. Ernst, A. M. and Smelik, P.: Site of action of dopamine and apomorphine on compulsive gnawing behavior in rats. *Experientia 22:* 837–838, 1966.
181. Euler, U. S., von and Lishajko, F.: Effect of adenine nucleotides on catecholamine release and uptake in isolated adrenergic nerve granules. *Acta Physiol. Scand. 59:* 454–461, 1963.
182. Euler, U.S. von, Lishajko, F., and Stjärne, L.: Catecholamines and adenosine triphosphate in isolated adrenergic nerve granules. *Acta Physiol. Scand. 59:* 495–496, 1963.
183. Euler, U.S. von and Lishajko, F.: Effect of drugs on the storage granules of adrenergic nerves. In Koelle, G., Douglas, W. W., and Carlsson, A. (Eds.): *Pharmacology of Cholinergic and Adrenergic Transmission* (Vol. 3). Oxford: Pergamon Press, 1965, pp. 245–259.
184. Everett, G. M. and Borcherding, J. W.: L-DOPA: Effect on concentrations in brains of mice. *Science 168:* 849–850, 1970.

185. Everett, G. M. and Wiegand, R. G.: Central amines and behavioral state: A critique and new data. Pharmacological analysis of the central nervous action. In Paton, W., and Lindgren, P. (Eds.): *Proc. 1st Internat. Pharmacol. Meeting* (Vol. 8). Oxford: Pergamon Press, 1962, pp. 85–92.
186. Evetts, K. D. and Iversen, L. L.: Effects of protriptyline on the depletion of catecholamines induced by 6-hydroxy-dopamine in the brain of the rat. *J. Pharm. Pharmacol. 22:* 540–543, 1970.
187. Fahn, S. and Côté, L. J.: Regional distribution of γ-aminobutyric acid (GABA) in brain of the Rhesus monkey. *J. Neurochem. 15:* 209–213, 1968.
188. Fahn, S., Rodman, J. S., and Côté, L. J.: Association of tyrosine hydroxylase with synaptic vesicles in bovine caudate nucleus. *J. Neurochem. 16:* 1293–1300, 1969.
189. Falck, B.: Observations on the possibilities of the cellular localization of monoamines by a fluorescence method. *Acta Physiol. Scand. 56* (Suppl. 197): 1–26, 1962.
190. Falck, B., Nordgren, E., and Rosengren, L.: Effect of haloperidol on monoamines in brain and heart. *Int. J. Neuropharmacol. 8:* 631–634, 1969.
191. Farnebo, L.-O. and Hamberger, B.: Drug-induced changes in the release of ^{3}H-monoamines from field stimulated rat brain slices. *Acta Physiol. Scand.* (Suppl. *371*): 35–44, 1971.
192. Farnebo, L.-O. and Hamberger, B.: Drug-induced changes in the release of ^{3}H-noradrenaline from field stimulation in rat iris. *Br. J. Pharmacol. 43:* 97–106, 1971.
193. Farnebo, L.-O. and Hamberger, B.: Release of norepinephrine from isolated rat iris by field stimulation. *J. Pharmacol. Exp. Ther. 172:* 332–341, 1970.
194. Faull, R. L. and Laverty, R.: Changes in dopamine levels in the corpus striatum following lesions in the substantia nigra. *Exp. Neurol. 23:* 332–340, 1969.
195. Faurbye, A.: The role of amines in the etiology of schizophrenia. *Compr. Psychiatr. 9:* 155–177, 1968.
196. Feltz, P.: Sensitivity to haloperidol of caudate neurons excited by nigral stimulation. *Eur. J. Pharmacol. 14:* 360–364, 1971.
197. Feltz, P.: Dopamine, amino acids and caudate unitary responses to nigral stimulation. *J. Physiol.* (Lond.) *205:* 8–9P, 1969.
198. Feltz, P. and Albe-Fessard, D.: A study of an ascending nigro-caudate pathway. *Electroenceph. Clin. Neurophysiol. 33:* 179–193, 1972.
199. Feltz, P. and De Champlain, J.: The postsynaptic effect of amphetamine on striatal dopamine-sensitive neurons. In Usdin, E. and Snyder, S. H. (Eds.): *Frontiers in Catecholamine Research.* Oxford: Pergamon Press, 1973, pp. 951–956.
200. Feltz, P. and De Champlain, J.: Persistence of caudate unitary responses to nigral stimulation after destruction and functional impairment of the striatal dopaminergic terminals. *Brain Res. 43:* 595–600, 1972.

201. Feltz, P. and De Champlain, J.: Enhanced sensitivity of caudate neurons to microiontophoretic injections of dopamine in 6-hydroxydopamine treated cats. *Brain Res. 43:* 601–605, 1972.
202. Feltz, P. and MacKenzie, J. S.: Properties of caudate unitary responses to repetitive nigral stimulation. *Brain Res. 13:* 612–616, 1969.
203. Fibiger, H. C., Pudritz, R. E., McGeer, P. L., et al.: Axonal transport in nigro-striatal and nigro-thalamic neurons: Effects of medial forebrain bundle lesions and 6-hydroxydopamine. *J. Neurochem. 19:* 1697–1708, 1972.
204. Fleckenstein, A. and Stöckle, D.: Zum Mechanismus der Wirkungs-Abschwächung sympathomimetischer Amine durch Cocain und andere Pharmaka. *Arch. Exp. Path. Pharmakol. 224:* 401–415, 1955.
205. Fog, R., Randrup, A., and Pakkenberg, H.: Lesions in corpus striatum and cortex of rat brains and the effect of pharmacologically induced stereotyped, aggressive and cataleptic behavior. *Psychopharmacologia 18:* 346–356, 1970.
206. Fog, R., Randrup, A., and Pakkenberg, H.: The aminergic mechanisms in corpus striatum and amphetamine-induced stereotyped behavior. *Psychopharmacologia 11:* 179–183, 1967.
207. Forman, O. and Ward, J. W.: Responses to electrical stimulation of caudate nucleus in cats in chronic experiments. *J. Neurophysiol. 20:* 230–244, 1957.
208. Frigyesi, T. L., Ige, A., Iulo, A., et al.: Denigration and sensorimotor disability induced by ventral tegmental injection of 6-hydroxydopamine in the cat. *Exp. Neurol. 33:* 78–87, 1971.
209. Frigyesi, T. L. and Purpura, D. P.: Electrophysiological analysis of reciprocal caudato-nigral relations. *Brain Res. 6:* 440–456, 1967.
210. Fukui, A. and Takagi, H.: Effect of morphine on the cerebral contents of metabolites of dopamine in normal and tolerant mice. *Br. J. Pharmacol. 44:* 45–51, 1972.
211. Furchgott, R. F. and Sanchez-Garcia, P.: Effects of inhibition of monoamine oxidase on the actions and interactions of norepinephrine, tyramine, and other drugs on guinea pig left atrium. *J. Pharmacol. Exp. Ther. 163:* 98–122, 1968.
212. Fuxe, K.: Evidence for the existence of monoamine neurons in the central nervous system. IV. Distribution of monoamine terminals in central nervous system. *Acta Physiol. Scand. 64* (Suppl. 247): 37–84, 1965.
213. Fuxe, K.: Cellular localization of monoamines in the median eminence and the infundibular stem of some animals. *Z. Zellforsch. 61:* 710–724, 1964.
214. Fuxe, K., Goldstein, M., and Lungdahl, A.: Anti-Parkinsonian drugs and central dopamine neurons. *Life Sci. 9:* 811– 1970.
215. Fuxe, K., Hamberger, B., and Malmfors, T.: The effect of drugs on the accumulation of monoamines in tuberoinfundibular neurons. *Eur. J. Pharmacol. 1:* 334–341, 1967.
216. Fuxe, K. and Jonsson, G.: The histochemical fluorescence method

for the demonstration of catecholamines. *J. Histochem. Cytochem.* 21: 293–311, 1973.
217. Fuxe, K. and Ungerstedt, U.: Histochemical, biochemical and functional studies on central monoamine neurons after acute and chronic amphetamine administration. In Costa, E. and Garattini, S. (Eds.): *Amphetamines and Related Compounds*. New York: Raven Press, 1970, pp. 258–288.
218. Fuxe, K. and Ungerstedt, U.: Histochemical studies on the effect of (+)-amphetamine, drugs of the imipramine group and tryptamine on central catecholamine and 5-hydroxytryptamine neurons after intraventricular injection of catecholamines and 5-hydroxytryptamine. *Eur. J. Pharmacol.* 4: 135–144, 1968.
219. Gauchy, C., Agid, Y., Glowinski, J., et al.: Acute effects of morphine on dopamine synthesis and release and tyrosine metabolism in the rat striatum. *Eur. J. Pharmacol.* 22: 311–319, 1973.
220. Gee, C.: *St. Bartholom, Hosp. Report* (Lond.) 5:215, 1869: quoted in: Eggleston and Hatcher, *J. Pharmacol. Exp. Ther.* 3: 551–580, 1912.
221. Geffen, L. B. and Livett, B. G.: Synaptic vesicles in sympathetic neurons. *Physiol. Rev.* 51: 98–157, 1971.
222. Gessa, A., Crabai, F., Vargiu, L., et al.: Selective increase of brain DA induced by γ-hydroxybutyrate: Study of the mechanism of action. *J. Neurochem.* 25: 377–381, 1968.
223. Gessa, G. L., Vargiu, L., Crabai, F., et al.: Selective increase of brain DA induced by γ-hydroxybutyrate. *Life Sci.* 5: 1921–1930, 1966.
224. Gey, K. F. and Pletscher, A.: Acceleration of turnover of ^{14}C-catecholamines in rat brain by chlorpromazine. *Experientia* 24:335–336, 1968.
225. Gloor, P., Murphy, J. T., and Dreifuss, J.-J.: Anatomical and physiological characteristics of the two amygdaloid projection systems to the ventromedial hypothalamus. In Hockman, C. H. (Ed.): *Limbic System Mechanisms and Autonomic Function*. Springfield, (Ill.): C.C. Thomas, 1972, pp. 60–77.
226. Glowinski, J.: Some new facts about synthesis, storage and release processes of monoamines in the central nervous system. In Snyder, S. H. (Ed.): *Perspectives in Neuropharmacology*. New York: Oxford University Press, 1972, pp. 349–404.
227. Glowinski, J.: Effects of amphetamine on various aspects of catecholamine metabolism in the central nervous system of the rat. In Costa, E. and Garattini, S. (Eds.): *Amphetamines and Related Compounds*. New York: Raven Press, 1970, pp. 301–317.
228. Glowinski, J. and Iversen, L.: Regional studies of catecholamines in the rat brain. III. Subcellular distribution of endogenous and exogenous catecholamines in various brain regions. *Biochem. Pharmacol.* 15:977–987, 1966.
229. Goldstein, M., Anagnoste, B., Battista, A. F., et al.: Studies of amines in the striatum in monkeys with nigral lesions. The disposition, biosynthesis and metabolites of [^{3}H] dopamine and [^{14}C] serotonin in the striatum. *J. Neurochem.* 16: 645–653, 1969.

230. Goldstein, M., Freedman, L. S., and Backstrom, T.: The inhibition of catecholamine biosynthesis by apomorphine. *J. Pharm. Pharmacol. 22:* 715–717, 1970.
231. Gonzalez-Vegas, J. A.: Actions of papaverine and bulbocapnine on synaptic transmission in nigro-striatal pathway in the rat. *J. Physiol.* (Lond.) *226:* 102P, 1972.
232. Gonzalez-Vegas, J. A.: Antagonism of catecholamine inhibition of brain stem neurons by mescaline. *Brain Res. 35:* 264–267, 1971.
233. Goodwin, F. K., Murphy, D. L., Brodie, H. K. H., et al.: L-dopa, catecholamines and behavior: A clinical and biochemical study in depressed patients. *Biol. Psychiatr. 2:* 341–366, 1970.
234. Griffith, J. D., Cavanaugh, J. H., Held, J., et al.: Dextroamphetamine: Evaluation of psychomimetic properties in man. *Arch. Gen. Psychiatr. 26:* 97–100, 1972.
235. Griffith, J. D., Cavanaugh, J. H., and Oates, J. A.: Psychosis induced by the administration of D-amphetamine to human volunteers. In Efron, D. H. (Ed.): *Psychotomimetic Drugs.* New York: Raven Press, 1970, pp. 287–298.
236. Grillner, S.: Supraspinal and segmental control of static and dynamic γ-motoneurons. *Acta Physiol. Scand.* (Suppl. *327*): 1–34, 1969.
237. Grofová, I. and Rinvik, E.: An experimental electron microscopic study on the striatonigral projection in the cat. *Exp. Brain Res. 11:*249–262, 1970.
238. Guinard, L.: Etude expérimental de pharmacodynamie comparée sur la morphine et l'apomorphine. Thèse méd. de Lyon, 1898, No. 107.
239. Guldberg, H. C. and Broch, O. J.: On the mode of action of reserpine on dopamine metabolism in the rat striatum. *Eur. J. Pharmacol. 13:* 155–167, 1971.
240. Gunne, L.-M., Jonsson, J., and Fuxe, K.: Effects of morphine intoxication on brain catecholamine neurons. *Eur. J. Pharmacol. 5:* 338–342, 1969.
241. Haefely, W., Huerlimann, A., and Thoenen, H.: A quantitative study of the effect of cocaine on the response of the cat nictating membrane to nerve stimulation and to injected noradrenaline. *Br. J. Pharmacol. 22:* 5–21, 1964.
242. Häggendal, J. and Malmfors, T.: Identification and cellular localization of the catecholamines in the retina and the choroid of the rabbit. *Acta. Physiol. Scand. 64:* 58–66, 1965.
243. Hamberger, B.: Reserpine-resistant uptake of catecholamines in isolated tissues of the rat. *Acta Physiol. Scand.* (Suppl. *295*): 1–56, 1967.
244. Hanson, L. C. F.: The disruption of conditioned avoidance response following selective depletion of brain catecholamines. *Psychopharmacologia 8:* 100–110, 1965.
245. Harik, S. I. and Morris, P. I.: The effect of lesions in the head of the caudate nucleus on spontaneous and L-DOPA induced activity in the cat. *Brain Res. 62:* 279–285, 1973.
246. Harnack, E.: Über die Wirkungen des Apomorphins am Säugetier und am Frosch. *Archiv. Exp. Pathol. Pharmakol. 2:* 254, 1874.

247. Harris, L. S.: Central neurohumoral systems involved with narcotic agonists and antagonists. *Fed. Proc. 29:* 28—32, 1970.
248. Hassler, R.: Zur Pathologie der Paralysis agitans und des postenzephalitischen Parkinsonismus. *J. Psychol. Neurol.* (Leipzig) *48*:387—476, 1938.
249. Heath, R. G., Nesslhof, W., Bishop, M. P., et al.: Behavioral and metabolic changes associated with administration of tetraethylthiuram disulfide. *Dis. Nerv. System 26:* 99—105, 1965.
250. Heikkila, R. and Cohen, G.: Inhibition of biogenic amine uptake by hydrogen peroxide: A mechanism for toxic effects of 6-hydroxydopamine. *Science 172:* 1257—1258, 1971.
251. Heikkila, R., Cohen, G., and Dembiec, D.: Tetrahydroisoquinoline alkaloids: Uptake by rat brain homogenates and inhibition of catecholamine uptake. *J. Pharmacol. Exp. Ther. 179:* 250—258, 1971.
252. Heimans, R. L. H., Rand, M. J., and Fennessi, M. R.: Effects of amantadine on uptake and release of dopamine by a particulate fraction of rat basal ganglia. *J. Pharm. Pharmacol. 24:* 875—895, 1972.
253. Heinrich, U., Lichtensteiger, W., and Langemann, H.: Effect of morphine on catecholamine content of midbrain nerve cell groups in rat and mouse. *J. Pharmacol. Exp. Ther. 179:* 259—267, 1971.
254. Hendley, E. D. and Snyder, S. H.: Stereoselectivity of catecholamine uptake in noradrenergic and dopaminergic peripheral organs. *Eur. J. Pharmacol. 19:* 56—66, 1972.
255. Hendley, E. D., Snyder, S. H., Fauley, J. J., et al.: Stereoselectivity of catecholamine uptake by brain synaptosomes: Studies with ephedrine, methylphenidate and phenyl-2-piperidyl carbinol. *J. Pharmacol. Exp. Ther. 183:* 103—116, 1972.
256. Herman, M. and Nagler, S. H.: Psychoses due to amphetamine. *J. Nerv. Ment. Dis. 120:* 268—272, 1954.
257. Hertting, D. and Suko, J.: Influence of neuronal and extraneuronal uptake on disposition, metabolism and potency of catecholamines. In Snyder, S. H. (Ed.): *Perspectives in Neuropharmacology.* New York: Oxford University Press, 1972, pp. 267—300.
258. Herz, A. and Zieglgänsberger, W.: Synaptic excitation of the corpus striatum inhibited by microiontrophoretically administered dopamine. *Experientia 22:* 839—842, 1966.
259. Hess, S. M., Connmacher, R. H., Ozaki, M., et al.: The effects of α-methyl-dopa and α-methyl-meta-tyrosine on the metabolism of norepinephrine and serotonin in vivo. *J. Pharmacol. Exp. Ther. 134:* 129—138, 1961.
260. Hill, R. G., Simmonds, M. A., and Straughan, D. W.: A comparative study of some convulsant substances as γ-aminobutyric acid antagonists in the feline cerebral cortex. *Br. J. Pharmacol. 49:* 37—51, 1973.
261. Hillen, F. C. and Noach, E. L.: The influence of 1-hydroxy-3-aminopyrrolidinone-2 (HA-966) on dopamine metabolism in the rat corpus striatum. *Eur. J. Pharmacol. 16:* 222—224, 1971.

262. Hockman, C. H., Lloyd, K. G., Farley, I. J., et al.: Experimental midbrain lesions: Neurochemical comparison between the animal model and Parkinson's disease. *Brain Res. 33:* 613—618, 1971.
263. Hökfelt, T.: *In vitro* studies on central and peripheral monoamine neurons at the ultrastructural level. *Z. Zellforsch. 91:* 1—74, 1968.
264. Hökfelt, T., Fuxe, K., Johansson, O., et al.: Pharmaco-histochemical evidence of the existence of dopamine nerve terminals in the limbic cortex. *Eur. J. Pharmacol. 25:* 108—112, 1974.
265. Hökfelt, T., Ljungdahl, A., Fuxe, K., et al.: Dopamine nerve terminals in the rat limbic cortex: Aspects of the dopamine hypothesis of schizophrenia. *Science 184:* 177—179, 1974.
266. Holtz, P., Heise, R., and Lüdtke, K.: Fermentativer Abbau von L-Dioxyphenylalanin (DOPA) durch Niere. *Arch. Exp. Pathol. Pharmacol. 191:* 87—118, 1938.
267. Holtz, P., Stock, K., and Westermann, E.: Formation of tetrahydropapaveroline from dopamine in vitro. *Nature 203:* 656—658, 1964.
268. Horn, A. S., Coyle, J. T., and Snyder, S. H.: Catecholamine uptake by synaptosomes from rat brain. *Mol. Pharmacol. 7:* 66—80, 1971.
269. Horn, A. S. and Snyder, S. H.: Chlorpromazine and dopamine: Conformational similarities that correlate with the antischizophrenic activity of phenothiazine drugs. *Proc. Nat. Acad. Sci. U.S.A. 68:* 2325—2328, 1971.
270. Hornykiewicz, O.: Dopamine and extrapyramidal motor function and dysfunction. In: *Neurotransmitters. Res. Publ. Assoc. Nerv. Ment. Dis. 50:* 390—415, 1972.
271. Hornykiewicz, O.: Dopamine: Its physiology, pharmacology and pathological neurochemistry. In Biel, J. H. and Abood, L. G. (Eds.): *Biological Amines and Physiological Membranes in Drug Therapy.* New York: Dekker, 1971, pp. 173—258.
272. Hornykiewicz, O.: Dopamine (3-hydroxytyramine) and brain function. *Pharmacol. Rev. 18:* 925—964, 1966.
273. Hornykiewicz, O.: Extrapyramidal side effects of neuro(psycho)tropic drugs. *Proc. Eur. Soc. for Study of Drug Toxicity,* Vol. VIII, Neurotoxicity of Drugs. Prague: Excerpta Medica Int. Congr. Ser. No. 118, 1966, pp. 122—135.
274. Hornykiewicz, O.: Zur Existenz "dopaminerger" Neurone im Gehirn. *Arch. Exp. Pathol. Pharmacol. 247:* 304, 1964.
275. Hornykiewicz, O.: Die topische Lokalisation und das Verhalten von Noradrenalin und Dopamin (3-Hydroxytyramine) in der Substantia nigra des normalen und Parkinson-kranken Menschen. *Wien. Klin. Wochenschr. 75:* 309—312, 1963.
276. Hughes, J.: Evaluation of mechanisms controlling the release and inactivation of the adrenergic transmitter in the rabbit portal vein and vas deferens. *Br. J. Pharmacol. 44:* 472—491, 1972.
277. Hull, C. D., Bernardi, G., and Buchwald, N. A.: Intracellular response of caudate neurons to brain stem stimulation. *Brain Res. 22:* 163—179, 1970.
278. Hutchins, D. A., Rayevsky, K. S., and Sharman, D. F.: The effect of

sodium γ-hydroxybutyrate on the metabolism of dopamine in the brain. *Br. J. Pharmacol. 46:* 409—415, 1972.
279. Ikeda, M., Fahien, L. A., and Udenfriend, S.: A kinetic study of bovine adrenal tyrosine hydroxylase. *J. Biol. Chem. 241:* 4452—4456, 1966.
280. Iversen, L. L.: Role of uptake mechanism in synaptic transmission. *Br. J. Pharmacol. 41:* 571—591, 1971.
281. Iversen, L. L.: Neuronal uptake processes for amines and amino acids. In Costa, E. and Jacobini, E. (Eds.): *Biochemistry of Simple Neuronal Models.* New York: Raven Press, 1970, pp. 109—132.
282. Iversen, L. L.: *The Uptake and Storage of Noradrenaline in Sympathetic Nerves.* New York: Cambridge University Press, 1967.
283. Janowsky, D. S., El-Youssel, M. K., Davis, J. M., et al.: Provocation of schizophrenia symptoms by intravenous methylphenidate. *Arch. Gen. Psychiatr. 28:* 185—191, 1973.
284. Janssen, P. A. J., Niemegeers, C. J. E., and Jageneau, A. H. M.: Apomorphine antagonism in rats. *Arzneim. Forsch. 10:* 1003—1005, 1960.
285. Janssen, P. A. J., Niemegeers, C. J. E., and Schellekens, K. H. L.: Is it possible to predict the clinical effects of neuroleptic drugs (major tranquillizers) from animal data? Part I. Neuroleptic activity spectra for rats. *Arzneim. Forsch. 15:* 104—117, 1965.
286. Janssen, P. A. J., Niemegeers, C. J. E., and Schellekens, K. H. L.: Is it possible to predict the clinical effects of neuroleptic drugs (major tranquillizers) from animal data? Part III. The subcutaneous and oral activity in rats and dogs of 56 neuroleptic drugs in the jumping box test. *Arzneim. Forsch. 16*:339—346, 1966.
287. Janssen, P. A. J., Niemegeers, C. J. E., Schellekens, K. H. L., et al.: Is it possible to predict the clinical effects of neuroleptic drugs (major tranquillizers) from animal data? Part IV. An improved experimental design for measuring the inhibitory effects of neuroleptic drugs on amphetamine- or apomorphine-induced "chewing" and "agitation" in rats. *Arzneim. Forsch. 17*:841—854, 1967.
288. Javoy, F. and Glowinski, J.: Dynamic characteristics of the "functional compartment" of dopamine in dopaminergic terminals of the rat striatum. *J. Neurochem. 18:* 1305—1311, 1971.
288a. Javoy, F., Agid, Y., Bouvet, D., et al.: Feedback control of dopamine synthesis in dopaminergic terminals of the rat striatum. *J. Pharmacol. Exp. Ther. 182:* 454—463, 1972.
289. Javoy, F., Hamon, M., and Glowinski, J.: Disposition of newly synthesized amine in cell bodies and terminals of central catecholaminergic neurons. I. Effect of amphetamine and thioproperazine on the metabolism of catecholamines in the caudate nucleus, the substantia nigra and the ventromedial nucleus of the hypothalamus. *Eur. J. Pharmacol. 10:* 178—188, 1970.
290. Jonason, J. and Rutledge, C. O.: Effects of reserpine, dopamine and nialamide on the synthesis of α-methyl-noradrenaline. *Eur. J. Pharmacol. 6:* 24—28, 1969.

291. Jones, B., Bobillier, P., and Jouvet, M.: Effet de la déstruction des neurones contenant des catecholamines du mésencéphale sur le cycle veille-sommeil du chat. *C. R. Soc. Biol. 163:* 176–180, 1969.
292. Jonsson, G.: Microfluorimetric studies on the formaldehyde-induced fluorescence of noradrenaline in adrenergic nerves of the rat iris. *J. Histochem. Cytochem. 17:* 714–723, 1969.
293. Jung, R. and Hassler, R.: The extrapyramidal motor system. In Field, J., Magoun, H. W., and Hall, V. E. (Eds.): *Handbook of Physiology, Neurophysiology* (Vol. 1). Washington, D.C.: Am. Physiol. Society, 1960, pp. 863–927.
294. Juorio, A. V. and Vogt, M.: The effect of prenylamine in the metabolism of catecholamines and 5-hydroxytryptamine in brain and adrenal medulla. *Br. J. Pharmacol. 24:* 566–573, 1965.
295. Kebabian, J. W., Petzold, G. L., and Greengard, P.: Dopamine-sensitive adenylate cyclase in caudate nucleus of rat brain, and its similarity to the "dopamine receptor." *Proc. Nat. Acad. Sci. U.S.A. 69:* 2145–2149, 1972.
296. Kebabian, J. W. and Greengard, P.: Dopamine-sensitive adenyl cyclase: Possible role in synaptic transmission. *Science 174:* 1346–1349, 1971.
297. Kehr, W., Carlsson, A., Lindqvist, M., et al.: Evidence for a receptor-mediated feedback control of striatal tyrosine hydroxylase activity. *J. Pharm. Pharmacol. 24:* 744–746, 1972.
298. Keller, P. J. and Lichtensteiger, W.: Stimulation of infundibular dopamine neurons and gonadotrophin secretion. *J. Physiol.* (Lond.) *219:* 385–401, 1971.
299. Kemp, J. M. and Powell, T. P. S.: The structure of the caudate nucleus of the cat: Light and electron microscopy. *Philos. Trans. R. Soc.* Lond. (Biol.) *262:* 383–401, 1971.
300. Kemp, J. M. and Powell, T. P. S.: The synaptic organization of the caudate nucleus. *Philos. Trans. R. Soc.* London (Biol.) *262:* 403–412, 1971.
301. Kemp, J. M. and Powell, T. P. S.: The site of termination of afferent fibers in the caudate nucleus. *Philos. Trans. R. Soc.* Lond. (Biol.) *262:* 413–427, 1971.
302. Kemp, J. M. and Powell, T. P. S.: The termination of fibers from the cerebral cortex and thalamus upon dendritic spines in the caudate nucleus: A study with the Golgi method. *Philos. Trans. R. Soc.* Lond. (Biol.) *262:* 429–439, 1971.
303. Kennear, J. H.: Interactions between central cholinergic agents and amphetamine in mice. *Psychopharmacologia 7:* 107–114, 1965.
304. Kerkut, G. A.: Catecholamines in invertebrates. *Br. Med. Bull, 29:* 100–104, 1973.
305. Kerkut, G. A., Horn, N., and Walker, R. J.: Long-lasting synaptic inhibition and its transmitter in the snail *Helix Aspersa*. *Comp. Biochem. Physiol. 30:* 1061–1074, 1969.
306. Kerkut, G. A., Pitman, R. M., and Walker, R. J.: Iontophoretic application of acetylcholine and GABA onto insect central neurons. *Comp. Biochem. Physiol. 31:* 611–633, 1969.

307. Killam, K. F., Jr.: Pharmacological aspects of limbic function. In Hockman, C. H. (Ed.): *Limbic System Mechanisms and Autonomic Function.* Springfield, (Ill.): C.C. Thomas, 1972, pp. 223–227.
308. Kim, J. S., Okada, Y., Hassler, R., et al.: The role of γ-aminobutyric acid (GABA) in extrapyramidal motor system: 2. Some evidence for the existence of a type of GABA-rich strionigral neurons. *Exp. Brain Res. 14:* 95–104, 1971.
309. Kirpekar, S. M. and Puig, M.: Effect of flow-stop on noradrenaline release from normal spleens and spleens treated with cocaine, phentolamine or phenoxybenzamine. *Br. J. Pharmacol. 43:* 359–369, 1971.
310. Klawans, H. C., Paulson, G. W., and Barbeau, A.: Predictive test for Huntington's chorea. *Lancet 2:* 1185–1186, 1970.
311. Knigge, K. M., Scott, D. E., and Weindl, A. (Eds.): *Brain-Endocrine Interaction. Median Eminence: Structure and Function.* Basel: Karger, 1972, pp. 181–223.
312. Kopin, I. J.: False adrenergic transmitters. *Ann. Rev. Pharmacol. 8:* 377–394, 1968.
313. Kornhuber, H. H.: Kleinhirn, Stammganglien und Grosshirnrinde in der motorischen Organisation von Primaten. *Verh. Dtsch. Zool Ges. 66:* 151–167, 1973.
314. Kornhuber, H. H.: Motor functions of the cerebellum and basal ganglia: The cerebello-cortical saccadic (ballistic) clock, the cerebellonuclear hold regulator, and the basal ganglia ramp (voluntary speed smooth movement) generator. *Kybernetik 8:* 157–162, 1971.
315. Kramer, S. G.: Dopamine: A retinal neurotransmitter. I. Retinal uptake, storage and light-stimulated release of ^{3}H-dopamine in vivo. *Invest. Ophthalmol. 10:* 438–452, 1971.
316. Kramer, S. G., Potts, A. M., and Magnall, Y.: Dopamine: A retinal neurotransmitter. II. Autoradiographic localization of ^{3}H-dopamine in the retina. *Invest. Ophthalmol. 10:* 617–624, 1971.
317. Krauthamer, G. and Albe-Fessard, D.: Inhibition of nonspecific sensory activities following striopallidal and capsular stimulation. *J. Neurophysiol. 28:* 100–124, 1965.
318. Kuczenski, R. T. and Mandell, A. J.: Regulatory properties of soluble and particulate rat brain tyrosine hydroxylase. *J. Biol. Chem. 247:* 3114–3122, 1972.
319. Kuhar, M. J., Roth, R. H., and Aghajanian, G. K.: Synaptosomes from forebrains of rats with midbrain raphé lesions: Selective reduction of serotonin uptake. *J. Pharmacol. Exp. Ther. 181:* 36–45, 1972.
320. Kuschinsky, K., and Hornykiewicz, O.: Personal communication, 1972.
321. Kuschinsky, K. and Hornykiewicz, O.: Morphine catalepsy in the rat: Relation to striatal dopamine metabolism. *Eur. J. Pharmacol. 19:* 119–122, 1972.
322. Laguzzi, R., Petitjean, F., Pujol, J.-E., et al.: Effets de l'injection intraventriculaire du 6-hydroxydopamine sur les états de sommeil

et les monoamines cérébrales du chat. *C. R. Soc. Biol.* (Paris) *165:* 1649—1651, 1971.
323. Lal, H., O'Brien, J., and Puri, S. K.: Morphine-withdrawal aggression: Sensitization by amphetamines. *Psychopharmacologia 22:* 217—223, 1971.
324. Lal, H., Sourkes, T. L., Missala, K., et al.: Effects of aporphine and emetine alkaloids on central dopaminergic mechanisms in rats. *Eur. J. Pharmacol. 20:* 71—79, 1972.
325. Lammers, A. J. J. C. and van Rossum, J. M.: Bizarre social behavior in rats induced by a combination of a peripheral decarboxylase inhibitor and DOPA. *Eur. J. Pharmacol. 5:* 103—105, 1968.
326. Langer, S. Z.: The metabolism of [^{3}H] noradrenaline released by electrical stimulation from the isolated nictitating membrane of the cat and from the vas deferens of the rat. *J. Physiol.* (Lond.) *208:* 515—546, 1970.
327. Langer, S. Z., Draskoczy, P. R., and Trendelenburg, U.: Time course of the development of supersensitivity to various amines in the nictitating membrane of the pithed cat after denervation or decentralization. *J. Pharmacol. Exp. Ther. 157:* 255—273, 1967.
328. Larochelle, L., Bédard, P., Poirier, L. J., et al.: Correlative neuroanatomical and neuropharmacological study of tremor and catatonia in the monkey. *Neuropharmacol. 10:* 273—288, 1971.
329. Laubie, M., Schmitt, H., and Le Donarec, J. C.: Cardiovascular effects of the 1-(2″-pyrimidyl)-4-piperonylpiperazine (ET 495). *Eur. J. Pharmacol. 6:* 75—82, 1969.
330. Laursen, A. M.: Corpus striatum. *Acta Physiol. Scand.* (Suppl. *211*): 1—105, 1963.
331. Laverty, R., Michelson, I. A., Sharman, D. F., et al.: The subcellular localization of dopamine and acetylcholine in the dog caudate nucleus. *Br. J. Pharmacol. 21:* 482—490, 1963.
332. Laverty, R. and Sharman, D. F.: The estimation of small quantities of 3,4-dihydroxyphenylethylamine in tissue. *Br. J. Pharmacol. 24:* 538—548, 1965.
333. Laverty, R. and Sharman, D. F.: Modification by drugs of the metabolism of 3,4-dihydroxyphenylethylamine and 5-hydroxytryptamine in the brain. *Br. J. Pharmacol. 24:* 759—772, 1965.
334. Lemmer, B., Kim, J. S., and Bak, I. J.: Effect of 5-hydroxydopamine on uptake and content of serotonin in rat striatum. *Experientia 28:* 439—441, 1972.
335. Levitt, M., Spector, S., Sjoerdsma, A., et al.: Elucidation of the rate-limiting step in norepinephrine biosynthesis in the perfused guinea pig heart. *J. Pharmacol. Exp. Ther. 148:* 1—8, 1965.
336. Libet, B. and Tosaka, T.: Dopamine as a synaptic transmitter and modulator in sympathetic ganglia: A different mode of synaptic action. *Proc. Natl. Acad. Sci. U.S.A. 67:* 667—673, 1970.
337. Liles, S. L. and Davis, G. D.: Interaction of caudate nucleus and thalamus in alteration of cortically induced movement. *J. Neurophysiol. 32:* 564—572, 1969.
338. Liles, S. L. and Davis, G. D.: Athetoid and choreiform hyperkinesias

produced by caudate lesions in the cat. *Science 164:* 195–197, 1969.
339. Lindmar, R., Löffelholz, K., and Muscholl, E.: A muscarinic mechanism inhibiting the release of noradrenaline from peripheral adrenergic nerve fibers by nicotinic agents. *Br. J. Pharmacol. 32:* 280–294, 1968.
340. Lindmar, R., Löffelholz, K., and Muscholl, E.: Unterschiede zwischen Tyramin und Dimethyl-phenyl-piperazin in der Ca^{++}-Abhängigkeit und im zeitlichen Verlauf der Noradrenalin-Freisetzung am isolierten Kaninchenherzen. *Experientia 23:* 933–934, 1967.
341. Lloyd, K. G., Hockman, C. H., Davidson, L., et al.: Kinetics of L-DOPA metabolism in the caudate nucleus of cats with midbrain lesions. (Submitted for publication.)
342. Lloyd, K. G. and Hornykiewicz, O.: Occurrence and distribution of aromatic L-amino acid (L-DOPA) decarboxylase in the human brain. *J. Neurochem. 19:* 1549–1559, 1972.
343. Lloyd, K. G., Stadler, H., and Bartholini, G.: Dopamine and acetylcholine neurons in striatal and limbic structures: Effect of neuroleptic drugs. In Usdin, E. and Snyder, S. H. (Eds.): *Frontiers in Catecholamine Research.* Oxford: Pergamon Press, 1973, pp. 777–779.
344. Lovenberg, W., Weissbach, H., and Udenfriend, S.: Aromatic L-amino acid decarboxylase. *J. Biol. Chem. 237:* 89–93, 1962.
345. Maj, J., Sowinska, H., Baran, L., et al.: The effect of clonidine on locomotor activity in mice. *Life Sci. 11:* 483–491, 1972.
346. Malmfors, T. and Sachs, C.: Degeneration of adrenergic nerves produced by 6-hydroxydopamine. *Eur. J. Pharmacol. 3:* 89–92, 1968.
347. Mantegazzini, P. and Glässer, A.: Action de la DL-3,4-dioxyphenylalanine (DOPA) et de la dopamine sur l'activité électrique du chat "cerveau isolé." *Arch. Ital. Biol. 98:* 367–374, 1960.
348. Marshall, J. F., Turner, B. H., and Teitelbaum, P.: Sensory neglect produced by lateral hypothalamic damage. *Science 174:* 523–525, 1971.
349. Martin, J. P.: *The Basal Ganglia and Posture.* London: Pitman, 1967.
350. Masuoka, D. T., Schott, H. F., and Petriello, L.: Formation of catecholamines by various areas of cat brain. *J. Pharmacol. Exp. Ther. 139:* 73–76, 1963.
351. Matussek, N., Bendert, O., Schneider, K., et al.: L-DOPA plus decarboxylase inhibitor in depression. *Lancet II:* 660–661, 1970.
352. Matussek, N., Pohlmeier, H., and Ruther, E.: Die Wirkung von DOPA auf gehemmte Depressionen. *Klin. Wochenschr. 44:* 727–728, 1966.
353. Mavrojannis, M.: L'action cataleptique de la morphine chez les rats. Contributions à la théorie toxique de la catalepsie. *C. R. Soc. Biol.* (Paris) *55:* 1092–1140, 1903.
354. McAfee, D. A., Schorderet, M., and Greengard, P.: Adenosine 3', 5'-monophosphate in nervous tissue: Increase associated with synaptic transmission. *Science 171:* 156–158, 1971.

355. McGeer, P. L., Bagchi, S. P., and McGeer, E. G.: Subcellular localization of tyrosine hydroxylase in beef caudate nucleus. *Life Sci. 4:* 1859–1867, 1965.
356. McGeer, E. G., Gibson, S., and McGeer, P. L.: Some characteristics of brain tyrosine hydroxylase. *Can. J. Biochem. 45:* 1557–1563, 1967.
357. McGeer, E. G., Gibson, S., Wada, J. A., et al.: Distribution of tyrosine hydroxylase activity in adult and developing brain. *Can. J. Biochem. 45:* 1943–1952, 1967.
358. McGeer, E. G. and McGeer, P. L.: In-vitro screen of inhibitors of rat brain tyrosine hydroxylase. *Can. J. Biochem. 45:* 115–131, 1967.
359. McGeer, E. G., McGeer, P. L., and Wada, J. A.: Distribution of tyrosine hydroxylase in human and animal brain. *J. Neurochem. 18:* 1647–1658, 1971.
360. McKenzie, G. M., Gordon, R. J., and Viik, K.: Some biochemical and behavioral correlates of a possible animal model of human hyperkinetic syndromes. *Brain Res. 47:* 439–456, 1972.
361. McKenzie, G. M. and Szerb, J. C.: The effect of dihydroxyphenylalanine, pheniprazine and dextro-amphetamine on the in vivo release of dopamine from the caudate nucleus. *J. Pharmacol. Exp. Ther. 162:* 302–308, 1968.
362. McKenzie, G. M. and White, H. L.: Evidence for the methylation of apomorphine by catechol-O-methyl transferase *in vivo* and *in vitro. Biochem. Pharmacol. 22:* 2329–2336, 1973.
363. McLennan, H.: The release of acetylcholine and of 3-hydroxytyramine from the caudate nucleus. *J. Physiol.* (Lond.) *174:* 152–161, 1964.
364. McLennan, H. and York, D. H.: The action of dopamine on neurones of the caudate nucleus. *J. Physiol.* (Lond.) *189:* 393–402, 1967.
365. McLennan, H.: The release of dopamine from the putamen. *Experientia 21:* 725–726, 1965.
366. Mettler, F. A.: Substantia nigra and Parkinsonism. *Arch. Neurol. 11:* 529–542, 1964.
367. Mettler, F. A.: Perceptual capacity, functions of the corpus striatum and schizophrenia. *Psychiatr. Q. 29:* 89–111, 1955.
368. Mettler, F. A. and Mettler, C. C.: Role of the neostriatum. *Am. J. Physiol. 133:* 594–601, 1949.
369. Miyamoto, E., Kuo, J. F., and Greengard, P.: Cyclic nucleotide-dependent protein kinases. Purification and properties of adenosine 3′,5′-monophosphate-dependent protein kinase from bovine brain. *J. Biol. Chem. 244:* 6395–6402, 1969.
370. Molinoff, P. B. and Axelrod, J.: Biochemistry of catecholamines. *Annu. Rev. Biochem. 40:* 465–500, 1971.
371. Montagu, K. A.: Catechol compounds in rat tissues and in brains of different animals. *Nature 180:* 244–245, 1957.
372. Moore, K. E., Carr, L. A., and Dominic, J. A.: Functional significance of amphetamine-induced release of brain catecholamines. In Costa, E. and Garattini, S. (Eds.): *Amphetamines and Related Compounds.* New York: Raven Press, 1970, pp. 371–385.

373. Moore, K. E., Bhatnagar, R. K., and Heller, A.: Anatomical and chemical studies of a nigro-neostriatal projection in the cat. *Brain Res. 30:* 119–135, 1971.
374. Mueller, R. A., Thoenen, H., and Axelrod, J.: Increase in tyrosine hydroxylase activity after reserpine administration. *J. Pharmacol. Exp. Ther. 169:* 74–79, 1969.
375. Munkvad, I., Pakkenberg, H., and Randrup, A.: Aminergic systems in basal ganglia associated with stereotyped hyperactive behavior and catalepsy. *Brain Behav. Evol. 1:* 89–100, 1968.
376. Murphy, D. L., Brodie, H. K. H., Goodwin, F. K., et al.: L-Dopa: Regular induction of hypomania in "bipolar" manic-depressive patients. *Nature 229:* 135–136, 1971.
377. Musacchio, J. M., D'Angelo, G. L., and McQueen, C. A.: Dihydropteridine reductase: Implication on the regulation of catecholamine biosynthesis. *Proc. Natl. Acad. Sci. U.S.A., 68:* 2087–2091, 1971.
378. Musacchio, J. M., Julou, L., Kety, S. S., et al.: Increase in rat brain tyrosine hydroxylase activity produced by electroconvulsive shock. *Proc. Nat. Acad. Sci. U.S.A. 63:* 1117–1119, 1969.
379. Nagatsu, T., Levitt, M., and Udenfriend, S.: The initial step in norepinephrine biosynthesis. *J. Biol. Chem. 239:* 2910–2917, 1964.
380. Nagatsu, T., Levitt, M., and Udenfriend, S.: Conversion of L-tyrosine to 3,4-dihydroxyphenylalanine by cell-free preparations of brain and sympathetically innervated tissues. *Biochem. Biophys. Res. Commun. 14:* 543–549, 1964.
381. Nagatsu, M., Mizutani, K., Nagatsu, I., et al.: Pteridines as cofactor or inhibitor of tyrosine hydroxylase. *Biochem. Pharmacol. 21:* 1945–1953, 1972.
382. Nagatsu, T. and Nagatsu, I.: Subcellular distribution of tyrosine and monoamine oxidase in the bovine caudate nucleus. *Experientia 26:* 722–723, 1970.
383. Nagatsu, T., Sudo, Y., and Nagatsu, I.: Tyrosine hydroxylase in bovine caudate nucleus. *J. Neurochem. 18:* 2179–2189, 1971.
384. Nagatsu, T. and Yamamoto, T.: Fluorescence assay of tyrosine hydroxylase activity in tissue homogenate. *Experientia 24:* 1183–1184, 1968.
385. Nathanson, J. A. and Greengard, P.: Octopamine-sensitive adenylate cyclase: Evidence for a biological role of octopamine in nervous tissue. *Science 180:* 308–310, 1973.
386. Nauta, W. J. H. and Mehler, W. R.: Projections of the lentiform nucleus in the monkey. *Brain Res. 1:* 3–42, 1966.
387. Nauta, W. J. H.: Central nervous organization and the endocrine motor system. In Nalbandov, A. V. (Ed.): *Advances in Neuroendocrinology.* Urbana: University of Illinois Press, 1963, pp. 5–21.
388. Naylor, R. J. and Olley, J. E.: Modification of the behavioral changes induced by haloperidol in the rat by lesions in the caudate nucleus, the caudate-putamen and globus pallidus. *Neuropharmacology 11:* 81–89, 1972.
389. Neff, N. H. and Costa, E.: Application of steady-state kinetics to the

study of catecholamine turnover after monoamine oxidase inhibition or reserpine administration. *J. Pharmacol. Exp. Ther. 160:* 40—47, 1968.
390. Neff, N. H. and Costa, E.: Effect of tricyclic antidepressants and chlorpromazine on brain catecholamine synthesis. In Garattini, S. and Dukes, M. N. G. (Eds.): *Antidepressant Drugs.* New York: Excerpta Medica Fdn., 1967, pp. 28—34.
391. Niemegeers, C. J. E., Verbruggen, F. J., and Janssen, P. A. J.: The influence of various neuroleptic drugs on shock avoidance responding in rats. II. Nondiscriminated Sidman avoidance procedure with alternate reinforcement and extinction periods and analysis of interresponse times (IRTs). *Psychopharmacologia 16:* 175—182, 1969.
392. Niemegeers, C. J. E. and Janssen, P. A. J.: A comparative study of the inhibitory effects of haloperidol and trifluperidol on learned shock-avoidance behavioral habits and on apomorphine-induced emesis in mongrel dogs and in beagles. *Psychopharmacologia 8:* 263—270, 1965.
393. Ng, L. K. Y., Chase, T. N., Colburn, R. W., et al.: Release of [^{3}H] dopamine by L-5-hydroxytryptophan. *Brain Res.* 499—500, 1972.
394. Ng, K. Y., Chase, T. N., Colburn, R. W., et al.: L-DOPA induced release of cerebral monoamines. *Science 170:* 76—77, 1970.
395. Nobin, A. and Björklund, A.: Topography of monoamine systems in the human brain as revealed in fetuses. *Acta Physiol. Scand.* (Suppl.) *388:* 1—40, 1973.
396. Nose, T., Segawa, T., and Takagi, H.: Subcellular localization of dopamine in the rat striatum *Jpn. J. Pharmacol. 22*:867—869, 1972.
397. Nyback, H., Schubert, J., and Sedvall, G.: Effect of apomorphine and pimozide on synthesis and turnover of labelled catecholamines in mouse brain. *J. Pharm. Pharmacol. 22*:622—624, 1970.
398. Nyback, H. and Sedvall, G.: Effect of nigral lesion on chlorpromazine-induced acceleration of dopamine synthesis from [^{14}C] tyrosine. *J. Pharm. Pharmacol. 23*:322—326, 1971.
399. Nyback, H. and Sedvall, G.: Further studies on the accumulation and disappearance of catecholamines formed from tyrosine - ^{14}C in the mouse brain. *Eur. J. Pharmacol. 10*:193—205, 1970.
400. Nyback, H. and Sedvall, G.: Regional accumulation of catecholamines formed from tyrosine-^{14}C in rat brain: Effect of chlorpromazine. *Eur. J. Pharmacol. 5*:245—252, 1969.
401. Nyback, H. Wiesel, F.-A., and Sedvall, G.: Receptor regulation of dopamine turnover. In Usdin, E. and Snyder, S. H. (Eds.): *Frontiers in Catecholamine Research.* Oxford: Pergamon Press, 1973, pp. 601—604.
402. Ohye, C., Bouchard, R., Boucher, R., et al.: Spontaneous activity of the putamen after chronic interruption of the dopaminergic pathway: Effect of L-DOPA. *J. Pharmacol. Exp. Ther. 175*:700—708, 1970.
403. Okada, Y., Nitsch-Hassler, C., Kim J. S., et al.: The role of γ-amino-

butyric acid (GABA) in extrapyramidal motor system. 1. Regional distribution of GABA in rabbit, rat and guinea-pig brain. *Exp. Brain Res. 13*:514–518, 1971.
404. O'Keefe, R., Sharman, D. F., and Vogt, M.: Effect of drugs used in psychoses on cerebral dopamine metabolism. *Br. J. Pharmacol. 38*:287–304, 1970.
405. Olivier, A., Parent, A., Simard, H., et al.: Cholinesterasic striatopallidal and striatonigral efferents in the cat and monkey. *Brain Res. 18*:273–282, 1970.
406. Olson, L., Nyström, B., and Seiger, Å.: Monoamine neuron systems in the normal and schizophrenic human brain: Fluorescence histochemistry of fetal, neurosurgical and post mortem material. In Usdin, E. and Snyder, S. H. (Eds.): *Frontiers in Catecholamine Research*. Oxford: Pergamon Press, 1973, pp. 1097–1100.
407. Olson, L., Seiger, Å., and Fuxe, K.: Heterogeneity of striatal and limbic dopamine innervation: Highly fluorescent islands in developing and adult brain. *Brain Res. 44*:283–288, 1972.
408. Oltmans, G. A. and Harvey, J. A.: LH syndrome and brain catecholamine levels after lesions of the nigrostriatal bundle. *Physiol. Behav. 8*:69–70, 1972.
409. Pakkenberg, H. and Brody, H.: Number of nerve cells in *substantia nigra* in paralysis agitans. *Acta Neuropathol. 5*:320–324, 1965.
410. Papavasiliou, P. S., Gellene, R., and Cotzias, G. C.: Modification of Parkinsonism: Diskinesias accompanying treatment with dopa. In Crane, G. E. and Gardner, R. (Eds.): *Psychotropic Drugs and Dysfunctions of the Basal Ganglia*. Publ. Health Serv. Pub. (U.S.A.), No. 1938, pp. 140–143, 1969.
411. Papeschi, R., Sourkes, T. L., Poirier, L. J., et al.: On the intracerebral origin of homovanillic acid of the cerebrospinal fluid of experimental animals. *Brain Res. 28*:527–533, 1971.
412. Parent, A. and Poirier, L. J.: Medial forebrain bundle (MFB) and ascending monoaminergic pathways in the cat. *Can. J. Physiol. Pharmacol. 47*:781–785, 1969.
413. Pellerin, J. and D'Iorio, A.: Metabolism of radioactive β-3, 4-dihydroxyphenylalanine-α-C^{14} in the albino rat. *Can. J. Biochem. 33*:1055–1061, 1955.
414. Persson, T. and Waldeck, B.: The interaction between different metabolic pathways of catecholamines in the brain studied by means of ^{3}H-dopa. *Acta Pharmacol. Toxicol. 27*:225–236, 1969.
415. Phillipu, A. and Heyd, W.: Release of dopamine from subcellular particles of the striatum. *Life Sci. 9*:361–373, 1970.
416. Pletscher, A., Brossi, A., and Gey, K. F.: Benzoquinolizine derivatives: A new class of monoamine decreasing drugs with psychotropic action. *Int. Rev. Neurobiol. 4*:275–306, 1962.
417. Pletscher, A., Gey, K. F., and Burkard, W. P.: Inhibitors of monoamine oxidase and decarboxylase of aromatic amino acids. In Eichler, O. and Farah, A. (Eds.): *Handbook of Experimental Pharmacology*. Vol. 19. Berlin: Springer, 1965, pp. 593–735.
418. Pletscher, A., Shore, P. A., and Brodie, B. B.: Serotonin as a mediator

of reserpine action in brain. *J. Pharmacol. Exp. Ther. 116*:84–89, 1956.
419. Poirier, L. J.: Experimental and histological study of midbrain dyskinesias. *J. Neurophysiol. 23*:534–551, 1960.
420. Poirier, L. J., Langelier, P., and Boucher, R.: Spontaneous and L-DOPA induced circus movements in cats with brain stem lesions. *J. Physiol.* (Paris) *66*: 735–754, 1973.
421. Poirier, L. J., Singh, P., Boucher, R., et al.: Effect of brain lesions on striatal monoamines in the cat. *Arch. Neurol. 17*:601–608, 1957.
422. Poirier, L. J., Singh, P., Sourkes, T. L., et al.: Effect of amine precursors on the concentration of striatal dopamine and serotonin in cats with and without unilateral brain stem lesions. *Brain Res. 6*:654–666, 1967.
423. Poirier, L. J. and Sourkes, T. L.: Experimentally induced Parkinsonism. In *Neurotransmitters. Res. Publ. Assoc. Nerv. Ment. Dis. 50*:416–433, 1972.
424. Poirier, L. J. and Sourkes, T. L.: Influence of the substantia nigra on the catecholamine content of the striatum. *Brain 88*:181–192, 1965.
425. Porter, C. C.: Aromatic amino acid decarboxylase inhibitors. *Fed. Proc. 30*:871–876, 1971.
426. Portig, P. J., Sharman, D. F., and Vogt, M.: Release by tubocurarine of dopamine and homovanillic acid from the superfused caudate nucleus. *J. Physiol.* (Lond.) *194*:565–672, 1968.
427. Portig, P. J. and Vogt, M.: Release into the cerebral ventricles of substances with possible transmitter function in the caudate nucleus. *J. Physiol.* (Lond.) *204*:687–715, 1969.
428. Precht, W. and Yoshida, M.: Monosynaptic inhibition of neurons of the substantia nigra by caudato-nigral fibers. *Brain Res. 32*:225–228, 1971.
429. Precht, W. and Yoshida, M.: Blockage of caudate-evoked inhibition of neurons in the substantia nigra by picrotoxin. *Brain Res. 32*:229–233, 1971.
430. Purpura, D. P., Frigyesi, T. L., and Malliani, A.: Intrinsic synaptic organization and relations of the corpus striatum. In Yahr, M. D. and Purpura, D. P. (Eds.): *Neurophysiological Basis of Normal and Abnormal Motor Activitivies.* (Proc. 3rd. Symposium of the Parkinson's Disease Information and Res. Ctr., Columbia Univ.). New York: Raven Press, 1967, pp. 177–214.
431. Randrup, A. and Munkvad, I.: Biochemical, anatomical and psychological investigations of stereotyped behavior induced by amphetamines. In Costa, E. and Garattini, S. (Eds.): *Amphetamines and Related Compounds.* New York: Raven Press, 1970, pp. 695–713.
432. Randrup, A. and Munkvad, I.: Stereotyped activities produced by amphetamine in several aminal species and man. *Psychopharmacologia 11*:300–310, 1967.
433. Randrup, A. and Munkvad, I.: Role of catecholamines in the amphetamine-excitatory response. *Nature 211*:540, 1966.
434. Rech, R. H. and Stolk, J. M.: Amphetamine-drug interactions that

relate brain catecholamines to behavior. In Costa, E. and Garattini, S. (Eds.): *Amphetamines and Related Compounds*. New York: Raven Press, 1970, pp. 385–415.
435. Reinert, H.: The depolarizing and blocking action of amphetamine in the cat's superior cervical ganglion. In Vane, J. R., Wolstenholme, G. E. W., and O'Connor, M. (Eds.): *Adrenergic Mechanisms* (A Ciba Foundation Symposium). Boston: Little, Brown, 1960, pp. 373–379.
436. Rekker, R. F., Engel, D. J. C., and Nys, G. G.: Apomorphine and its dopamine-like action. *J. Pharm. Pharmacol.* 24:589–591, 1972.
437. Riddell, D. and Szerb, J. C.: The release *in vivo* of dopamine synthesized from labeled precursors in the caudate nucleus of the cat. *J. Neurochem.* 18:989–1006, 1971.
438. Rinvik, E. and Grofová, I.: Observations on the fine structure of the substantia nigra in the cat. *Exp. Brain Res.* 11:229–248, 1970.
439. Robins, E. and Hartman, B. K.: Biochemical theories of mental disorders. In Albers, W. R., Siegel, G. J., Katzman, R., et al.: (Eds.): *Basic Neurochemistry*. Boston: Little, Brown, 1972, pp. 607–644.
440. Robison, G. A., Butcher, R. W., and Sutherland, E. W.: *Cyclic AMP*. New York: Academic Press, 1971.
441. Roffler-Tarlov, S., Sharman, D. F., and Tegerdine, P.: 3,4-Dihydroxyphenylacetic acid and 4- hydroxy-3-methoxy-phenylacetic acid in the mouse striatum: A reflection of intra- and extraneuronal metabolism of dopamine. *Br. J. Pharmacol.* 46:409–415, 1972.
442. Roos, B.-E.: Decrease in homovanillic acid as evidence for dopamine receptor stimulation by apomorphine in the neostriatum of the rat. *J. Pharm. Pharmac.* 21:263–264, 1969.
443. Rosengren, E.: On the role of monoamine oxidase for the inactivation of dopamine in brain. *Acta Physiol. Scand.* 49:370–375, 1960.
444. Rosenman, S. J. and Smith, C. B.: ^{14}C-catecholamine synthesis in mouse brain during withdrawal. *Nature* 204:153–155, 1972.
445. Roth, R. H.: Quoted in Bustos, G., Kuhar, M. J., and Roth, R. H.: Effect of gamma-hydroxybutyrate and gamma-butyrolactone on dopamine synthesis and uptake by rat striatum. *Biochem. Pharmacol.* 21:2649–2652, 1972.
446. Roth, R. H.: Effect of anesthetic doses of γ-hydroxy-butyrate on subcortical concentrations of homovanillic acid. *Eur. J. Pharmacol.* 15:52–59, 1971.
447. Roth, R. H. and Surh, Y.: Mechanism of the γ-hydroxy-butyrate-induced increase in brain dopamine and its relationship to "sleep." *Biochem. Pharmacol.* 19:3001–3012, 1970.
448. Roth, R. H., Walters, J. R., and Aghajanian, G. K.: Effect of impulse flow on the release and synthesis of dopamine in the rat striatum. In Usdin, E. and Snyder, S. H. (Eds.): *Frontiers in Catecholamine Research*. Oxford: Pergamon Press, 1973, pp. 567–574.
449. Rosvold, H. E. and Delgado, J. M. R.: The effect on delayed alternation test performance of stimulating or destroying electrically structures within the frontal lobes of the monkey's brain. *J. Comp. Physiol. Psychol.* 45:565–575, 1952.

450. Sandler, M., Carter, S. B., Hunter, K. R., et al.: Tetrahydroisoquinoline alkaloids, *in vivo* metabolites of L-Dopa in man. *Nature 241*: 439–443, 1973.
451. Sano, I., Gamo, T., Kakimoto, Y., et al.: Distribution of catechol compounds in human brain. *Biochem. Biophys. Acta. 32*:586–587, 1959.
452. Sasame, H. A., Perez-Cruet, J., DiChiara, G., et al.: Evidence that methadone blocks dopamine receptors in the brain. *J. Neurochem. 19*:1953–1957, 1972.
453. Scheel-Krüger, J.: Comparative studies of various amphetamine analogues demonstrating different interactions with the metabolism of the catecholamines in the brain. *Eur. J. Pharmacol. 14*:47–59, 1971.
454. Scheel-Krüger, J.: Central effects of anticholinergic drugs measured by the apomorphine gnawing test in mice. *Acta Pharmacol. Toxicol. 28*:1–16, 1970.
455. Scheel-Krüger, J. and Randrup, A.: Stereotyped hyperactive behavior produced by dopamine in the absence of noradrenaline. *Life Sci. 6*:1389–1398, 1967.
456. Schelkunow, E. L.: Integrated effect of psychotropic drugs on the balance of cholino-, adreno-, and serotoninergic processes in the brain as a basis of their gross behavioral and therapeutic actions. *Activitas Nerv. Supp.* (Praha) *9*:207–217, 1967.
457. Schildkraut, J. J.: The catecholamine hypothesis of affective disorders: A review supporting evidence. *Am. J. Psychiatr. 122*:509–522, 1965.
458. Schildkraut, J. J. and Kety, S. S.: Biogenic amines and emotion. *Science 156*:21–30, 1967.
459. Schwab, R. S., Amador, L. V., and Letvin, J. Y.: Apomorphine in Parkinson's disease. *Trans. Am. Neurol. Assoc. 76*:251–253, 1951.
460. Schwab, R. S., England, A. C., and Peterson, E.: Akinesia in Parkinson's disease. *Neurology 9*:65–72, 1959.
461. Sedgwick, E. M. and Williams, T. D.: The response of single units in the caudate nucleus to peripheral stimulation. *J. Physiol.* (Lond.) *189*:281–298, 1967.
462. Seeman, P.: Personal communication, 1974.
463. Segal, D. S., Sullivan, J. L., Kuczenski, R. T., et al.: Effects of long-term reserpine treatment on brain tyrosine hydroxylase and behavioral activity. *Science 173*:847–849, 1971.
464. Segundo, J. P. and Machne, X.: Unitary responses to afferent volleys in lenticular nucleus and claustrum. *J. Neurophysiol. 19*:325–339, 1956.
465. Seiden, L. S. and Carlsson, A.: Temporary and partial antagonism by L-dopa of reserpine-induced suppression of a conditioned avoidance response. *Psychopharmacologia 4*:418–423, 1963.
466. Seiden, L. S. and Martin, T. W., Jr.: Potentiation of effects of L-dopa on conditioned avoidance behavior by inhibition of extracerebral dopa decarboxylase. *Physiol. Behav. 6*:453–458, 1971.
467. Seiden, L. S. and Petersen, D. D.: Blockade of L-dopa reversal of

reserpine-induced conditional avoidance response suppression by disulfiram. *J. Pharmacol. Exp. Ther. 163*:84–89, 1968.
468. Sharman, D. F.: Changes in the metabolism of 3,4-dihydroxyphenylethylamine (dopamine) in the striatum of the mouse induced by drugs. *Br. J. Pharmacol. 28*:153–163, 1966.
469. Shaskan, E. and Snyder, S. H.: Kinetics of serotonin accumulation into slices from rat brain: Relationship to catecholamine uptake. *J. Pharmacol. Exp. Ther. 175*:404–418, 1971.
470. Shaw K. N. F., McMillan, A., and Armstrong, M. D.: The metabolism of 3,4-dihydroxyphenylalanine. *J. Biol. Chem. 226*:255–266, 1957.
471. Shimizu, N. and Ohnishi, S.: Demonstration of nigrostriatal tract by degeneration silver method. *Exp. Brain Res. 17*:133–138, 1973.
472. Shore, P. A., Silver, S. L., and Brodie, B. B.: Interaction of reserpine, serotonin, and lysergic acid diethylamide in brain. *Science 122*:284–285, 1955.
473. Shute, C. C. O. and Lewis, P. R.: The ascending cholinergic reticular system: Neocortical, olfactory and subcortical projections. *Brain 90*:497–520, 1967.
474. Sims, K. L. and Bloom, F. E.: Rat brain L-3,4-dihydroxyphenylalanine and L-hydroxytryptophan decarboxylase activities: Differential effect of 6-hydroxydopamine. *Brain Res. 49*:165–175, 1973.
475. Singh, M. and Smith, J. M.: Reversal of some therapeutic effects of haloperidol in schizophrenia by anti-Parkinson drugs. *Pharmacologist 13*:207, 1971.
476. Smelik, P. G. and Ernst, A. M.: Role of nigro-neostriatal dopaminergic fibers in compulsive gnawing behavior in rats. *Life Sci. 5*:1485–1488, 1966.
477. Smith, C. B., Villarreal, J. E., Bednarczyk, J. H., and Sheldon, M. I.: Tolerance to morphine-induced increases in (^{14}C)-catecholamine synthesis in mouse brain. *Science 170*:1106–1107, 1970.
478. Snyder, S. H.: Amphetamine psychosis: A "model" schizophrenia mediated by catecholamines. *Am. J. Psychiatr. 130*:61–67, 1973.
479. Snyder, S. H. and Coyle, J. T.: Regional differences in ^{3}H-norepinephrine and ^{3}H-dopamine uptake into rat brain homogenates. *J. Pharmacol. Exp. Ther. 165*:78–86, 1969.
480. Snyder, S. H., Green, A. I., and Hendley, E. D.: Kinetics of ^{3}H-norepinephrine accumulation into slices from different regions of the rat brain. *J. Pharmacol. Exp. Ther. 164*:90–102, 1968.
481. Solomon, P., Mitchell, R. S., and Prinzmetal, M.: The use of benztropine sulfate in postencephalitic Parkinson's disease. *J. Am. M. A. 108*:1765–1770, 1937.
482. Sourkes, T. L.: Possible new metabolites mediating actions of L-DOPA. *Nature 229*:413–414, 1971.
483. Sourkes, T. L.: Dopa decarboxylase: Substrates, coenzyme, inhibitors. *Pharmacol. Rev. 18*:53–60, 1966.
484. Sourkes, T. L.: Inhibition of dihydroxyphenylalanine decarboxylase by derivatives of phenylalanine. *Arch. Biochem. Biophys. 51*:444–456, 1954.

485. Spano, P. F. and Neff, N. H.: Metabolic fate of caudate nucleus dopamine. *Brain Res. 42*:139—145, 1972.
486. Spector, S., Gordon, R., Sjoerdsma, A., et al.: End-product inhibition of tyrosine hydroxylase as a possible mechanism for regulation of norepinephrine synthesis. *Mol. Pharmacol. 3*:555—559, 1967.
487. Spector, S., Sjoerdsma, A., and Udenfriend, S.: Blockade of endogenous norepinephrine synthesis by α-methyl-tyrosine, an inhibitor of tyrosine hydroxylase. *J. Pharmacol. Exp. Ther. 147*:86—95, 1965.
488. Starke, K.: Influence of extracellular noradrenaline on the stimulation evoked secretion of noradrenaline from sympathetic nerves: Evidence for an α-receptor-mediated feedback inhibition of noradrenaline release. *Naunyn Schmiedebergs Arch. Pharmacol. 275*:1—23, 1972.
489. Steg, G.: Striatal cell activity during systemic administration of monoaminergic and cholinergic drugs. In Gillingham, F. J. and Donaldson, I. M. L. (Eds.): *Third Symposium on Parkinson's Disease.* London: Livingstone, 1969, pp. 26—29.
490. Steg, G.: Efferent muscle innervation and rigidity. *Acta Physiol. Scand. 61* (Suppl. 225):1—53, 1964.
491. Stein, L. and Wise, C. D.: Mechanism of the facilitating effects of amphetamine on behavior. In Efron, D. H. (Ed.): *Psychotomimetic Drugs.* New York: Raven Press, 1970, pp. 123—149.
492. Stern, G.: The effects of lesions in the substantia nigra. *Brain 89*:449—478, 1966.
493. Stille, G.: Zur Pharmakologie katatonigener Stoffe. 4. Mitteilung: Die Wirkung von Morphin. *Arzneim Forsch. 21*:650—654, 1971.
494. Stille, G.: Zur Pharmakologie katatonigener Stoffe. 3. Mitteilung: Die Wirkung von Bulbokapnin. *Arzneim Forsch. 21*:528—535, 1971.
495. Stille, G. und Lauener, H.: Zur Pharmakologie katonigener Stoffe. 1. Mitteilung: Korrelation zwischen neuroleptischer Katalepsie und Homovanillinsäuregehalt im C. striatum. *Arzneim Forsch. 21*:252—255, 1971.
496. Stille, G. and Sayers, A.: Effect of striatal spreading depression on the pharmacological catatonia. *Neuropharmacol. 8*:181—189, 1969.
497. Stjärne, L.: Studies of catecholamine uptake, storage and release mechanisms. *Acta Physiol. Scand. 62* (Suppl.):1—97, 1964.
498. Stjärne, L. and Euler, U. S. von: Stereospecificity of amine uptake mechanism in nerve granules. *J. Pharmacol. Exp. Ther. 150*:335—340, 1965.
499. Stock, G., Magnusson, T., and Andén, N.-E.: Increase in brain dopamine after axotomy or treatment with gammahydroxybutyric acid due to elimination of the nerve impulse flow. *Naunyn Schmiedebergs Arch. Pharmacol. 278*:347—361, 1973.
500. Stone, C. A., Porter, C. C., Stavorski, J. M., et al.: Antagonism of certain effects of catecholamine-depleting agents by antidepressant and related compounds. *J. Pharmacol. Exp. Ther. 144*:196—204, 1964.

501. Struyker Boudier, H. A. J., Gielen, W., and van Rossum J. M.: Analysis of dopamine specific, excitatory and inhibitory actions on neurons of the snail. In Usdin, E. and Snyder, S. H. (Eds.): *Frontiers in Catecholamine Research.* Oxford: Pergamon Press, 1973, pp. 673–674.
502. Svensson, T. H.: On the role of central norepinephrine in the regulation of motor activity and body temperature in the mouse. *Naunyn Schmiedebergs Arch. Pharmacol. 271*:111–120, 1971.
503. Svensson, T. H. and Waldeck, B.: On the role of brain catecholamines in motor activity: Experiments with inhibitors of synthesis and of monoamine oxidase. *Psychopharmacologia 18*:357–365, 1970.
504. Symchowicz, S., Korduba, C. A., and Veals, J.: Inhibition of dopamine uptake into synaptosomes of rat corpus striatum by chlorpheniramine and its structural analogs. *Life Sci. 10*:35–42, 1971.
505. Szabo, J.: Projections from the body of the caudate nucleus in the Rhesus monkey. *Exp. Neurol. 27*:1–15, 1970.
506. Szabo, J.: Topical distribution of striatal efferents in the monkey. *Exp. Neurol. 5*:21–36, 1962.
507. Tagliamonte, A., Tagliamonte, P., and Gessa, G. L.: Reserpine-like action of chlorpromazine on rabbit basal ganglia. *J. Neurochem. 17*:733–738, 1970.
508. Takatsuka, K., Segawa, T., and Takagi, H.: Regional specificity of 5-hydroxytryptamine uptake in rabbit brain stem. *Neurochem. 17*:695–696, 1970.
509. Taugner, G.: The membrane of catecholamine storage vesicles of adrenal medulla. Uptake and release of noradrenaline in relation to the pH and concentration and steric configuration of the amine present in the medium. *Naunyn Schmiedebergs Arch. Pharmacol. 274*:299–314, 1972.
510. Taylor, K. M. and Laverty, R.: The metabolism of tritiated dopamine in regions of the rat brain *in vivo*. II. The significance of the neutral metabolites of catecholamines. *J. Neurochem. 16*:1367–1376, 1969.
511. Taylor, K. M. and Snyder, S. H.: Differential effects of D- and L-amphetamine on behavior and on catecholamine disposition in dopamine and norepinephrine-containing neurons of rat brain. *Brain Res. 28*:295–309, 1971.
512. Timsit, J.: Sur l'activité cataleptigène de quelques dérivés de la butyrophenone. *Thérapie 21*:1453–1471, 1966.
513. Thoenen, H.: Chemical sympathectomy: A new tool in the investigation of the physiology and pharmacology of peripheral and central adrenergic neurons. In Snyder, S. H. (Ed.): *Perspectives in Neuropharmacology.* New York: Oxford University Press, 1972, pp. 203–238.
514. Thoenen, H.: Induction of tyrosine hydroxylase in peripheral and central adrenergic neurones by cold-exposure of rats. *Nature 228*:861–862, 1970.

515. Thoenen, H., Huerlimann, A., and Haefely, W.: Cation dependence of the noradrenaline-releasing action of tyramine. *Eur. J. Pharmacol.* 6:29–37, 1969.
516. Thudichum, J. W. L.: *A Treatise on the Chemical Constitution of the Brain*. London: Balliere, Tindall and Cox, 1884.
517. Trendelenburg, U., Draskoczy, P. R., and Graefe, K. H.: The influence of intraneuronal monoamine oxidase on neuronal net uptake of noradrenaline and on sensitivity to noradrenaline. *Adv. Biochem. Psychopharm.* 5:371–377, 1972.
518. Trendelenburg, U., Maxwell, R. A., and Pluchino, S.: Methoxamine as a tool to assess the importance of intraneuronal uptake of l-norepinephrine in the cat's nictitating membrane. *J. Pharmacol. Exp. Ther.* 172:91–99, 1970.
519. Tseng, L. P. and Walaszek, E. J.: Blockade of dopamine depressor response by bulbocapnine. *Fed. Proc.* 29:741, 1970.
520. Udenfriend, S., Zaltzman-Nirenberg, P., and Nagatsu, T.: Inhibitors of purified beef adrenal tyrosine hydroxylase. *Biochem. Pharmacol.* 14:837–845, 1965.
521. Ungerstedt, U.: Stereotaxic mapping of the monoamine pathways in the rat brain. *Acta Physiol. Scand.* 82 (Suppl. 367):1–48, 1971.
522. Ungerstedt, U.: Histochemical studies on the effects of intracerebral and intraventricular injections of 6-hydroxy-dopamine on monoamine neurons in the rat brain. In Malmfors, T. and Thoenen, H. (Eds.): *6-Hydroxydopamine and Catecholamine Neurons*. Amsterdam: North-Holland, 1971, pp. 101–127.
523. Ungerstedt, U.: Adipsia and aphagia after 6-hydroxydopamine induced degeneration of the nigrostriatal dopamine system. *Acta Physiol. Scand.* 82 (Suppl. 367):95–122, 1971.
524. Ungerstedt, U.: Postsynaptic supersensitivity after 6-hydroxydopamine induced degeneration of the nigrostriatal dopamine system in the rat brain. *Acta Physiol. Scand.* 82 (Suppl. 367):69–93, 1971.
525. Ungerstedt, U.: Striatal dopamine release after amphetamine of nerve degeneration revealed by rotational behavior. *Acta Physiol. Scand.* 82 (Suppl. 367):49–68, 1971.
526. Ungerstedt, U.: 6-Hydroxy-dopamine induced degeneration of central monoamine neurons. *Eur. J. Pharmacol.* 5:1067–110, 1968.
527. Ungerstedt, U. and Arbuthnott, G. W.: Quantitative recording of rotational behavior in rats after 6-hydroxy-dopamine lesions of the nigrostriatal dopamine system. *Brain Res.* 24:485–493, 1970.
528. Ungerstedt, U., Butcher, L. L., Butcher, S. G., et al.: Direct chemical stimulation of dopaminergic mechanisms in the neostriatum of the rat. *Brain Res.* 14:461–471, 1969.
528a. Uretsky, N. J., Chase, G. J., and Lorenzo, A. V.: *J. Pharmacol. Exp. Ther.* 193:73–87, 1975.
529. Uretsky, N. J. and Iversen, L. L.: Effects of 6-hydroxy-dopamine on catecholamine-containing neurons in the rat brain. *J. Neurochem.* 17:269–278, 1970.

530. Valenstein, E. S., Cox, V. C., and Kakolenski, J. W.: Reexamination of the role of the hypothalamus in motivation. *Psychol. Rev.* 77:16–31, 1970.
531. van Rossum, J. M.: The significance of dopamine receptor blockade. In Brill, H., Cole, J. O., Deniker, P., et al. (Eds.): *Neuro-Psycho-Pharmacology*. Amsterdam: Excerpta Medica Foundation, 1967, pp. 321–329.
532. van Rossum, J. M. and Hurkmans, J. A. T. H. M.: Mechanism of action of psychomotor stimulant drugs. *Int. J. Neuropharmacol.* 3:227–239, 1964.
533. Vedernikow, Yu. P.: The influence of single and chronic morphine administration on some central effects of amphetamine and apomorphine. *Psychopharmacologia* 17:283–288, 1970.
534. Vogel, W. H., Orfei, V., and Century, G.: Activities of enzymes involved in the formation and destruction of biogenic amines in various areas of the human brain. *J. Pharmacol. Exp. Ther.* 165:196–203, 1969.
535. Vogt, C. and Vogt, O.: Sitz und Wesen der Krankheiten im Lichte der topistischen Hirnforschung und des Variierens der Tiere. *J. Psychol. Neurol.* 47:237–457, 1937.
536. Vogt, M.: Catechol amines in brain. *Pharmacol Rev.* 11:483–489, 1959.
537. Vogt, M.: The concentration of sympathin in different parts of central nervous system under normal conditions and after the administration of drugs. *J. Physiol.* (Lond.) 123:451–481, 1954.
538. Voneida, T. J.: An experimental study of the course and destination of fibers arising in the head of the caudate nucleus in the cat and monkey. *J. Comp. Neurol.* 115:75–87, 1960.
539. Von Uexküll, T.: Ein Beitrag zur Pathologie und Klinik der Bereitstellungsregulationen. *Verh. Dsch. Ges. Inn. Med.* 59:104–107, 1953.
540. Von Voigtlander, P. F. and Moore, K. E.: *In vivo* release of ^{3}H-dopamine from the brain during the administration of haloperidol. *Res. Comm. Chem. Path. Pharmacol.* 5:223–232, 1973.
541. Von Voigtlander, P. F. and Moore, K. E.: Involvement of nigrostriatal neurons in the *in vivo* release of dopamine by amphetamine, amantadine and tyramine. *J. Pharmacol. Exp. Ther.* 184:542–552, 1973.
542. Von Voigtlander, P. F. and Moore, K. E.: The release of ^{3}H-dopamine from cat brain following electrical stimulation of the substantia nigra and caudate nucleus. *Neuropharmacology* 10:733–741, 1971.
543. Von Voigtlander, P. F. and Moore, K. E.: Nigrostriatal pathway: Stimulation-evoked release of [^{3}H]dopamine from caudate nucleus. *Brain Res.* 35:580–583, 1971.
544. Walker, R. J. and Woodruff, G. N.: Structure activity studies on the Helix dopamine receptor. *Br. J. Pharmacol.* 35:359P, 1969.

545. Walters, J. R. and Roth, R. H.: Effect of gamma-hydroxy-butyrate on dopamine and dopamine metabolites in the rat striatum. *Biochem. Pharmacol. 21*:2111–2121, 1972.
546. Way, E. L. and Shen, F. H.: Catecholamines and 5-hydroxytryptamine. In Clouet, D. H. (Ed.): *Narcotic Drugs. Biochemical Pharmacology*. New York: Plenum Press, 1971, pp. 229–253.
547. Weight, F. F. and Padjen, A.: Slow postsynaptic inhibition and sodium inactivation in frog sympathetic ganglion cells. *Vth Int. Congr. Pharmacol.*, Abstract No. 1489.
548. Weight, F. F. and Votava, J.: Slow synaptic excitation in sympathetic ganglion cells: Evidence for synaptic inactivation of potassium conductance. *Science 170*:755–758, 1970.
549. Weil-Malherbe, H. and Bone, A. D.: Intracellular distribution of catecholamines in the brain. *Nature 180*:1050–1051, 1957.
550. Weissman, A. and Finger, K. F.: Effects of benzquinamide on avoidance behavior and brain amine levels. *Biochem. Pharmacol. 11*: 871–880, 1962.
551. Weissman, A., Koe, B. K., and Tenen, S. S.: Anti-amphetamine effects following inhibition of tyrosine hydroxylase. *J. Pharmacol. Exp. Ther. 151*:339–352, 1966.
552. Werdinius, B.: Prenylamine and brain monoamine metabolism. *Acta Pharmacol Toxicol. 25*:1–8, 1967.
553. Werman, R.: Criteria for identification of a central nervous system transmitter. *Comp. Biochem. Physiol. 18*:745–766, 1966.
554. Wirz-Justice, A.: The influence of various amines and amino acids on K^+-stimulated release of dopamine and serotonin from rat striatum. *Eur. J. Pharmacol. 19*:281–284, 1972.
555. Wise, C. D. and Stein, L.: Facilitation of brain self-stimulation by central administration of norepinephrine. *Science 163*:299–301, 1969.
556. Wolf, H. H., Rollins, D. E., Rowland, C. R., et al.: The importance of endogenous catecholamines in the activity of some C. N. S. stimulants. *Int. J. Neuropharmacol. 8*:319–328, 1969.
557. Woodruff, G. N.: Dopamine receptors. *Comp. Gen. Pharmacol. 2*:439–455, 1971.
558. Wurtman, R. J. and Fernstrom J. D.: L-tryptophan, L-tyrosine, and the control of brain monoamine biosynthesis. In Snyder, S. H. (Ed.): *Perspectives in Neuropharmacology*. New York: Oxford University Press, 1972, pp. 143–193.
559. Yahr, M. D.: L-DOPA in neurological disease: Current status. In Kopin, I. J. (Ed.): *Neurotransmitters. Res. Publ. Assoc. Nerv. Ment. Dis. 50*:494–511, 1972.
560. Yahr, M. D., Duvoisin, R. C., Schear, M. J., et al.: Treatment of Parkinsonism with Levodopa. *Arch. Neurol. 21*:343–354, 1969.
561. Yakovlev, P. I.: A proposed definition of the limbic system. In Hockman, C. H. (Ed.): *Limbic System Mechanisms and Autonomic Function*. Springfield (Ill.): C.C. Thomas, 1972, pp. 241–283.
562. York, D. H.: Dopamine receptor blockade—a central action of chlorpromazine on striatal neurons. *Brain Res. 37*:91–99, 1972.

563. York, D. H.: Possible dopaminergic pathway from substantia nigra to putamen. *Brain Res. 20*:233–249, 1970.
564. York, D. H.: The inhibitory action of dopamine on neurons of the caudate nucleus. *Brain Res. 5*:263–266, 1967.
565. Yoshida, M., Rabin, A., and Anderson, M.: Monosynaptic inhibition of pallidal neurons by axon collaterals of caudato-nigral fibers. *Exp. Brain Res. 15*:333–347, 1972.
566. Zetler, G.: Cataleptic state and hypothermia in mice caused by central cholinergic stimulation and antagonized by anticholinergic and antidepressant drugs. *Int. J. Neuropharmacol. 7*:325–335, 1968.

CHAPTER SIX
Norepinephrine and Central Neurons

Barry J. Hoffer and Floyd E. Bloom

The actions of norepinephrine (NE) on central neurons have recently been reviewed (*15*). We shall concentrate on an extensive analysis of this fund of knowledge and attempt to point out those gaps that currently prevent generalized statements on the actions of this very interesting catecholamine. It is our view that the proper characterization of physiological central noradrenergic receptors depends on a cytochemical foundation in which cells are positively identified as being postsynaptic to NE-containing synapses. We will therefore briefly review the central anatomy of the NE-projection systems (see also [*95*]) and then indicate what we believe are the best methods for localizing these synaptic boutons at the electron microscopic level (*12, 13*). With these background details in mind we can then turn to an examination of the iontophoretic experiments that directly reflect on the physiological actions of NE on central neurons.

ANATOMICAL CONSIDERATIONS

Our present understanding of the locations of NE-containing central neurons and of the distribution of their axonal arborizations derives directly from the Falck-Hillarp method of fluorescence histochemistry (*36*). When applied to the central nervous system (CNS), the method demonstrates that central NE cells are concentrated into a half dozen nuclei within the pons, medulla, and mesencephalon of most mammals (*32*), and that from these nuclei, monosynaptic projections extend to the entire brain and to the spinal cord (*94*). Recent observations with discrete lesions have aided in the determination of specific terminal projections of these nuclei (*58, 94*) and indicate that the locus ceruleus (LC) is the source of the majority of NE axons projecting to both diencephalon and other telencephalic structures. Ungerstedt's work has shown that the axons of the LC form two separate bundles which course forward: 1) a dorsal bundle that sends axon collaterals to the entire neocortex, limbic cortex, and cerebellar cortex; and 2) the ventral bundle, composed of fibers from the locus subceruleus and other pontine NE nuclei, which innervates the hypothalamus and the basal telencephalon (94, 95). The dorsal and ventral NE bundles come together in the caudal diencephalon, join dopamine fibers from the substantia nigra (SN) and ventral tegmentum and 5-HT fibers from the raphé nuclei, and eventually reach the diencephalon and cortex by way of the medial forebrain bundle (see section on "Self-Stimulation" in this chapter). Thus the medial forebrain bundle contains a variety of ascending monoamine axon projections.

These anatomical studies also suggest that the LC offers a more selective opportunity for activation of NE axons than does the medial forebrain bundle. Recent detailed studies of the projections from the LC of the rat indicate that the dorsal nucleus is subdivided (25, 94, 95); the posterior lateral group sends a heavy projection to the cerebellum via the superior cerebellar peduncle, whereas the telencephalic projections of the LC arise mainly from the anterior pole of the nucleus. Both groups appear, however, to overlap in their terminal fields.

Electron Microscopic Localization

The varicosities of the axons demonstrated by fluorescence histochemistry indicate presumed sites of transmitter release. However, because of the limited resolution of the optical microscope relative to the complexly interrelated cellular processes of the neuropil, electron microscopic methods are needed to determine precisely which neurons in a given region receive synaptic contact from NE-containing axons. Although NE is a very small molecule in comparison to other cellular constituents that have been specifically localized by electron microscopy (EM)—e.g., DNA, RNA, glycogen, and collagen—certain empirical and pharmacological methods have been developed in pursuit of this objective. To determine the success of these methods it is useful to consider a set of operational criteria for the localization of any transmitter substance (12, 13).

Electron Microscopic Identification

To establish that a cytochemical reaction can specifically identify a substance as small as the neurotransmitter NE, the following objectives should be considered:

1. The fixation techniques required for structural analysis should neither disturb the cellular location of the transmitter substance nor impair its

cytochemical reactivity. Either before or after fixation the transmitter should be shown to react specifically with cytochemical reagents to yield an electron-opaque deposit. This reaction should be verified by model experiments in vitro.

2. The distribution of the reactive sites in the tissue should agree with the grosser estimates obtained by light microscopy and by cellular biochemical analyses of the transmitter. Almost all quantitative ultrastructural evaluations of staining methods for monoamines, e.g., are predicated on data obtained from formaldehyde-induced fluorescence histochemistry.

3. The amount of demonstrable substances should vary in proportion to the biochemically measurable levels of the substance when drugs have been administered that deplete or elevate the transmitter concentration in the tissue.

4. Selective destruction of the cells of origin, or of the tract of preterminal axons believed to contain the substance being localized, should remove the cytochemically reactive material. In the earliest cytological response to the destruction of the tract, the topological distribution of the degenerating terminals should agree with the cytochemical location of the transmitter.

Criteria for Localization

No single electron microscopic histochemical method has yet achieved the consistency and selectivity of localization desired for analysis of NE-transmitting synapses. Permanganate fixation methods (45–47, 72) offer the most direct approach to the successful visualization of small granular synaptic vesicles which seem identical morphologically and pharmacologically to the storage vesicles of NE in peripheral sympathetic nerve terminals. Technical problems, however, such as poor penetration yielding small usable tissue samples, limit this method to regions with a high density of NE axons, e.g., pons or hypothalamus.

We have found most useful for our purposes a combination of two methods: 1) autoradiographic localization of processes that accumulate tracer amounts of ^{3}H-NE in vivo (1) or in vitro (64), and 2) the acute degeneration that occurs in NE terminals within eight to 48 hr after injection of 6-hydroxydopamine (6-OHDA) into the cerebrospinal fluid (CSF) (11).

Both methods must, however, be applied cautiously. Several types of experiments demonstrate that the selective uptake of NE by noradrenergic processes is a saturable process (51, 83). When tissues are exposed to NE (or to any NE congeners that are to act as ultrastructural markers, such as α-methyl NE, 5-hydroxydopamine, or 6-OHDA) in concentrations above the saturation point for the selective uptake, marker can enter cells that do not normally contain NE. Furthermore, although 6-OHDA is known to produce damage only in central catecholamine neurons (*51*), the acute degeneration is often undetectable by EM (*11, 16*). It is also possible that ultrastructural effects may occur in cell groups for which no neurochemical tests are available.

We have found the most satisfactory localizations of NE-containing synaptic terminals to be based on complementary results obtained from multiple cytochemical methods. Recently we applied the experimental method of autoradiography in a different fashion (*27*), in an effort to localize NE synaptic projections independently of cytochemical tests for the transmitter molecule. This is accomplished by injecting small volumes of radioactive precursors of polypeptides or glycoproteins into the nucleus the projections of which are to be traced. In this context we injected ^{3}H-proline (*70a*) into the LC. The amino acid is incorporated by the neuronal perikarya, and the resultant labeled polypeptide is transported axopetally, labeling the entire axonal arborization within 24 hr. These results confirm the distribution of NE-projections obtained from fluorescence microscopy and are also capable of providing electron microscopic resolution of the labeled terminals by autoradiography (*70a*).

PROCEDURES AND STUDIES

Different procedures have been employed to study the effect of NE on central neurons. Injection of precursors parenterally (*75*), e.g., has been reported to alter both cortical slow waves and unit potentials. The most useful technique, however, for evaluating the effects of NE on central neurons utilizes microiontophoretic application from multi-barrel micropipettes, thus circumventing many of the temporal, chemical, and structural restrictions encountered in other test procedures (*74*).

Microiontophoresis

In contrast to earlier studies which indicated negligible NE effects on cortical (*58*) and spinal (*31*) neurons, recent experiments indicate that NE can affect nerve cells at virtually all levels of the neuraxis (Table 1).

The critical parameters determining the presence and qualitative nature of response to NE have been clarified by several recent studies. Thus in the cerebral cortex the response to iontophoresis of NE depends on the type of anesthesia: excitatory responses are more prevalent with halothane or in unanesthetized preparations, whereas inhibitory actions are usually seen with barbiturate anesthesia (*53*). The pH of drug solution also appears critical: cortical neurons are reported to be excited by NE ejected from

Table 1. Catalog of qualitative CNS responses to microelectrophoretic administration of NE

Brain Region	Cell Type[a]	Prep.[b]	Responses[c] & drug studies[d]	References
Cortex				
Cerebral (general)	U	1,3,5	3	58
		1,2,3	1,2; depending on anesthesia	53
		1,2,3	1,2; excitation blocked by α and β blockers, depression not blocked by either	52
		1	1,2; depending on pH of drug solution	38
Polysensory	I	1,3,4	1	67
Piriform	U	2x	1	63
Limbic system				
Hippocampus	I (pyramidal)	2	1	41
	I (pyramidal)	2	1	9
Septum	U	5	1	40, 84
Olfactory bulb	I	1	1	74
		5	1; B: dibenamine & LSD	73
Amygdala		2,4	1	89
Caudate nucleus	U	5	1	14
		1	1	42
Diencephalon				
Lateral geniculate	U	3	1	69
	I (principal)	3	1	69
	I (principal)	2,4	5	29
	I (principal)	5	2	77
Medial geniculate	U	3	1	93
Hypothalamus	I (perifornical and ventromedial)	1	4	57
	I (supraoptic)	2,3,5	1; B:MJ-1999; P:DMI[e]	6
	U	2,3	1,2	18

continued.

Table 1. continued

Brain Region	Cell Type[a]	Prep.[b]	Responses[c] & drug studies[d]	References
Brainstem				
Paramedian reticular nucleus	I	5x	2,3; B:chlorpromazine	5, 22
Inferior colliculus	U	2	5	30
Reticulospinal nucleus	I	5	2	49
Pontine raphé	I	2	1,2	26
Unidentified	U	3	4; amphetamine sensitivity correlated with NE response	19
	U	3	1,2; excitation blocked by α-methyl NE	21
Spinal cord				
	I (motoneuron)	5	1	98
	I (motoneuron)	2x	5	31
	I (interneuron)	5	3	97
	I (Renshaw)	1x	1	7
	I (Renshaw)	1x	1; B:phenoxybenzamine	8
	I (interneuron)	1	3; no blockade by parenteral or local phenoxybenzamine, phentolamine, or propranolol	35
	I (motoneuron)	3		
	I (Renshaw)	3		
Cerebellum				
	Purkinje	1,3,5	1; B:MJ-1999, prostaglandin E₁, nicotinate; P:DMI, methyl-xanthines, papaverine	43

[a]I, identified; U, unidentified.
[b]1, unanesthetized but with local anesthesia and neuromuscular blockade; 2, barbiturate anesthesia; 3, ether or halothane gaseous anesthesia; 4, chloralose; 5, unanesthetized decerebrate; x, extensive neurosurgical manipulation was performed.
[c]1, mainly depression; 2, mainly facilitation; 3, more responsive neurons depressed than facilitated; 4, more neurons facilitated than depressed; 5, few tested neurons responsive.
[d]P or B, drugs that potentiate or block response.

solutions with pH less than 4.0 and inhibited by NE from solutions with a pH greater than 4.0 (38). We have not observed a strict pH dependency for NE responses in other brain areas (15). It is also suggested that excitatory responses to NE are secondary to vasoconstriction (88), although this view has been vigorously challenged (20).

Identity of Action

That many neurons respond to microiontophoresis of NE is not sufficient per se to prove that the sensitive cells receive an NE-containing afferent input. The most crucial experiment requires selective stimulation of the NE-containing afferent pathway and the demonstration that the synaptic activation via NE application produces similar responses. These considerations lead to the conclusion that NE synaptic input can only be defined if identifiable postsynaptic elements known to receive NE inputs (see following section) are studied.

DEFINED POSTSYNAPTIC NEURONS

Norepinephrine-containing synapses have been demonstrated by a combination of fluorescence histochemistry, autoradiography after ^{3}H-NE, and acute degeneration after 6-OHDA in diverse structures in which neurons have also been tested with NE by iontophoresis. Supraoptic (6) and olfactory bulb (14) neurons are reproducibly inhibited by NE, but NE-receptor blockade did not completely eliminate the recurrent type of synaptic inhibition in either case. Cells of the medial geniculate body are also inhibited by NE (93) as are cat motoneurons (98). In neither of these cases, however, is it possible to activate the noradrenergic pathway selectively.

Much interest in central adrenergic afferents is derived from the clinical experience that psychoactive drugs that can either induce or ameliorate behavioral and perceptual dysfunction can also affect catecholamine metabolism (78). Such studies suggest that NE may be important within higher cortical centers for perceptual integration. However, previous studies of the cerebral cortex have shown a diversity of NE effects (Table 1). Recently we studied polysensory neurons in the frontal association cortex of the squirrel monkey, i.e., neurons that can be identified by their responsiveness to different sensory modalities (67).

Cytochemical studies have shown that polysensory cortex contains an extensive network of fine NE-containing axons which establish both axosomatic and axodendritic synapses on these neurons (67). Moreover, NE administered by microiontophoresis depressed spontaneous or induced discharge in well over 90% of the neurons (67). The NE responses were usually of short latency and had very low thresholds. The few excitatory responses that were seen could be readily reversed to inhibition by concurrent iontophoresis of desmethylimipramine (13). Further analysis of these cortical projections requires knowledge of the source of the NE cortical afferent terminals. As to be described subsequently, the source of these fibers to cerebellar and hippocampal cortex has been identified, and extension of these studies to the cerebral cortex is now in progress.

Rat Cerebellar Purkinje Cells

The most serious obstacle to physiological analysis of the central noradrenergic pathways has been the inability to activate the pathway selectively. The noradrenergic projection to rat cerebellar Purkinje cells has, however, been recently studied by both electrophysiological and cytochemical methods.

Cytochemical Studies The NE-containing axons of the cerebellar cortex can be localized, at the light microscopic level, using either formaldehyde-induced fluorescence in normal animals (16) or in vitro incubation with catecholamine analogues (48). These techniques may be combined with 6-OHDA pretreatment to insure that NE-containing rather than serotonin-containing fibers are being visualized (17). The thin fluorescent fibers branch extensively and manifest multiple varicosities in the molecular layer of the cerebellar cortex, giving off branches that run in both frontal and sagittal planes.

Norepinephrine-containing synapses can be localized at the electron microscopic level by degeneration after 6-OHDA exposure or autoradiography of sites taking up ^{3}H-NE. These ultrastructural studies indicate that NE-containing fibers synapse on the Purkinje cell dendritic tree, in the mid- to outer molecular layer (16). Recent techniques for facilitating visualization of these fibers (44) suggest that they are of sufficient density to permit contact with each Purkinje cell.

Purkinje Cell Adrenergic Receptor When NE is applied to Purkinje cells from micropipettes, there is a uniform and powerful depression of spontaneous discharge (43). Spike interval histograms show that NE produces no effect on climbing fiber bursts or on the most probable single spike interval, but rather that NE specifically increases the probability of long pauses when compared with control Purkinje cell firing.

Several lines of evidence suggest that a β-receptor is involved in the Purkinje cell response. Epinephrine and isoproterenol, e.g., produce changes in mean rate and in the interspike interval histogram analogous to those of NE (43). Moreover, iontophoretic administration of MJ-1999, a specific β-adrenergic blocking agent (43), antagonizes NE effects.

Evidence from the peripheral nervous system suggests that the sympathomimetic amines amphetamine and tyramine act by release of NE from presynaptic terminals. Yet iontophoresis of these amines slows Purkinje cell discharge even when adrenergic synaptic terminals are selectively destroyed by prior injection of 6-OHDA (*43*). This suggests that amphetamine and tyramine may have direct postsynaptic actions on Purkinje cells. Similar conclusions have been reached concerning the mechanism of action of amphetamine on neurons of the caudate nucleus (*37*).

It is also possible to record intracellularly from Purkinje cells during extracellular application of NE (*82*); NE produces hyperpolarization with either no change or an increase in membrane resistance. Similar transmembranous changes after iontophoresis of NE have recently been described in cat motoneurons by Engberg and Marshall (*34*).

The hyperpolarization with increased resistance seen with NE is in direct contrast to changes seen with classic inhibitory postsynaptic potentials (*33*) (IPSPs) or during iontophoresis of GABA (*82*). The classic inhibitory pathways and amino acids are thought to operate exclusively through mechanisms that increase conductance to ionic species whose equilibrium potentials are more negative than the resting membrane potential. In such cases the hyperpolarization is associated with decreased membrane resistance. On the contrary, hyperpolarization produced by NE may be attributable to a decrease in conductance to some ion such as sodium or calcium, or to activation of an electrogenic pump (*82*).

Activation of Adrenergic Pathway Experiments utilizing changes in formaldehyde-induced fluorescence of neurons and axons, in animals with lesions of the ascending NE bundles (see section on "Anatomy of Central NE Neurons" in this chapter), or cerebellar peduncles, show that the cerebellar adrenergic projection arises from the NE-containing neurons of the LC (*44, 68*).

Purkinje cells showed remarkably uniform inhibitory responses to stimulation of the LC with trains of pulses: 94 of 102 cells (20 animals) recorded extracellularly displayed depression of spontaneous discharge rate (*24*). This summated response was greatest at relatively low stimulus frequencies (3 to 50 Hz) and was markedly diminished at faster rates. Complete cessation of discharge outlasting the stimulation period by four to 65 sec (mean 21 sec) could be obtained with 20 to 100 pulses at 10 Hz). At this frequency, threshold currents ranged from 0.03 to 1.2 mA (mean 0.35 mA).

Although the response to a single LC stimulus often escaped detection on direct visual inspection, construction of poststimulation time histograms revealed a reduced probability of spike discharge over prolonged intervals of 60 to 470 msec (mean 293 msec) with a long latent period of 50 to 290 msec (mean 148 msec).

Intracellular recording from some Purkinje cells during stimulation of the LC with single shocks revealed late hyperpolarizations (not directly related to climbing fiber responses) that were usually very small. Signal averaging techniques better delineated these late hyperpolarizing responses (0.5 to 2 mV), which are comparable in latency and duration to the late prolonged periods of spike suppression seen in poststimulus histograms.

With trains of pulses, large hyperpolarizations extending well beyond the stimulation period and averaging 14 mV (range, 2 to 39 mV) were recorded. An index of membrane resistance was obtained by measuring 1) the size of climbing fiber excitatory postsynaptic potentials (EPSPs), and 2) the potential deflections produced by hyperpolarizing currents (0.5 to 1 nA, 40 msec duration) passed through the recording micropipette in conjunction with a Wheatstone bridge circuit. In all cases input resistance, as measured by these two parameters, either increased (ten cells) or did not change (two cells) during the LC-evoked hyperpolarizations. Thus LC stimulation exactly mimics the action of exogenously applied NE, which also produces hyperpolarization without a decrease in membrane resistance.

The anatomical discreteness of brainstem loci evoking this characteristic inhibitory response is underscored by the finding that the incidence of depression of Purkinje cell discharge falls abruptly when the stimulating electrode is placed slightly outside of the LC. In many cases individual cells, which responded when the electrode was in the LC, failed to do so when the stimulation electrode was displaced from the LC by as little as 0.5 mm; recovery of response ensued with repositioning in the LC. Furthermore, when catecholamine-containing pathways were selectively and chronically destroyed by intracisternally injected 6-OHDA, only five of 60 cells were inhibited when the LC was stimulated directly.

Although the effects of LC stimulation produce the same qualitative effect on Purkinje cells as does the iontophoretic administration of NE, additional studies were undertaken to confirm the noradrenergic nature of the LC effects. If the effects of LC stimulation were attributable to release of NE from the nerve terminals already demonstrated to synapse with Purkinje cells, pharmacological depletion of NE should seriously impair the LC inhibitory response. Indeed, when animals are acutely pretreated with reserpine (1.5 mg/kg, IV) and α-methyl-p-tyrosine (α-MpT) (100 to 200 mg/kg, IP), the loss of the LC inhibitory effects, whether to single or multiple shocks, correlates closely with the regional loss of NE (*39*), as does the subsequent recovery.

Inhibitions of neurons in other brainstem areas have been found with LC stimulation. Thus cells in the spinal trigeminal nucleus (76) and lateral geniculate nucleus (66) show a decreased probability of discharge following LC activation.

Recently the hippocampal cortex, a brain region known to receive an extensive input of NE-containing fibers (9, 41), has been examined (78a). We have confirmed the presence of these terminals by fluorescence histochemistry, 6-OHDA-induced degeneration, and autoradiography after microinjection of labeled precursors into the LC (70a). These studies also indicate that the hippocampal NE projections on pyramidal cells function in a fashion markedly similar to the effects of the LC on cerebellar Purkinje cells: 1) the LC and NE slow pyramidal cell discharge with long latency and long duration actions; 2) the receptor is blocked with MJ-1999 and by prostaglandins of the E series; and 3) the action of the pathway is totally blocked by chronic pretreatment with 6-OHDA or acute pretreatment with reserpine and α-MpT. Preliminary evidence suggests that in the hippocampus, as in the cerebellum, the NE inhibitory actions may be mediated postsynaptically by the formation of cyclic adenosine monophosphate (AMP).

CYCLIC ADENOSINE 3′,5′-MONOPHOSPHATE AS A MEDIATOR

In many peripheral organs receiving adrenergic input, the biochemical mechanism of the response of sympathetically innervated tissue involves activation of adenylcyclase by NE. Adenylcyclase then catalyzes the formation of cyclic 3′,5′-AMP from ATP (91). The cyclic nucleotide then triggers the cellular events mediated by the sympathetic innervation. Cyclic AMP is subsequently catabolized by a specific phosphodiesterase (91).

There is strong evidence in the CNS that at least one adrenergic pathway, the adrenergic input to cerebellar Purkinje cells, involves generation of cyclic AMP. The cerebellum has one of the highest regional levels of both cyclic AMP and adenylcyclase (99). In cerebellar slices, treatment with NE results in increased cyclic AMP content (55). These biochemical findings prompted neuropharmacological and electrophysiological studies which indicate that the inhibitory effect of NE and LC stimulation on rat cerebellar Purkinje cells is mediated by cyclic AMP (44, 80).

In brief, this hypothesis is based on the following points: 1) Both NE and cyclic AMP slow the discharge of Purkinje cells by prolongation of the pauses between single spikes. 2) In many cases the response and recovery latencies with cyclic AMP are briefer than with NE. 3) The duration and magnitude of the response to NE and to cyclic AMP are increased by either parenteral or electrophoretic administration of methylxanthines, such as theophylline or aminophylline, or of papaverine, compounds that are known to inhibit phosphodiesterase (59). 4) The response to NE can be completely blocked by electrophoretic administration of nicotinate and of prostaglandins of the E series (80); all of these substances are known to block the ability of NE to elevate levels of cyclic AMP in peripheral tissues (24). 5) In addition the transmembrane responses both to NE and to cyclic AMP involve a novel type of hyperpolarization in which there is no increase in the passive membrane conductance to ionic flow such as occurs in the response to other inhibitory substances like GABA (20) (see section on "Activation of Adrenergic Pathway" in this chapter). 6) Moreover, the response to NE and to cyclic AMP and its potentiation by methylxanthines can be seen in animals pretreated with 6-OHDA to remove the endogenous catecholamine nerve terminals; these results strongly support a direct postsynaptic activation of adenylcyclase as one step in the molecular mechanism that causes the inhibition by NE (80). 7) Furthermore, when prostaglandin E_1 or E_2 is administered to Purkinje cells during stimulation of LC, the inhibitory effects of LC stimulation are reproducibly and reversibly blocked (44). Of all of the transmitters tested in the rat cerebellar cortex, NE is the only substance to be antagonized by prostaglandins (80). 8) Finally, microiontophoretic administration of papaverine not only potentiates Purkinje cell responses to NE and cyclic AMP but also enhances the inhibition produced by LC stimulation. These recent results suggest that the LC gives rise to the NE—cyclic AMP-mediated inhibition of Purkinje cells.

In the brainstem many neurons respond to iontophoresis of cyclic AMP with a depression of discharge (2). The cyclic AMP depression is correlated with a reduction in discharge following iontophoresis of NE. Cells that are excited by NE usually show no effect to the iontophoresis of cyclic AMP. Prostaglandin E_1, however, antagonizes NE depressions in only about 50% of the neurons tested (2). Furthermore, effects of methylxanthines appear relatively independent of effects on the cyclic AMP system.

Cyclic AMP may not mediate the effects of NE in other regions of the neuraxis. Thus in the cerebral cortex NE acts to depress more cells than does cyclic AMP (61). In this area the effects of the iontophoresis of methylxanthines and prostaglandins have also been interpreted to be independent of adenylcyclase and cyclic AMP (61). But see (88a).

Determination of the positive or negative nature of iontophoretic drug administrations is a complicated matter in itself which we have discussed in some detail elsewhere (15). In the cerebral cortex the precise distribution of NE terminals is not completely known at the ultrastructural level, the phosphodiesterase content of the tissue is several fold higher than in the cerebellum (99), and possibly artifactual NE responses may compound the interpretation of NE receptivity (38, 53). Conversely, in the hippocampal cortex, where the fibers and the postsynaptic cells can be more accurately determined, NE and cyclic AMP produce identical actions, and phosphodiesterase inhibitors potentiate cyclic AMP, the actions of LC stimulation, and the effects of applied NE. Thus, a conservative interpretation at present would be that several transmitters may be able to activate certain postsynaptic neuronal receptors which trigger the formation of cyclic AMP.

Regardless of the transmitter in activating the cyclase, methylxanthines—and perhaps prostaglandins—would be expected to interact with the postsynaptic cells, even if these cells do not receive NE synapses or exhibit responsivity to NE. Recent experiments indicate, however, that the LC input to several neuronal structures may function by the intermediation of an adenylate cyclase-dependent step. However, we still do not know if the ability to activate the adenylate cyclase of the Purkinje cell or of other specific postsynaptic elements reflects special properties of 1) the receptive cell; 2) the nucleus the axons of which make the synapses onto the cell; or 3) the transmitter molecule. Both dopamine (DA) and 5-HT can also activate the adenylate cyclase of aplysia ganglia, but in this case the effects of 5-HT on cyclic AMP synthesis appear to be independent of transmitter release or of changes in synaptic potentials.

Evaluation of the role of cyclic AMP in central noradrenergic function has recently been facilitated by the development of a histochemical technique for localization of intracellularly bound cyclic AMP in unfixed frozen tissues (96). With this technique it is possible to show that topical NE application or LC stimulation selectively elevates cyclic AMP cytochemical staining in cerebellar Purkinje cells (79). Other putative cerebellar transmitters, such as GABA, ACh, 5-HT, or histamine, do not augment staining (79). This technique should provide a valuable adjunct to the more classic electrophysiological and biochemical studies on cyclic AMP roles in the CNS.

PLASTICITY AND REGENERATION

Norepinephrine-containing neurons in the CNS have recently been shown to sprout new collateral axon and terminal arborizations. The stimuli for this structural plasticity are unknown but appear to involve diverse factors, including injury to the NE-containing cell bodies or fibers themselves. (10, 56, 70), presence of "foreign" but normally adrenergically innervated tissue (*86*), and the spatial and topographical relationship to injured nonadrenergic synapses (*65*). In general, the regeneration and proliferation of central NE-containing fibers can be grouped as homotypic (injured NE terminals are replaced, with or without overproliferation, by other NE fibers) or heterotypic (adrenergic terminals replace lesioned nonadrenergic afferents). Homotypic responses have been found with midbrain (*56*) and spinal cord lesions (*10*) and with damage to the superior cerebellar peduncle (*70*). Heterotypic responses have been seen in the septum (*65*) and lateral geniculate body (*87*). With respect to heterotypic regeneration, it is important to note that the topographical and spatial relationships between the lesioned afferents and adrenergic terminals are important, since destruction of some afferents is a far better stimulus to the induction of NE-fiber sprouting than similar lesions in other afferents (*87*).

The postsynaptic target cells also play an important role in central noradrenergic regeneration. Peripheral tissues, implanted directly into the brain, are reinnervated in direct proportion to their normal sympathetic innervation (*86*). Thus the iris shows a large catecholamine fiber ingrowth, the mitral valve less so, and the transplanted diaphragm or uterus few ingrowing fibers. The pattern of central adrenergic reinnervation also appear similar to normal sympathetic input.

Reis and Ross (*71*) have studied biochemical changes in noradrenergic terminals during and after injury and have emphasized the widespread nature of the response. After unilateral lesions of noradrenergic fibers in the medial forebrain bundle, large changes in the level of dopamine-β-hydroxylase, exclusive to noradrenergic terminals, are found at many different levels of the neuraxis. The contralateral cortex, cerebellum, and brainstem all show profound reductions in enzyme activity 14 days after the lesion. Thus biochemical as well as histofluorescence evidence suggests a marked response in adrenergic neurons to homotypic or heterotypic manipulation. The relationship of these phenomena to psychological "plasticity" remains for future investigation.

SELF-STIMULATION

One of the striking correlations that has been derived from the recent extensive mapping of NE pathways in the brain is the relationship of NE cells and fibers to self-stimulation sites. This correlation has been strengthened by the finding that electrodes implanted in the area of the LC produce positive self-stimulation (*28*). In an extension of this study (*4*) it was shown that the ventral NE bundle was most critical for self-stimulation, and that the high rates of stimulation produced by electrodes in this bundle elicited decreased NE histofluorescence in the ipsilateral supraoptic and paraventricular nuclei, preoptic area, and ventral part of the striae terminalis. In view of possible current spread, a critical evaluation of this problem necessitates use of 6-OHDA to cause selective destruction of NE terminals. The effects of 6-OHDA pretreatment on self-stimulation, however, have been widely variable from one laboratory to the next. Wise and Stein (*100*) and Breese et al. (*23*) found that 6-OHDA pretreatment produces a permanent decrement in self-stimulation. However, Taylor and Laverty (*62, 92*) have shown that the effects of 6-OHDA on behavior are relatively transient, lasting only two to three weeks, whereas 6-OHDA treatment produces a permanent decrement in the NE content of the brain. Moreover, Taylor and Laverty (*92*) have shown that when desmethylimipramine is used to protect the animal from permanent NE depletion by 6-OHDA, the transient behavioral effects of the latter are not affected.

On the basis of the facilitation of self-stimulation by central administration of NE (*100*) and a decrement in this behavior after pretreatment with 6-OHDA (*85*), Wise and Stein have speculated that damage to the "noradrenergic reward system" may play a crucial role in the etiology of schizophrenia. However, this hypothesis has been challenged on both technical (*3*) and conceptual (*90*) levels.

NOREPINEPHRINE-CONTAINING NEURONS AND SLEEP

The NE-containing neurons of the LC have been causally related to the regulation of sleep and waking behavior in the cat, on the basis of sleep disturbances that follow 1) electrolytic destruction of the LC or 2) treatment with drugs that alter NE synthesis or storage (*54*). Laguzzi et al. (*60*) have described disturbances in sleep, pontogeniculate-occipital (PGO) activity, and temperature regulation after intraventricular administration of 6-OHDA in the cat. Moreover, Chu and Bloom (*25*), recording from LC neurons in unrestrained unanesthetized cats, have shown that the discharge pattern is closely correlated with the level of arousal. In quiet waking and slow wave sleep, neuronal discharge is slow and regular. During paradoxical sleep the LC neurons show characteristic high frequency bursts correlated with PGO spikes. When the cat is awake and attentive to its environment, the neuronal discharge is again elevated. This approach may eventually lead to a correlation of LC discharge with other behaviors, such as feeding, exploration, or aggression.

Addendum Since this chapter was prepared, a number of studies have appeared which link noradrenalin pathways to arousal, exploratory behavior, and reinforcement. Two excellent recent accounts on the role of central noradrenergic mechanisms in behavior have also been published (Lidbrink, P.: *Studies on Central Catecholamine Neurons with Special Reference to their Role in Wakefulness*. Stockholm: Karolinska, 1974; Snyder, S. H., and Usdin, E. (Eds.): *Frontiers in Catecholamine Research*. Oxford: Pergamon Press, 1974).

LITERATURE CITED

1. Aghajanian, G. K. and Bloom, F. E : Electron microscopic localization of tritiated norepinephrine in rat brain: Effects of drugs. *J. Pharmacol. Exp. Ther. 156*:407–416, 1967.
2. Anderson, E., Haas, H., and Hösli, L.: Comparison of effects of noradrenaline and histamine with cyclic AMP on brain stem neurons. *Brain Res. 49*:471–475, 1973.
3. Antelman, S., Lippa, A., and Fisher, A.: Technical comment. *Science 175*:919–920, 1972.
4. Arbuthnott, G., Fuxe, K., and Ungerstedt, U.: Central catecholamine turnover and self-stimulation. *Brain Res. 27*:406–413, 1971.
5. Avanzino, G. L., Bradley, P. B., and Wolstencroft, J. H.: Pharmacological properties of neurones of the paramedian reticular nucleus. *Experientia 22*:410–411, 1966.
6. Baker, J., Crayton, J., and Nicoll, R.: Noradrenalin and acetylcholine responses of supraoptic neurosecretory cells. *J. Physiol.* (Lond.) *218*:19–32, 1971.
7. Biscoe, T. J. and Curtis, D. R.: Noradrenaline and inhibition of Renshaw cells. *Science 151*:1230–1231, 1966.
8. Biscoe, T. J. and Krnjevic, K.: Chloralose and the activity of Renshaw cells. *Exp. Neurol. 8*:395–405, 1963.
9. Biscoe, T. J. and Straughan, D. W.: Microelectrophoretic studies of neurons in the cat hippocampus. *J. Physiol.* (Lond.) *183*:341–359, 1966.
10. Björklund, A., Katzman, R., Stenevi, U., et al.: Development and growth of axonal sprouts from NA and 5-HT neurons in the rat spinal cord. *Brain Res. 31*:21–33, 1971.
11. Bloom, F. E.: Fine structural changes in rat brain after intracisternal injection of 6-hydroxydopamine. In Malmfors, T. and Thoenen, H. (Eds.): *6-Hydroxydopamine and Catecholamine Neurons*. Amsterdam: North-Holland, 1971, p. 135.
12. Bloom, F. E.: Electron microscopy of catecholamine-containing structures. *Handbook Exp. Pharmacol. 33*:46, 1972.
13. Bloom, F. E.: Ultrastructural identification of catecholamine-containing central synaptic terminals. *J. Histochem. Cytochem. 21*:333–348, 1973.
14. Bloom, F. E., Costa, E., and Salmoiraghi, G. C.: Anesthesia and the responsiveness of individual neurons of the caudate nucleus of the cat to acetylcholine, norepinephrine and dopamine administered by

microelectrophoresis. *J. Pharmacol. Exp. Ther. 150*244–252, 1965.
15. Bloom, F. E. and Hoffer, B. J.: Norepinephrine as a central synaptic transmitter. In Usdin, E. and Snyder, S. H. (Eds.): *Frontiers in Catecholamine Research.* Oxford: Pergamon Press, 1973, pp. 637–642.
16. Bloom, F. E., Hoffer, B. J., and Siggins, G. R.: Studies on norepinephrine-containing afferents to Purkinje cells of rat cerebellum: I. Localization of the fibers and their synapses. *Brain Res. 25*: 501–521, 1971.
17. Bloom, F. E., Hoffer, B. J., and Siggins, G. R.: Norepinephrine mediated synapses: A model system for neuropsychopharmacology. *Biol. Psychiatry 4*:157–177, 1972.
18. Bloom, F. E., Oliver, A. P., and Salmoiraghi, G. C.: The responsiveness of individual hypothalamic neurons to microelectrophoretically administered endogenous amines. *Int. J. Neuropharmacol. 2*:181–193, 1963.
19. Boakes, R. F., Bradley, P., and Candy, J. M.: A neuronal basis for the alerting action of amphetamine. *Br. J. Pharmacol. 45*:391–403, 1972.
20. Boakes, R. F., Bradley, P., Candy, J., et al.: Noradrenaline. *Nature* (New Biol.) *239*:151–152, 1972.
21. Boakes, R. F., Candy, J., and Wolstencroft, J. H.: Agonistic and antagonistic effects of alphamethyl noradrenaline at central receptors. *Brain Res. 11*:450–452, 1968.
22. Bradley, P. B., Wolstencroft, J. H., Hösli, L., et al.: Neuronal basis for the central action of chlorpromazine. *Nature 212*:1425–1427, 1966.
23. Breese, G., Howard, J., and Leaky, J. P.: Effect of 6-hydroxydopamine on electrical self-stimulation of the brain. *Br. J. Pharmacol. 43*:255–257, 1971.
24. Butcher, R. and Baird, C.: Effects of prostaglandins on adenosine 3'5'-monophosphate levels in fat and other tissue. *J. Biol. Chem. 243*:1713–1717, 1968.
25. Chu, N.-S. and Bloom, F. E.: Norepinephrine-containing neurons: Changes in spontaneous discharge patterns during unrestrained sleeping and waking. *Science 179*:908–910, 1973.
26. Couch, J.: Responses of neurons in the raphé nuclei to serotonin, norepinephrine and acetylcholine and their correlation with excitatory synaptic input. *Brain Res. 19*:137–150, 1970.
27. Cowan, M. W., Gottlieb, D. I., Hendrickson, A. E., et al.: The autoradiographic demonstration of axonal connections in the central nervous system. *Brain Res. 37*:21–51, 1972.
28. Crow, T. J., Spear, P., and Arbuthnott, G.: Intracranial self-stimulation with electrodes in the region of the locus coeruleus. *Brain Res. 36*:275–287, 1972.
29. Curtis, D. R. and Davis, R.: Pharmacological studies upon neurons of the lateral geniculate nucleus of the cat. *Br. J. Pharmacol. 18*:217–246, 1962.

30. Curtis, D. R. and Koizumi, D.: Chemical transmitter substances in the brain stem of the cat. *J. Neurophysiol. 24*:80–90, 1961.
31. Curtis, D. R., Phillis, J. W., and Watkins, J. C.: Cholinergic and non-cholinergic transmission in the mammalian spinal cord. *J. Physiol.* (Lond.) *158*:296–323, 1961.
32. Dahlström, A. and Fuxe, K.: Evidence for the existence of monoamine neurons in the central nervous system. I. Demonstration of monoamines in the cell bodies of brain stem neurons. *Acta. Physiol. Scand. 62* (Suppl. 232):1–55, 1964.
33. Eccles, J., Ito, M., and Szentagothai, J.: *The Cerebellum as a Neuronal Machine.* New York: Springer, 1967.
34. Engberg, I. and Marshall, K.: Mechanism of noradrenaline hyperpolarization in spinal cord motoneurons of the cat. *Acta Physiol. Scand. 83*:142–144, 1971.
35. Engberg, I. and Ryall, R.: The inhibitory action of noradrenaline and other monoamines on spinal neurons. *J. Physiol.* (Lond.) *185*: 298–322, 1966.
36. Falck, B., Hillarp, N.-Å., Thieme, G., et al.: Fluorescence of catecholamines and related compounds condensed with formaldehyde. *J. Histochem. Cytochem. 10*:348–354, 1962.
37. Feltz, P. and De Champlain, J.: Enhanced sensitivity of caudate neurons to microiontophoretic injections of dopamine in 6-hydroxydopamine-treated cats. *Brain Res. 43*:601–605, 1972.
38. Frederickson, R., Jordan, L., and Phillis, J.: The action of noradrenaline on central neurons. *Brain Res. 35*:556–560, 1971.
39. Glowinski, J. and Baldessarini, R.: Metabolism of norepinephrine in the central nervous system. *Pharmacol. Rev. 18*:1201–1238, 1966.
40. Herz, A. and Gogolak, G.: Mikroelektrophoretische Untersuchungen am Septum des Kaninchens. *Pflügers Arch. 285*:317–330, 1965.
41. Herz, A. and Nacimiento, A. C.: Über die Wirkung von Pharmaka auf Neurone des Hippocampus nach mikroelektrophoretischer Verabfolgung. *Naunyn Schmiedebergs Arch. Pharmacol. 251*:295–314, 1965.
42. Herz, A. and Zieglgänsberger, W.: Synaptic excitation in the corpus striatum inhibited by microelectrophoretically administered dopamine. *Experientia 22*:839–840, 1966.
43. Hoffer, B. J., Siggins, G. R., and Bloom, F. E.: Studies on norepinephrine-containing afferents to Purkinje cells of rat cerebellum: II. Sensitivity of Purkinje cells to norepinephrine and related substances administered by microiontophoresis. *Brain Res. 25*:523–534, 1971.
44. Hoffer, B. J., Siggins, G. R., Oliver, A. P., et al.: Activation of the pathway from locus coeruleus to rat cerebellar Purkinje neurons: Pharmacological evidence of noradrenergic central inhibition. *J. Pharmacol. Exp. Ther. 184*:553–569, 1973.
45. Hökfelt, T.: Electron microscopic studies on brain slices from regions rich in catecholamine nerve terminals. *Acta Physiol. Scand. 69*: 119–120, 1967.

46. Hökfelt, T.: In vitro studies on central and peripheral monoamine neurons at the ultrastructural level. *Z. Zellforsch. Mikrosk. Anat.* *91*:1–74, 1968.
47. Hökfelt, T.: Ultrastructural localization of intraneuronal monoamines. Some aspects on methodology. *Prog. Brain Res. 34*: 213–222, 1971.
48. Hökfelt, T. and Fuxe, F.: Cerebellar monoamine nerve terminals, a new type of afferent fiber to the cortex cerebelli. *Exp. Brain Res.* *9*:63–72, 1969.
49. Hösli, L, Tebēcis, A., and Schonwelter, H.: A comparison of the effects of monoamines on neurons of the bulbar reticular formation. *Brain Res. 25*:357–370, 1971.
50. Iversen, L. L.: Role of transmitter uptake mechanisms in synaptic neurotransmission. *Br. J. Pharmacol. 41*:571–591, 1971.
51. Iversen, L. L. and Uretsky, N. J.: Biochemical effects of 6-hydroxydopamine on catecholamine-containing neurons in the rat central nervous system. In Malmfors, T. and Thoenen, H. (Eds.): *6-Hydroxydopamine and Catecholamine Neurons.* Amsterdam: North-Holland, 1971, p. 171.
52. Johnson, E., Roberts, M., Sobrezek, A., et al.: Noradrenaline cells in cat cerebral cortex. *Int. J. Neuropharmacol. 8*:549–566, 1969.
53. Johnson, E., Roberts, M., and Straughan, D.: The responses of cortical neurons to monoamines under differing anesthetic conditions. *J. Physiol.* (Lond.) *203*:261–280, 1969.
54. Jouvet, M.: Biogenic amines and the states of sleep. *Science 163*:32–41, 1969.
55. Kakiuchi, S. and Rall, T.: Influence of chemical agents on the accumulation of adenosine 3'5'-phosphate in slices of rabbit cerebellum *Mol. Pharmacol. 4*:367–378, 1968.
56. Katzman, R., Björklund, A., Owman, C., et al.: Evidence for regenerative axon sprouting of central catecholamine neurons in the rat mesencephalon following electrolytic lesions. *Brain Res. 25*:579–596, 1971.
57. Krebs, H. and Bindra, D.: Noradrenaline and "chemical coding" of hypothalamic neurons. *Nature* (New Biol.) *229*:178–180, 1971.
58. Krnjević, K. and Phillis, J. W.: Iontophoretic studies of neurons in the mammalian cerebral cortex. *J. Physiol.* (Lond.) *165*:274–304, 1963.
59. Kukovetz, W. and Poch, G.: Inhibition of cyclic 3'5'-nucleotide phosphodiesterase as a possible mode of action of papaverine and similarly acting drugs. *Naunyn Schmiedebergs Arch. Pharmacol.* *267*:189–194, 1970.
60. Laguzzi, R., Petitjean, F., Pujol, J., et al.: Effets de l'injection intraventriculaire de 6-hydroxydopamine. *Brain Res. 48*:295–310, 1972.
61. Lake, N., Jordan, L., and Phillis, J. W.: Mechanism of noradrenaline action in cat cerebral cortex. *Nature* (New Biol.) *240*:249–250, 1972.

62. Laverty, R. and Taylor, K.: Effects of intraventricular 6-hydroxydopamine on rat behavior and brain catecholamine metabolism. *Br. J. Pharmacol.* 40:836–846, 1970.
63. Legge, K. F., Randic, M., and Straughan, D. W.: The pharmacology of neurons in the pyriform cortex. *Br. J. Pharmacol.* 26:87–107, 1966.
64. Lenn, N. J.: Localization of uptake of tritiated norepinephrine by rat brain in vivo and in vitro using electron microscopy. *Am. J. Anat.* 120:377–389, 1967.
65. Moore, R., Björklund, A., and Stenevi, U.: Plastic changes in the adrenergic innervation of the rat septal area in response to denervation. *Brain Res.* 33:13–35, 1971.
66. Nakai, Y. and Takaori, S.: Influence of catecholamine on the lateral geniculate neuron activity. *Proc. V. Int. Congr. Pharmacol.* 983, 1972.
67. Nelson, C., Hoffer, B., Chu, N.-S., et al.: Cytochemical and pharmacological studies on the primate frontal cortex. *Brain Res.* 62:115–133, 1973.
68. Olson, L. and Fuxe, K.: On the projections from the locus coeruleus noradrenaline neurons. *Brain Res.* 28:165–171, 1971.
69. Phillis, J. W., Tebēcis, A. K., and York, D. H.: The inhibitory actions of monoamines on lateral geniculate neurons. *J. Physiol.* (Lond.) 190:563–581, 1967.
70. Pickel, V., Krebs, H., and Bloom, F. E.: Proliferation of norepinephrine-containing axons in rat cerebellar cortex after peduncle lesions. *Brain Res.* 59:169–179, 1973.
70a. Pickel, V., Segal, M., and Bloom, F. E.: A radioautographic study of the efferent pathways of the locus coeruleus. *J. Comp. Neurol.* 155:15–42, 1974.
71. Reis, D. J. and Ross, R. A.: Dynamic changes in brain dopamine β-hydroxylase activity during anterograde and retrograde reactions to injury of central noradrenergic axons. *Brain Res.* 57:307–326, 1973.
72. Richardson, K. C.: Electron microscopic identification of autonomic nerve endings. *Nature* 210:756–757, 1966.
73. Salmoiraghi, G. C.: Electrophoretic administration of drugs to individual nerve cells. *Neuropsychopharmacol.* 3:219, 1964.
74. Salmoiraghi, G. C., Bloom, F. E., and Costa, E.: Adrenergic mechanisms in rabbit olfactory bulb. *Am. J. Physiol.* 207:1417–1424, 1964.
75. Salmoiraghi, G. C. and Stefanis, C.: Central synapses and suspected transmitters. *Int. Rev. Neurobiol.* 10:1–30, 1967.
76. Sasa, S. and Takaori, S.: Relationship between morphine and regulation mechanisms on the trigeminal nucleus from the locus coeruleus. *Proc. V. Int. Congr. Pharmacol.* 1202, 1972.
77. Satinsky, D. and Salmoiraghi, G. C.: Pharmacology of lateral geniculate neurons. *Fed. Proc.* 26:323, 1967.
78. Schildkraut, J. and Kety, S.: Biogenic amines and emotion. *Science* 156:21–30, 1967.

78a. Segal, M. and Bloom, F. E.: The action of norepinephrine in the rat hippocampus. I. Iontophoretic Studies. *Brain Res.* 72:79–98, 1974.
79. Siggins, G. R., Battenberg, E. F., Hoffer, B. J., et al.: Noradrenergic stimulation of cyclic adenosine monophosphate in rat Purkinje neurons: An immunocytochemical study. *Science* 179:585–588, 1973.
80. Siggins, G. R., Hoffer, B. J., and Bloom, F. E.: Studies on norepinephrine-containing afferents to Purkinje cells of rat cerebellum: III. Evidence for mediation of norepinephrine effects by cyclic 3'5'-adenosine monophosphate. *Brain Res.* 25:535–553, 1971.
81. Siggins, G. R., Hoffer, B. J., Oliver, A. P., et al.: Activation of a central noradrenergic projection to cerebellum. *Nature* 233:481–483, 1971.
82. Siggins, G. R., Oliver, A. P., Hoffer, B. J., et al.: Cyclic adenosine monophosphate and norepinephrine: Effects on transmembrane properties of cerebellar Purkinje cells. *Science* 171:192–194, 1971.
83. Snyder, S. H., Kuhar, M. J., Green, A. I., et al.: Uptake and subcellular localization of neurotransmitters in the brain. *Int. Rev. Neurobiol.* 13:127–158, 1970.
84. Stefanis, C.: Hippocampal neurons: Their responsiveness to microelectrophoretically administered endogenous amines. *Pharmacologist* 6:171, 1964.
85. Stein, L. and Wise, C. D.: Possible etiology of schizophrenia: Progressive damage to the noradrenergic reward system by 6-hydroxydopamine. *Science* 171:1032–1036, 1971.
86. Stenevi, U.: Sprouting and plasticity of central monoamine with microspectrofluorometric characterization of their intraneuronal fluorophores. Lund, Sweden: University of Lund, pp. 1–26, 1971.
87. Stenevi, U., Björklund, A., and Moore, R. Y.: Growth of intact central adrenergic axons in the denervated lateral geniculate body. *Exp. Neurol.* 35:290–299, 1972.
88. Stone, T.: Are noradrenaline excitations artifacts? *Nature* 234:145–146, 1971.
88a. Stone, T. W., Taylor, D. A., and Bloom, F. E.: Cyclic AMP and cyclic GMP may mediate opposite neuronal responses in the rat cerebral cortex. *Science* 187:845–847, 1975.
89. Straughan, D. and Legge, K.: The pharmacology of amygdala neurons. *J. Pharm. Pharmacol.* 17:675–677, 1966.
90. Strauss, J. and Carpenter, W.: Technical comment. *Science* 175:921, 1972.
91. Sutherland, E. W., Robison, G. A., and Butcher, R.: Some aspects of the biological role of adenosine 3'5'-monophosphate. *Circulation* 37:279–306, 1968.
92. Taylor, K. and Laverty, R.: The effects of drugs on the behavioral and biochemical actions of 6-hydroxydopamine. *Eur. J. Pharmacol.* 17:16–24, 1972.
93. Tebēcis, A.: Effects of monoamines and amino acid on medial geniculate neurons of the cat. *Neuropharmacology*, 9:381–390, 1970.

94. Ungerstedt, U.: Stereotaxic mapping of the monoamine pathways in rat brain. *Acta Physiol. Scand.* (Suppl. 367):1–122, 1971.
95. Ungerstedt, U.: Functional dynamics of central monoamine pathways. In Schmitt, F. O. (Ed.): *The Neurosciences, Third Study Program.* Cambridge: MIT Press, 1974, pp. 979–988.
96. Wedner, H. J., Hoffer, B. J., Battenberg, E., et al.: A method for detecting intracellular cyclic adenosine monophosphate by immunofluorescence. *J. Histochem. Cytochem. 20*:293–299, 1972.
97. Weight, F. F. and Salmoiraghi, G. C.: Responses of spinal cord interneurons to acetylcholine, norepinephrine and serotonin administered by microelectrophoresis. *J. Pharmacol. Exp. Ther. 153*:420–427, 1966.
98. Weight, F. F. and Salmoiraghi, G. C.: Motoneuron depression by norepinephrine. *Nature 213*:1229–1230, 1967.
99. Weiss, B. and Kidman, A. D.: Neurobiological significance of cyclic 3'5'-adenosine monophosphate. *Adv. Biochem. Psychopharmacol. 1*:131–164, 1969.
100. Wise, C. D. and Stein, L.: Facilitation of brain self-stimulation by central administration of norepinephrine. *Science 163*:299–301, 1969.

CHAPTER SEVEN
Serotonin and the Central Nervous System

Thaddeus J. Marczynski

In 1953 Twarog and Page (*411*) reported that 5-hydroxytryptamine (5-HT or serotonin) was a natural constituent in the mammalian central nervous system (CNS), and a few years later it was hypothesized that this indoleamine modulated synaptic activity (*276*). Within the past two decades investigations into the role of 5-HT have grown to encompass a wide range of neural phenomena, and this research has stimulated its own growth by abetting the development of a host of new methodologies. A wealth of data, often contradictory, defies attempts to formulate a unifying concept on the role of serotonin in the brain. Even for specific brain functions—such as temperature regulation, gonadotropin secretion, integration of pain perception, and sleep mechanisms—no firm and specific conclusions can be reached about the contributions of the 5-HT system, since the neural circuitry subserving each of these functions is not well defined. For the same reason the vagueness of current working hypotheses does little to validate the proposal that, in most target areas of serotonin fiber projections, the primary role of 5-HT is to dampen or inhibit neuronal activity, as suggested by numerous microelectrophoretic studies in the neocortex, thalamus, hypothalamus, and limbic system (*46*).

In this chapter no attempt has been made to give a comprehensive treatment to the anatomical, metabolic, and electrophysiological studies on the 5-HT system because of the availability of excellent reviews on these topics (*29, 44, 82, 157, 347*). Nonetheless, in several instances a detailed account of the data has been given in order to show that a simplistic concept of 5-HT function would not generate a working hypothesis of potential heuristic value.

TOPOGRAPHY AND NEUROCHEMISTRY OF BRAIN SEROTONIN NEURONS

Shortly after the discovery of 5-HT in the mammalian brain (*8, 411*), it became clear that the distribution of this amine is very uneven. The highest concentrations, in declining order, are found in the brainstem structures, limbic system, hypothalamus, basal ganglia, thalamus, and meso- and neocortex (*48*). In general, the early data based on biochemical and fluorimetric methods are in agreement with the more recent studies based on fluorescence histochemical techniques which permit a visualization of cell bodies, their axons, and their terminals (*105, 106*).

Ascending Pathways

Although most of the ascending pathways originate in the raphé nuclei, there are a few clusters of 5-HT cells in the ventromedial reticular formation of the pons and the mesencephalon that make a significant contribution to the projections. It must be pointed out, however, that there are cell bodies in several raphé nuclei that do not contain endogenous 5-HT, such as the nucleus linearis caudalis of the mesencephalon which contains a considerable number of dopamine (DA) nerve cells. Ascending 5-HT pathways can be subdivided into several bundles: 1) A *medial pathway* which primarily innervates the hypothalamus and the preoptic region originates in the mesencephalic raphé nuclei (locus B7, B8 of Fuxe and Dahlström), and the pontine raphé region (locus B5 and B6); it provides a relatively dense innervation to the following areas of the telencephalon and diencephalon: nucleus suprachiasmaticus, certain amygdaloid nuclei, ventral part of the lateral geniculate body, septal area, preoptic area, anterior colliculi, hippocampal formation in which discrete concentrations of 5-HT-containing terminals can be found in the stratum lacunosum moleculare, and in area 31 and 26 of the gyrus hippocampi and subiculum. 2) A *lateral ascending* 5-HT pathway primarily innervates the mesocortex and neocortex and originates mainly from the mesencephalic raphé cell groups designated as B7 and B9. 3) A minor, *far lateral pathway* projects to the extrapyramidal motor nuclei (neostriatum) and originates partly from the mesencephalic raphé cell groups (B7, B8, B9). 4) The *cerebellar innervation* originates mostly from the mesencephalic raphé nuclei (B7, B8) and possibly from cell groups in the pons (B5, B6).

Microspectrofluorometric analysis (*39*) of intracellular monoamine fluorophores has shown that the yellow fluorescent cells and axons, originally described a decade ago (*105, 106*) as the 5-HT system, do actually belong to two distinct indole-amine systems. One neuron type shows all of the characteristics of the 5-HT fluorophore, such as 1) rapid photodecomposition on UV irradiation; 2) disappearance of fluorescence after treatment of the animal with the tryptophan-5-hydroxylase inhibitor, *p*-chlorophenylalanine (PCPA), or with reserpine; and 3) increased fluorescence after treatment with the MAO inhibitor, nialamide. The other yellow fluorescent neuron type shows fluorophores with 1) different excitation and emission maxima; 2) a rather low photodecomposition rate; and 3) susceptibility to reserpine; however, in contrast to the 5-HT system, it is not affected by PCPA or nialamide. Although the latter type of cells is intermingled with the 5-HT cell bodies in the raphé nuclei, the bulk of the cells containing the unknown monoamine was found in the nucleus *linearis caudalis* and in the *nucleus raphé medianus*. Anteriorly both types of axons run into the medial forebrain bundle where they intermingle. It is suggested that the unknown monoamine may be 5,6-dihydroxytryptamine or 5-methoxytryptamine (*39*). In the cat, the histofluorescent technique showed that the topographical distribution of the 5-HT-containing cells and their axons is, with certain differences, similar to that of the rat (*334*).

Studies of the retrograde cellular changes in the raphé nuclei following various lesions in the feline brain, cerebellum and spinal cord reveal that the long ascending fibers are quantitatively the most important of the efferent projections (*53*). These fibers originate from all raphé nuclei, the middle group of nuclei (nucleus raphé magnus and nucleus raphé pontis) sending a relatively smaller contingent of axons than do the caudal and rostral ones. The nucleus linearis rostralis which does not seem to have a fluorescent counterpart in the rat, sends fibers mostly to the region of the globus pallidus, putamen, and either caudate nucleus or anterior thalamus or both. The caudal nuclei of the raphé system send fibers primarily to the mesencephalon. (For a more detailed description, the reader may consult [4, 105, 145, 148].)

Descending Bulbospinal System

Using the fluorescence technique, Swedish investigators (*72, 107*) have shown that in the *rat* the descending pathways originate mainly from cell bodies in the nucleus raphé obscurus and the nucleus raphé pallidus as well as from scattered cell bodies surrounding the pyramidal tract at the level of the medulla oblongata. The descending axons give rise to large numbers of collaterals to the gray matter of the spinal cord. The lumbosacral portion of the spinal cord shows a particularly dense network of 5-HT-containing fibers.

Considering the morphological relationships between the 5-HT- and the norepinephrine (NE)-containing nerve terminals and the functional implications of such relationships, it should be pointed out that many of the 5-HT cell bodies in the medulla oblongata appear to be innervated by NE-containing nerve terminals which probably exert tonic inhibitory influences.

A study of distribution of the retrograde cellular changes indicates that the descending fibers originate mostly from the nucleus raphé magnus in the *cat*. A relatively smaller contingent comes from the nucleus raphé obscurus and the nucleus raphé pontis. These fibers descend in part in the dorsal half of the lateral funiculus of the spinal cord. In general, the descending are less numerous than are the ascending fibers.

Neurons Innervating Lower Brainstem

It is not clear yet if the 5-HT-containing nerve terminals in the brainstem area derive from short axons or from collaterals of long axons descending to the spinal cord (*148*).

Afferent Connections

Studies of the preterminal and terminal degeneration of fibers after various lesions of the brain reveal that the afferents from the cerebral cortex, spinal cord, and cerebellum (fastigial nucleus) are relatively few (*54*) and, by inference, the bulk of afferents presumably comes from structures in the brainstem. In four raphé nuclei (obscurus, dorsalis, linearis intermedius, and linearis rostralis) no afferent terminals could be demonstrated from sources located outside the brainstem Of the remaining four nuclei, the pallidus and magnus receive only a moderate input from the spinal cord and the sensorimotor cortex, and the raphé pontis, magnus, and centralis superior receive moderate input from the cerebellum. None of the raphé nuclei showed degenerating terminals following a lesion of the occipital, temporal, or basal regions of the cortex.

Metabolism

There is a consensus that brain 5-HT is synthesized within the neurons in which it is normally localized by histofluorescence technique. This amine is synthesized from L-tryptophan (TR), which passes from the blood plasma into the neural parenchyma; TR is first hydroxylated to 5-hydroxytryptophan (5-HTP), then to be acted upon by aromatic L-amino acid decarboxylase and converted to 5-HT. There is convincing evidence that the same enzyme catalyzes the decarboxylation of both L-DOPA and

L-5-HTP. However, it has recently been proposed that in certain brain areas two separate subtypes of the enzyme may exist, displaying a discrete substrate specificity toward either of these amino acids. (See Ch. 5.)

Although TR hydroxylation is regarded as a rate-limiting step in 5-HT synthesis, observations that suggest that the amount of TR available to the brain actually controls 5-HT synthesis include: 1) daily and parallel rhythm in the brain 5-HT and TR; 2) a high Michaelis constant for TR hydroxylase relative to the whole brain TR concentration; and 3) a great increase of brain TR levels and its metabolite, 5-HIAA, after a larger systemic dose of TR (*435*). Unlike tyrosine hydroxylase, TR hydroxylase has a Michaelis constant (K_m) for its substrate much higher than the concentration of TR present in the whole mammalian brain (*250*). Hence, it is not surprising that 5-HT synthesis depends on the plasma levels of free TR, its active transport into the brain, and its concentration in the serotoninergic neurons. Thus, in contrast to noradrenergic neurons in which the dopamine-β-hydroxylase activity is regarded as a rate-limiting step in the synthesis of NE, in the tryptaminergic system the transport of TR into the serotoninergic neurons may actually provide a rate-limiting step in 5-HT synthesis (*435*). Peripheral injection (*296*) or intracerebroventricular administration (*96*) of TR markedly elevates 5-HT concentration and its metabolite 5-HIAA in the brain. Animals fed a TR-free diet show reduced CNS levels of 5-HT (*152*).

Amino acids (neutral, acidic, or basic) are transported into the brain by specific carrier systems, and they compete for common transport sites. Therefore the amount of TR in the plasma may control 5-HT synthesis in the brain in a fairly predictable manner, provided that the plasma levels of the competing amino acids, such as tyrosine, phenylalanine, leucine, isoleucine, and valine, are low (*435*). If the competing amino acids are not excluded from the diet and, e.g., rats consume casein—which results in an increase of plasma TR and other amino acids—then the brain levels neither of TR nor of 5-HT are found to be elevated. Interestingly in rats, a hypoglycemic dose of insulin or the consumption of a large amount of carbohydrates after a 15-hr fast, cause a selective mobilization of relatively large amounts of TR from undefined pools, a response that is coupled with low plasma levels of other amino acids and leads to a marked increase in brain 5-HT (*435*).

The issue of feedback control of TR hydroxylase as a rate-limiting step is still controversial, since evidence for and against end-product inhibition has been reported (*173*). If a feedback control operates in the mammalian brain via stimulation of the serotoninergic receptors, such a mechanism is far less apparent than in the catecholaminergic system. Although the conversion of TR to 5-HT is reduced when intrinsic levels of 5-HT are raised by MAO inhibitors in vivo (*252*), in order to demonstrate a significant inhibition of TR hydroxylase in mouse brain, as measured by diminished accumulation of 5-HTP, end-product levels have to be raised by 60% if MAO inhibitors, such as pargyline or nialamide, are used (*76, 77*). Receptor-mediated feedback regulation of 5-HT synthesis may explain the action of hallucinogenic drugs, such as *N,N*-dimethyltryptamine and LSD-25, which stimulate central 5-HT receptors in proper concentrations and partially inhibit 5-HTP formation in the mouse brain (*77*). The tricyclic antidepressant, chlorimipramine, which shows a certain degree of selectivity in blocking transmitter reuptake at serotonin fiber terminals, may cause an accumulation of 5-HT in the synaptic cleft, stimulating 5-HT receptors, and thereby inhibit TR hydroxylase, as suggested by retarded formation of 5-HTP in the rat brain (*295*).

Several observations indicate that the flow of neuronal impulses in the 5-HT-containing fibers exerts a strong stimulating effect on TR hydroxylase activity in nerve terminals. This is demonstrated by the observation that electrical stimulation of 5-HT-containing neurons accelerates 5-HT synthesis (*378*). On the other hand, acute transection of 5-HT pathways depresses the TR hydroxylase activity. A spinal transection that separates rostral 5-HT perikarya from their axons, e.g., causes a 50% decrease in 5-HTP formation in distal segments. A similar effect, although less conspicuous, can be obtained in the forebrain after axotomy of rostrally projecting pathways. The reduction in the activity of TR hydroxylase can already be observed one to two hr after transection and must, therefore, be attributed to the cessation of impulse traffic, rather than to the loss of enzyme molecules. It is interesting that the DA system reacts in a manner opposite to that of the 5-HT system, i.e., the acute axotomy causes a marked increase in the tyrosine hydroxylase activity in dopaminergic nerve terminals and has no significant effect on this enzyme in NE terminals. The aforementioned data support the concept that a feedback mechanism, operating via the degree of postsynaptic receptor activation at the terminals far away from cell soma, may control the synthesis of DA, but it has no recognizable role in the NE and 5-HT systems (*75*). In the 5-HT system, feedback regulation of TR hydroxylase probably occurs via 5-HT inhibitory receptors located presynaptically, not at the distant axon terminals but rather on the cell bodies in the raphé nuclei. When applied microintophoretically to raphé neurons, 5-HT and all drugs known to inhibit its synthesis by repressing TR hydroxylation show a strong inhibitory effect on their spontaneous firing rate. If the 5-HT presynaptic inhibitory receptors are evenly distributed on the 5-HT-containing neurons including their terminals, it is very possible that they play a certain role in local feedback regulation of 5-HT synthesis and release, analogous to that postulated for dopaminergic neurons (*216*).

Although the major catabolic pathway of 5-HT and tryptamine is the oxidative deamination to 5-HIAA and indole acetic acid, respectively, other routes are of potential significance in normal function of the mammalian brain. The existence of an alternative route by 6-hydroxylation has been demonstrated (*244*), and whether or not this metabolite may interfere with normal function of the 5-HT-containing terminals remains to be explored. Among other routes, the *N*-methylation of tryptamine may assume greater importance after drug-induced inhibition of MAO activity and a resulting higher-than-normal concentration of tryptamine, thus leading to the synthesis of a substantial amount of *N*-methyltryptamine and *N*-dimethyltryptamine, which have psychotomimetic properties (*258, 362*). In addition *N*-methylation of 5-HT may occur and will result in a relatively significant amount of bufotenine, which is claimed to represent a pathogenetic factor in generating schizophrenia-like symptoms (*259*) (see section on "Schizophrenia" in this chapter). The *N*-methylating enzyme has been detected in chick, sheep, and human brain (*258, 301*) and in rabbit and human lung (*22*).

Of potential importance is the metabolic route of 5-HT that leads to the synthesis of melatonin in the pineal gland and its immediate vicinity where a unique example of interaction exists between the NE and the indolamine systems. In this region NE terminals from the superior cervical ganglion regulate the activity of the *N*-acetyltransferase, which converts 5-HT to *N*-acetylserotonin. The latter in turn is acted on by the hydroxyindole-*O*-methyltransferase to form melatonin (*221*). Synthesis of melatonin shows a clear circadian rhythm in mammals. Since melatonin appears to have an influence on the 5-HT system and on sleep mechanisms (see section on "Sleep Mechanisms and . . ." in this chapter), the relevant question is whether or not this circadian rhythm can be entrained by light-dark cycles, sleep, or food consumption.

Release and Uptake

There is evidence that 5-HT is released in the CNS both under resting conditions and as a result of neuronal activity. The release of 5-HT and related substances was first demonstrated in studies of the superfused frog spinal cord (*12*). Electrical stimulation of afferent nerves of the isolated frog and mouse spinal cord induced a release of 5-HT and an increase in its synthesis (*9*). Ventricular perfusion experiments in anesthetized cats showed that the basal release of 5-HT from the caudate nucleus and septal region varied from 0.25 to 4.0 µg in 25 min. Electrical stimulation of the nucleus linearis intermedius or the nucleus linearis of the raphé system caused an enhanced release of 5-HT. Low frequency stimuli, 0.5 Hz, were more effective per stimulus than were high-frequency trains in releasing the indoleamine. No stimulus-bound release was obtained if the stimulating electrode was positioned in structures outside the linear nuclei (*187*). In unanesthetized monkeys spontaneous release of 5-HT from the superior colliculus perfused with a push-pull cannula was also reported (*36*). Electrical stimulation of the midbrain raphé nuclei caused a release of 5-HT and 5-HIAA from the neocortex (*120, 326*).

There is convincing evidence that 5-HT released in response to physiological stimuli is taken up into nerve terminals by what appears to be an active, high-affinity, saturable uptake process, a finding that accords with the general concept of inactivation or recycling of neurotransmitters, or both. This process is very selective and, if synaptosomes from the forebrain of raphé-lesioned animals are studied and proper concentrations of putative neurotransmitters used, only the uptake of 5-HT is affected, whereas the uptake of TR, DA, and NE is unchanged (*240*).

Synaptic Action

A large body of data suggesting the identity of 5-HT as an inhibitory transmitter in the mammalian CNS has been accumulated from microelectrophoretic studies of neurons that most likely receive 5-HT-containing terminals, and from the electrical stimulation of the raphé nuclei. On the basis of these studies it can be argued that the 5-HT raphé system exerts a tonic inhibitory influence on postsynaptic neurons as follows: 1) Electrical stimulation of the raphé nuclei results in inhibition of postsynaptic neurons in the telencephalon, e.g., in the suprachiasmatic nuclei of the cat hypothalamus, a region very densely supplied with serotonin terminals; electrophoretic application of 5-HT to these neurons also causes inhibition of firing. (Microelectrophoretic application of 5-HT to other neuronal populations in the telencephalon, presumed to receive 5-HT terminals, causes inhibition of firing rate, e.g., the frontal association cortex of the squirrel monkey [*45, 46*].) 2) Analogously, neurons in the corticomedial nucleus of the amygdala, ventral lateral geniculate, and ventral hippocampus show an inverse relationship between their firing rate and that of the raphé nuclei. This can be demonstrated by systemic administration of a small dose (20 µg/kg, i.v.) of LSD-25 which selectively depresses raphé neurons, a phenomenon associated with concomitant acceleration of firing rate of the postsynaptic neurons (*171*). 3) Microelectrophoretic application of 5-HT causes hyperpolarization of the neurons in the mammalian

CNS. In the spinal cord motoneurons, such hyperpolarization is associated with a decrease in amplitude of both EPSPs and IPSPs, as well as with blockade of both synaptic and antidromic invasion of the neuron, i.e., phenomena characteristic of the action of an inhibitory transmitter (*333*). 4) Intracellular recording from cortical cells, the firing rate of which is reduced or blocked by a close intraarterial injection of 5-HT, shows hyperpolarization and a decrease in membrane conductance, i.e., a type of inhibition similar to that induced by NE in the Purkinje cells of the cerebellum (*383*).

Attempts to link actual serotonin release with a specific biological action have perhaps been most successful in investigations of the role of 5-HT in thermoregulation. In donor-recipient experiments utilizing push-pull perfusion of discrete hypothalamic regions in unanesthetized macaques, an inverse relationship between release rates of 5-HT and levels of body temperature was established. A rise in 5-HT levels in the perfusate to the recipient, whether owing to augmented output from the donor or exogenously applied amine, produced qualitatively indistinguishable temperature responses in the recipient (*312*). Thus, 5-HT appears to be a physiological mediator in neural circuits subserving the activation of heat production (see section on "Thermoregulation" in this chapter).

NEUROPHARMACOLOGY

Responses to Drugs and Functional Dynamics

Several MAO inhibitors administered systemically inhibit unit firing in the raphé nuclei. Since this effect is attenuated by prior treatment with PCPA, an inhibitor of 5-HT synthesis, it can be assumed that there is an inverse relationship between the 5-HT concentration in the brain and the rate of firing of raphé cells, a phenomenon that may be based on a chemical feedback mechanism. In agreement with this concept, inhibition of raphé unit firing associated with an increase in cellular fluorescence is observed after parenteral or intracerebroventricular administration of TR (*1*). The specificity of this inhibitory effect is validated by the fact that the catecholamine precursors, L-tyrosine and L-DOPA, do not alter raphé cell fluorescence or firing rate.

Contrary to expectations, L-5-HTP, which produces a rapid elevation of brain 5-HT (*412*), has no significant effect on the firing of raphé cells (*3*). Also contrary to expectation, PCPA, the specific TR hydroxylase inhibitor, does not seem to impair the ability of TR to enhance raphé cell fluorescence, but the nonspecific inhibitor of aromatic amino acid hydroxylase–α-propyldopacetamide *(71)*–or reserpine (a nonspecific monoamine-depleting drug) prevents the TR-induced increase in raphé fluorescence. These observations raise serious doubts about the assumption that 5-HT is the only indoleamine present within the raphé system and suggest that an unidentified (presumably hydroxylated) TR derivative may be present in the raphé neurons in addition to 5-HT.

Several investigators have reported that the nucleus linearis caudalis receives many terminals from the 5-HT-containing raphé neurons *(1, 145)*. The postsynaptic neurons in this area are very sensitive to low doses of LSD administered systemically, as evidenced by the increase in their firing rate in response to a dose of this hallucinogen as low as 10 µg/kg. Since the same dose of LSD concomitantly inhibits the firing rate of raphé neurons, it is suggested that there is negative feedback from the area of the nucleus linearis caudalis to the 5-HT-containing raphé system *(1)*. It can be assumed that LSD stimulates cells in the nucleus linearis caudalis directly, since it does not block the uptake of 5-HT by brain tissue *(358)*, and even appears to enhance the binding of 5-HT by nerve terminals, thus decreasing the availability of 5-HT at receptor sites *(141)*. By contrast, tricyclic antidepressant compounds inhibit raphé cell firing *(1)* by blocking serotonin uptake, thus increasing the availability of extraneuronal 5-HT at synapses *(69)*.

Microelectrophoretic studies clearly show that LSD blocks the excitatory effect of 5-HT on brainstem neurons, whereas the inhibitory effects of 5-HT remain unaffected *(47)*. It is suggested that LSD inhibits the firing rate of 5-HT-containing neurons of the raphé system by negative feedback, which operates by activating neurons of the nucleus linearis caudalis *(1)*, presumably composed of cells containing an as yet unknown indole derivative the characteristics of which are different from those of 5-HT *(39)*. Alternatively, the concept has been entertained that the LSD-induced decrease in the 5-HT turnover rate is attributable to a negative feedback from 5-HT synapses at which LSD acts as a 5-HT agonist *(10)*. Neither explanation is easily reconcilable with the LSD-blocking action on 5-HT excitatory effects *(47)* or with the LSD-induced enhancement of PGO activity *(302)*, if one presumes the latter to reflect LSD antagonism at 5-HT receptors.

Additional evidence indicating that cells in the raphé nuclei are not homogeneous in their sensitivity to drugs and neurohormonal substances comes from a study that shows that systemic administration of mescaline (2 to 8 µg/kg, i.v.) produces an inhibition of raphé cells; however, there is no correlation between the population of cells depressed by microiontophoretic administration of mescaline and that depressed by systemic injection *(172)*. Again, an indirect effect on raphé cells produced by an excitatory action on a subpopulation of cells offers a plausible explanation, since a small percentage of cells in and immediately outside of the raphé system is excited by systemic administration of mescaline *(2)*. Systemic administration of 2,5-dimethoxy-4-methylamphetamine produces unit responses similar to those seen after treatment with mescaline. Large doses of atropine or scopolamine administered systemically, and phencyclidine—a psychotogen of the anticholinergic group—have no effect on raphé cells and do not prevent inhibition by LSD. Pretreatment of the animals with chlorpromazine does not abolish the inhibitory effect of LSD *(2)*.

Responses of raphé neurons driven synaptically by electrical stimulation of the nucleus paragigantocellularis of the reticular formation were compared with responses evoked by microiontophoresis of 5-HT, NE, and ACh; 98% of the raphé cells receiving an excitatory input from this nucleus were also excited by 5-HT, and 87% of the raphé cells inhibited by the input from the reticular nucleus were also inhibited by 5-HT. In contrast the effects of NE and ACh showed no such correlation.

In the cat, a lesion of the NE-containing fibers ascending at the level of the pontomesencephalic isthmus causes an increase of endogenous 5-HIAA in mesencephalon and telencephalon *(331)*. An increase in endogenous 5-HIAA in the telencephalon of rats is also found 43 hr to one week after intracisternal administration of 6-hydroxydopamine (6-OHDA), a compound that selectively destroys NE- and DA-containing nerve terminals when given intraventricularly *(52, 413)*. In such experiments, the level of 5-HT is unchanged but that of NE is decreased by 46 to 64%. These results suggest that destruction of NE terminals by 6-OHDA induces an increase in 5-HT synthesis and utilization. Interestingly, at the same time, no increase in 5-HIAA is observed in the hypothalamus, indicating that the functional relationships between the NE- and 5-HT-containing neurons in this region are different *(43)*. In the rat, a possibility of a tonic inhibitory influence of the NE-containing neurons originating in the locus ceruleus is indicated by a study of monoamine terminals in which there is a projection of NE-containing terminals to the 5-HT neurons of the raphé nuclei *(202, 321)*.

Large doses of L-DOPA given to mice produce a marked increase in brain DA, no change in NE, and a very significant decrease in brain 5-HT *(128)*. Systemic administration of DOPA (together with an appropriate dose of an AAAD inhibitor) causes a marked and rapid fall in the intraneuronal level of 5-HT, and the serotoninergic raphé neurons show substantial quantities of DA as evidenced by strong green fluorescence *(31,*

66). It is not known whether or not the depletion of 5-HT resulting from a displacement by the newly formed DA has any therapeutic or untoward consequences in patients during treatment with L-DOPA. The conversion of L-DOPA to dopamine (DA) in serotoninergic neurons seemingly may be responsible for 5-HT depletion (*128, 317*). DA concentrations as low as 10^{-7} M in the incubation medium markedly inhibit the synthesis of 5-HT from TR in brain slices of the striatum and brainstem (*160*). However, synthesis is not inhibited when 5-HTP is a precursor, indicating that DOPA inhibits TR hydroxylation. L-DOPA may accumulate in 5-HT-containing nerve terminals (*163, 318*) and cause a decrease in 5-HT turnover (*318, 328*). On the other hand, the level of DA in the caudate nucleus can be markedly reduced by larger doses of 5-HTP, the immediate precursor of 5-HT (*166*). Nerve terminals containing 5-HT project to the substantia nigra (SN) (*328*) and probably make synaptic contacts with the cell bodies containing DA. Although little is known about the physiological interaction between the serotoninergic and the dopaminergic systems, the nature of 5-HT influences is probably a tonic inhibitory one, since the cataleptogenic effect of neuroleptic drugs in rats can be reduced by inhibition of 5-HT synthesis with PCPA or by a raphé lesion (*238*).

FUNCTIONAL CONSIDERATIONS

Sleep Mechanisms and Raphé Neuron System

Lesions Subtotal (80 to 90%) coagulation of the feline raphé nuclei causes a decrease in brain 5-HT and a continuous EEG arousal lasting three to four days. The electrocortical low-voltage, fast activity is accompanied by mydriasis and agitation, the latter being evidenced by almost continuous running movements of the forepaws. After three days the total percentage of time spent in slow-wave sleep (SWS) does not exceed 10%. One of the most important findings is that paradoxical sleep (PS) is totally suppressed (*203, 353*), despite the fact that the animals show almost continuous (30 to 40/min) pontogeniculooccipital spikes (PGOS) characteristic of rapid eye movement (REM) sleep (*112, 220*). The enhancement of PGOS in the raphé-lesioned cats appears to be a specific effect caused by removal or impairment of the serotoninergic system, since pharmacologically induced 5-HT deficiency with PCPA (*112*) also enhances PGOS, and 5-HT precursors suppress these phenomena (*204*). Larger doses of 5-HTP, administered to raphé-lesioned cats, restore EEG synchronization, but the animals remain awake (*346, 353*). In cats that show no pronounced changes in the level of brain catecholamines (CA) after a selected lesion of raphé nuclei, a significant correlation was found between the extent of the lesion, the level of 5-HT, and the amount of sleep (*203*).

Results from in vitro studies of 5-HT metabolism in brain tissue from raphé-lesioned cats strongly support the view that the insomnia and enhanced occurrence of PGOS are related to the impairment of normal function of the serotoninergic system. At the peak of insomnia the activity of the TR hydroxylase and the synthesis of 5-HT from labeled TR are markedly decreased (*346*). Although no changes in endogenous 5-HT or 5-HIAA can be found in brain tissue explants, the spontaneous release of 5-HT in vitro from brain tissue taken from raphé-lesioned and insomniac cats shows a very significant decrease, indicating that functional impairment of serotoninergic terminals precedes the decline of 5-HT levels. Only ten days after the raphé lesion, there is a marked decrease in endogenous 5-HT and 5-HIAA, paralleled by a decrease in labeled 5-HT synthesized in vitro from labeled TR. However, brain tissue explants from raphé-lesioned cats show little or no impairment of 5-HT synthesis from labeled 5-HTP (*346*), a finding clearly indicating that 1) other monoaminergic neurons presumably take up 5-HTP, which is then converted to 5-HT by nonspecific AAAD; 2) a lesion confined to the raphé nuclei selectively eliminates the serotoninergic system, leaving CA neurons functionally intact; and 3) 5-HTP, owing to its propensity to undergo heteroneuronal metabolism, is not very suitable for studies of 5-HT neuronal functions.

The behavioral and EEG effects of raphé lesions in rats are comparable to those observed in the cat. In the rat, however, the decrease in forebrain 5-HT and 5-HIAA after a raphé lesion seems to parallel more closely the suppression of behavioral and EEG signs of sleep than in the cat (*236*).

Electrical Stimulation of Raphé Nuclei In the rat, electrical stimulation of the raphé nuclei blocks the habituation processes, as evidenced by persistence of startle responses to repetitive auditory stimuli (*379*). Since reserpine and PCPA abolish this effect of raphé stimulation (*95*), the active inhibition of habituation processes may be linked to an activation of serotonin synapses. This interpretation is supported by observations that electrical stimulation of the raphé nuclei causes a release of 5-HT in the forebrain nuclei (*187, 235, 379*) and from the cortex (*121*).

The induction of behavioral and EEG sleep in the rat with low-frequency electrical stimulation of the raphé nuclei associated with enhanced 5-HT metabolism (*162, 235, 237*) is seemingly incompatible with the blocking effect of raphé stimulation on habituation processes. The hypnogenic effect of low-frequency electrical stimulation of raphé nuclei (and the pontine reticular formation) was confirmed in the cat in which the stimulation was much more effective after than before administration of 5-HTP and was antagonized by LSD-25 (*234*). Evidence from these experiments is insufficient to establish the direct involvement of 5-HT as a hypnogenic factor, since low-frequency electrical stimulation of the medulla (tractus solitarius and reticular formation) also produces EEG synchronization and sleep (*253*), and since stimuli applied outside the raphé system do not release 5-HT (*379*).

A decrease in firing rate of dorsal raphé units during SWS when compared with the waking state (*284*), and the aforementioned disruption of habituation processes by both raphé stimulation and systemic administration of LSD (*219*), are not compatible with the current "serotonin hypothesis" of sleep (*203, 204*). The modulating influences of the 5-HT system on the brainstem reticular formation and the filtering processes of sensory input may be indirect and much more complex than implied by investigators who first surmised the possible hypnogenic role of 5-HT in normal brain function (*51, 56, 205, 226, 264–266, 270–272, 279, 298, 353, 359, 392*).

Local Chemical Stimulation In freely moving cats the application of 20 to 30 µg of crystalline 5-HT to the hypnogenic basal forebrain region, lateral preoptic area of the hypothalamus, and intralaminar nuclei of the thalamus consistently produced SWS. Conversely, a comparable amount of NE placed in the corresponding anatomic loci—with the exception of the thalamus—induces behavioral arousal, restlessness, and low-voltage desynchronized electrocortical patterns associated with regular theta activity in the dorsal hippocampus (*263, 271, 439, 440*). The synchronizing effect of 5-HT applied to the nonspecific thalamic nuclei was studied recently with multiple microelectrodes (*418*). Application of 5-HT to the mesencephalic reticular formation is less effective and produces no significant change in the sleep-wakefulness profile of the normal cat when applied to the raphé nuclei or serotoninergic systems (*263*). However, if the animal is pretreated with PCPA and rendered insomniac, 5-HT placements in these areas fully restore the normal sleep profile (*302*). The specificity of the 5-HT effect is indicated by very limited spread in the brain tissue, as shown by the fluorescence technique (*91, 385*) and by the differential effect of NE administered to the nonspecific thalamic nuclei where it produces sleep with enhanced REM episodes, in contrast to the arousing effect in both the hypothalamus and mesencephalic reticular formation (*263*).

In all loci in which 5-HT produces sleep, bufotenine (the N-dimethyl derivative of 5-HT) causes arousal, increased sympathetic tone as reflected by mydriasis and tachycardia, restlessness, and electrocortical low voltage, fast activity (*263, 271*). These results have been corroborated in studies on the behavioral effects of peripherally administered 5-HT and bufotenine to neonatal chicks with an immature blood-brain barrier (259, 260). The action of bufotenine, contrasted with that of 5-HT, can partially be explained by different pharmacological properties of this 5-HT derivative. In contrast to 5-HT, bufotenine does not fulfill requirements for a putative neurotransmitter. Owing to the presence in its molecule of two methyl groups attached to the nitrogen of the side chain, bufotenine is deaminated and oxidized at a rate approximately six times slower than 5-HT (42, 165). Hence it can be assumed that even a relatively low concentration of this derivative may produce an effect comparable to that of an excess of free 5-HT.

The excess of free 5-HT in the brain that can be achieved by various pharmacological manipulations causes behavioral excitation, an effect opposite to that of moderate increase of free 5-HT. Such a dual action of 5-HT can be demonstrated by systemic administration of a small dose of 5-HTP, causing sedation. However, larger doses of 5-HTP (or a smaller dose of this 5-HT precursor administered in conjunction with an MAO inhibitor) cause behavioral excitation (261), an effect that is attributed to a release of NE (55).

Studies of molecular models (387) indicate that the gap between the nucleosides of the receptors into which the side chain of 5-HT has to be fitted in order to bind the N^+ group to the negative atoms of phosphate and ribose oxygen is very narrow, and two methyl groups attached to the nitrogen of the side chain would prevent the entry. Hence, the molecule of bufotenine as well as that of dimethyltryptamine, a known hallucinogen, would tend to bind to the 5-HT receptor in an inappropriate position and thus prevent the access of the 5-HT molecule. In connection with the hypnogenic action of 5-HT and the lack of this effect shown by bufotenine, oxidative metabolites of 5-HT (254), such as 3-indoleacetaldehyde (and possibly also 5-hydroxyindoleacetaldehyde), cause inhibition of succinoxidase activity, whereas the tertiary amines such as N,N-dimethyltryptamine or its 5-hydroxy derivative, i.e., bufotenine, which are not readily oxidized by the MAO system, fail to inhibit activity of succinoxidase (254). In homogenates from brain tissue of animals pretreated with an MAO inhibitor, 5-HT and tryptamine fail to inhibit succinoxidase activity, whereas these amines suppress this enzyme in tissue from nonpretreated animals, indicating that 5-HT metabolites may play a crucial role in the hypnogenic action by causing reduction of energy derived from oxidative metabolism.

The suppressant effect of MAO inhibitors on PS may be due to a decrease in 5-HT turnover and, subsequently, to low levels of 5-HT catabolites such as 5-hydroxyindoleacetaldehyde, which may act as a "priming" agent in the occurrence of PS (*204*). Since PS is totally suppressed when the daily amount of SWS decreases below 15% in cats treated with PCPA (*227*), it can be postulated that the "priming" catabolite never reaches the threshold concentration and that the NE system, which is believed to be responsible for the "executive" mechanism in the occurrence of PS, is not sufficient for the development of PS (*204*).

Intracerebroventricular administration of 5-HT, after initial EEG desynchronization, produces sedation and EEG synchronization (*51*). It is difficult to assess how specific these effects are in terms of selective activation of serotoninergic receptor sites, since intraventricularly administered 5-HT is accumulated in part by catecholaminergic neurons in the caudate nucleus (*149*). Moreover, it was recently pointed out (*375*) that most investigators who administer 5-HT intraventricularly to study thermoregulation or sleep employ concentrations of 5-HT that would be in excess of 5×10^{-6} M if distributed evenly throughout the brain. Under such circumstances it is predicted that large quantities of 5-HT would enter the catecholaminergic nerve terminals (*375*). However, there is suggestive evidence that the hypnogenic effect of intracerebroventricularly administered 5-HT is caused by a specific action on receptors located in the area postrema, since topical application of 5-HT to this region induces suppression of the brainstem reticular activating system and elicits cortical synchronization that can be quantified by the enhancement of cortical recruiting responses to electrical stimulation of the nonspecific thalamic nuclei (*226*).

Cauterization of the area postrema or local application of the 5-HT antagonists, methysergide or LSD-25, abolishes the synchronizing effect of 5-HT. On the basis of these experiments (*226*), it is proposed that the area postrema is endowed with 5-HT receptors the stimulation of which increases the tonus of the hypnogenic and synchronizing region of the reticular formation known to be located in the area of the nucleus of the tractus solitarius (*49, 253*). This interpretation is supported by observations that fibers from the plexus around the area postrema run toward the region of the nucleus of the solitarius (*300*). The synchronizing and hypnogenic effect of 5-HT by an action on the area postrema was recently confirmed by investigators (*58*) who based their conclusion on the analysis of frequency-density spectra and power-density spectra from the temporal and frontal electrocorticograms of cats. The specificity of the 5-HT receptors in the area postrema was demonstrated by showing that no significant changes in the EEG activity could be produced by application of NE, acetylcholine (ACH), or the local anesthetic, Xylocaine.

Microinjections of melatonin (5-methoxy-N-acetyltryptamine) into the preoptic region of freely moving cats produce behavioral and EEG effects qualitatively similar to those induced by 5-HT (272). Several synthetic derivatives of 5-HT (containing a morpholine, piperidine, or pyrrolidine ring attached to the side chain and a methoxy or hydroxyl group in position 5 of the indole nucleus) that exhibit a weak but long-lasting serotonin-like stimulating effect on the isolated heart of a clam, *Anodonta cygnea*, produce sedation, drowsiness, and SWS when applied in microamounts (15 to 20 µg) through chronically implanted cannulas into the preoptic region of the hypothalamus of cats (262, 270).

The compound 5,6-dihydroxytryptamine (5,6-DHT) is a strongly reducing 5-HT congener and, when injected intracerebroventricularly, is taken up mainly by 5-HT-containing neurons and causes a long-lasting reduction in brain and spinal cord 5-HT levels (33). The damage of 5-HT axons and terminals which can be shown at the ultrastructural level (33) and with the fluorescence technique (32, 40) is not permanent, since signs of recovery of 5-HT levels have been noticed during the third and fourth weeks after administration of the 5,6-DHT (40, 41). The usefulness of this compound as a tool in investigating the role of brain 5-HT system depends on the selectivity of the neuronal damage that remains to be ascertained. Judging from a preliminary study in which an attempt was made to correlate the destructive effect of 5,6-DHT on the 5-HT system with behavior in rats (419), it can be predicted that this 5-HT congener causes a severe disruption of sleep mechanisms.

Systemic Effects Sleep-like behavior associated with ECoG synchronization is produced by systemic administration of 5-HT in neonatal chicks (390, 391) in which the blood-brain barrier is not yet fully developed (174). The evidence for the central hypnogenic influences of circulating 5-HT in adult animals with a fully developed blood-brain barrier was provided in experiments in which 5-HT was injected into the artery supplying the area postrema, which is devoid of blood-brain barrier, and, when stimulated by 5-HT, triggers abrupt ECoG synchronization (226).

Since MAO inhibitors block REM sleep and the occurrence of PGOS, it is postulated that metabolites of monoamines may be responsible for triggering REM and PGOS (203). During the last few years several 5-HT metabolites have been tested for their hypnogenic action. Tryptophol, 5-hydroxytryptophol, and the corresponding aldehydes produce, in neonatal chicks (with immature blood-brain barrier) and in adult mice, "postural depression" and "sleep-like state" (in chicken, defined as roosting posture and eyelid closure) (361, 363). Some investigators have been unable to show the hypnogenic effect of 5-hydroxytryptophol injected intracerebroventricularly *(351)*. More recently, in an electrophysiological analysis of feline sleep profiles, it was demonstrated that systemic or intracerebroventricular administration of tryptophol, 5-hydroxytryptophol, or 5-hydroxyindoleacetic acid produced severe motor depression and autonomic disturbances without any hypnogenic effect. On the contrary, these compounds abolished both SWS and REM sleep *(303)*.

Another 5-HT metabolite, 5-methoxy-*N*-acetyltryptamine (melatonin), produced only by the pineal *(21)* seems to conform more closely to Jouvet's postulate *(203)*, since it is reported to enhance SWS in newborn chicks *(184)* and to promote both SWS and REM in man. In adult rabbits, an i.v. injection of melatonin produces a marked sedation and EEG synchronization after a latency of approximately 30 min *(395, 396)*. When this substance is administered systemically to adult cats (0.5 mg/kg i.p.), it increases the total amount of sleep without changing its general profile (Marczynski, unpublished results).

The rate of synthesis of melatonin depends on the activity of hydroxyindole-*O*-methyltransferase (HIOMT), an enzyme the activity of which shows a daily rhythm corresponding to the photoperiod and is present only in the pineal. The activity of HIOMT is lowest at the end of the daily light period and increases approximately 5 hr after the onset of darkness *(434)*. Thus the pineal can be regarded as a neuroendocrine transducer which converts a reduction in retinal photoreceptor output to an enhanced synthesis and output of melatonin *(13, 24, 433)*. The information about reduced retinal photoreceptor output reaches the pineal via a polysynaptic pathway involving the accessory optic tract which joins the medial forebrain bundle and terminates in the medial terminal nucleus located lateral to the mammillary peduncle and dorsomedial to the SN. Subsequently the afferent pathway involves the superior cervical ganglion, but there is no information to indicate what pathway may carry information about environmental lighting from the medial terminal nucleus to the thoracic intermediolateral cell column *(299)*.

During the night not only is the activity of HIOMT increased but the synthesis of the pineal enzyme responsible for *N*-acetylation of 5-HT is also enhanced 50-fold. Exposure of the animal to light at night or administration of a β-adrenergic blocking agent causes a pronounced inhibition of enzyme synthesis, indicating that an adrenergic mechanism is involved in the modulation of melatonin synthesis *(109)*. Administered systemically, melatonin readily crosses the blood-brain barrier *(434)*, and in rats a dose of 30 μg/kg i.p. causes a release of 5-HT from the cortex after 20 to 60 min, increasing simultaneously the levels of 5-HT in hypothalamus and midbrain. After 180 min, 5-HT levels in the cortex and hypothalamus return to normal, whereas the midbrain concentration continues to rise (*14*). The loci in which melatonin causes the greatest increase in 5-HT are also sites at which the circulating ^{3}H-melatonin is most rapidly taken up and accumulated (*17*).

Since melatonin markedly enhances the activity of brain pyridoxal kinase and increases the levels of GABA in the cortex and hypothalamus, these changes may be brought about by the effect of melatonin on the availability of pyridoxal phosphate, a cofactor in the activities of 5-hydroxytryptophan and glutamic carboxylases (*13, 16*). An interesting aspect of the melatonin effect on the level of 5-HT in the brain is that doses as low as 1.0 μg/kg administered i.v. increase 5-HT in the midbrain, whereas doses larger than 100 μg/kg are less effective than those of the order of magnitude of 50 μg/kg (*14*).

When melatonin is administered to healthy volunteers in i.v. doses of 0.25 to 1.25 mg/kg, it enhances the ECoG α activity, causes marked sedation, and often induces sleep during which PS and REM episodes are enhanced (*15*). After approximately 45 min subjects can be awakened easily, and they report a sensation of well-being or even elation and can recall vivid dream episodes. None of the treated subjects showed any behavioral or mental disturbances. On the contrary, parkinsonian patients treated during four weeks with daily oral doses of 1.2 g of melatonin show marked improvement: the α rhythm increases in regularity and amplitude over the parietooccipital cortex, the β rhythm becomes more regular over the frontal cortex, and the tremor and rigidity disappear. Substitution of melatonin with a placebo is followed by a gradual deterioration which was plainly evident two weeks after withdrawal of melatonin.

A derivative of serotonin, 5-methoxy-*N*-methyltryptamine, is a naturally occurring alkaloid isolated from perennial grass *Phalaris arundinacea* (*428*). This 5-HT congener can probably be synthesized in the mammalian brain, since monomethyltryptamine has been isolated from rat and human brain (*361*), and the hydroxyindole-*O*-methyltransferase is present in the pineal tissue (*20*). When administered systemically (1 mg/kg, i.v.) to freely moving rabbits, it produces sedation and, up to 24 hr, a lasting hypersynchronization of the EEG that resembles SWS patterns. This effect contrasts with the restlessness and EEG desynchronization produced by 5-hydroxy-*N*-dimethyltryptamine, i.e., bufotenine (s,265). Since 5-methoxy-*N*-monomethyltryptamine shows weak but prolonged serotonin-like stimulating effect on the isolated heart of the clam *Anodonta cygnea* (*264, 265*), it is assumed that the hypnogenic effect is mediated by prolonged stimulation of the 5-HT receptor sites (*266*) in agreement with the hypothesis that the hypothalamic 5-HT modulates the tonus of the trophotropic and hypnogenic neuronal substrates (56). When taken orally (0.5 mg/kg) by healthy adult volunteers, this 5-HT derivative produces a feeling of well-being and drowsiness, without hallucinations and major autonomic disturbances (Marczynski, unpublished results).

Because of their long-lasting serotonin-like effects, several synthetic tryptamine derivatives with a morpholine or piperidine ring attached to the side chain, and with or without a methoxy group in position 5 of the indole nucleus, e.g., 3-beta-morpholineethylindole, 3-beta-piperidineethylindole, and their 5-methoxy derivatives, have been examined (262). All of these compounds produce sedation and an EEG synchronization that closely resembles SWS patterns when administered IV in doses of 1 to 2 mg/kg to freely moving cats and rabbits; these derivatives of 5-HT also block hippocampal theta waves (an effect comparable to that of 5-HTP and tryptamine [394]) and raise the threshold for EEG desynchronization to electrical stimulation of the brainstem reticular formation. All of these effects are observed while the animals remain awake and are able to respond to external stimuli with eye, head, or limb movements.

In addition, the 5-HT derivatives block the conditioned avoidance—escape responses in rats, an action characteristic of major tranquilizers. Parenthetically, in cats with intact neuraxis (under urethan-chloralose anesthesia), the 5-HT derivatives, in doses 50 to 100 times smaller than the LD_{50}, block the monosynaptic patellar reflex (and its polysynaptic facilitation) without causing significant cardiovascular changes (267), a finding both compatible with the view that the descending 5-HT system exerts tonic inhibitory influences on spinal cord reflexes (11, 125), and consistent with the depressant effect of iontophoretically administered 5-HT to spinal cord motoneurons (46). An electrophysiological study of feline sleep profiles showed that the aforementioned 5-HT derivatives increase the total amount of sleep, but the percentage of SWS, PS, and REM sleep remains unchanged (Marczynski, unpublished results).

Precursors

L-*Tryptophan* Substantial evidence indicates that TR hydroxylase is the rate-limiting step in the biosynthesis of 5-HT (200). Several investigators found that this enzyme is not saturated with the substrate under normal conditions and therefore concluded that the rate of 5-HT synthesis may depend on availability of TR (296). It is also reported that i.v. administration of TR to dogs produces a more "physiological" increase of brain 5-HT than that caused by 5-HTP, since TR hydroxylase is in higher concentrations only in those areas that receive projections from 5-HT neurons (296). Hence despite the large increases of 5-HT and 5-HIAA after TR administration, the patterns and distribution of these changes closely correspond to the physiological patterns in untreated animals. On the other hand, administration of 5-HTP not only abolishes the physiological patterns of distribution of 5-hydroxyindoles, but also generates a different pattern that corresponds closely to the distribution of AAAD activity (48). A good example is the caudate nucleus which normally shows a low rate of TR hydroxylation and a low level of 5-HT but, after 5-HTP loading, shows large rises of 5-HIAA and 5-HT owing to the high level of the nonspecific decarboxylase (296). In the caudate nucleus, this enzyme is presumably associated with dopaminergic nerve terminals, and it may produce relatively high concentrations of 5-HT that could act as a false transmitter (18). Tryptophan is actively transported by catalytic mechanisms (or permeases) from the circulation to the intracellular site of the TR hydroxylase. Overwhelming evidence indicates that TR, and not 5-HTP, is the natural precursor of brain 5-HT. Results from in vivo studies show that cerebral tissue can hydroxylate TR (154). Moreover, the rise in cerebral 5-HT in rats after i.p. administration of TR is not affected by evisceration that removes most of the peripheral organs containing tryptophan hydroxylase (422). Perfusion of a cerebral hemisphere with a TR solution increases 5-HT concentration on the perfused side only, indicating that no extracerebral source of 5-HTP is necessary for the synthesis of 5-HT in the brain.

In untreated phenylketonuric patients, blood 5-HT and its major catabolite, 5-HIAA, are present in subnormal concentrations (327). Several studies support the view that the excess of phenylalanine in these patients results in competitive inhibition of transport of TR into the brain (286, 442). In necropsied brains of untreated phenylketonuric patients, 5-HT levels (as well as those of DA and NE) were found to be reduced to 30 to 40% of normal values, and the level of TR was decreased 40 to 50% of normal. These amine deficits could be reversed by the administration of TR and a diet in which the intake of phenylalanine was restricted (285). The degree to which the deficit in brain 5-HT contributes to behavioral and sleep disturbances in phenylketonuric patients remains to be investigated.

A study of the effect of α_2-globulin, isolated from plasma of schizophrenic patients, on active transport of amino acids across cellular membrane shows that the influx (and to a lesser degree, the efflux) of the following three important precursors of neurohormonal substances is greatly enhanced, namely, glutamic acid (a precursor of GABA), TR, and 5-HTP (142). The effect appears to be specific since tyrosine was barely altered, and there was no change in the transport of phenylalanine, serine, methionine, histidine, lysine, α-aminoisobutyric acid, or aspartic acid. Although it is difficult to predict what would be the net effect of such changes in the membrane transport on synaptic transmission involving GABA or 5-HT, existing data seem to indicate that an impairment may occur in the function of the serotoninergic system (67). In patients with a high level of the α_2-globulin, SWS (and particularly Stage IV) is markedly suppressed or even abolished, and there is a negative correlation between the level of this globulin and the amount of Stage IV sleep (67).

In rats, systemic administration of TR (50 mg/kg, i.p.) increases free TR in plasma and brain by approximately 400 to 500% after 30 to 60 min, and these increases are associated with elevation of brain 5-HT (18%) and 5-HIAA (80%), indicating that brain 5-HT turnover is enhanced (397), which, in turn, may imply that the "tonus" of the serotoninergic system is increased. An excessive amount of TR in the diet of rats induces a small reduction in the total duration of sleep, but this effect is associated with a marked enhancement of the frequency of PS episodes. On the other hand, a TR-free diet reduces such episodes (176). Oral administration of 6 to 10 g of TP to *normal* human subjects at bedtime enhances onset of sleep and shortens the interval of the initial occurrence of a REM episode. This effect is abolished by methysergide, a central and peripheral 5-HT blocking agent (177, 325, 426). An increase in SWS (Stages III and IV) was observed in normal adults with doses of 6 to 10 g of TR but there were no significant changes in REM sleep. Conversely, a decrease in REM sleep has been reported (438).

Patients with idiopathic narcolepsy often develop REM sleep almost immediately after falling asleep, a phenomenon not normally observed in healthy humans and animals. When TR (5 g) is administered orally to narcoleptic patients 15 min before bedtime, the average duration of the initial REM period is doubled (127). It is also reported that in man moderate loads of TR (7.5 g) cause sedation, a marked increase in SWS, and shortening of the periods between REM episodes (429). With larger doses (12 g) a marked sedation is observed; however, the conspicuous enhancement of SWS is delayed until the recovery night, 24 hr later.

The amount of REM sleep during the first night after TR administration is greatly increased, as if the demand for REM had priority over the demand for SWS. When administered to patients suffering from chronic insomnia, TR (7.5 g) produces a significant increase in total sleep, diminution in early morning awakenings, a decrease in intermittent awakening, and only a slight increase in REM sleep (438). Approximately 20 min after taking TR orally, most subjects complain of drowsiness and sometimes fall asleep, or they complain of mild feelings of disorientation and drunkenness; few of them experience euphoria and compare this state to a psychedelic experience. In some subjects shortly after they receive TR (as well as after 5-HTP), their conversation tends to become lewd *(325, 429)*.

5-Hydroxytryptophan Ever since 5-HTP was shown to cross the blood-brain barrier readily and raise the levels of brain 5-HT *(412)*, the amino acid has been used extensively in studies on sleep mechanism and behavior, the L-isomer of 5-HTP being the immediate precursor of 5-HT *(420)*. Systemic administration of 5-HTP (30 to 50 mg/kg) to waking cats *(234, 251, 298)*, rabbits, and monkeys induces sedation and ECoG synchronization, lasting several hours. In *normal* human subjects i.v. administration of 5-HTP (40 to 50 mg) shortens the interval from sleep onset to the first REM period *(324, 325)* and increases the percentage of time spent in REM sleep from 22 to 31% *(257)*. Oral administration of 5-HTP (600 mg) to normal subjects over a five-night period causes an increase in REM sleep from 5 to 53% of placebo baseline, whereas the non-REM sleep decreases slightly, probably compensating for the increased amount of REM sleep *(438)*. No changes are observed in SWS Stages III and IV *(436)*.

The effect of 5-HTP was tested in a patient who became *chronically hyposomniac* after brain injury that resulted in a permanent brainstem lesion, presumably involving the raphé nuclei as evidenced by abnormally low levels of 5-hydroxyindoleacetic acid (5-HIAA) in the CSF, the level of homovanillic acid being normal *(168)*. Administration of 1.0 to 1.5 g of 5-HTP restored sleep and the CSF level of 5-HIAA to normal. Larger doses of 5-HTP, however, suppressed REM sleep, presumably by interfering with the function of the adrenergic system, which appears to be necessary for REM sleep *(203)*. There is evidence that 5-HTP in larger doses not only activates serotoninergic receptor sites by increasing the level of 5-HT, but also causes release of catecholamines from nerve terminals, thus eliciting a biphasic response, i.e., sedation followed by arousal, or a hypnogenic effect that differs from physiological sleep.

In the cat, during the 5-HTP-induced sleep-like behavior, despite the ECoG synchronization, the nictitating membrane remains retracted, and PGOS and PS are abolished for several hours. Afterward a rebound of PS is observed *(111)*. With higher doses of 5-HTP, a clear biphasic effect is produced; the initial sedation is followed by excitation *(103)*, an effect that may be produced by a release through displacement of transmitters from catecholaminergic nerve terminals *(55)*. This contention is supported by early observations that large systemic doses of 5-HTP do not induce excitation in animals depleted of stores of NE *(78)*. Moreover *(247)*, histochemical evidence indicates that systemic administration of 5-HTP results in accumulation of 5-HT in central dopaminergic neurons. In brain slices, 5-HTP, a substrate of unspecific AAAD, is converted to 5-HT in serotoninergic and noradrenergic nerve terminals, thus causing a substantial release of NE from such preparations. In view of these data, any interpretation of central effects of 5-HTP should be made with caution.

Relatively large doses of 5-HTP, administered to animals pretreated with an MAO inhibitor, result in a significant production of bufotenine (*260, 390*) that, as mentioned earlier, when applied to the basal forebrain hypnogenic area or to the preoptic region of freely moving cats, causes behavioral and ECoG effects opposite to those of 5-HT, i.e., an increased sympathetic tone, mydriasis, behavioral excitation, restlessness, and desynchronized ECoG patterns, an effect comparable to that produced by NE (*262, 263*). It is also reported that a relatively low level of bufotenine in the brain as compared to other organs is associated with bizarre forms of behavioral excitation (*367*). In this connection, increased levels of bufotenine in the urine of schizophrenic patients treated with an MAO inhibitor (*183*) are associated with exacerbation of the schizophrenic syndrome (*403*).

The enzyme responsible for N-methylation of 5-HT has been demonstrated in the lung of rabbit and of man (*22*), and the highest levels of bufotenine are present in the lung of chicks after a combined treatment with 5-HTP and MAO inhibitor (*260*). This has led to the suggestion that the brain may be the passive victim of N-methylated tryptamine derivatives that are transported by the blood. However, more recent evidence indicates that there may be an indole-ethylamine N-methyltransferase in the human brain that may be responsible for N-methylation of 5-HT (*256*). The optimum pH for human brain indole-ethylamine N-methyltransferase activity very closely resembles that of the enzyme obtained from the lung (*22*). Since the enzyme obtained from the brain has relatively high K_m, one might argue that brain 5-HT never reaches a sufficiently high concentration for the enzyme to function under physiological conditions. However, during treatment with MAO inhibitors or amphetamine, some workers (*256*) believe that N-methyltransferase may be functional, thus raising the levels of brain bufotenine and N-dimethyltryptamine, both of which are strong hallucinogenic compounds.

α-Methyl-5-hydroxytryptophan, administered orally in daily doses of 0.5 to 2 g for seven to 17 days to patients with hypertension, is converted to α-methylserotonin (*436*) which, unlike 5-HT, is a poor substrate for MAO (*416*). An increase as high as 50% in REM sleep is found with doses of 0.75 g or higher. Since this compound presumably is not readily metabolized, it provides some evidence that the triggering of REM sleep does not depend on a 5-HT metabolite, as has been suggested (204).

Synthesis Inhibitors

p-Chlorophenylalanine When administered systemically, *p*-chlorophenylalanine (PCPA) selectively decreases the level of 5-HT in the brain and other tissues without significantly altering levels of catecholamines by inhibiting the hydroxylation of TR, the first rate-limiting step in the biosynthesis of 5-HT (*155, 200, 224*), whereas its synthesis from 5-HTP remains normal (*346*). In the pineal, TR hydroxylase is relatively resistant to inhibition with PCPA when compared with other parts of the brain, indicating that the pineal enzyme has distinctive molecular properties. After treatment with PCPA, enzyme activity in the rat brainstem or cortex, assayed at 10 μM TR, decreases to 5% of control, whereas enzyme activity in the pineal remains at 60 to 75% of that in control rats (*110*).

After a latency of not more than 24 hr, a single injection of PCPA (200 mg/kg, i.p.) reduces sleep in the cat to a daily average of less than 10% of the normal sleep time, with maximal suppression occurring approximately 48 to 72 hr after administration of the drug. The ratio of PS to total sleep remains approximately constant, except when total sleep approaches extremely low levels during which there is a relative increase in PS. Administration of 5-HTP (30 mg/kg, i.p.) to an insomniac cat 48 hr after injection of PCPA causes long-lasting SWS within 10 min. After PCPA administration the levels of 5-HT in the diencephalon, medulla, and midbrain are closely correlated with the total amount of sleep. In general, a similar correlation between the amount of sleep and 5-HT levels has been demonstrated in the rat during prolonged treatment with PCPA (*304*) but, after the cessation of PCPA treatment, the recovery of brain 5-HT level lags considerably behind the recovery of sleep (see section on "Precursors" under "Functional Considerations" in this chapter).

In the cat 48 to 72 hr after administration of PCPA, there is a marked reduction of the "deep slow-wave sleep" with little or no effect on either "light slow-wave sleep" or REM sleep (*414*). During recovery from a large single dose of PCPA, approximately 40 hr after drug administration, cats show markedly enhanced PGO activity in the lateral geniculate and occipital cortex. Full recovery of normal sleep patterns is observed after approximately 200 hr (*346*). In macaques, i.p. doses of 330 to 1,000 mg/kg of PCPA significantly reduce SWS but have no effect on REM sleep. These alterations in sleep patterns are correlated with a 31 to 46% decrease in brain 5-HT content. Interestingly, the concentration of cerebellar 5-HT increases by 44% (*425*). The administration of PCPA to the baboon *Papio Papio* also markedly suppresses SWS. It also reduces the PS; however, the latter effect appears to be secondary to the suppression of SWS (*36a*).

In man, oral doses of PCPA (2 to 4 g/24 hr) for several weeks or even months markedly reduce REM sleep after a latency of two to three weeks. Although the total amount of non-REM sleep shows no significant change, the occurrence of sleep spindles in Stage II sleep is reduced. Administration of DL-5-HTP (400 to 800 mg) to subjects taking PCPA, at the time when REM is maximally suppressed, causes an almost complete return of REM sleep. The non-REM sleep is not affected by 5-HTP (*436*).

Fenfluramine The value of PCPA as a 5-HT depletor and as a tool for investigating the relationship between brain 5-HT level and SWS has been challenged by investigators (*202*), who found that fenfluramine (an anorexiant), like PCPA, selectively depletes brain 5-HT (*118, 322*) but markedly enhances SWS in both normal and 5-HT-deficient cats pretreated with PCPA. In interpreting these data, however, one should bear in mind that fenfluramine increases the turnover of 5-HT whereas PCPA decreases it (*224*).

Theoretical Implications and Critique of Sleep Hypothesis In agreement with the hypothesis that a deficit in the serotoninergic system may be the cause of schizophrenia (*431*), some workers believe that impairment of the serotoninergic gating mechanism for PGOS, which prevents our dreaming while we are awake, may be responsible for at least some of the symptoms of schizophrenia (*114*). This conjecture is based on the following parallels drawn between the gross behavior, sleep patterns, and PGOS characteristics in 5-HT-deficient animals treated with PCPA and schizophrenics: 1) In schizophrenics, prior to and during the onset of a psychotic episode, and in PCPA-treated animals, there is a severe insomnia which usually lasts several days. 2) Drive functions are severely disrupted both in serotonin-deficient animals treated with PCPA and in schizophrenics. 3) REM-deprived, serotonin-deficient animals (*114*), and schizophrenics after REM deprivation, unlike most normal persons do not show REM rebound (*241, 443*). 4) Electrophysiological and behavioral observations indicate that PGO potentials, electrophysiologically identical with those of REM sleep, intrude into the waking state in serotonin-deficient animals treated with PCPA and trigger bizarre behavior characterized by frequent orientations to nonexistent stimuli which is reminiscent of hallucination (*135, 284*). It is thus conceivable that a similar intrusion of PGO activity may occur in schizophrenics with all of the perceptual and emotional consequences that such an event comprises. Psychotic symptomatology has been reported in man after administration of PCPA (*125, 437*) and after maintenance on a TR-deficient diet (*246*).

Large potentials, termed *lambda waves, occur in man over the occipital cortex during eye movements* (*30, 167*) and in animals (*192*) both during wakefulness and in REM sleep. During REM sleep in the cat, the high-voltage EEG potentials in the pons, lateral geniculate, and occipital cortex, i.e., the PGOS, are regarded as essential components of REM sleep (*61, 206*). There is electrophysiological evidence from the study of the H reflexes in man that phasic events, comparable to the PGO activity, are manifested during dream episodes (*335*), and they can be recorded from subcortical nuclei (*430*). Despite the similarity between the PGO waves associated with eye movements during wakefulness and REM sleep, several differences have been noted (*62, 199*): 1) PGOS in the lateral geniculate during wakefulness are always much smaller in amplitude than those recorded during REM sleep from the same electrodes. 2) Positive cortical PGO waves during REM are often preceded by an initial surface negativity never present in PGO waves during wakefulness. 3) Most importantly, cortical PGO waves during wakefulness are attenuated in the dark and abolished by bilateral optic nerve section, whereas during REM sleep they are not modulated by retinal input. Moreover, the administration of reserpine changes the patterns of PGO waves characteristic of wakefulness to those of REM sleep in the waking animal.

Reserpine enhances dramatically PGOS in animals that remain awake, and a similar enhancement has been observed after treatment with PCPA (*112, 113*). Moreover, as mentioned earlier, an almost continuous PGO activity occurs in cats rendered sleepless after a selective destruction of the raphé nuclei (*203*). One could argue that the PGOS are simply related to exploratory eye movements in a sleepless animal. This interpretation, however, does not seem to be correct in view of the fact that electrophysiological characteristics of PGOS recorded during wakefulness after depletion of 5-HT with reserpine closely resemble PGOS normally observed during REM sleep in a non-premedicated animal (*62*).

In connection with the hypothesis of a serotoninergic gating mechanism of PGOS, one should mention that LSD, the most powerful of the hallucinogens, 1) depresses the firing rate of raphé units both after topical or systemic administration (*2*); 2) antagonizes the excitatory effect of 5-HT on brainstem reticular neurons when applied iontophoretically (*47*); and 3) markedly enhances PGOS during both wakefulness and sleep, whereas bromo-LSD, known to be devoid of hallucinogenic properties, is without effect on PGOS (*302*).

The following arguments can be marshaled to emphasize the importance of research on the PGO phenomena and to support the contention that the disinhibited PGOS intruding into the waking state may be responsible for severe perceptual disturbances in the visual and other modalities. There is overwhelming evidence that both saccadic and pursuit eye movements—which can be defined as goal-directed eyeball rotations that achieve foveal fixation—are not randomly occurring phenomena. Instead they represent active and well-preprogrammed performances of the brain. The time of onset, as well as location of the goal of fixation, must be predetermined by complex computation and integration of the prior input. This process involves coordinated prelocation, timing, and cancellation of the previous image. To obtain effective localization, the patterns of the object to be recognized must not only be monitored but also continuously matched with the expected patterns retrieved from the "memory engram" (*207*).

There is also strong evidence that incorrect programming of eye movements (even though the primary visual system, per se, remains intact) may lead to "psychic blindness" or to considerable difficulties in recognizing familiar objects. When efferent impulses go to the eye muscles, other impulses go simultaneously to the lateral geniculate, visual cortex, and other brain areas to compensate for movement of the image. The impulses from the eye motor system to the sensory system are referred to as "corollary discharge" (*389*), which may be represented by PGO waves. The propagation of the PGO wave from the area of its presumable origin in the pontine reticular formation near the oculomotor nuclei (*90*) to the lateral geniculate causes strong presynaptic inhibition of retinal input to this nucleus (*37, 215*). The so-called saccadic suppression of visual input occurs during eye movements both in man (*445*) and in the cat (*294*). At the cortical level the PGO waves are the consequence of postsynaptic depolarization in the region of the soma of neurons located in the deeper cortical laminae (*60*). The corollary discharge during eye movements not only causes as much as 70% reduction in cortical responses to flash when the eyes are driven optokinetically, but also markedly reduces or modifies responses to auditory stimuli (*119*), thus interfering with integration of heterosensory input.

Indirect support for the hypothesis formulated by Dement and his colleagues (*114*) also comes from the study of eye-tracking patterns in schizophrenic as compared to normal individuals and nonschizophrenic patients (*188*). The experimental task requires the subject to watch an oscillating pendulum and eye movements are recorded by monitoring changes in the field of the standing corneoretinal potential. Although the number of saccades disrupting the pursuit movements is markedly enhanced in schizophrenics, they can be suppressed after alerting the subjects. However, the number of velocity arrests during pursuit movements is less subject to voluntary control, and in a large population of schizophrenics this interference was much higher than in nonschizophrenic patients and in normal individuals, reflecting a dysfunction in the smooth pursuit eye movements. Although the possibility that this dysfunction is caused by changes in the eye muscles or neuromuscular junction cannot be excluded, impairment of smooth pursuit movements may also be generated by the release of PGO-like activity that interferes with the central programming of pursuit eye movements.

Despite 1) parallels between 5-HT-deficient animals and schizophrenic patients; 2) clinical observations that the administration of L-TR to phenothiazine-resistant, chronic undifferentiated schizophrenic patients results in improvement (*438*); and 3) aggravation of schizophrenic symptoms caused by lowering 5-HT levels with alpha-methyldopa (*181*), clinical implications of the hypothesis of Dement and his colleagues on the serotoninergic gating mechanism of PGO activity cannot be easily reconciled with the fact that several antipsychotic drugs *more* potent than reserpine, e.g., chlorpromazine, haloperidol, and methiothepin (a member of the new class of dibenzthiepins with potent antiserotonin and neuroleptic activity [*329*]) have been found to potentiate PGOS in waking cats pretreated with PCPA. Methysergide, a central 5-HT antagonist, was reported to abolish PGOS (*297*).

On the basis of previously discussed data, it can be hypothesized that the total amount of sleep depends on both the integrity of the serotoninergic raphé system and the level and availability of 5-HT in the nerve terminals (*203, 204*), the general implication being that the serotoninergic system activates the hypnogenic pathways that suppress the waking system.

The serotonin-sleep hypothesis has been challenged by investigators (*284*) who recorded firing patterns of raphé units in chronic cats. When the electrodes were lowered into the dorsal raphé nucleus, 80% of the units exhibited a remarkably rhythmic firing pattern at a frequency of 0.5 to 5.0 spikes/sec closely resembling that described in the rat (*1*). And when these freely moving animals were exposed to various environmental stimuli, these units did not display the phasic changes in activity characteristic of neurons involved in sensorimotor pathways. Instead they displayed a very stable firing frequency in a particular state of wakefulness, and only gradually did they slow their rate during transition from wakefulness to SWS and PS, as though they were involved in a tonic modulation of brain activity related to the level of arousal, drive, or affective state. The firing rate consistently decreased during transition from wakefulness to SWS and remained reduced by 54% during SWS. It dropped even further, up to 92% when compared with the waking state, prior to and during PS, and most units were inhibited prior to and during bursts of REM. The latter observation supports the contention that a tonic serotoninergic mechanism is involved in inhibition of the neuronal population that generates the PGOS. On the other hand, the reduced firing rate during SWS is not compatible with the "serotonin-sleep hypothesis" in its present form. These data (*284*) are of particular relevance, since it can be argued that the firing of raphé neurons is directly related to the release of 5-HT from nerve terminals. First, depletion of 5-HT after inhibition of synthesis depends on the flow of nerve impulses, since such a depletion can be prevented by transection of serotonergic pathways (*146*). Second, the depletion of 5-HT following inhibition of synthesis can be markedly enhanced by electrical stimulation of raphé nuclei (*108*). Third, as already mentioned, electrical stimulation of the raphé nuclei is known to induce a release of 5-HT from brain regions that receive raphé projections (*5, 237*). Recent studies on the content of CSF in cats reveal that there is no correlation between the levels of 5-HIAA and SWS or REM sleep (*348*), despite the fact that 5-HIAA levels faithfully mirror such factors as immobilization stress during sleep deprivation (Radulovacki, unpublished results).

Another criticism of the serotonin-sleep hypothesis comes from a study in which cats were chronically treated with PCPA (*113*). By the eighth day of drug administration the total amount of sleep returned to about 70% of the baseline values, although the brain 5-HT level remained as low as during the period of almost total sleep suppression. Moreover, in rats there is no correlation between the recovery of brain 5-HT and the recovery of sleep after the withdrawal of chronically administered PCPA: the recovery of the 5-HT level consistently lagged behind the recovery of SWS by approximately 30% (*304*).

One could argue against this lack of correspondence between the 5-HT level and the amount of sleep by assuming that the recovery of sleep was made possible by a prolonged depletion of 5-HT leading to supersensitivity of receptors, thus allowing lower concentrations of 5-HT in synaptic clefts to restore physiological function. This contention is supported by observations that chronic depletion of monoamines in the peripheral nervous system, e.g., with reserpine, results in a receptor supersensitivity to exogenous monoamines (*404, 410*). Denervation supersensitivity in the CNS is also suggested by studies on nigroneostriatal dopamine-containing neurons undergoing degeneration following topical injection of 6-OHDA into the

SN or after treatment of experimental animals with reserpine. In such animals the effects of apomorphine, amphetamine, or DOPA are markedly potentiated (*148*).

An interesting challenge to the serotonin-sleep hypothesis also comes from work that has shown that drastic reduction of sleep in the cat, lasting from three to five weeks, can be produced by a bilateral longitudinal pontine split which spares the raphé nuclei (*255*). Unfortunately, it has not been ascertained if such sections cut across some of the rostrally projecting 5-HT-containing fibers originating in the raphé nuclei.

The effect of manipulating brain 5-HT with PCPA has not been uniform across species (*351*), i.e., in rats only moderate sleep decrements are obtained with 90% depletion of brain 5-HT, whereas in nontreated rats the sleep-wakefulness cycle is observed with spontaneous brain 5-HT fluctuation only over an 18% range (*371, 437*). When PCPA was administered to four patients with carcinoid, no change in the amount of non-REM sleep was observed; one patient indeed even showed an increase in non-REM sleep.

It is possible that the very large changes in brain 5-HT level caused by the administration of either large doses of a 5-HT precursor or a depletor are necessary to influence sleep mechanism because the total brain level of 5-HT is of less importance than a regional change in the function of the 5-HT-containing neurons and their terminals. The regional distribution of hypnogenic 5-HT receptors in the area postrema (*226*) and the projections from this area to the "synchronizing" bulbopontine reticular formation could be regarded as a potential "hypnogenic trigger zone" if the 5-HT levels in the CSF or blood plasma could be correlated with the sleep-wakefulness cycle. Since no such correlation can be demonstrated in the CSF or blood plasma (*158*; Isaac, personal communication), the area postrema is unlikely to play a major role in normal sleep mechanisms.

With push-pull cannulas it is possible to map out a topographical distribution of serotoninergic regions in the hypothalamus that are involved in the regulation of body temperature, and to differentiate between this area and surrounding regions that also contain 5-HT terminals, which presumably are not involved in temperature regulation, since a steady 5-HT release is not affected by thermal stress (*312*) (see section on "Thermoregulation" in this chapter). It is possible that part of the 5-HT system involved in sleep mechanisms is as well differentiated as that which subserves temperature regulation, and that it can be topographically delineated, first at the level of 5-HT-containing terminals and later, by retrograde degeneration studies, at the level of origin of these fibers in discrete areas of the raphé nuclei.

It appears that the topographical distribution of hypnogenic loci from which sleep can be produced by topical microinjection of 5-HT gives certain, although rather crude, indications of where to look for regional changes in the function of 5-HT-containing nerve terminals involved in the regulation of sleep. On the basis of 1) latency, intensity, and duration of the hypnogenic effect of 5-HT; 2) the specificity of such responses as indicated by the lack of effect of NE; and 3) the lack of effect of substances intended to provide osmotic stimulation, such as glucose or sodium chloride, it appears that the preoptic region of the hypothalamus and the more rostrally located basal forebrain hypnogenic area are the most likely regions in which localized release of 5-HT may produce sleep (*263*).

Thermoregulation

Fever produced by a bacterial pyrogen administered into the cerebral ventricle of a cat is markedly reduced or even totally abolished by NE or epinephrine (E) administered via the same route (*130, 131*). By the same token, the administration of 5-HT triggers a rise of body temperature accompanied by shivering. The 5-HT-induced hyperthermia can likewise be abolished by NE or E (*132*). The site of action of these putative neurotransmitters has been localized in the anterior hypothalamus with microinjection techniques (*132*). Since not only pyrogen-induced but also the 5-HT-induced hyperthermia can be abolished by NE or E, Feldberg and Myers advanced the hypothesis that a monoaminergic mechanism regulated body temperature, and they postulated that in the cat hypothalamus NE and 5-HT play antagonistic roles, the former subserving a mechanism of heat loss and the latter one of heat production. Intraventricularly injected 5-HT also reverses the hypothermia associated with barbiturate anesthesia, whereas NE or E enhances both anesthesia and hypothermia.

Subsequent comparative studies (*307*) have confirmed a similar antagonistic effect of 5-HT and NE in the chicken, dog, and monkey, whereas in the rat, goat, sheep, and ox these amines produce effects opposite to those observed in the cat and monkey, i.e., 5-HT triggers heat loss and NE, heat production (*307*). In the rabbit, several investigators observed rises in body temperature after 5-HT or 5-HTP injection into the cisterna magna (*68*), after systemic administration of 5-HT (*26*), and 5-HTP (*90, 197*). On the other hand, marked hypothermia has also been observed following administration of 5-HT into either the lateral cerebral ventricle or hypothalamus. A careful reinvestigation of the effects of 5-HT and 5-HTP on body temperature in the rabbit was undertaken by investigators

(*198*) who found that biphasic responses are produced: 5-HT or DL-5-HTP, administered by systemic, intracisternal or intracerebroventricular routes, induce hypothermia which is gradually superseded by hyperthermia. Pretreatment of the animals with a MAO inhibitor, phenylisopropylhydrazine, potentiates the biphasic character of the responses to both 5-HT and 5-HTP injected into the CSF, whereas only the hyperthermic response is potentiated by the MAO inhibitor after systemic administration of these substances. These workers concluded that the 5-HT neuronal thermoregulatory system in the rabbit may have a dual action, depending on the actual concentration of the free 5-HT at the receptor sites.

When push-pull cannulas are implanted in various hypothalamic loci in unanesthetized monkeys, two distinct factors are released into the perfusate: one when the animal is exposed to high temperature, and another when the animal is kept in a cool environment. If the perfusate collected from the anterior hypothalamus of a warmed monkey is transfused via a push-pull cannula to a homologous hypothalamic locus of another normothermic monkey kept at room temperature, a marked hypothermia develops in the recipient. However, if the donor is cooled, the normothermic recipient develops hyperthermia and shivering. The amount of 5-HT in a perfusate collected from the anterior preoptic region of a cooled monkey increases 4- to 24-fold above the resting release. The warming of the donor monkey suppresses the normal output of 5-HT by 50 to 100% and, since the administration of exogenous 5-HT to the anterior hypothalamus mimics the effect of the perfusate collected from the donor monkey, it can be assumed that the release of 5-HT activates the heat production pathway (*312*). A more systematic anatomical mapping of 5-HT release after cooling of the monkey revealed that the region of greatest 5-HT release encompasses the caudal-most portion of the anterior hypothalamus and extends rostrally to include the preoptic area. In the midhypothalamic area, 5-HT was also present in the perfusate, but its output showed no change during thermal stress (*308*). Hence it can be assumed that not all 5-HT nerve terminals in this area are involved in the regulation of body temperature.

In the monkey, 5-HT is not the only factor involved in heat production since ACh is released during either warming or cooling of the monkey from discrete loci in the mid- and posterior hypothalamus, and microinjections of ACh into such loci produce either hyper- or hypothermia (*313*). Hence in the posterior hypothalamus a different thermoregulatory mechanism is present. The functional relationship between the serotoninergic and the cholinergic mechanisms remains to be investigated.

Recent investigation of the effect of barbiturate-induced feline hypothermia revealed that a rise in 5-HIAA in the CSF is associated with a reduction of body temperature resulting from exposure to cold, since prevention of the hypothermia also prevents the increase in 5-HIAA in the CSF (*196*). These data are in harmony with the concept of a serotoninergic mechanism in heat production; they also show that barbiturates per se do not influence the serotoninergic system subserving heat production. The nature of the neurohormonal factors released from the hypothalamus during warming of the monkey and possibly responsible for heat loss is unknown, although a marked increase in the release of the nonesterified fatty acids (NEFA) in the perfusate from several hypothalamic loci is reported (*311*).

As mentioned earlier, the serotoninergic system in the rat, in contrast to primates and cats, is probably involved in the heat loss mechanism. In agreement with this role an elevated ambient temperature in the rat causes an increase both in the firing rate of the 5-HT-containing neurons in the raphé nuclei (*423*) and in 5-HT turnover (*101, 352*). On the other hand, hyperpyrexia induced by 1,1,1-trichloro-2,2-bis(*p*-chlorophenyl) ethane (*p,p*-DDT) is abolished by the administration of PCPA or DL-6-fluorotryptophan, inhibitors of 5-HT synthesis (*330*), an observation supporting the concept of a serotoninergic heat-producing mechanism.

A rise of 1° to 1.5°C in body temperature in rats is consistently observed after a half-hour stimulation of the caudal raphé nucleus but not of adjacent structures (*379*), a procedure associated with release of 5-HT and accumulation of its metabolite, 5-HIAA, in the basal forebrain (*5, 236*). Hence it is conceivable that in the rat, depending on the level and actual distribution of free 5-HT in brain structures manipulated by various drugs or electrical stimulation, the heat-producing function of 5-HT neurons may override the 5-HT mechanism of heat loss. This makes the rat somewhat comparable to the rabbit, a much more difficult model in which to study the mechanisms of thermoregulation than the cat or the primate.

Modification of Epileptiform Activity

In general, drugs that increase the level of free 5-HT in the brain also elevate seizure threshold, whereas drugs that reduce brain 5-HT have the opposite effect. For two reasons such an influence of 5-HT is not surprising: 1) iontophoretic application of this amine to single neurons inhibits spontaneous or driven discharges of the majority of neurons in various brain structures, including the cortex (*46*); and 2) convulsive seizures are one of the clinical manifestations of phenylketonuria (*222*) in which a reduced level of brain 5-HT has been reported (*432*).

Most drugs that increase the levels of NE, DA, and 5-HT at brain receptor sites exert marked anticonvulsant influences (282). In Table I only those drugs or drug combinations have been listed that seem to influence the 5-HT system more or less selectively, except for reserpine, which is known to block the storage mechanism for both catecholamines and 5-HT without affecting their synthesis (55, 382). An enhancement of 5-HT release from nerve terminals and an increased level of free 5-HT at the receptor sites probably account for the anticonvulsant action of p-chloramphetamine and related compounds, provided that doses of this drug which do not seem to affect the catecholamine system are used (144). A more selective increase in free 5-HT can be produced by systemic administration of 5-HTP, resulting in a significant elevation of seizure threshold to electric shock or to pentylenetetrazol (223, 345).

A strong anticonvulsant effect can be obtained by the administration of 5-HTP with a MAO inhibitor or reserpine, procedures that are believed to cause a high level of 5-HT at receptor sites. On the other hand, systemic administration of PCPA with the resulting inhibition of 5-HT synthesis lowers the threshold to convulsions induced in rats by electroshock and pentylenetetrazol (7, 360), auditory stimuli (372), fluorothyl (344), and withdrawal of barbiturates (319). It should be noted, however, that in addition to affecting brain 5-HT, PCPA also produces phenylketonuria in rats (248) and alterations in brain lipids and demyelination (139), a condition that predisposes an organism to convulsive seizures. Thus, despite the selective depletion of 5-HT, the evidence for its inhibitory role based solely on the effect of PCPA remains uncertain.

Indirect evidence for the anticonvulsant action of 5-HT also comes from clinical observations on the antiepileptic effect of oral administration of the 5-HT metabolite, melatonin, which is known to increase brain 5-HT and GABA (15).

Efforts to correlate *genetically determined susceptibility* to audiogenic seizures with levels of brain monoamines in certain inbred strains of mice reveal that the susceptible mice have significantly lower concentrations of both 5-HT and NE at the age of maximal seizure sensitivity (372). It has also been demonstrated that susceptible mouse strains have not only lower 5-HT but also an increased turnover of 5-HT and reduced MAO activity, as measured by the production of 5-HIAA after administration of 5-HTP (217).

A correlation between endogenous levels of both brain NE and 5-HT and the *circadian rhythm* of susceptibility to seizures has been investigated in genetically susceptible and less susceptible strains of mice (373), and a distinct nocturnal rhythm of greater susceptibility to both sound and

Table 1. Effect on seizure susceptibility of drugs that theoretically influence free serotonin at brain receptors

Drug	Free serotonin	Seizure susceptibility	Methods[a]	References
p-Chloramphetamine	↑ ?	→	AS, PST	242, 332
5-HTP	↑	→	MES, PST	223, 345
MAO inhibitor plus 5-HTP	↑	→	MES, PST	223, 245, 345
Reserpine plus 5-HTP	↑	→	MES	166
Reserpine	→	↑	EST, MES, PST	83, 360
PCPA	→	↑	MES, PST	7
PCPA	→	↑	PST	360
PCPA	→	↑	AS	372
PCPA	→	↑	FL	344
PCPA	→	↑	BW	319

[a] AS, audiogenic seizures; PST, pentylenetetrazol seizure threshold; MES, maximal electroshock seizure; FL, flurothyl; BW, barbiturate withdrawal.
[b] PCPA, p-Chlorophenylalanine.

electrically induced seizures in three tested genotypes of mice has been demonstrated. The increased susceptibility is associated with significantly lower 5-HT levels in all parts of the brain tested when compared with values obtained during the day when susceptibility was reduced. Since no day-night differences in the levels of NE were observed, the inverse relationship between 5-HT levels and seizure susceptibility seems to indicate that the 5-HT system is more intimately involved in the regulation of the general excitability of brain neurons than is the CA system. It was shown that the genetically susceptible strain of mice has a faster rate of 5-HT synthesis, higher fractional rate constant of disappearance of 5-HT and 5-HIAA, a markedly lower endogenous 5-HT level, and reduced MAO activity (217). The low 5-HT content coupled with the higher rate of 5-HT synthesis suggests a release from inhibition (via negative feedback) of the TR hydroxylase, the rate-limiting step in 5-HT synthesis. All of these changes may be brought about by an inefficient 5-HT system that, in turn, might result in a compensatory increase in neuronal activity and an enhanced release and turnover of 5-HT.

Neuroendocrine Function

Gonadotropin Release During the last five years, substantial indirect evidence has been accumulated indicating that serotoninergic nerve terminals in the region of the median eminence exert an inhibitory control over hypothalamic processes regulating the release of pituitary luteinizing hormone (LH). The cell bodies that send axons to the median eminence are probably located in the raphé nuclei, since electrical stimulation of the latter inhibits ovulation (79). Thus the serotonin nerve terminals appear to play a role antagonistic to that of the tuberoinfundibular dopaminergic system which is believed to be responsible for triggering the release of LH (146, 229, 230). Pharmacological evidence supporting the concept of 5-HT inhibitory influences on ovulation can be marshaled as follows: systemic administration of 5-HT to immature rats just prior to the "critical period" of ovulation control (during the short interval at the end of proestrus during which the ovulating amount of pituitary LH is released) blocks ovulation (323).

Similar effects can be obtained after administration of 5-HTP, the precursor of 5-HT (231). Moreover, selective blockade of 5-HT synthesis by PCPA markedly stimulates ovulation, and the intensity of this "provoked" ovulation is inversely proportional to the 5-HT level in the hypothalamus (228). However, the enhanced ovulation can only be observed when PCPA is administered just prior to the "critical period": if administered 20 hr earlier, it has an opposite effect in that it almost totally suppresses ovulation (*231*). In a recent review of the monoaminergic control of gonadotropin release, it was suggested that the dual action of the PCPA-induced impairment of the serotoninergic system implies that there are at least two functionally different serotonin pathways involved in the complex processes of LH secretion, one operating during the longer interval of pre-ovulatory processing of information necessary for preparing the "critical period," and the other involved in inhibition of the information transfer from the hypothalamus to the hypophysis, resulting in LH secretion and ovulation.

The facilitatory or inhibitory effect of PCPA on ovulation, contingent on the time of administration, shows that a long-term inhibition of 5-HT synthesis may impair various—and sometimes even antagonistic—functions of the serotoninergic system and should thus always be interpreted with great caution (*230*). Specific localization of serotoninergic receptors responsible for inhibition of the ovulatory response has been mapped using microinjections of MAO inhibitors into the hypothalamus of rats with or without proper blockade of 5-HT synthesis. The general area of the median eminence turned out to be the only one in which increased levels of 5-HT caused suppression of the ovulatory response (*228, 232*).

Prolactin Intracerebroventricular administration of 5-HT or melatonin to rats enhances the release of prolactin and inhibits the release of follicle-stimulating hormone (FSH); however, these effects cannot be obtained when 5-HT or melatonin is infused into the anterior pituitary by a cannulated portal vessel, suggesting that these indolamines act on the hypothalamic prolactin-inhibiting and FSH-releasing factors (*208*).

ACTH Topical application of 5-HT into discrete areas of the hypothalamus causes a release of ACTH as evidenced by a sharp rise of plasma corticosteroid levels (*239, 316*). There are indications that the hypothalamic 5-HT system is involved in the negative feedback effects of plasma glucocorticoids on ACTH secretion, since systemic administration of hydrocortisone or dexamethasone to rats normalizes the brain 5-HT turnover if it was reduced after adrenalectomy (*25, 147*). The important link between glucocorticoids and the inhibition of ACTH secretion is probably the stimulating effect of adrenal steroids on the activity of TR hydroxylase (*25*).

Growth Hormone There is indirect evidence that catecholamines, and particularly DA, are involved in the release of growth hormone (GH) through an action on the secretion of the hypothalamic-releasing factor (*278*). It appears that the complex circuitry that leads to the release of GH also involves 5-HT. Several investigators have reported a correlation between the initial phase of SWS, GH secretion, and onset of REM sleep

(368, 369, 401). Deprivation of SWS prevents the release of GH in man *(368, 369)*. The hypothesis on the role of 5-HT in GH release is supported not only by the role of 5-HT in SWS but also by the fact that intracerebroventricular administration of this indoleamine causes GH release in the rat *(94)*, and oral administration of 5-HTP causes a rise of plasma GH in man *(195)*.

Sexual Behavior

Female rat copulatory behavior such as the lordosis response, enacted as the animal in heat is about to be mounted by a male, is abolished after castration, but it can readily be restored by sequential administration of estrogen and progesterone, secretion of which from the ovaries is normally controlled by the pituitary hormones FSH and LH. The lordosis response can be totally suppressed by the combined systemic administration of a MAO inhibitor, pargyline, and a 5-HT precursor, 5-HTP *(288, 290, 291)*. Pargyline alone also tends to suppress estrous behavior. This effect cannot be prevented by prior catecholamine depletion with alpha-methyl-metatyrosine but can be prevented with alpha-propyldopacetamide, a drug that prevents the accumulation of 5-HT induced by pargyline *(288)*. Hence, it is postulated that 5-HT-containing neurons exert an inhibitory influence on estrous behavior *(288)*: a low level of 5-HT releases estrous behavior, whereas a high 5-HT level suppresses or terminates this behavioral pattern.

It appears that the 5-HT system is responsible for inhibition of the LH release, i.e., an action antagonistic to that of the dopaminergic system in the median eminence region of the hypothalamus (see review by Kordon and Glowinski [230]). Although estrogen treatment is necessary to restore the lordosis response in ovariectomized rats, the 5-HT synthesis inhibitor PCPA can act as a substitute for progesterone, a finding that suggests that progesterone may behave as an antagonist with regard to the serotoninergic system normally involved in suppression of the lordosis response *(293)*. Monoamine depletors such as reserpine or tetrabenazine also mimic the effect of progesterone.

In support of the aforementioned hypothesis, investigators have demonstrated that not only the inhibition of 5-HT synthesis with PCPA but also systemic administration of the 5-HT receptor-blocking agent, methysergide, to ovariectomized, adrenalectomized, and estrogen-primed rats induces a strong lordosis response comparable to that elicited by progesterone *(444)*. Moreover, topical application of 10 μg of methysergide to discrete loci in the medial preoptic and posterior hypothalamus is also effective in producing the lordosis response, whereas phentolamine, an α-adrenergic receptor blocking agent, is ineffective. It should be pointed out that no treatment aimed at blocking the hypothalamic 5-HT system was able to activate the solicitory component of the estrous behavior in the rat, such as darting, hopping, and earwiggling, thus indicating that activation of the total pattern of estrus probably involves transmitters other than 5-HT.

An interesting aspect of the physiological role of the serotoninergic system in sexual behavior is revealed by the observation that in the *male* rat, castrated as an adult, the lordosis response is produced by treatment with estrogens followed by either progesterone or monoamine depletors (reserpine or tetrabenazine), suggesting that in the male, a monoaminergic (perhaps serotoninergic) mechanism is involved in the tonic suppression of female sexual behavior (*292*).

Several groups of investigators have found that administration of PCPA, or PCPA plus the MAO inhibitor, pargyline—a procedure resulting in a presumably high level of free CA and a low level of 5-HT—markedly enhances homosexual mounting behavior in the adult rats (*159, 377, 381, 398, 399*), cats (*135*), and rabbits (*400*). Systemic administration of 5-HT promptly abolishes this abnormal behavior (*398*). However, it is reported that the heterosexual interactions in adult male rats treated with PCPA and pargyline, such as frequency of mounting, intromissions, and ejaculations, are not enhanced (*427*). Data from such studies suggest that such a manipulation of brain 5-HT and CA does not actually increase sexual motivation, and that the induced homosexual interactions reported by other workers are triggered by disturbances in perception and by the male's inability to distinguish adequately appropriate sexual partners. This contention is supported by the observation (*135*) that cats which display homosexual interactions after treatment with PCPA also appear to be perceptually disoriented. Chronic administration of PCPA to cats or monkeys produces bizarre behavior, and they display "pseudohallucinatory activity," i.e., they appear to be watching moving objects that the human observer cannot see. Such behavior is associated with the appearance of PGOS in the electroencephalogram which is characteristic of REM sleep (see section on "Lesions" under "Functional Considerations" in this chapter).

Neonatal administration of PCPA to female rats is reported to cause a delay in vaginal opening even though the cytological cycles are normal. In addition, as adults these females show significantly less earwiggling activity during both spontaneous and hormone-induced copulatory behavior. Males treated neonatally with PCPA show marked changes in mating behavior: they required significantly fewer mounts and intromissions prior to ejaculation and shorter refractory periods and intervals between intromissions.

However, hypothalamic and forebrain levels of 5-HT and 5-HIAA cannot be correlated with changes in mating behavior (*194*). A single injection of reserpine to neonatal male rats also reduces the number of intromissions needed for ejaculation in adults (*243*), and a similar action of reserpine is observed in adult animals (*116, 388*).

Aggressive Behavior

Stress-induced Aggressiveness in mice and rats is known to increase markedly after the animals have been isolated for four to six weeks. The behavioral changes are associated with changes in brain biochemistry. A decrease of 5-HIAA content and a reduced 5-HT turnover rate, e.g., have been demonstrated in mice that were rendered aggressive (*156*). These changes appear to be specific, since they are found only in those animals that develop aggressive behavior and are not observed in animals whose behavior is markedly influenced by isolation (*156*). In male rats, isolated for six weeks, the turnover rate of 5-HT decreases only in those animals that are rendered aggressive with respect to the mouse ("muricidal") or overly inhibited ("indifferent") after isolation, and that show lack of interest when confronted with a mouse (*415*). By contrast, those animals the behavioral score of which can be defined as "friendly" (characterized by sniffing, licking, grooming, nest-preparing for the mouse, and a whole pattern of playful behavior, including picking up, carrying around, jockeying, and jumping) show a marked increase in brain 5-HT turnover. Simultaneously, brainstem NE turnover decreases in "friendly" and in "muricidal" animals but does not change in "indifferent" rats. In addition, in "friendly" rats, the turnover rate of striatal DA decreases, but it remains unchanged in "muricidal" and in "indifferent" animals.

Drug-induced A behavioral deficit can be produced in various strains of mice by chronic administration of *amphetamine* and *amphetamine-like drugs*, a procedure that gradually leads to high levels and increased turnover of brain 5-HT and to low levels of NE. The behavioral deficit, ascribed to an increased level of 5-HT, consists of depression of exploration, reduction of aggressive behavior, and increase in stereotypicity. Other specific behavioral and ethological changes induced by these drugs do not relate clearly to the neurotransmitters. Certain types of "stress" also induce stereotypic and emotional behavior, increased levels and turnover of 5-HT, decreased NE levels, and variable changes in the levels of ACh (*214*). It appears that several transmitter or transmitter-like substances are affected by such stress as isolation, footshock, and application of cold.

In agreement with the contention that stress-induced or genetically determined low level of brain 5-HT with regard to CA is associated with increased aggressiveness, *PCPA-induced depletion of brain 5-HT leads to a* markedly enhanced muricidal behavior in rats (*377*) and to rat killing by cats (*135*). PCPA-treated cats that normally ignore laboratory rats begin to kill them with a rapidity and concentrated savagery that evoke "images of the jungle" (*114*). Administration of 5-HTP reverses the PCPA-induced aggressiveness (*377, 381*). The specificity of PCPA-induced aggressiveness is supported by the fact that a different approach to suppressing the normal function of the 5-HT system consisting of intraventricular injection of *5,6-* or *5,7-dihydroxytryptamine,* which causes selective degeneration of 5-HT-containing terminals, also markedly enhances aggressive behavior in rats (*419*).

A study of fighting behavior in rats induced by electric shocks delivered from a grid floor demonstrated that PCPA in doses that lower brain 5-HT to 10% of the control levels (NE concentrations remaining at 75% of normal levels) has no effect on fighting behavior (*95*). It is therefore possible that brain 5-HT influences only certain types of aggressive behavior such as predatory and spontaneous aggression, but not irritable or shock-induced aggression, which may be triggered by different integrative processes involving different neurohormonal mechanisms.

Genetically Determined Several investigators have extensively studied brain 5-HT and other putative transmitters in numerous strains and genera of mice and other small rodents and have attempted to relate the levels and turnover of these substances to the behavioral repertoire of the animals, as well as to brain excitability as demonstrated by the electroshock threshold and latency (*211*). While brain 5-HT levels differed by 300 to 400% between the various rodent types, ACh, NE, and DA also varied. Furthermore all of these substances generally changed in parallel (*210, 211*). It was hypothesized that relatively low levels of 5-HT and other transmitters are related to the ethological and behavioral syndrome of relatively high aggression, effective learning, efficient exploratory behavior, low stereotypicity, and relatively high brain excitability, whereas high levels of monoamines coincided with emotional, stereotypic behavior, with relatively poor problem solving, as well as with low brain excitability (*212, 213, 374*). It was stressed that certain experiential and learning factors were as important in establishing a behavioral syndrome as were the neurochemical changes. *In summary,* aggressive behavior was coupled with an increase in the activity of the catecholaminergic neuronal system relative to the serotoninergic system whereas the reverse was true for the "friendly" animals.

Consummatory Behavior

Administration of PCPA to rats and the resulting depletion of brain 5-HT do not significantly change the animals' consumption of water and food

(*57, 322, 377, 405*). However, PCPA-treated rats appear to be more sensitive or finicky than controls if the taste of food or water is adulterated with alcohol (*86, 322, 377*), saccharin (*314*), dextrose, or quinine (*57*), indicating that PCPA somewhat increases gustatory and olfactory acuity.

Volitional Alcohol Intake

It can be assumed that a genetic predisposition or aversion to ethanol exists since rats of some strains will select alcohol even on first exposure in a free-choice milieu (*126, 275*). Such an inclination to alcohol in a particular strain of rats can be significantly reduced by lowering the brain 5-HT with PCPA (*417*). On the other hand, lowering catecholamines with α-methyl-paratyrosine has no effect on preference for ethanol. The suppression of alcoholic preference by PCPA is very persistent and can be demonstrated in rats even two months after the last dose of this drug, when the brain 5-HT levels have returned to normal. It is thus suggested that cerebral 5-HT is involved in the complex processes that determine preference for ethanol (*417*). The specificity of these results can be questioned since PCPA-treated animals exhibit unusual irritability, and the ordinarily aversive gustatory properties of ethanol may be intensified as a result of increased sensitivity to sensory stimuli during PCPA treatment (*306*).

However, results from recent experiments (*6*) support the notion that increased brain 5-HT is associated with preference for alcohol, since brain 5-HT and 5-HIAA levels in rats that have been selected for 17 generations for preferring alcohol over water were 15 to 20% higher than in water-selecting animals (*6*). Moreover, when the rats were given a free choice between ethanol and water for one month, the brain 5-HT in the alcohol-habituated rats was 31% and the 5-HIAA content 10% higher when compared with the water-selecting group. Unfortunately, simultaneous changes in catecholamines were not measured in these experiments. A selectivity of PCPA action appears to be consistent with the observation that another 5-HT depletor, *p*-chloroamphetamine, also reduces rodent preference for ethanol (*140*). PCPA has an effect on rats whose preference for ethanol is a result of either inherent predisposition, prolonged exposure to alcohol (*86*), or environmental stress (*309*).

Paradoxically, 5-HTP administered systemically also causes a significant and long-lasting suppression of preference for ethanol for as long as four months after cessation of 5-HTP treatment, an effect that is not mirrored by an equally long-lasting increase in brain 5-HT (*310*). These observations raise doubts about the specificity of PCPA-induced suppression of alcoholic intake. To complicate the problem, the enhancement of ethanol intake produced by chronic intracerebroventricular administration of low doses of either ethanol, acetaldehyde, paraldehyde, or 5-hydroxytryptophol is significantly reduced by 5-HTP injected by the same route. Systemic administration of PCPA also reduces ethanol intake in ethanol-, acetaldehyde-, or 5-hydroxytryptophol-infused rats, but not in paraldehyde-infused animals. On the other hand, PCPA enhances ethanol preference in 5-HTP-infused rats (*310*).

The following two explanations have been offered for the paradoxically similar inhibitory action of 5-HTP and PCPA on the animals' preference for ethanol. First, it is possible that reduced preference for ethanol is engendered by the peripheral potentiation of 5-HT toxicity by ethanol (*402*) caused by an ethanol-induced alternate pathway of 5-HT metabolism in the liver to 5-hydroxytryptophol (*122*). Second, it is possible that 5-HTP enhances a feedback inhibition of TR hydroxylase, an action similar to that of PCPA. The plausibility of both explanations can be questioned. Clearly, more specific experimental approaches are desirable, such as discrete electrolytic lesions in selected parts of the raphé system or administration of a specific precursor of 5-HT, TR, instead of 5-HTP, which is taken up and decarboxylated to 5-HT not only by serotonin neurons but also by CA neurons, thus interfering with their physiological function.

Morphine Analgesia

In general, studies designed to assess the role of monoamines in both the complex processes of integration of pain sensation and mechanisms of analgetic action of morphine suggest that the neuronal system containing 5-HT as a putative transmitter plays an important, if not a key role. Although convincing arguments can be marshaled to support the view that NE, DA, and ACh are also involved in the mechanisms of morphine analgesia (*88*), a substantial, although indirect, body of evidence has been obtained during the last five years to indicate that the 5-HT system may play a critical role. Rats treated with PCPA show a marked increase in sensitivity to pain (*406*). When the flinch-jump response is used as the experimental variable, the loss of 5-HT is found to coincide with a decrease in morphine analgesia (*405*).

In keeping with these observations, a lesion in the medial forebrain bundle in rats or administration of PCPA, both of which result in a decrease in brain 5-HT, significantly lowers the threshold to electrical footshock (*178*). The lesion in the medial forebrain bundle causes reduction of 5-HT only in the telencephalon, indicating that this part of the brain participates in the integration of pain and that 5-HT modulates these functions. The control values of both the telencephalic 5-HT and pain threshold can be promptly restored by i.p. administration of 75 mg/kg of 5-HTP. With the aid of the hot plate procedure it has furthermore been established that the decrease of brain 5-HT caused by a lesion in the nucleus raphé medianus is accompanied by a reduction in morphine analgesia. Intraventricular injection of 5-HT enhanced morphine analgesia (*364*).

Electrical stimulation of the nucleus raphé medianus causes a release of 5-HT in the forebrain (*5, 162, 237*), in the general region where topical application of microamounts of 5-HT in unrestrained animals induces EEG synchronization and sleep (*263*). Hence activation of the 5-HT system by electrical stimulation of the median raphé can be expected to interfere with integration of pain sensation and to potentiate morphine analgesia. Indeed this idea receives support from analgesimetric studies on the rat (in which the hot plate technique and squeal response to electrical stimulation of the base of the tail were employed); thus, electrical stimulation of the nucleus raphé medianus markedly potentiates morphine analgesia, whereas pretreatment of the animals with PCPA abolishes the effect of raphé stimulation (*365*). The potentiating effect of raphé stimulation on morphine analgesia was observed after a latency of 30 min when the forebrain level of 5-HIAA was maximally increased. Electrical stimulation of the *lateral* dorsal raphé nuclei failed to increase the release of 5-HT or accumulation of 5-HIAA and did not enhance morphine analgesia. However, electrical stimulation of the dorsal raphé nuclei proper produces analgesia that can be blocked by pretreatment of the animals with PCPA (*281*).

Analgesimetric tests differ significantly in their potential to produce and measure modes of pain encountered in human disease. Clinical observations and evidence obtained from studies with experimental animals strongly indicate that the analgetic action of morphine can be attributed primarily to a dampening of the emotional, threat-motivated, "reactive" components of pain and the experience of "suffering," i.e., complex processes that involve polysensory thalamocortical and limbic systems. In other words, under the influence of the narcotic agent, subjective interpretation of actual or perceived injury to the self becomes dissociated from affective contents, whereas sensory thresholds for the detection of pain remain unaltered (*35, 81*). This unique action is probably best illustrated by many anecdotal reports from patients receiving morphine who often state: "My pain is the same, but it doesn't hurt me anymore" (*35*). Thus in animal experiments, the analgesimetric procedures based on responses that reflect to a high degree *affective* involvement, such as cry of pain (squeak or squeal responses) on electrical stimulation of a peripheral sensory nerve, appear more suitable than those that involve a tail-flick response to radiant heat or a hot plate. Results from recent work indicate that a lesion in the raphé nucleus of rats attenuates the "squeal-threshold elevating" effect of morphine, suggesting that morphine analgesia is produced by *activation of the central 5-HT system (366)*. Such an interpretation of morphine action is supported by the previously discussed experiments showing that electrical stimulation of the raphé system per se can produce analgesia and enhance the effect of morphine, and that administration of 5-HTP counteracts the effect of 5-HT deficit caused by PCPA or a raphé lesion.

More recent, unpublished observations (*441*) show that raphé-lesioned rats in which the brain 5-HT content is reduced by 77% exhibit a markedly elevated squeal threshold and a diminished analgetic responsiveness to morphine. However, the lesioned animals did not differ from controls in pain threshold or morphine analgesia as determined by the tail-flick procedure. Although these data confirm that in raphé-lesioned rats the squeal threshold elevating effect of morphine is reduced (*366*), they fail to support the concept that morphine analgesia involves stimulation of the serotoninergic raphé system. On the contrary, these data favor the notion that the elevation of squeal thresholds, as well as the interaction between lesions and morphine, is attributable to the disruption of the serotoninergic punishment system. For morphine this is achieved by a blockade of synaptic serotonin receptors; and with raphé lesions by destruction of the presynaptic portion of the pathway. Both groups of investigators found that chronic deficits of brain NE and DA (produced by the intranventricular injection of 6-OHDA) were accompanied by lower squeal thresholds and enhanced analgetic action of morphine. Hence, the data are consistent with the "reward-punishment" concept as outlined by Stein (*393*) and by others.

Several observations support the concept that morphine blocks brain serotoninergic receptors. Gaddum and Vogt (*151*) found that morphine and 5-HT act antagonistically when administered into the feline lateral ventricle: 5-HT normally produces marked sedation; after morphine the same dose induces an excitatory state. In rats, analgesia produced by intraventricular injection of morphine is abolished when morphine is administered together with 5-HT (*421*).

An excess of free 5-HT produced by the administration of the MAO inhibitor, harmaline, or large doses of 5-HTP increase the pain threshold in rats as measured by the vocalization paradigm, whereas small doses of 5-HTP markedly lower the pain threshold. Methysergide, a 5-HT receptor antagonist, has an effect similar to that produced by an excess of free

5-HT. Hence, it is proposed that the analgetic action of morphine partially depends on its blocking action on central 5-HT receptors, a concept that is supported by studies of peripheral 5-HT receptors. Using the isolated guinea-pig ileum in which 5-HT is believed to initiate peristaltic reflexes (*150, 269*), several investigators found that 5-HT antagonism in a large group of central analgesic drugs parallels almost perfectly their analgetic potency (*170, 233, 287*).

In man, chemical or electrical stimulation of presumptive reward substrates, i.e., those brain regions that support self-stimulation in the rat, produces intense feelings of pleasure and a total obliteration of severe pain (*179, 180*). Induction of pleasurable feelings by external stimuli—or even discussion of pleasurable experiences—causes high-voltage discharges in the same brain areas. Moreover, systemic administration of euphoriant drugs, such as opium alkaloids, is also associated with EEG discharges in reward substrates (*180*). These observations are consistent with those in unanesthetized monkeys (*Macaca speciosa*) which show bursts of single unit activity in discrete loci of the anterior and lateral hypothalamus after systemic administration of 5 to 10 mg/kg of morphine sulfate. These effects of morphine are accompanied by a behavioral state of inebriation. Repeated administration of morphine for days and weeks is less effective in producing neuronal discharges so that increasing doses are required to induce a given effect, a phenomenon that can be related to the development of tolerance (*123*).

Iontophoretic administration of 5-HT to neurons in the hypothalamus, amygdala and septum, i.e., areas presumed to contain reward substrates, causes marked depression of spontaneous and glutamate-induced activity. Moreover these inhibitory responses are mimicked by electrical stimulation of the median raphé nucleus, indicating that inhibition is the primary physiological role of serotoninergic terminals in these structures (*46*). Hence, it can be assumed that morphine removes tonic inhibitory or moderating influences impinging on reward substrates by blocking the 5-HT receptors. This interpretation is in harmony with the concept of reciprocal antagonism between the adrenergic reward system and the serotoninergic punishment system (*338, 393*), and with the postulated homeostatic or buffering role of the 5-HT system in affective processes (*218*). (See also Chs. 5 and 6.)

Physical Dependence on Morphine

It is proposed that 5-HT plays a part in the genesis of morphine tolerance and physical dependence. The three-dimensional molecular models of morphine and 5-HT exhibit structural similarities (*421*) sufficiently close as to suggest the possibility of competitive interactions at receptor sites. The administration of 5-HT or its precursors, TR and 5-HTP, attenuates certain signs of the abstinence syndrome (*192*). Nonetheless certain morphine withdrawal symptoms are similar to those produced by the central action of 5-HTP (*92*). The turnover of brain 5-HT in tolerant mice is more than doubled when compared to nontolerant animals (*421*). The increased rate of 5-HT turnover during chronic administration of morphine and the development of tolerance and physical dependence can be prevented by simultaneous administration of cycloheximide, an inhibitor of protein synthesis (*249*). Moreover, in mice the concurrent administration of morphine and naloxone, a specific antagonist of morphine, prevents the acceleration of 5-HT turnover as well as the development of tolerance and dependence. Approximately two weeks after withdrawal, the elevated brain 5-HT turnover in tolerant animals returns to a normal level (*277, 380*). Although they succeeded in producing a withdrawal syndrome in mice (piloerection, micturition, diarrhea, and characteristic jumping), other workers could not confirm an increased rate of brain 5-HT synthesis (*277*). Furthermore, no change in 5-HT turnover is detectable in animals rendered tolerant to and dependent on morphine (*85*). For a detailed discussion of the possible sources of discrepancies, such as the different methods of measuring the 5-HT turnover and their pitfalls, the reader is referred to the reviews by Hitzemann et al. (*185*) and by Cheney and Costa (*84*).

If the 5-HT system plays a crucial role in the development of morphine tolerance and dependence, then PCPA, an inhibitor of 5-HT synthesis, should greatly reduce these processes. Indeed this appears to be the case (*421*). The incomplete blockade of morphine tolerance and dependence by PCPA can be tentatively explained by the possible involvement of CA which also shows an increased turnover rate (*386*), and by the fact that PCPA does not completely reduce brain 5-HT levels to zero.

The bulk of the evidence regarding the effect of single doses of morphine on brain 5-HT steady-state levels in a variety of animal species indicates that no significant change can be obtained even with a dose as high as 300 mg/kg (*283, 441*). This is also true for repeated administration of morphine in doses of 40 to 70 mg/kg for as long as 45 days (*85, 89, 283, 305*). Apparently it is much more meaningful to evaluate the rate of 5-HT turnover than it is to rely on measurements of steady-state levels.

Biochemical Teratology and the 5-HT Neuronal System

Pharmacological manipulation of neuronal enzyme systems in the developing fetus may cause permanent discrete changes in neuronal systems as well as behavioral alterations without gross anatomical malformations (*164*). Research in this area is of great importance because of clinical and social implications.

In rats, permanent behavioral changes have been demonstrated in pups of dams treated with 5-HTP i.p., 100 mg/kg or 50 mg/kg daily from the eighth through the 14th day of gestation. The weight of the offspring was less than that of the control litters, the differences becoming more apparent as the animals grew older. No deficit was observed in maturation of motor and learning ability as tested by maze performance and conditioned avoidance learning. However, a significant increase in "emotionality" was observed as evidenced by the open field behavior. Furthermore, there was a much greater susceptibility to audiogenic seizures and a markedly increased level of activity on the "inclined plane." Unfortunately, neither levels nor turnover of brain 5-HT were determined in this study.

Phencyclidine is a hallucinogenic drug known to increase brain 5-HT in mature rats on acute administration (*409*). The 5-HT and 5-HIAA concentrations in various rat brain areas of offspring the dams of which had been chronically treated with phencyclidine during gestation and suckling have been studied (*408*). The offspring, sacrificed nine months after weaning, showed highly significant increases in 5-HT concentrations in the striatum, thalamus, midbrain, pons-medulla, and area of the nucleus amygdalae. The brain 5-HIAA was elevated in the hypothalamus and pons-medulla, indicating an enhanced 5-HT turnover. Unfortunately, no parallel behavioral study has been conducted.

PATHONEUROBIOLOGY

Schizophrenia

Almost twenty years ago, Harley-Mason suggested that aberrant transmethylation could give rise to the formation of psychotogenic CA derivatives. Meanwhile, an analogous theory implicating methylated indoleamines is gaining ground (*280*).

An early report indicated that paper chromatograms of urines from unmedicated schizophrenic patients contained an *N*-methylated 5-HT derivative, namely, a bufotenine-like substance, in contradistinction to negative findings in nonschizophrenic patients (*136*). These results have been confirmed (*64*) and extended by other investigators (*137*). Excretion of a relatively large amount of bufotenine is detectable in 75% of untreated acute schizophrenics, but only in 37% of chronic or medicated schizophrenics was the compound detectable. No bufotenine was detected in a large population of nonschizophrenic patients. According to more recent work, nonschizophrenic patients excrete small amounts of bufotenine (approximately 2.9 µg/100 ml of urine), whereas untreated acute schizophrenics excrete 15.5 µg/100 ml (*384*). Changes in diet have no significant effect on the urinary level of bufotenine.

The administration of a MAO inhibitor, tranylcypromine, alone or with L-cystein as a methyl donor, increases urinary excretion of three psychotomimetic compounds, including *N*-dimethyltryptamine, bufotenine, and 5-methoxy-*N*-dimethyltryptamine. These substances appear in higher concentration shortly before or during exacerbation of schizophrenic symptoms, and their concentration declines in parallel with amelioration of the clinical picture. The changes in thought content, characteristic of schizophrenia, correlated closely with variations in the level of the three *N*-dimethyltryptamine derivatives (*183*). A similar treatment of normal controls with tranylcypromine and L-cystein failed to produce any psychotomimetic effect. Although both the schizophrenic patient and normal subjects showed rises of tryptamine during the drug administration, it is noteworthy that only the former excreted the three derivatives of *N*-dimethyltryptamine. Hence, it appears that normal subjects cannot produce and accumulate the potentially hallucinogenic compounds, despite the pharmacological treatment aimed at enhancing the *N*-methylation processes by specific enzymes present in the human brain and peripheral organs (*301, 361*). Administration of other methyl donors, such as methionine and betain, also exacerbate the schizophrenic syndrome (*63*) in a manner similar to that observed following administration of L-cystein.

In combination with a MAO inhibitor, TR also markedly aggravates psychotic symptoms (*336*). The normally insignificant *N*-methylation processes are believed to yield a substantial amount of bufotenine and dimethyltryptamine if the normal oxidative degradation of indolealkylamines is inhibited by a MAO inhibitor (*256*). In this connection, the indolethylamine-*N*-methyltransferase in serum of normal controls or nonschizophrenic patients is not detectable when 5-HT or 5-methoxytryptamine is used as a substrate. However, this enzyme is clearly detectable in both chronic and acute schizophrenics (*315*). All of these findings, coupled with data indicating that schizophrenic patients may have an additional enzymatic abnormality that permits excess TR to enter the brain (*142*), support the contention that enhanced *N*-methylation of indoleamines may play a causative role in the pathogenesis of schizophrenia. (See section on "Systemic Effects" under "Functional Considerations" in this chapter.)

Finally, it should be recalled that the N-dimethylation distorts the configuration of the indolethylamine molecule in a manner such that it can only partially enter the 5-HT receptor site, thus impairing the normal function of the 5-HT system (see section on "Sleep Mechanism and Raphé Neuron System" in this chapter). Moreover, methysergide, one of the most potent and specific 5-HT antagonists (*117, 186*), when administered to autistic and schizophrenic children, markedly aggravates the syndrome (*138*). All of these observations are in agreement with the hypothesis of Woolley and Shaw (*431*) and of Dement and his colleagues (*114*) (see section on "Precursors" under "Functional Considerations" in this chapter) suggesting that the 5-HT deficiency is partially responsible for schizophrenic symptoms.

Affective Illness

Several observations support the contention that the 5-HT system is involved in the development of mental depression. Depressed patients show a decreased formation of 5-HT, and administration of TR potentiates the antidepressant effect of MAO inhibitors (*100*). Furthermore, depressed patients who commit suicide have significantly lower levels of 5-HT and 5-HIAA in the brain (*376*) and a lower concentration of 5-HT metabolites in the CSF than have normal subjects (*342*).

In patients who complain of depression, the concentration of 5-HIAA in CSF is significantly lowered, and its fluctuation parallels the severity of depression. Moreover, remissions are accompanied by a rise in 5-HIAA (*19*). These data have been confirmed by investigators who also reported that, despite the alterations in the 5-HIAA levels, there were no changes in the content of homovanillic acid (HVA) (*115*), suggesting that there is no major involvement of the dopaminergic system. Probenecid (p-[di-N-propylsulfamyl]-benzoic acid), which inhibits the active transport of organic acids from CSF to bloodstream, has been used to assess cerebral 5-HT metabolism mirrored by CSF levels of 5-HIAA (*340*). There appears to be a diminished accumulation of 5-HIAA in CSF, in the presence of probenecid, in patients suffering from endogenous depression when compared to a nondepressive control group. Similar results have been reported by other investigators (*356*).

More recent studies in which a larger population of patients received i.v. injections of probenecid demonstrate that the apparently heterogeneous depressive group can be subdivided into patients with a normal and a subnormal accumulation of 5-HIAA in the CSF, indicating that endogenous depression encompasses two biochemically different categories, i.e., with and without disturbances in the metabolism of brain 5-HT (*340*). In the latter, the deficiency in the serotoninergic system may be caused by decreased activity of TR 5-hydroxylase (*50*). This assumption is supported by the observation that 5-HTP is effective in the treatment of this type of depressive patients (*341*).

Antidepressant tricyclic amines, such as imipramine and chlorimipramine, selectively block 5-HT uptake by brain tissue (*70, 74*). This may result in enhanced stimulation of 5-HT receptors by a relatively smaller amount of 5-HT released from nerve terminals.

The discovery of at least four molecular forms of MAO in different areas of the human brain offers a challenge for the restitution of a deficient 5-HT system (*93*). These enzymes show distinct substrate specificity (*93*), and their blockade by specific MAO inhibitors may permit selective manipulation of a particular monoamine level (*143*), a finding that may have far-reaching clinical implications.

In addition to the aforementioned approaches, attempts are being made to develop drugs that will easily cross the blood-brain barrier and selectively stimulate 5-HT receptors. A smooth muscle stimulant, quipazine (2-[1-piperazinyl]-quinoline) with a spectrum of 5-HT-like effects that can be blocked by methysergide, BOL-148, or cyproheptadine, exhibits a typical "antidepressant profile" of actions in animal tests. Quipazine is comparable to tricyclic antidepressants in several actions: It antagonizes reserpine or tetrabenazine-induced sedation, reverses reserpine hypothermia, and inhibits the mouse-killing behavior of rats. However, unlike imipramine, quipazine counteracts tetrabenazine-induced sedation even in catecholamine-depleted animals, suggesting that its action does not involve an adrenergic mechanism.

Investigations designed to examine the correlation between TR turnover and the manic-depressive syndrome revealed a significant increase in conversion of TR to kynurenine during a depressive phase, associated with a rise in 17-hydroxycorticosteroids. Alternative metabolic pathways of TR, reflected by the levels of 5-HIAA, indolylacroylglycine, and N-acetyltryptophan, were unaffected (*260*). A similar pattern of enhanced TR-kynurenine conversion is produced by the administration of ACTH. This has prompted the suggestion that physical or mental stress, which causes a release of hydrocortisone, may greatly influence the metabolic fate of TR and other monoamines and may perhaps explain the intermittency of "attacks" of mental disturbances precipitated by various forms of life stress (*260*).

A recent proposal concerning the role of the 5-HT system is compatible with most of the clinical data as well as with biochemical, electrophysiological, and behavioral results obtained in animals (*218*). According to this hypothesis, 5-HT neurons fulfill the function of a homeostatic or buffering system responsible for dampening the swings in or stabilizing the "tonus" of the noradrenergic system. Central 5-HT deficiency would thus permit either mania or depression triggered by increased or decreased activity of the adrenergic system, respectively. This hypothesis is supported by several observations and reconciles seemingly contradictory findings: 1) Low CSF levels of 5-HIAA occur both in mania and in depression (*98*). 2) Methysergide worsens mania (*99*). 3) TR improves the clinical picture of patients with affective disorders (*65, 343*).

Furthermore, the homeostatic theory of a 5-HT role explains numerous observations derived from various experimental approaches that have already been discussed and which indicate that a serotonin deficiency is associated with hyperresponsiveness to a wide range of environmental stimuli: 1) Serotonin-depleted animals show a considerably lower threshold to painful stimuli as well as to audiovisual, gustatory, olfactory, and (particularly) social stimuli. The last-mentioned is manifested by sexual hyperreactivity and aggressiveness. 2) Insomnia or hyposomnia in 5-HT-depleted animals may be regarded as one of the many consequences of the removal of the buffering 5-HT system. 3) The pharmacological studies of the EEG correlates of positive reinforcement in cats trained to press a lever for milk reward also support this hypothesis. In well-trained cats, a lever press, resulting in a delivery of 0.5 to 1 cc of milk reward, triggers a burst of high-amplitude, 6 to 9 c/sec, alpha-like activity over the striate and parastriate cortex, the so-called postreinforcement synchronization (PRS) (*87, 268*). This phenomenon is associated with a large epicortical positive steady-potential shift, termed "Reward Contingent Positive Variation" (RCPV) (*273, 274*). Since these phenomena in the alert cat are indistinguishable from those observed during the onset of sleep, one might expect that administration of 5-HT precursors, such as 5-HTP or TR, would enhance the PRS-RCPV responses. These drugs not only fail to enhance the PRS-RCPV phenomena, they actually suppress them, indicating that the effect of biological reward is reduced or even totally "buffered" (Marczynski, unpublished data). On the other hand, shortly after a single dose, α-methylmetatyrosine potentiates the effect of reward, presumably by increasing the availability of NE by displacing it from presynaptic stores. Subsequently there is a total suppression of the PRS responses, presumably reflecting depletion of NE and its reduced synaptic availability (*357*). In this connection, it should be mentioned that the concept of an adrenergic mechanism in self-stimulation behavior (*337, 338*), and the facilitation and suppression of this behavior by selectively lowering or elevating brain 5-HT, respectively (*393*), can be regarded as compatible with the aforementioned hypothesis (*218*).

LITERATURE CITED

1. Aghajanian, G. K.: Influence of drugs on the firing of serotonin-containing neurons in brain. *Fed. Proc. 31*(1):91–96, 1972.
2. Aghajanian, G. K., Foote, W. E., and Sheard, M. H.: Action of psychotogenic drugs on single midbrain raphé neurons. *J. Pharmacol. Exp. Ther. 171*:178–187, 1970.
3. Aghajanian, G. K., Graham, A. W., and Sheard, M. H.: Serotonin-containing neurons in brain: Depression of firing by monoamine oxidase inhibitors. *Science 169*:1100–1102, 1970.
4. Aghajanian, G. K., Kuhar, M. J., and Roth, R. H.: Serotonin-containing neuronal perikarya and terminals: differential effects of p-chlorophenylalanine. *Brain Res. 54*:85–101, 1973.
5. Aghajanian, G. K., Rosencrans, J. A., and Sheard, M. H.: Serotonin: Release in the forebrain by stimulation of midbrain raphé. *Science 156*:402–403, 1967.
6. Ahtee, L. and Eriksson, K.: 5-Hydroxytryptamine and 5-hydroxyindoleacetic acid content in brain of rat strains selected for their alcohol intake. *Physiol. Behav. 8*:123–126, 1972.
7. Alexander, G. J. and Kopeloff, L. M.: Metrazol seizures in rats: Effect of p-chlorophenylalanine. *Brain Res. 22*:231–235, 1970.
8. Amin, A. H., Crawford, T. B. B., and Gaddum, J. H.: The distribution of substance P and 5-hydroxytryptamine in the central nervous system of the dog. *J. Physiol.* (Lond.) *126*:596–618, 1954.
9. Andén, N.-E., Carlsson, A., Hillarp, N.-Å., et al. 5-Hydroxytryptamine release by nerve stimulation of the spinal cord. *Life Sci. 3*:473–478, 1964.
10. Andén, N.-E., Fuxe, K., and Hökfelt, T.: The importance of the nervous impulse flow for the depletion of the monoamines from central neurons by some drugs. *J. Pharm. Pharmacol. 18*:630–632, 1966.
11. Anderson, E. G.: Bulbospinal serotonin-containing neurons and motor control. *Fed. Proc. 31*(1):107–112, 1972.
12. Angelucci, L.: Experiments with perfused frog's spinal cord. *Br. J. Pharmacol. 11*:1761–170, 1956.
13. Anton-Tay, F.: Pineal-brain relationships. In Wolstenholme, G. E. W. and Knight, J. (Eds.): *The Pineal Gland*. Edinburgh: Churchill Livingstone, 1971, pp. 213–227.
14. Anton-Tay, F., Chou, Ch., Anton, S., et al.: Brain serotonin concentration: Elevation following intraperitoneal administration of melatonin. *Science 162*:277–278, 1968.

15. Anton-Tay, F., Diaz, J. L., and Fernandez-Guardiola, A.: On the effect of melatonin upon human brain. Its possible therapeutic implications. *Life Sci. 10* (1):841–850, 1971.
16. Anton-Tay, F., Sepulveda, J., and Gonzales, S.: Increase of brain pyridoxal phosphokinase activity following melatonin administration. *Life Sci. 9* (2):1283–1288, 1970.
17. Anton-Tay, F. and Wurtman, R. J.: Regional uptake of ^{3}H-melatonin from blood or cerebrospinal fluid by rat brain. *Nature 221*: 474–475, 1969.
18. Aprison, M. H., Kariya, T., Hingtgen, J. N., et al.: Neurochemical correlates of behavior, changes in acetylcholine, norepinephrine and 5-hydroxytryptamine concentrations in several discrete brain areas of the rat during behavioral excitation. *J. Neurochem. 15*: 1131–1139, 1968.
19. Ashcroft, G. W., Crawford, T. B. B., Eccleston, D., et al.: 5-Hydroxyindole compounds in the cerebrospinal fluid of patients with psychiatric or neurological disease. *Lancet 2*:1049–1052, 1966.
20. Axelrod, J. The mode of action of psychotomimetic drugs. In: Metabolic Fate of the Drugs. *Neurosci. Res. Program Bull. 8*(1):16–21, 1970.
21. Axelrod, J.: Neural control of indoleamine metabolism in the pineal. In Wolstenholme, G. E. W. and Knight, J. (Eds.): *The Pineal Gland.* Edinburgh: Churchill Livingstone, 1971, pp. 35–52.
22. Axelrod, J.: The enzymatic N-methylation of serotonin and other amines. *J. Pharmacol. Exp. Ther. 138*:28–33, 1962.
23. Axelrod, J., MacLean, P. D., Albers, R. W., et al.: *Regional Neurochemistry.* Oxford: Pergamon Press, 1961, p. 307.
24. Axelrod, J., Shein, H. M., and Wurtman, R. J.: Stimulation of C^{14}-melatonin synthesis from C^{14}-tryptophan by noradrenaline in rat pineal in organ culture. *Proc. Nat. Acad. Sci. U.S.A. 62*: 544–549, 1969.
25. Azmitia, E. C., Algeri, S., and Costa, E.: *In vivo* conversion of ^{3}H-L-tryptophan into ^{3}H-serotonin in brain areas of adrenalectomized rats. *Science 169*:201–203, 1970.
26. Bachtold, H. and Pletscher, A.: Einfluss von Isonikotinsäurehydraziden auf den Verlauf der Körpertemperatur nach Reserpin, Monoaminen und Chlorpromazin. *Experientia 13*:163–165, 1957.
27. Baldessarini, R. J.: Release of aromatic amines from brain tissue of the rat in vitro. *J. Neurochem. 18*:2509–2518, 1971.
28. Barchas, J. D.: Relation of melatonin to sleep. *Proc. West. Pharmacol. Soc. 11*:22–23, 1968.
29. Barchas, J. and Usdin, E. (Eds.): *Serotonin and Behavior.* New York: Academic Press, 1972.
30. Barlow, J. S. and Ciganek, L.: Lambda responses in relation to visual evoked responses in man. *Electroencephalogr. Clin. Neurophysiol. 26*:183–192, 1969.
31. Bartholini, G., Da Prada, M., and Pletscher, A.: Decrease of cerebral 5-hydroxytryptamine by 3,4-dihydroxyphenylalanine after inhibi-

tion of extracerebral decarboxylase. *J. Pharm Pharmacol.* 20: 228—229, 1968.
32. Baumgarten, H. G., Björklund, A., Holstein, A. F., et al.: Chemical degeneration of indoleamine axons in rat brain by 5,6-dihydroxytryptamine, an ultrastructural study. *Z. Zellforsch. 129*:256—271, 1972.
33. Baumgarten, H. G., Björklund, A., Lachenmayer, L., et al.: Long-lasting selective depletion of brain serotonin by 5,6-dihydroxytryptamine. *Acta Physiol. Scand.* (Suppl.) 373:1—15, 1971.
34. Baumgarten, H. G., Lachenmayer, L., and Schlossenberger, H. G.: Evidence for a degeneration of indole-containing nerve terminals in rat brain induced by 5,6-dihydroxytryptamine. *Z. Zellforsch. 125*: 553—569, 1972.
35. Beecher, H. K.: The measurement of pain. *Pharm. Rev. 9*:59—210, 1957.
36. Beleslin, D. B. and Myers, R. D.: The release of acetylcholine and 5-hydroxytryptamine from the mesencephalon of the unanesthetized rhesus monkey. *Brain Res. 23*:437—442, 1970.
36a. Bert, J.: Action de la p-chlorophenylalanine sur le sommeil du babuin Papio papio. *Electroencephalogr. Clin. Neurophysiol. 33*:99—103, 1972.
37. Bizzi, E.: Changes in the orthodromic and antidromic responses of optic tract during the eye movements of sleep. *J. Neurophysiol. 29*:861—870, 1966.
38. Björklund, A., Falck, B., and Stenevi, U.: Microfluorometric characterization of monoamines in the central nervous system: Evidence for a new neuronal monoamine-like compound. *Prog. Brain Res. 34*:49—59, 1971.
39. Björklund, A., Falck, B., and Stenevi, U.: Classification of monoamine neurones in the rat mesencephalon: Distribution of a new monoamine neurone system *Brain Res. 32*:269—285, 1971.
40. Björklund, A., Nobin, A., and Stenevi, U.: Regeneration of central serotonin neurons after axonal degeneration induced by 5,6-dihydroxytryptamine. *Brain Res. 50*:214—220, 1973.
41. Björklund, A., Nobin, A., and Stenevi, U.: Effect of 5,6-dihydroxytryptamine on nerve terminal serotonin and serotonin uptake in the rat brain. *Brain Res. 50*:214—220, 1973.
42. Blaschko, H. and Philpot, F. J.: Enzymatic oxydation of tryptamine derivatives. *J. Physiol.* (Lond.) *122*:403—408, 1953.
43. Blondaux, Ch., Juge, A., Sordet, F., et al.: Modification du metabolisme de la sérotonine (5-HT) cérébrale induite chez le rat par administration de 6-hydroxydopamine. *Brain Res. 50*:101—114, 1973.
44. Bloom, F. E.: Serotonin neurons: Localization and possible physiological role. *Adv. Biochem. Pharmacol. 1*:27—47, 1969.
45. Bloom F. E., Hoffer, B. J., Nelson, C., et al.: The physiology and pharmacology of serotonin mediated synapses. In Barchas, J. and Usdin, E. (Eds.): *Serotonin and Behavior*, New York: Academic Press, 1973, pp. 249—266.

46. Bloom F. E., Hoffer, B. J., Siggins, G. R., et al.: Effects of serotonin on central neurons: Microiontophoretic administration. *Fed. Proc.* *31*(1):97–106, 1972.
47. Boakes, R. J., Bradley, P. B., Briggs, I., et al.: Antagonism of 5-hydroxytryptamine by LSD-25 in the central nervous system: A possible neuronal basis for the actions of LSD-25. *Br. J. Pharmacol.* *40*:202–218, 1970.
48. Bogdanski, D. F., Weissbach, H., and Udenfriend, S.: The distribution of serotonin, 5-hydroxytryptophan decarboxylase, and monoamine oxidase in brain. *J. Neurochem. 1*:272–278, 1957.
49. Bonvallet, M. and Bloch, V.: Bulbar control of cortical arousal. *Science 133*:1133–1134, 1961.
50. Bowers, M. B.: Cerebrospinal fluid 5-hydroxyindoles and behavior after L-tryptophan and pyridoxine administration to psychiatric patients. *Neuropharmacology 9*:599–604, 1970.
51. Bradley, P. B.: The effect of 5-hydroxytryptamine on the electrical activity of the brain and on behavior in the conscious cat. In Lewis, G. P. (Ed.): *5-Hydroxytryptamine.* London: Pergamon Press, 1958.
52. Breese, G. R. and Taylor, T. D.: Effect of 6-hydroxydopamine on brain norepinephrine and dopamine: Evidence for selective degeneration of catecholamine neurons. *J. Pharmacol. Exp. Ther. 174*: 413–420, 1970.
53. Brodal, A., Taber, E., and Walberg, F.: The raphé nuclei of the brain stem in the cat. II. Efferent connections. *J. Comp. Neurol. 114*: 239–259, 1960a.
54. Brodal, A., Walberg, F., and Taber, E.: The raphé nuclei of the brain stem in the cat. III. Afferent connections. *J. Comp. Neurol. 114*: 261–281, 1960b.
55. Brodie, B. B., Comer, M. S., Costa, E., et al.: The role of brain serotonin in the mechanisms of central action of reserpine. *J. Pharmacol. Exp. Ther. 152*:340–350, 1966.
56. Brodie, B. B. and Shore, P. A.: A concept for the role of serotonin and norepinephrine as chemical mediators in the brain. *Ann. N.Y. Acad. Sci. 66*:631–642, 1957.
57. Brody, J. F., Jr.: Behavioral effects of serotonin depletion and of p-chlorophenylalanine (a serotonin depletor) in rats. *Psychopharmacologia 17*:14–33, 1970.
58. Bronzino, J. D., Morgane, P. J., and Stern, W. C.: EEG synchronization following application of serotonin to area postrema. *Am. J. Physiol. 223*:376–383, 1972.
59. Brooks, D. C.: Waves associated with eye movements in the awake and sleeping cat. *Electroencephalogr. Clin. Neurophysiol. 24*: 532–541, 1968.
60. Brooks, D. C.: Localization and characteristics of the cortical waves associated with eye movement in the cat. *Exp. Neurol. 22*: 603–613, 1968.
61. Brooks, D. C. and Bizzi, E.: Brain stem electrical activity during deep sleep. *Arch. Ital. Biol. 101*:648–666, 1963.
62. Brooks, D. C. and Gershon, M. D.: Eye movement potentials in the

oculomotor and visual systems of the cat: A comparison of reserpine induced waves with those present during wakefulness and rapid eye movement sleep. *Brain Res. 27*:223—239, 1971.
63. Brune, G. G. and Himwich, H. E.: Effects of methionine loading on the behavior of schizophrenic patients. *J. Nerv. Ment. Dis. 134*: 447—450, 1962.
64. Brune, G. G., Hohl, H. H., and Himwich, H. E.: Urinary excretion of bufotenin-like substance in psychotic patients. *J. Neuropsychol. 5*:14, 1963.
65. Bunney, W. E., Brodie, K. H., Murphy, D. L., et al.: Studies of alpha-methyl-paratyrosine, L-dopa, and L-tryptophan in depression and mania. *Am. J. Psychiatr. 127*:872—881, 1971.
66. Butcher, L., Engel, J., and Fuxe, K.: L-Dopa induced changes in central monoamine neurons after peripheral decarboxylase inhibition. *J. Pharm. Pharmacol. 22*:313—316, 1970.
67. Caldwell, D. and Domino, E.: Electroencephalographic and eye movement patterns during sleep in chronic schizophrenic patients. *Electroencephalogr. Clin. Neurophysiol. 22*:414—420, 1967.
68. Canal, N. and Ornesi, A.: Serotonina encefalica e ipertermia da vaccino. *Atti Acad. Med. Lomb. 16*:69—73, 1961.
69. Carlsson, A.: Structural specificity for inhibition of ^{14}C-5-hydroxytryptamine uptake by cerebral slices. *J. Pharm. Pharmacol. 22*: 729—732, 1970.
70. Carlsson, A., Corrodi, H., Fuxe, K., et al.: Effect of antidepressant drugs on the depletion of intraneuronal brain 5-hydroxytryptamine stores caused by 4-methyl-α-ethyl-metatyramine. *Eur. J. Pharmacol. 5*:357—361, 1969.
71. Carlsson, A., Corrodi, H., and Waldeck, B.: α-Substituierte Dopacetamide als Hemmer der Catechol-O-Methyltransferase und der enzymatischen Hydroxylierung aromatischer Aminosäuren. In den Catecholamin-Metabolismus eingreifende Substanzen. *Helv. Chim. Acta 46*:2270—2285, 1963.
72. Carlsson, A., Falck, B., Fuxe, K., et al.: Cellular localization of monoamines in the spinal cord. *Acta Physiol. Scand. 60*:112—119, 1964.
73. Carlsson, A., Hillarp, N.-Å., and Waldeck, B.: Analysis of the Mg^{++}-ATP dependent mechanism in the amine granules of the adrenal medulla. *Acta Physiol. Scand. 59*(Suppl. 215):1—38, 1963.
74. Carlsson, A., Jonason, J., and Lindqvist, M.: On the mechanism of 5-hydroxytryptamine release by thymoleptics. *J. Pharm Pharmacol. 21*:769—773, 1969.
75. Carlsson, A., Kehr, W., Lindqvist, M., et al.: Regulation of monoamine metabolism in the central nervous system. *Pharmacol. Rev. 24*:371—384, 1972.
76. Carlsson, A. and Lindqvist, M.: In vivo measurements of tryptophan and tyrosine hydroxylase activities in mouse brain. *J. Neural Transm. 34*:79—91, 1973.
77. Carlsson, A. and Lindqvist, M.: The effect of L-tryptophan and some

psychotropic drugs on the formation of 5-hydroxytryptophan in the mouse brain in vivo. *J. Neural Trans. 33*:23–43, 1972.
78. Carlsson, A., Rosengren, E., Bertler, Å., et al.: Effect of reserpine on the metabolism of catecholamines. In Garattini, S. and Ghetti, V. (Eds.): *Psychotropic Drugs.* Amsterdam: Elsevier, 1957, pp. 363–372.
79. Carrer, H. F. and Taleisnik, S.: Effect of mesencephalic stimulation on the release of gonadotropins. *J. Endocrinol. 48*:527–533, 1970.
80. Cazzullo, C. L., Mangoni, A., and Mascherpa, G.: Tryptophan metabolism in affective psychoses. *Br. J. Psychiatry 112*:157–162, 1966.
81. Chapman, L. F., Dingman, H. F., and Ginzberg, S. P.: Failure of systemic analgesic agents to alter the absolute sensory threshold for simple detection of pain. *Brain 88*:1011–1022, 1965.
82. Chase, T. N. and Murphy, D. L.: Serotonin and central nervous system function. *Ann. Rev. Pharmacol. 13*:181–197, 1973.
83. Chen, G., Ensor, C. R., and Bohner, B.: A facilitation action of reserpine on the central nervous system *Proc. Soc. Exp. Biol. Med. 86*:507–510, 1954.
84. Cheney, D. L. and Costa, E.: Narcotic tolerance and dependence and serotonin turnover. *Science 178*:647, 1972.
85. Cheney, D. L., Goldstein, A., Algeri, S., et al.: Narcotic tolerance and dependence: Lack of relationship with serotonin turnover in the brain. *Science 171*:1169–1170, 1971.
86. Cicero, T. J. and Hill, S. Y.: Ethanol self-selection in rats: A distinction between absolute and 95% ethanol. *Physiol. Behav. 5*: 787–791, 1970.
87. Clemente, D. C., Sterman, M. B., and Wyrwicka, W.: Post-reinforcement EEG synchronization during alimentary behavior. *Electroencephalogr. Clin. Neurophysiol. 16*:355–365, 1964.
88. Clouet, D. H. and Ratner, M.: Catecholamine biosynthesis in brains of rats treated with morphine. *Science 168*:854–855, 1970.
89. Cochin, J. and Axelrod, J.: Biochemical and pharmacological changes in the rat following chronic administration of morphine, nalorphine, and normorphine. *J. Pharmacol. Exp. Ther. 125*:105–110, 1959.
90. Cohen, B. and Feldman, M.: Relationship of electrical activity in pontine reticular formation and lateral geniculate body to rapid eye movements. *J. Neurophysiol. 31*:806–817, 1968.
91. Cohen, D. L. and Sladek, J. R.: Evidence for the limited spread of intracerebrally applied serotonin. *Brain Res. 45*:630–634, 1972.
92. Collier, H. O. J.: A general theory of the genesis of drug dependence by induction of receptors. *Nature 205*:181–182, 1965.
93. Collins, G. G. S., Sandler, M., Williams, E. D., et al.: Multiple forms of human brain mitochondrial monoamine oxidase. *Nature 225*: 817–820, 1970.
94. Collu, R., Faschini, F., and Visconti, P.: Adrenergic and serotoninergic control of growth hormone secretion in adult male rats. *Endocrinology 90*:1231–1237, 1972.

95. Conner, R. L., Stolk, J. M., Barchas, J. D., et al.: The effect of parachlorophenylalanine (PCPA) on shock-induced fighting behavior in rats. *Physiol Behav.* 5:1221–1224, 1970.
96. Consolo, S., Garattini, S., Gielmetti, R., et al.: The hydroxylation of tryptophan in vivo by brain. *Life Sci.* 4:625–630, 1965.
97. Cooper, K. E., Cranston, W. I., and Honour, A. J.: Effects of intraventricular and intrahypothalamic injections of noradrenaline and 5-hydroxytryptamine on body temperature in conscious rabbits. *J. Physiol.* (Lond.) 181:852–864, 1965.
98. Coppen, A.: The biochemistry of affective disorders. *Br. J. Psychiatr.* 113:1237, 1967.
99. Coppen, A., Prange, A. J., Jr., Whybrow, P. C., et al.: Methysergide in mania, a controlled trial. *Lancet* 2:338, 1969.
100. Coppen, A., Shaw, D. M., Malleson, A., et al.: Tryptamine metabolism in depression. *Br. J. Psychiatry* 111:993–995, 1965.
101. Corrodi, H., Fuxe, K., and Hökfelt, T.: A possible role played by central monoamine neurons in thermoregulation. *Acta Physiol. Scand.* 71:224–232, 1967.
102. Costa, E.: The role of serotonin in neurobiology. *Neurobiology* 2:175–227, 1960.
103. Costa, E., Pscheidt, G. R., van Meter, W. G., et al.: Brain concentrations of biogenic amines and EEG patterns of rabbits. *J. Pharmacol. Exp. Ther.* 130:81–88, 1960.
104. Couch, J. R.: Responses of neurons in the raphé nuclei to serotonin, norepinephrine and acetylcholine and their correlation with an excitatory synaptic input. *Brain Res.* 19:137–150, 1970.
105. Dahlström, A. and Fuxe, K.: A method for the demonstration of monoamine-containing nerve fibres in the central nervous system.. *Acta Physiol. Scand.* 60:293–294, 1964.
106. Dahlström, A. and Fuxe, K.: Evidence for the existence of monoamine-containing neurons in the central nervous system. *Acta Physiol. Scand.* 62(Suppl. 232):1–55, 1965a.
107. Dahlström, A. and Fuxe, K.: Evidence for the existence of monoaminergic neurons in the cerebral nervous system. II. Experimentally induced changes in the intraneuronal amine levels of bulbospinal neuron systems. *Acta Physiol. Scand.* 64(Suppl. 247):7–36, 1965b.
108. Dahlström, A., Fuxe, K., Kernell, D., et al.: Reduction of the monoamine stores in the terminals of bulbospinal neurons following stimulation in the medulla oblongata. *Life Sci.* 4:1207–1212, 1965.
109. Deguchi, T. and Axelrod, J.: Control of circadian change of serotonin-N-acetyltransferase activity in the pineal organ by the beta-adrenergic receptor. *Proc. Nat. Acad. Sci.* U.S.A. 69:2547–2550, 1972.
110. Deguchi, T. and Barchas, J.: Effect of p-chlorophenylalanine on hydroxylation of tryptophan in pineal and brain of rats. *Mol. Pharmacol.* 8:770–779, 1972.
111. Delorme, F.: Monoamines et sommeils, étude polygraphique, neuro-

pharmacologique et histochimique des états de sommeil chez le chat. Thèse Université de Lyon, Imprimerie LMD, 1966, 168 pp.
112. Delorme, F., Froment, J. L., and Jouvet, M.: Suppression du sommeil par la p-chloromethamphetamine et la p-chlorophenylalanine. *C. R. Soc. Biol.* (Paris) *160*:2347–2351, 1966.
113. Dement, W. C.: The biological role of REM sleep. In Kales, A. (Ed.): *Sleep, Physiology and Pathology*. Philadelphia: Lippincott, 1969, pp. 245–265.
114. Dement, W. Zarcone, V., Ferguson, J., et al.: Some parallel findings in schizophrenic patients and serotonin-depleted cats. In Siva Sankar, D. V. (Ed.): *Schizophrenia, Current Concepts and Research*. Hicksville, N.Y.: PJD Publications, Ltd., 1969, p. 775.
115. Dencker, S. J., Malm, M., and Roos, B. E.: Acid monoamine metabolites of CSF in mental depression and mania. *J. Neurochem. 13*:1545–1548, 1966.
116. Dewsbury, D. A. and Davis, H. N.: Effects of reserpine on the copulatory behavior of male rats. *Physiol. Behav. 5*:1331–1333, 1970.
117. Doepfner, W. and Cerletti, A.: Comparison of lysergic acid derivatives and antihistamines as inhibitors of the edema provoked in the rat's paw by serotonin. *Int. Arch. Allergy Appl. Immunol. 12*:89–97, 1958.
118. Duhault, J. and Verdavainne, C.: Modification du taut de serotonine cérebrale chez le rat par le trifluoromethylphenyl-2-ethyl aminopropane (fenfluramine 768 S). *Arch. Intern. Pharmacodyn. 170*: 276–286, 1967.
119. Ebersole, J. S. and Galambos, R.: Modification of the cortical click-evoked response during eye movement in cats. *Electroencephalogr. Clin. Neurophysiol. 26*:273–279, 1969.
120. Eccleston, D. A., Padjen, A., and Randic, M.: Release of 5-hydroxytryptamine and 5-hydroxyindol-3-acetic acid in the forebrain by stimulation of midbrain raphé. *J. Physiol.* (Lond.) *201*:22–23, 1969.
121. Eccleston, D., Randic, M., Roberts, M. H. T., et al.: In Hooper, G. (Ed.): *Metabolism of Amines in the Brain*. London: MacMillan, 1968, p. 29.
122. Eccleston, D., Reading, W. H., and Ritchie, I. M.: 5-Hydroxytryptamine metabolism in brain and liver slices and the effects of ethanol. *J. Neurochem. 16*:274–276, 1969.
123. Eidelberg, E. and Barstow, C. A.: Effects of morphine on hypothalamic electrical activity in monkeys. In *Communications on Problems of Drug Dependence*. Washington, 1971, pp. 767–777.
124. Engberg, I., Lundberg, A., and Ryall, R. W.: Is the tonic decerebrate inhibition of reflex paths mediated by monoaminergic pathways? *Acta Physiol. Scand. 72*:123–133, 1968.
125. Engelman, K., Lovenberg, W., and Sjoerdsma, A.: Inhibition of serotonin synthesis by para-chlorophenylalanine in patients with carcinoid syndrome. *New Engl. J. Med. 277*:1103, 1967.

126. Eriksson, K.: Genetic selection for voluntary alcohol consumption in the albino rat. *Science 159*:739—741, 1968.
127. Evans, J. I. and Oswald, I.: Some experiments in the chemistry of narcoleptic sleep. *Br. J. Psychiatry 112*:401—404, 1966.
128. Everett, G. M. and Borcherding, J. W.: L-Dopa: Effect on concentrations of dopamine, norepinephrine and serotonin in brains of mice. *Science 168*:849—850, 1970.
129. Feer, H. and Wirz-Justice, A.: The effect of 5-hydroxytryptophan on the efflux of noradrenaline from brain slices. *Experientia 27*: 885—886, 1971.
130. Feldberg, W. and Myers, R. D.: A new concept of temperature regulation by amines in the hypothalamus. *Nature 200*:1325—1326, 1963.
131. Feldberg, W. and Myers, R. D.: Effects on temperature of amines injected into the cerebral ventricles. A new concept of temperature regulation. *J. Physiol.* (Lond.) *173*:226—237, 1964.
132. Feldberg, W. and Myers, R. D.: Changes in temperature produced by microinjections of amines into the hypothalamus of cats. *J. Physiol.* (Lond.) *177*:239—245, 1965.
133. Feldstein, A., Chang, F. H., and Kucharski, J. M.: Tryptophol, 5-hydroxytryptophol and 5-methoxytryptophol-induced sleep in mice. *Life Sci. 9*:323—329, 1970.
134. Fennessy, M. R. and Lee, J. R.: Modification of morphine analgesia by drugs affecting adrenergic and tryptaminergic mechanisms. *J. Pharm. Pharmacol. 22*:930—935, 1970.
135. Ferguson, J., Henriksen, S., Cohen, H., et al.: Hypersexuality and changes in aggressive and perceptual behavior caused by chronic administration of parachlorophenylalanine in cats. *Science 168*: 499—501, 1970.
136. Fischer, E., Fernandez-Lagravere, T. A., Vazquez, A. J., et al.: Bufotenine-like substance in the urine of schizophrenics. *J. Nerv. Ment. Dis. 133*:441, 1961.
137. Fischer, E. and Spatz, H.: Quantitativer Nachweis einer vermehrten Bufotenin-Ausscheidung im Urin der Schizophrenen. *Arch. Psychiatr. Z. Ges. Neurol. 211*:241, 1968.
138. Fish, B., Campbell, M., Shapiro, T., et al.: Schizophrenic children treated with methysergide (Sansert). *Dis. Nerv. Syst. 30*:534—540, 1969.
139. Foote, J. L. and Tao, R. V. P.: The effects of p-chlorophenylalanine and alanine on brain esterbound fatty acids of developing rats. *Life Sci. 7*:1187—1192, 1968.
140. Frey, H. -H., Magnussen, M. P., and Nielsen, K.: The effect of p-chloramphetamine on the consumption of ethanol by rats. *Arch. Int. Pharmacodyn. Ther. 183*:165—172, 1970.
141. Friedman, D. X.: Effect of LSD-25 on brain serotonin. *J. Pharmacol. Exp. Ther. 134*:160—166, 1961.
142. Frohman, C. E., Warner, K. A., Barry, C. T., et al.: Amino acid transport and the plasma factor in schizophrenia. *Biol. Psychiatry 1*:201—207, 1969.

143. Fuller, R. W.: Influence of substrate in the inhibition of rat liver and brain monoamine oxidase. *Arch. Int. Pharmacodyn.* 174:32–36, 1968.
144. Fuller, R. W., Hines, C. W., and Mills, J.: Lowering of brain serotonin level by chloramphetamines. *Biochem. Pharmacol.* 14:483–488, 1965.
145. Fuxe, K.: The distribution of monoamine terminals in the central nervous system. *Acta Physiol. Scand.* 64(Suppl. 247):1–84, 1965.
146. Fuxe, K. and Hökfelt, T.: Further evidence for the existence of tuberoinfundibular dopamine neurons. *Acta Physiol. Scand.* 66: 245–246, 1966.
147. Fuxe, K., Hökfelt, T., and Jonsson, G.: Participation of central monoaminergic neurons in the regulation of anterior pituitary secretion. In Martini, L. and Meites, J. (Eds.): *Neurochemical Aspects of Hypothalamic Function.* New York: Academic Press, 1970, pp. 61–83.
148. Fuxe, K., Hökfelt, T., and Ungerstedt, U.: Morphological and functional aspects of central monoamine neurons. *Int. Rev. Neurobiol.* 13:93–126, 1970.
149. Fuxe, K. and Ungerstedt, U.: Localization of 5-hydroxytryptamine uptake in rat brain after intraventricular injection. *J. Pharm. Pharmacol.* 19:335–337, 1967.
150. Gaddum, J. H. and Picarelli, Z. P.: Two kinds of tryptamine receptors. *Br. J. Pharmacol.* 12:323–328, 1957.
151. Gaddum, J. H. and Vogt, M.: Some central actions of 5-hydroxytryptamine and various antagonists. *Br. J. Pharmacol.* 11:175–179, 1956.
152. Gal, E. M. and Drewes, P. A.: Studies on the metabolism of 5-HT. II. Effect of tryptophan deficiency in rats. *Proc. Soc. Exp. Biol. Med.* 110:368–371, 1962.
153. Gal, E. M., Morgan, M., and Marshall, F. D.: Studies on the metabolism of 5-hydroxytryptamine (Serotonin). IV. The effects of various drugs on the *in vivo* hydroxylation of tryptophan by the brain tissue. *Life Sci.* 4:1765–1772, 1965.
154. Gal, E. M., Poczik, M., and Marshall, F. D., Jr.: Hydroxylation of tryptophan to 5-hydroxytryptophan by brain tissue *in vivo*. *Biochem Biophys. Res. Commun.* 12:39–43, 1963.
155. Gal, E. M., Raggeveen, A. E., and Millard, S. A.: DL-(2-^{14}C)-p-chlorophenylalanine as an inhibitor of tryptophan 5-hydroxylase. *J. Neurochem.* 17:1221–1235, 1970.
156. Garattini, S., Giacalone, E., and Valzelli, L.: Biochemical changes during isolation-induced aggressiveness in mice. In Garattini, S. and Sigg, E. G. (Eds.): *Aggressive Behavior.* Amsterdam: Excerpta Medica, 1969, pp. 179–187.
157. Garattini, S. and Shore, P. A. (Eds.): *Biological Role of Indolealkylamine Derivatives. Advan. Pharmacol.* 6A, 6B, New York: Academic Press, 1968, pp. 301, 421.
158. Garattini, S. and Valzelli, L.: *Serotonin.* Amsterdam: Elsevier, 1965, p. 392.

159. Gessa, G. L., Tagliamonte, A., Tagliamonte, P., et al.: Essential role of testosterone in the sexual stimulation by p-chlorophenylalanine in male animals. *Nature* 227:616–617, 1970.
160. Goldstein, M. and Frenkel, R.: Inhibition of serotonin synthesis by dopa and other catechols. *Nature* 233:179–180, 1971.
161. Goldstein, S., Himwich, W. A., Leiner, K., et al.: Psychoactive agents in dogs with bilateral lesions in subcortical structures. *Neurology* (Minneap.) 21:847–852, 1971.
162. Gomulka, W., Marcucci, F., and Samanin, R.: Effect of lesion and electrical stimulation of midbrain raphé on the level of N-acetyl-L-aspartic acid in the rat brain. *Eur. J. Pharmacol.* 13:364–366, 1971.
163. Goodwin, F. K., Dunner, D. L., and Gershon, E. S.: Effects of L-DOPA treatment on brain serotonin in depressed patients. *Life Sci.* 10:751–759, 1971.
164. Gottlieb, J. S., Frohman, C. E., and Havlena, J. M.: The effect of antimetabolites on embryonic development. *J. Mich. Med. Soc.* 57:364–366, 1958.
165. Govier, W. M., Hoves, G., and Gibbons, A. J.: The oxidative deamination of serotonin and other 3(beta-aminoethyl)-indoles by monoamine oxidase and the effect of these compounds on the deamination of tyramine. *Science* 118:596–597, 1953.
166. Gray, W. D., Rauh, C. E., and Shanahan, R. W.: The mechanisms of the antagonistic action of reserpine on the anticonvulsant effect of inhibitors of carbonic anhydrase. *J. Pharmacol. Exp. Ther.* 139:350–360, 1963.
167. Green, J.: Some observations on lambda waves and peripheral stimulation. *Electroencephalogr. Clin. Neurophysiol.* 9:691–704, 1957.
168. Guilleminault, C., Cathala, J. P., and Castaigne, P.: Effects of 5-hydroxytryptophan on sleep of a patient with a brain stem lesion. *Electroencephalogr. Clin. Neurophysiol.* 34:177–184, 1973.
169. Gunne, L.-M.: Catecholamines and 5-hydroxytryptamine in morphine tolerance and withdrawal. *Acta Physiol. Scand.* 58(Suppl. 204): 1963.
170. Gyand, E. A., Kosterlitz, H. W., and Lees, G. M.: The inhibition of autonomic neuro-effector transmission by morphine-like drugs and its use as a screening test for narcotic analgesic drugs. *Arch. Exp. Pathol. Pharmacol.* 248:231–246, 1964.
171. Haigler, H. J. and Aghajanian, G. K.: A comparison of effect of D-lysergic acid diethylamide (LSD) and serotonin on pre- and postsynaptic cells in the serotonin system. *Fed. Proc.* 32(3):303, 1973.
172. Haigler, H. J. and Aghajanian, G. K.: Mescaline and LSD: Direct and indirect effects on serotonin-containing neurons in brain. *Eur. J. Pharmacol.* 21:53–60, 1973.
173. Hamon, M., Bourgoin, S., Morot-Gaudry, Y., et al.: End-product inhibition of serotonin synthesis in the rat striatum. *Nature (New Biol.)* 237:184–187, 1972.
174. Hanig, J. P., Aiello, E., and Seifter, J.: Permeability of the blood-

brain barrier to parenteral 5-hydroxytryptamine in the neonate chick. *Eur. J. Pharmacol. 12*:180–182, 1970.
175. Hart, E. R., Rodriguez, J. M., and Marrazzi, A. S.: Carotid vagal afferents and drug action on transcallosally evoked cortical potentials. *Science 134*:1696–1697, 1961.
176. Hartmann, E.: On the pharmacology of dreaming sleep (the D State). *J. Nerv. Ment. Dis. 146*:165–173, 1968.
177. Hartmann, E. and Chung, R.: Sleep-inducing effects of L-tryptophan. *J. Pharm. Pharmacol. 24*:252–253, 1972.
178. Harvey, J. A., Lints, C. E., and Carlton, E.: Lesions in the medial forebrain bundle: Relationship between pain sensitivity and telencephalic content of serotonin. *J. Comp. Physiol. Psychol. 16*:298–302, 1971.
179. Heath, R. G.: Pleasure response of human subjects to direct stimulation of the brain: Physiologic and psychodynamic considerations. In Heath, R. G. (Ed.): *The Role of Pleasure in Behavior*. New York: Hoeber, Harper and Row, 1964, pp. 219–243.
180. Heath, R. G.: Pleasure and brain activity in man. *J. Ment. Nerv. Dis. 154*:3–18, 1972.
181. Herkert, E. E. and Keup, W.: Excretion patterns of tryptamine, indoleacetic acid, and 5-hydroxyindoleacetic acid, and their correlation with mental changes in schizophrenic patients under medication with alpha-methyldopa. *Psychopharmacologia 15*:48–59, 1969.
182. Herold, M. and Chan, J.: The possible role of serotonin in affective component of pain behavioral reaction in the rat. In: *Third International Pharmacology Meeting* (Vol. 9). London: Pergamon Press, 1968, pp. 87–99.
183. Himwich, H. R., Narasimhachair, N., Helier, B., et al.: Comparative behavioral and urinary studies on schizophrenics and normal controls. In Bowman, R. E. and Datta, S. P. (Eds.): *Biochemistry of Brain and Behavior*. New York: Plenum Press, 1970, pp. 207–221.
184. Hisikawa, Y., Cramer, H., and Kuhlo, W.: Natural and melatonin induced sleep in young chickens. A behavioral and electrographic study. *Exp. Brain Res. 7*:84–88, 1969.
185. Hitzemann, R. J., Ho, I. K., and Loh, H. H.: Narcotic tolerance and dependence and serotonin turnover. *Science 178*:645–647, 1972.
186. Hoffman, A.: Recent developments in ergot alkaloids. *Australas. J. Pharmacy 42*:7–18, 1961.
187. Holman, R. B. and Vogt, M.: Release of 5-hydroxytryptamine from caudate nucleus and septum. *J. Physiol.* (Lond.) *223*:243–254, 1972.
188. Holzman, P. S., Proctor, L. R., and Hughes, D. W.: Eye tracking patterns in schizophrenia. *Science 181*:179–181, 1973.
189. Hong, E., Sancilio, L. F., Vargas, R., et al.: Similarities between the pharmacological actions of quipazine and serotonin. *Eur. J. Pharmacol. 6*:274–280, 1969.
190. Horita, A. and Gogerty, J. H.: The pyrogenic effect of 5-hydroxy-

tryptophan and its comparison with that of LSD. *J. Pharmacol. Exp. Ther. 122*:195–200, 1958.
191. Huang, Ch. C., and Marrazzi, A. S.: Analysis of drug block of LSD/5-HT/GABA by monitoring neuronal membrane changes. *Fed. Proc. 32*(3):303, 1973.
192. Huges, J. R.: Responses from the visual cortex of unanesthetized monkeys. *Int. Rev. Neurobiol. 7*:99–152, 1964.
193. Huidobro, F., Contreras, E., and Croxatto, R.: Studies on morphine. I. Effects of nalorphine and levallorphan in mice implanted with pellets of morphine. *Arch. Int. Pharmacodyn. Ther. 144*:196–205, 1963.
194. Hyyppa, M., Lampinen, P., and Lehtinen, P.: Alteration in the xual behavior of male and female rats after neonatal administration of p-chlorophenylalanine. *Psychopharmacologia 25*:152–161, 1972.
195. Imura, H., Nakai, Y., and Yoshimi, T.: Effect of 5-hydroxy- *J. Clin. Endocrinol. Metab. 36*:204–206, 1973.
196. Isaac, L.: Temperature alterations of monoamine metabolites in the cerebrospinal fluid. *Nature* (New Biol.) *243*:269–271, 1973.
197. Jacob, J., Robert, J. M., and Lafille, C.: Activités hyperthermisantes de diverses tryptamines administrées par voie intraveneuse et cisternale chez le lapin. *C. R. Séanc. Acad. Sci.* (Paris) *262*:209–212, 1966.
198. Jacob, J., Girault, J. M., and Peindaries, R.: Actions of 5-hydroxytryptamine and 5-hydroxytryptophan injected by various routes on the rectal temperature of the rabbit. *Neuropharmacology 11*:1–16, 1972.
199. Jeannerod, M. and Saki, K.: Occipital and geniculate potentials related to eye movements in the unanesthetized cat. *Brain Res. 19*:361–377, 1970.
200. Jequier, E., Lovenberg, W., and Sjoerdsma, A.: Tryptophan hydroxylase inhibition: The mechanism by which p-chlorophenylalanine depletes rat brain serotonin. *Mol. Pharmacol. 3*:274–279, 1967.
201. Johnson, D. N., Funderburk, W. H., Ruckart, R. T., et al.: Contrasting effects of two 5-hydroxytryptamine-depleting drugs on sleep patterns in cats. *Eur. J. Pharmacol. 20*:80–84, 1972.
202. Johnson, G. A., Kim, E. G., and Boukma, S. S.: 5-Hydroxyindole levels in rat brain after inhibition of dopamine beta-hydroxylase. *J. Pharmacol. Exp. Ther. 180*:539–546, 1972.
203. Jouvet, M.: Biogenic amines and the states of sleep. *Science 163*: 32–41, 1969.
204. Jouvet, M.: The role of monoamines and acetylcholine-containing neurons in the regulation of the sleep-waking cycle. *Ergeb. Physiol. 64*:167–307, 1972.
205. Jouvet, M., Bobillier, P., Pujol, J. F., et al.: Effets des lésions du systeme du raphé sur le sommeil et la sérotonine cérebrale. *C. R. Soc. Biol. 160*:2343, 1966.
206. Jouvet, M. and Michel, F.: Correlations electromyographiques du

sommeil chez le chat décortiqué et mésencéphalique chronique. *C. R. Soc. Biol.* (Paris) *153*:422–425, 1959.
207. Jung, R.: Neurophysiological and psychophysical correlates in vision research. In Karczmar, A. G. and Eccles, J. C. (Eds.): *Brain and Human Behavior.* Berlin: Springer, 1972, pp. 209–258.
208. Kamberi, I. A., Mical, R. S., and Porter, J. C.: Effects of melatonin and serotonin on the release of FSH and prolactin. *Endocrinology 88*:1288–1293, 1971.
209. Karczmar, A. G.: Is the central cholinergic nervous system over-exploited? In Karczmar, A. G. (Ed.): *Symposium on Central Cholinergic Transmission and Its Behavior Aspects. Fed. Proc. 28*:147–157, 1969.
210. Karczmar, A. G.: Neurophysiological, behavioral and neurochemical correlates of the central cholinergic synapses. In Vinar, O., Votava, Z., and Bradley, P. B. (Eds.): *Advances in Neurosychopharmacology.* Amsterdam: North Holland, 1971, pp. 455–480.
211. Karczmar, A. G.: Les synapses cholinergiques et leurs enchainements In Nahas, G. G., Salamagne, J. C., Viars, P., et al. (Eds.): *Le Systeme Cholinergique.* Paris: Librairie Arnette, 1972, pp. 1313–156.
212. Karczmar, A. G. and Scudder, C. L.: Behavioral responses to drugs and brain catecholamine levels in mice of different genera and strains. In Cafruny, E. J. (Ed.): *Proceedings of an International Symposium on Comparative Pharmacology. Fed. Proc. 26*: 1186–1191, 1967.
213. Karczmar, A. G. and Scudder, C. L.: Learning and effect of drugs on learning of related mice genera and strains. In Karczmar, A. G. and Koella, W. P. (Eds.): *Neurophysiological and Behavioral Aspects of Psychotropic drugs.* Springfield, (Ill.): C. C. Thomas, 1960, pp. 133–160.
214. Karczmar, A. G., Scudder, C. L. and Richardson, D.: Interdisciplinary approach to the study of behavior in related mice types. *Neurosci. Res. 5,* in press.
215. Kawamura, H. and Marchiafava, P. L.: Excitability changes along visual pathways during eye tracking movements. *Arch. Ital. Biol. 106*:141–156, 1968.
216. Kehr, W., Carlsson, A., Lindqvist, T., et al.: Evidence for receptor-mediated feed-back control of striatal tyrosine hydroxylase activity. *J. Pharm Pharmacol. 24*:744–747, 1972.
217. Kellogg, C.: Serotonin metabolism in the brains of mice sensitive or resistant to audiogenic seizures. *J. Neurobiol. 2*:209–219, 1971.
218. Kety, S.: Brain amines and affective disorders. In Ho, B. T. and Melsaac, W. M. (Eds.): *Brain Chemistry and Mental Disease.* New York: Plenum Press, 1971, pp. 237–244.
219. Key, B. J.: Effect of lysergic acid diethylamide on potentials evoked in the specific sensory pathways. *Br. Med. Bull. 21*:30–35, 1965.
220. Kiyono, S. and Jeannerod, M.: Relations entre l'activité géniculaire

phasique et les mouvements oculaires chez le chat normal et sous réserpine. *C. R. Soc. Biol.* (Paris) *161*:1607—1611, 1967.
221. Klein, D. C.: The role of serotonin N-acetyltransferase in the adrenergic regulation of indole metabolism in the pineal gland. In Barchas, J. and Usdin, E. (Eds.): *Serotonin and Behavior.* New York: Academic Press, 1973, pp. 109—119.
222. Knox, W. E.: Phenylketonuria. In Stanbury, J. B., Wyngaarden, J. B., and Frederickson, D. S. (Eds.): *The Metabolic Basis of Inherited Disease.* New York: Blakiston-McGraw-Hill, 1966, pp. 258—294.
223. Kobinger, W.: Beeinflussung der Cardiazolkrampfschwelle durch veränderten 5-Hydroxytryptamingehalt des Zentralnervensystems. *Arch. Exp. Path. Pharmacol. 233*:559—566, 1958.
224. Koe, B. K. and Weissman, A.: p-Chlorophenylalanine, a specific depletor of brain serotonin. *J Pharmacol. Exp. Ther. 154*:499—516, 1966.
225. Koella, W. P.: The physiology and pharmacology of sleep. In Mendeles, J. (Ed.): *Biological Psychiatry.* New York: Wiley, 1973, pp. 263—296.
226. Koella, W. P. and Czicman, J. S.: Mechanism of the EEG-synchronizing action of serotonin. *Am. J. Physiol. 211*:926—934, 1966.
227. Koella, W. P., Feldstein, A., and Czicman, J. S.: The effect of parachlorophenylalanine on the sleep of cats. *Electroencephalogr. Clin. Neurophysiol. 25*:481—490, 1968.
228. Kordon, C.: Role des monoamines hypothalamiques dans les régulations adénohypophysaires. In: *Neuroendocrinologie.* Paris: Edition du CNRS, 1969, p. 73.
229. Kordon, C. and Glowinski, J.: Selective inhibition of superovulation by blockade of dopamine synthesis during the "critical period" the immature rat. *Endocrinology 85*:924—931, 1969.
230. Kordon, C. and Glowinski, J.: Role of hypothalamic monoaminergic neurones in the gonadotrophin release-regulating mechanisms. *Neuropharmacology 11*:153—162, 1972.
231. Kordon, C., Javoy, F., Vassent, G., et al.: Blockade of superovulation in the immature rat by increased brain serotonin. *Eur. J. Pharmacol. 4*:169—174, 1968.
232. Kordon, C. and Vassent, G.: Effet de microinjections intrahypothalamiques d'un inhibiteur de la monoamine-oxidase sur l'ovulation provoquée chez la ratte impubère. *C. R. Hebd. Séanc. Acad. Sci.* (Paris): *266*:2473—2476, 1968.
233. Kosterlitz, H. W. and Lees, G. J.: Pharmacological analysis of intrinsic intestinal reflexes. *Pharmacol. Rev. 16*:301—339, 1964.
234. Kostowski, W.: The effects of some drugs affecting brain 5-HT on electrocortical synchronization following low frequency stimulation of brain. *Brain Res. 31*:151—157, 1971.
235. Kostowski, W. and Giacalone, E.: Effect of psychotropic drugs on the release of serotonin induced by electrical stimulation of midbrain raphé in rats. C.I.N.P. VI. International Congress (Excerpta Medica Foundation, Amsterdam), 1968.

236. Kostowski, W., Giacalone, E., Garattini, S., et al.: Studies on behavioral and biochemical changes in rats after lesion of midbrain raphé. *Eur. J. Pharmacol. 4*:371–376, 1968.
237. Kostowski, W., Giacalone, E., Garattini, S., et al.: Electrical stimulation of midbrain raphé: biochemical, behavioral and bioelectric effects. *Eur. J. Pharmacol. 7*:170–175, 1969.
238. Kostowski, W., Gomulka, W., and Czlonkowski, A.: Reduced cataleptogenic effects of some neuroleptics in rats with lesioned midbrain raphé and treated with p-chlorphenylalanine. *Brain Res. 48*:443–446, 1972.
239. Krieger, H. P. and Krieger, D. T.: Chemical stimulation of the brain: Effect on adrenal corticoid release. *Am. J. Physiol. 218*:1632–1641, 1970.
240. Kuhar, M. J., Roth, R. H., and Aghajanian, G. K.: Synaptosomes from forebrains of rats with midbrain raphé lesions: Selective reduction of serotonin uptake. *J. Pharmacol. Exp. Ther. 181*:36–45, 1972.
241. Kupfer, D. J., Wyatt, R. J., Scott, J., et al.: Sleep disturbance in acute schizophrenic patients. *Am. J. Psychiatr. 126*:1213, 1970.
242. Lehmann, A.: Audiogenic seizures data in mice supporting new theories of biogenic amines mechanisms in the central nervous system. *Life Sci. 6*:1423–1431, 1967.
243. Lehtinen, P., Hyyppa, M., and Lampinen, P.: Sexual behavior of adult rats after the neonatal single injection of reserpine. *Psychopharmacologia 23*:171–179, 1972.
244. Lemberger, L., Axelrod, J., and Kopin, I. J.: The disposition and metabolism of tryptamine and the in vivo formation of 6-hydroxytryptamine in the rabbit. *J. Pharmacol. Exper, Ther. 177*:169–176, 1971.
245. Lessin, A. W. and Parkes, M. W.: The effects of reserpine and other agents upon leptazol convulsions in mice. *Br. J. Pharmacol. 14*:108–111, 1959.
246. Lester, B. K., Chanes, R. E., and Condit, P. T.: A clinical syndrome and EEG-sleep changes associated with amino acid deprivation. *Am. J. Psychiatr. 126*(2):185–190, 1969.
247. Lichtensteiger, W., Mutzner, V., and Langemann, H.: Uptake of 5-HT and 5-HTP by neurons of the central nervous system normally containing catecholamines. *J. Neurochem. 14*:489–497, 1967.
248. Lipton, M. A., Gordon, R., Guroff, G., et al.: p-Chlorophenylalanine-induced chemical manifestations of phenylketonuria in rats. *Science 156*:248–250, 1967.
249. Loh, W. H., Shen, F., and Way, E. L.: Inhibition of morphine tolerance and physical dependence development and brain serotonin synthesis by cycloheximide. *Biochem. Pharmacol. 18*:2711–2721, 1969.
250. Lovenberg, W., Jequier, E., and Sjoerdsma, A.: Tryptophan hydroxylation in mammalian systems. *Adv. Pharmacol. 6A*:21–36, 1968.
251. Maccitelli, F. J., Fischetti, D., and Montararelli, N.: Changes in

behavior and electrocortical activity in the monkey following administration of 5-hydroxytryptophan. *Psychopharmacologia 9*: 447—456, 1966.
252. Macon, J. B., Sokoloff, L., and Glowinski, J.: Feed-back control of rat brain 5-hydroxytryptamine synthesis. *J. Neurochem. 18*:323—331, 1971.
253. Magnes, J., Moruzzi, G., and Pompeiano, O.: Synchronization of the EEG produced by low frequency electrical stimulation of the region of the solitary tract. *Arch. Ital. Biol. 99*:33—67, 1961.
254. Mahler, D. J. and Humoller, F. L.: Effects of serotonin on brain metabolism *Int. J. Neuropsychiatr. 3*:229—233, 1967.
255. Mancia, M.: Electrophysiological and behavioral changes owing to splitting of the brain stem in cats. *Electroencephalogr. Clin. Neurophysiol. 27*:487—502, 1969.
256. Mandell, A. J., Buckingham, B., and Segal, D.: Behavioral, metabolic and enzymatic studies of a brain indoleethylamine N-methylating system. In Ho, B. T. and McIsaac, W. M. (Eds.): *Brain Chemistry and Mental Disease.* New York: Plenum Press, 1972, pp. 37—60.
257. Mandell, M. P., Mandell, A. J., and Jacobson, A.: Biochemical and neurophysiological studies of the paradoxical sleep. *Rec. Adv. Biol. Psychiatr. 7*:115—122, 1965.
258. Mandell, A. and Morgan, M.: Indole (ethyl) amine N-methyltransferase in human brain. *Nature (New Biol.) 230*:85—87, 1971.
259. Mandell, A. J. and Spooner C. E.: An N, N-indole transmethylation theory of the mechanism of MAOI-indole amino acid load behavioral activation. In Siva Sankar, D. V. (Ed.): *Schizophrenia. Current Concepts and Research.* Hicksville, N.Y.:PJD Publications, 1969, pp. 496—507.
260. Mandell, A. J., Spooner, C. E., and Brunet, D.: Whither the "sleep transmitter." *Biol. Psychiatry 1*:13—30, 1969.
261. Mantegazzini, P.: Pharmacological actions of indolealkylamines and precursor amino acids on the central nervous system. In Erspamer, V. (Ed.): *Handbook of Experimental Pharmacology* (Vol. 19). 5-Hydroxytryptamine and related indolealkylamines. Berlin: Springer, 1966, pp. 424—470.
262. Marczynski, T. J.: Pharmacology and mechanisms of action of some new tryptamine derivatives. *Folia Med. Cracov. 6*:1—86, 1964.
263. Marczynski, T. J.: Topical application of drugs to subcortical brain structures and selected aspects of electrical stimulation. *Ergeb. Physiol. 59*:86—159, 1967.
264. Marczynski, T. J.: The fresh water clam *Anodonta cygnea L.* as a test object for serotonin and related compounds. *Bull. Acad. Pol. Sci.* (Biol.) *VI7*:147—150, 1959a.
265. Marczynski, T. J.: Pharmacological properties of 5-methoxy-N-methyltryptamine. The fresh water clam as a test for serotonin. *Diss. Pharm.* (Cracow) *11*:297—313, 1959b.
266. Marczynski, T. J.: Some pharmacological properties of a recently isolated alkaloid, 5-methoxy-N-methyltryptamine. *Bull. Acad. Pol. Sci. VI7*:151—154, 1959c.

267. Marczynski, T. J.: Pharmacological properties of some new tryptamine derivatives. I. The effect on spinal cord reflexes. *Diss. Pharm.* (Cracow) *14*:247–258, 1962.
268. Marczynski, T. J., Rosen, A. J., and Hackett, J. T.: Postreinforcement electrocortical synchronization and facilitation of cortical auditory evoked potentials in appetitive instrumental conditioning. *Electroencephalogr. Clin. Neurophysiol. 24*:227–241, 1968.
269. Marczynski, T. J. and Vetulani, J.: Observations on the effect of serotonin and bufotenine on some autonomic reflexes. *Diss. Pharm.* (Cracow) *13*:11–25, 1961.
270. Marczynski, T. J. and Yamaguchi, N.: Central depressant effects of some new tryptamine derivatives. *Biochem Pharmacol. 12*:Suppl. 209, 1963.
271. Marczynski, T. J., Yamaguchi, N., and Ling, G. M.: Effects of bufotenine and melatonin applied in crystalline form to the preoptic region or nucleus centralis medialis in unrestrained cats. *Pharmacologist 4*:175, 1962.
272. Marczynski, T. J., Yamaguchi, N., Ling, G. M., et al.: Sleep induced by the administration of melatonin, 5-methoxy-N-acetyltryptamine, to the hypothalamus in unrestrained cats. *Experientia 26*: 435, 1964.
273. Marczynski, T. J., York, J. L., and Hackett, J. T.: Steady potential correlates of positive reinforcement: Reward Contingent Positive Variation. *Science 163*:301–304, 1969.
274. Marczynski, T. J., York, J. L., Allen, S. L., et al.: Steady potential correlates of positive reinforcement and sleep onset in the cat; "Reward Contingent Positive Variation" (RCPV). *Brain Res. 26*: 305–332, 1971.
275. Mardones, J., Segovia-Requelme, N., Hedebra, A., et al.: Effect of some self-selection conditions on the voluntary alcohol intake of rats. *Q. J. Stud. Alcohol 16*:425–437, 1955.
276. Marrazzi, A. S. and Hart, E. R.: Relationship of hallucinogens to adrenergic cerebral neurohumors. *Science 121*:365–367, 1955.
277. Marshall, I. and Grahame-Smith, D. G.: Evidence against a role of brain 5-hydroxytryptamine in the development of physical dependence upon morphine in mice. *J. Pharmacol. Exp. Ther. 173*:634–641, 1971.
278. Martin, J. B.: Neural regulation of growth hormone secretion. *New Engl. J. Med. 288*:1384–1393, 1973.
279. Matsumoto, J. and Jouvet, M.: Effects de réserpine, DOPA et 5-HTP sur les deux états de sommeil. *C. R. Soc. Biol. 158*:2137–2140, 1964.
280. Matthysse, S., Smith, E. L., Puck, T. T., et al.: Prospects for research on schizophrenia. *Neurosci. Res. Prog. Bull. 10*:446–450, 1972.
281. Mayer, D. J., Wolfe, T. L., Akil, H., et al.: Analgesia from electrical stimulation of the brain. *Science 174*:1351–1354, 1971.
282. Maynert, E. W.: The role of biochemical and neurohumoral factors in the laboratory evaluation of antiepileptic drugs. *Epilepsia 10*: 145–162, 1969.

283. Maynert, E. W., Klingman, G. I., and Kaji, H. K.: Tolerance to morphine. II. Lack of effects on brain 5-hydroxytryptamine and gamma-aminobutyric acid. *J. Pharmacol. Exp. Ther. 135*:296–299, 1962.
284. McGinty, D. J., Harper, R. M., and Fairbanks, M. K.: 5-HT-containing neurons: Unit activity in behaving cats. In Barchas, J. (Ed.): *Serotonin and Behavior.* New York: Academic Press, 1973.
285. McKean, C. M.: The effects of high phenylalanine concentrations on serotonin and catecholamine metabolism in the human brain. *Brain Res. 47*:469–476, 1972.
286. McKean, C. M., Schanberg, S. M., and Giarman, N. J.: A mechanism of the indole defect in experimental phenylketonuria. *Science 137*: 604–605, 1962.
287. McMenamy, R. H., Lund, C. C., and Oncley, J. L.: Unbound amino acid concentrations in human blood plasma. *J. Clin. Invest. 36*: 1672–1679, 1957.
288. Medakovic, M.: Comparisons of antagonistic potencies of morphine-like analgesics towards 5-hydroxytryptamine on the guinea pig ileum. *Arch. Int. Pharmacodyn. Ther. 114*:201–209, 1958.
289. Meyerson, B. J.: Central nervous monoamines and hormone induced estrus behavior in the spayed rat. *Acta Physiol. Scand. 63* (Suppl. 241):1–32, 1964a.
290. Meyerson, B. J.: The effect of neuropharmacological agents on hormone-activated estrus behavior in ovariectomized rats. *Arch. Int. Pharmacodyn. 150*:4–33, 1964b.
291. Meyerson, B. J.: Estrus behavior in spayed rats after estrogen or progesterone treatment in combination with reserpine or tetrabenazine. *Psychopharmacologia 6*:210–218, 1964c.
292. Meyerson, B. J.: Female copulatory behavior in male and androgenized female rats after oestrogen-amine depletor treatment. *Nature 217*:683–684, 1968.
293. Meyerson, B. J. and Lewander, T.: Serotonin synthesis inhibition and estrus behavior in female rats. *Life Sci. 9*:661–671, 1970.
294. Michael, J. A. and Stark, L.: Interactions between eye movements and the visually evoked response in the cat. *Electroencephalogr. Clin. Neurophysiol. 21*:478–488, 1966.
295. Modigh, K.: The effect of chlorimipramine on the rate of tryptophan hydroxylation in the intact and transected spinal cord. *J. Pharm. Pharmacol. 25*:928–929, 1973.
296. Moir, A. T. B. and Eccleston, D: The effects of precursor loading in the cerebral metabolism of 5-hydroxyindoles. *J. Neurochem 15*: 1093–1108, 1968.
297. Monachon, M. A., Burkard, W. P., Jalfre, M., et al.: Blockade of central 5-hydroxytryptamine receptors by methiothepin. *Naunyn Schmiedebergs Arch. Pharmacol. 274*:192–197, 1972.
298. Monnier, M. and Tissot, R.: Action de la réserpine et des ses médiateurs (5-HTP, serotonin, and DOPA-noradrénaline) sur le comportement et le cerveau du lapin. *Helv. Physiol. Pharmacol. Acta 16*:255–267, 1958.

299. Moore, R. Y., Heller, A., Bhatnagar, R. K., et al.: Central control of the pineal gland: Visual pathways. *Arch. Neurol. 18*:208–218, 1968.
300. Morest, D. K.: A study of the structure of the area postrema with Golgi method. *Am. J. Anat. 107*:291–303, 1960.
301. Morgan, M. and Mandell, M.: Indole (ethyl) amine N-methyltransferase in the brain. *Science 165*:492, 1969.
302. Morgane, P. J. and Stern, W. C.: Relationship of sleep to neuroanatomical circuits, biochemistry and behavior. *Ann. N. Y. Acad. Sci. 193*:95–111, 1972.
303. Morgane, P. J. and Stern, W. C.: Effects of serotonin metabolites on sleep-waking activity in cats. *Brain Res. 50*:205–213, 1973.
304. Mouret, J. R., Bobillier, P. and Jouvet, M.: Insomnia following parachlorophenylalanine in the rat. *Eur. J. Pharmacol. 5*:17–22, 1968.
305. Muruyama, Y., Hayashi, G., and Takemori, A. E.: Relation between 5-hydroxytryptamine (5-HT) and tolerance and physical dependence in mice. *Pharmacologist 12*:231, 1970.
306. Myers, R. D.: Voluntary alcohol consumption in animals: Peripheral and intracerebral factors. *Psychosom. Med. 28*:484–497, 1966.
307. Myers, R.: The role of hypothalamic transmitter factors in the control of body temperature. In Hardy, J. D. et al. (Eds.): *Physiological and Behavioral Temperature Regulation*. Springfield, (Ill.): C. C. Thomas, 1970. pp. 648–666.
308. Myers, R. D. and Beleslin, D. B.: Changes in serotonin release in hypothalamus during cooling or warming of the monkey. *Am. J. Physiol. 220*:1746–1754, 1971.
309. Myers, R. D. and Cicero, T. J.: Effects of serotonin depletion on the volitional alcohol intake of rats during a condition of psychological stress. *Psychopharmacologia 15*:373–381, 1969.
310. Myers, R. D., Evans, J. E., and Yaksh, T. L.: Ethanol preference in the rat: Interactions between brain serotonin and ethanol, acetaldehyde, paraldehyde, 5-HTP and 5-HTOL. *Neuropharmacology 11*: 539–549, 1972.
311. Myers, R. D., Kava, A., and Beleslin, D. B: Evoked release of 5-HT and NEFA from the hypothalamus of the conscious monkey during thermoregulation. *Experientia 25*:705–706, 1969.
312. Myers, R. D. and Sharpe, L. G.: Temperature in the monkey: Transmitter factors released from the brain during thermoregulation. *Science 161*:572–573, 1968.
313. Myers, R. D. and Yaksh, T. L.: Control of body temperature in the unanesthetized monkey by cholinergic and aminergic systems in the hypothalamus. *J. Physiol.* (Lond.) *202*:483–500, 1969.
314. Nachman, J., Lester, D., and LeMagnen, J.: Alcohol aversion in the rat: Behavioral assessment of noxious drug effect. *Science 168*: 1244–1246, 1970.
315. Narasimhachari, N., Plant, J. M., and Himwich, H. E.: Indolethylamine-N-methyltransferase in serum samples of schizophrenics and normal controls. *Life Sci. 11*(2):221–227, 1972.

316. Naumenko, E. V.: Hypothalamic chemoreactive structures and the regulation of pituitary-adrenal function. Effects of local injections of norepinephrine, carbachol and serotonin into the brain of guinea pigs. *Brain Res. 11*:1–10, 1968.
317. Ng, K. Y., Chase, T. N., Colburn, R. W., et al.: L-Dopa-induced release of cerebral monoamines. *Science, 170*:76–77, 1970.
318. Ng, K. Y., Chase, T. N., Colburn, R. W., et al.: Dopamine stimulation induced release from central neurons. *Science 172*:487–489, 1972.
319. Norton, P. R. E.: The effects of drugs on barbiturate withdrawal convulsions in the rat. *J. Pharm Pharmacol. 22*:763–766, 1970.
320. Oka, T., Nozaki M., and Hosoya, E.: Effects of p-chlorophenylalanine and cholinergic antagonists on body temperature changes induced by the administration of morphine to nontolerant and morphine-tolerant rats. *J. Pharmacol. Exp. Ther. 180*:136–143, 1972.
321. Olson, L. and Fuxe, K.: On the projections from the locus coeruleus noradrenaline neurons: The cerebellar innervation. *Brain Res. 28*: 165–171, 1971.
322. Opitz, K.: Anorexigene Phenylalkylamine and Serotoninstoffwechsel. *Arch. Pharmakol. Exp. Pathol. 259*:56–65, 1967.
323. O'Steen, W. K.: Serotonin suppression of luteinization in gonadotropin-treated immature rats. *Endocrinology 74*:885–888, 1964.
324. Oswald, I.: Drugs and sleep. *Pharmacol. Rev. 20*:273–303, 1968.
325. Oswald, I., Ashcroft, G. W., Berger, R. J., et al.: Some experiments in the chemistry of normal sleep. *Br. J. Psychiatry 112*:391–399, 1966.
326. Padjen, A. and Randic, M. Some factors influencing the release of 5-hydroxyindol-3-acetic acid in the forebrain. *Br. J. Pharmacol. 39*:1–8, 1970.
327. Pare, C. M., Sandler, M., and Stacey, R. S.: 5-Hydroxytryptamine deficiency in phenylketonuria. *Lancet 1*:551–553, 1957.
328. Parizek, J., Hassler, R., and Bak, I. J.: Light and electron microscopic autoradiography of substantia nigra of rat after intraventricular administration of tritium labelled norepinephrine, serotonin, dopamine and the precursors. *Z. Zellforsch. 115*:137–148, 1971.
329. Pelz, K., Jirkowsky, I., Adlerova, E., et al.: Neurotrope and psychotrope Substanzen XXV. Über die in 8-Stellung durch die Methyl-tertbuty-methoxy-methylthio- and methansulfonylgruppe substituierten 10-(4-methylpiperazin)10,11-dihydrobenzo [b,f] thiepinderivate. *Collect. Czechoslov. Chem Commun. 33*:1895–1910, 1968.
330. Peters, D. A. V., Hrdina, P. D., Singhal, R. L., et al.: The role of brain serotonin in DDT-induced hyperpyrexia. *J. Neurochem. 19*:1131–1136, 1972.
331. Petitjean, F. and Jouvet, M.: Hypersomnie et augmentation de l'acide 5-hydroxyindole acétique cérebrale par lésion isthmique chez le chat. *C. R. Soc. Biol. 164*:2288–2293, 1970.
332. Pfeifer, A. K. and Galambos, E.: The effect of p-chloramphetamine on the susceptibility to seizures and the monoamine level in brain

and heart of mice and rats. *J. Pharm Pharmacodyn. 165*:201—211, 1967.
333. Phillis, J. W., Tebēcis, A. K., and York, D. H. Depression of spinal motoneurons by noradrenaline, 5-hydroxytryptamine and histamine. *Eur. J. Pharmacol.* 7:471—475, 1968.
334. Pin, C., Jones, B., and Jouvet, M.: Topographie des neurones monoaminergiques du tronc cérebral du chat; étude par histofluorescence. *C. R. Soc. Biol.* (Paris) *162*:2136—2141, 1968.
335. Pivik, T. and Dement, W.: Phasic changes in muscular and reflex activity during non-REM sleep. *Exp. Neurol.* 27:115—124, 1970.
336. Pollin, W., Cardon, P. V., and Kety, S. S.: Effects of amino acid feedings in schizophrenic patients treated with iproniazid. *Science 133*:104—105, 1961.
337. Poschel, B. P. H. and Ninteman, F. W.: Norepinephrine: A possible excitatory neurohormone of the reward system. *Life Sci.* 2:782—788, 1963.
338. Poschel, B. P. H. and Ninteman, F. W.: Intracranial reward and the forebrain's serotonergic mechanisms: Studies employing para-chloro-phenylalanine and para-chloramphetamine. *Physiol. Behav.* 7:39—46, 1971.
339. Praag, van, H. M.: Indoleamines and the central nervous system. *Psychiatr. Neurol. Neurochir. 73*:9—36, 1970.
340. Praag, van, H. M., Korf, J., Dols, L. C. W., et al.: A pilot study of the predictive value of the probenecid test in application of 5-hydroxytryptophan as antidepressant. *Psychopharmacologia 25*:14—21, 1972.
341. Praag, van H. M. and Korf, J.: In Barchas, J. and Usdin, E. (Eds.), *Serotonin and Behavior.* New York: Academic Press, 1972.
342. Praag, van, H. M., Korf, J., and Puite, J.: 5-Hydroxyindoleacetic acid levels in the cerebrospinal fluid of depressive patients treated with probenecid. *Nature 225*:1259—1260, 1970.
343. Prange, A. J., Sisk, J. L., Wilson, I. C., et al.: In Barchas, J. and Usdin, E. (Eds.), *Serotonin and Behavior.* New York: Academic Press. 1973. pp. 539—548.
344. Prichard, J. W. and Guroff, G.: Increased cerebral excitability caused by p-chlorophenylalanine in young rats. *J. Neurochem. 18*:153—160, 1971.
345. Prockop, D. J., Shore, P. A., and Brodie, B. B.: Anticonvulsant properties of monoamine oxidase inhibitors. *Ann. N. Y. Acad. Sci. 80*:643—651, 1959.
346. Pujol, J. F., Buguet, A., Froment, J. L., et al.: The central metabolism of serotonin in the cat during insomnia: A neurophysiological and biochemical study after p-chlorophenylalanine or destruction of the raphé system. *Brain Res. 29*:195—212, 1971.
347. Quay, W. B.: Catecholamines and tryptamines. *J. Neurovisc. Relat.* Suppl. *9*:212—235, 1969.
348. Radulovacki, M.: 5-Hydroxyindoleacetic acid in cerebrospinal fluid:

Measurements in wakefulness, slow-wave and paradoxical sleep in cats. *Brain Res. 50*:484—488, 1973.
349. Ramanamurthy, P. S. U. and Srikantia, S. G.: Effects of leucine on brain serotonin. *J. Neurochem. 17*:27—32, 1970.
350. Rechtschaffen, A., Ledecky-Janecek, S., and Lovell, R.: The effect of 5-hydroxyindoleacetic acid (5-HIAA) on sleep in the cat. *Psychophysiology 5*:211, 1968.
351. Rechtschaffen, A., Lovell, R. A., Freedman, D. X., et al.: The effect of para-chlorophenylalanine on sleep in the rat: Some implications for the serotonin-sleep hypothesis. In Barchas, J. and Usdin, E. (Eds.): *Serotonin and Behavior.* New York: Academic Press, 1973. pp. 401—418.
352. Reid, W. D., Volicer, L., Smookler, H., et al.: Brain amines and temperature regulation. *Pharmacology 1*:329—344, 1968.
353. Renault, J.: Monoamines et sommeils. III. Role du système du Raphé, et de sérotonine cérebrale dans l'endormissement. Thèse de Médecine, Lyon. Imp. Beaux-Arts, p. 144, 1967.
354. Richardson, D., Karczmar, G., and Scudder, C. L.: Intergeneric behavioral differences among methamphetamine treated mice. *Psychopharmacologia 25*:347—375, 1972.
355. Rodriguez, R. and Pardo, E. G.: Quipazine, a new type of antidepressant agent. *Psychopharmacologia 21*:89—100, 1971.
356. Roos, B. E. and Sjöström R.: 5-Hydroxyindoleacetic acid (and homovanillic acid) levels in the CSF after probenecid application in patients with manic-depressive psychosis. *Pharmacol. Clin. 1*:153—155, 1969.
357. Rosen, A. J. and Marczynski, T. J.: Effects of alpha-methyl-metatyrosine on operant behavior, auditory evoked potentials and postreinforcement electrocortical synchronization in the cat. *Physiol. Behav. 2*:413—419, 1967.
358. Ross, S. B. and Renyi, A. L.: Accumulation of tritiated 5-hydroxytryptamine in brain slices. *Life Sci. 6*:1407—1415, 1967.
359. Rothballer, A. B.: The effect of phenylephrine, methamphetamine, cocaine and serotonin upon the adrenaline sensitive component of the reticular activating system. *Electroencephalogr. Clin. Neurophysiol. 9*:409—414, 1957.
360. Rudzik, A. D. and Johnson, G. A.: Effect of amphetamines and amphetamine analogues on convulsive thresholds. In Costa, E. and Garattine (Eds.): *Amphetamines and Related Compounds.* New York: Raven Press, 1970, pp. 715—728.
361. Saavedra, J. and Axelrod, J.: Psychotomimetic N-methylated tryptamines: Formation in brain in vivo and in vitro. *Science 175*: 1365—1366, 1972.
362. Sabelli, H. C. and Giardina, W. J.: Tryptaldehydes (indoleacetaldehydes) in serotonergic sleep in newly hatched chicks. *Arzneim. Forsch. 20*:74—80, 1970.
363. Sabelli, H. C., Giardina, W. J., Alivasatos, S. G. A., et al.: Indoleacetaldehydes: Serotonin-like effects on the central nervous system. *Nature 223*:73—74, 1969.

364. Samanin, R., Gomulka, W., and Valzelli, L.: Reduced effect of morphine in midbrain raphé lesioned rats. *Eur. J. Pharmacol.* *10*:339–343, 1970.
365. Samanin, R. and Valzelli, L.: Increase of morphine-induced analgesia by stimulation of the nucleus raphé dorsalis. *Eur. J. Pharmacol.* *16*:298–302, 1971.
366. Samanin, R. and Bernasconi, S.: Effects of intraventricularly injected 6-OH dopamine or midbrain raphé lesion on morphine analgesia in rats. *Psychopharmacologia 25*:175–182, 1972.
367. Sanders, E. and Bush, M. T.: Distribution, metabolism and excretion of bufotenine in the rat with preliminary studies of its O-methyl derivative. *J. Pharmacol. Exp. Ther. 158*:340–352, 1967.
368. Sassin, J. F., Parker, D. C., and Johnson, L. C.: Effects of slow wave sleep deprivation on human growth hormone release in sleep; preliminary study. *Life Sci. 8*(1):1299–1307, 1969.
369. Sassin, J. F., Parker, D. C., and Mace, J. W.: Human growth hormone release: Relation to slow wave sleep and sleep-waking cycles. *Science 165*:513–515, 1969.
370. Schain, R. J., Copenhaver, J. H., and Carver, M. J.: Inhibition by phenylalanine of the entry of 5-hydroxytryptophan-1-C^{14} into cerebrospinal fluid. *Proc. Soc. Exp. Biol. 118*:184–186, 1965.
371. Scheving, L. E., Harrison, W. R., Gordon, P., et al.: Daily fluctuation (circadian and ultradian) in biogenic amines of the rat brain. *Am. J. Physiol. 214*:166–173, 1968.
372. Schlesinger, K., Boggan, W. O., and Freedman, D. X.: Genetics of audiogenic seizures: I. Relation to brain serotonin and norepinephrine in mice. *Life Sci. 4*:2345–2351, 1965.
373. Schreiber, R. A. and Schlesinger, K.: Circadian rhythms and seizure susceptibility: Relation to 5-hydroxytryptamine and norepinephrine in brain. *Physiol. Behav. 6*:635–640, 1971.
374. Scudder, C. L., Karczmar, A. G., Everett, G. M., Brain catechol and serotonin levels in various strains and genera of mice and a possible interpretation for the correlations of amine levels with electroshock latency and behavior. *Int. J. Neuropharmacol. 5*:343–351, 1966.
375. Shaskan, E. G. and Snyder, S. H.: Kinetics of serotonin accumulation into slices from rat brain: Relationship to catecholamine uptake. *J. Pharmacol. Exp. Ther. 175*:404–418, 1970.
376. Shaw, D. M., Camps, F. E., and Eccleston, E.: 5-Hydroxytryptamine in the hindbrain of depressive suicides. *Br. J. Psychiatry 113*:1407, 1967.
377. Sheard, M. H.: The effect of p-chlorophenylalanine on behavior in rats: Relation to brain serotonin and 5-hydroxyindoleacetic acid. *Brain Res. 15*:524–528, 1969.
378. Sheard, M. H. and Aghajanian, G. K.: Stimulation of the midbrain raphé. Effect on serotonin metabolism. *J. Pharmacol. Exp. Ther. 163*:425–430, 1968.
379. Sheard, M. H. and Aghajanian, G. K.: Stimulation of midbrain raphé neurons: behavioral effects of serotonin release. *Life Sci. 7*:19–25, 1968.

380. Shen, F. H., Loh, H. H., and Way, E. L.: Brain serotonin turnover in morphine tolerant and dependent mice. *J. Pharmacol. Exp. Ther.* *175*:427–434, 1970.
381. Shillito, E. E.: The effect of parachlorophenylalanine on social interactions of male rats. *Br. J. Pharmacol. 38*:305–315, 1970.
382. Shore, P. A.: Release of serotonin and catecholamines by drugs. *Pharmacol. Rev. 14*:531–550, 1962.
383. Siggins, G. R., Oliver, A. P., Hoffer, B. J., et al.: Cyclic adenosine monophosphate and norepinephrine: Effects on transmembrane properties of cerebellar Purkinje cells. *Science 171*:192–194, 1971.
384. Sirex, D. W. and Marini, F. A.: Bufotenine in human urine. *Biol. Psychiatry 1*:189–191, 1969.
385. Sladek, J. R., Jr. and Cohen, D. L.: Histochemical fluorescence of intracerebrally implanted serotonin. *Exp. Neurol. 36*:539–548, 1972.
386. Smith, C. B., Villarreal, J. E., Bednarczyk, J. H., et al.: Tolerance to morphine-induced increases in [^{14}C] catecholamine synthesis in mouse brain. *Science 170*:1106–1107, 1970.
387. Smythies, J. R.: The chemical nature of the receptor site. *Int. Rev. Neurobiol. 13*:181–222, 1970.
388. Soulairac, M. L.: Étude experimentale des régulations hormonon-nerveuses du comportement sexuel du rat mâle. *Ann. Endocrinol.* (Paris):1–98, 1963.
389. Sperry, L. W.: Neural basis of the spontaneous optokinetic response produced by visual inversion. *J. Comp. Physiol. Psychol. 43*: 482–489, 1950.
390. Spooner, C. E., Mandell, A. J., Winters, W. D., et al.: Pharmacological and biochemical correlates of 5-hydroxytryptamine entry into the central nervous system during maturation. *Proc. West. Pharmacol. Soc. 11*:98–105, 1968.
391. Spooner, C. E. and Winters, W. D.: Evidence for a direct action of monoamines on the chick central nervous system. *Experientia 21*: 256–258, 1965.
392. Spooner, C. E. and Winters, W. D.: Neuropharmacological profile of the young chick. *Int. J. Neuropharmacol. 5*:217–236, 1966.
393. Stein, L.: Neurochemistry of reward and punishment: Some implications for the etiology of schizophrenia. *J. Psychiatr. Res. 8*:345–361, 1971.
394. Stumpf, C.: Drug action on the electrical activity of the hippocampus. *Int. Rev. Neurobiol. 8*:77–138, 1965.
395. Supniewski, J. V., Marczynski, T. J., and Misztal, S.: Biological properties of melatonin, 5-methoxy-N-acetyltryptamine. *Bull. Acad. Pol. Sci.* VI, *8*:483–487, 1960.
396. Supniewski, J. V., Misztal, S., and Marczynski, T. J.: Synthesis and biological properties of melatonin. *Diss. Pharm.* (Cracow) *13*: 205–217, 1961.
397. Tagliamonte, A., Biggio, G., Vargin, L., et al.: Free tryptophan in serum controls brain tryptophan level and serotonin synthesis. *Life Sci.* II, *12*:277–278, 1973.

398. Tagliamonte, A., Tagliamonte, P., Gessa, G., et al.: Compulsive sexual activity induced by p-chlorophenylalanine in normal and pinealectomized male rats. *Science 166*:1433—1435, 1969.
399. Tagliamonte, A., Tagliamonte, P., and Gessa, G.: Reversal of pargyline induced inhibition of sexual behavior in male rats by p-chlorophenylalanine. *Nature 230*:244—245, 1971.
400. Tagliamonte, D., Perez-Cruet, J., Tagliamonte, A., et al.: Differential effect of p-chlorophenylalanine (PCPA) on the sexual behavior and on the electrocorticogram of male rabbits. *Pharmacologist 12*:205, 1970.
401. Takahashi, Y., Kipnis, D. M., and Daughaday, W. H.: Growth hormone secretion during sleep. *J. Clin. Invest. 47*:2079—2090, 1968.
402. Tammisto, T. and Ylitalo, P.: Tachyphylaxis and toxicity of 5-HT in anaesthetized rats. *Acta Pharmacol. Toxicol. 27*:461—468, 1969.
403. Tanimukai, H., Ginter, R., Spaide, J., et al.: Psychotogenic N,N-dimethylated indole amines and behavior in schizophrenic patients. *Rec. Adv. Biol. Psychiatr. 10*:6—15, 1968.
404. Taylor, J. and Green, R. D.: Analysis of reserpine-induced supersensitivity in aortic strips of rabbits. *J. Pharmacol. Exp. Ther. 177*:127—135, 1971.
405. Tenen, S. S.: The effects of para-chlorophenylalanine, a serotonin depletor, on avoidance acquisition, pain sensitivity and related behavior in the rat. *Psychopharmacologia 10*:204—219, 1967.
406. Tenen, S. S.: Antagonism of the analgesic effect of morphine and other drugs by p-chlorophenylalanine, a serotonin depletor. *Psychopharmacologia 12*:278—285, 1968.
407. Thoenen, H., Hürlimann, A., and Haefely, W.: Mechanism of amphetamine accumulation in the isolated perfused heart of the rat. *J. Pharm. Pharmacol. 20*:1—11, 1968.
408. Tonge, S. R.: Neurochemical teratology: 5-Hydroxyindole concentrations in discrete areas of rat brain after pre- and neonatal administration of phencyclidine and imipramine. *Life Sci. 12*:481—486, 1973.
409. Tonge, S. R. and Leonard, B. E.: The effect of hallucinogenic drugs upon the metabolism of 5-hydroxytryptamine in the brain. *Life Sci. 8*:805—814, 1969.
410. Trendelenburg, U. and Weiner, N.: Sensitivity of the nictitating membrane after various procedures and agents. *J. Pharmacol. Exp. Ther. 136*:152—161, 1962.
411. Twarog, B. M. and Page, I. H.: Serotonin content of some mammalian tissues and urine and a method for its determination. *Am. J. Physiol. 175*:157—161, 1953.
412. Udenfriend, H., Weissbach, H., and Bogdanski, D. F.: Increase in tissue serotonin following administration of its precursor 5-hydroxytryptophan. *J. Biol. Chem. 224*:803—810, 1957.
413. Uretsky, N. J. and Iversen, L. L.: Effects of 6-hydroxydopamine on catecholamine neurons in the rat brain. *J. Neurochem. 17*:269—278, 1970.
414. Ursin, R.: Differential effect of para-chlorophenylalanine on the two

slow wave sleep stages in the cat. *Acta Physiol. Scand.* 86:278–285, 1972.
415. Valzelli, L. and Garattini, S.: Biochemical and behavioral changes induced by isolation in rats. *Neuropharmacol.* 11:17–22, 1972.
416. Vane, J.: The relative activities of some tryptamine analogues on the isolated rat stomach strip preparation. *Br. J. Pharmacol. Chemother.* 14:87–98, 1959.
417. Veale, W. L. and Myers, R. D.: Decrease in ethanol intake in rats following administration of p-chlorophenylalanine. *Neuropharmacol.* 9:317–326, 1970.
418. Verzeano, M. and Mahnke, J. H.: Serotonin and thalamic synchronization. *Physiol. Behav.* 9:649–653, 1972.
419. Victor, S. J., Baumgarten, H. G., and Lovenberg, W.: Effect of intraventricular administration of 5,6 and 5,7-dihydroxytryptamine on regional tryptophan hydroxylase activity in rat brain. *Fed. Proc.* 32(3):1954, 1973.
420. Wada, J. A. and McGeer, E. G.: Central aromatic amines and behavior. *Arch. Neurol.* 14:129–143, 1966.
421. Way, E. L.: Role of serotonin in morphine effects. *Fed. Proc.* 31(1):113–120, 1972.
422. Weber, L. J. and Horita, A.: A study of 5-hydroxytryptamine formation from L-tryptophan in the brain and other tissues. *Biochem. Pharmacol.* 14:1141–1149, 1965.
423. Weiss, L. and Aghajanian, G. K.: Activation of brain serotonin metabolism by heat: role of midbrain raphé neurons. *Brain Res.* 26:37–48, 1971.
424. Weissman, A. and Harbert, C. A.: Recent developments relating serotonin and behavior. In Barchas, J. and Usdin, E. (Eds.), *Serotonin and Behavior*. New York: Academic Press, 1973, pp. 47–58.
425. Weitzman, E. D., Rapport, M. M., McGregor, P., et al.: Sleep patterns of the monkey and brain serotonin concentrations: Effect of p-chlorophenylalanine. *Science* 160:1361–1363, 1968.
426. Werboff, J., Gottlieb, J. S., Havlena, J., et al.: Behavioral effects of prenatal drug administration in the white rat. *Pediatrics* 27: 318–324, 1961.
427. Whalen, R. E. and Luttge, G. W.: p-Chlorophenylalanine methylester: An aphrodisiac? *Science* 169:111–1001, 1970.
428. Wilkinson, S.: 5-Methoxy-N-methyltryptamine: A new indole alkaloid from *Phalaris arundinacea L.J. Chem. Soc.* 2:2079–2081, 1958.
429. Williams, H. L.: The new biology of sleep. *J. Psychiatr. Res.* 8: 445–478, 1971.
430. Wilson, W. P. and Nashold, B. S.: The sleep rhythms of subcortical nuclei: Some observations in man. *Biol. Psychiatry* 1:289–296, 1969.
431. Woolley, D. W. and Shaw, E. N.: A biochemical and pharmacological suggestion about certain mental disorders. *Proc. Nat. Acad. Sci. U.S.A.* 40:228, 1954.
432. Woolley, D. W. and van der Hoeven, T.: Serotonin deficiency in

infancy as a cause of a mental defect in experimental phenylketonuria. *Int. J. Neuropsychiatry* 1:529–544, 1965.
433. Wurtman, R. J. and Anton-Tay, F.: The mammalian pineal as a neuroendocrine transducer. *Rec. Prog. Horm. Res.* 25:493–514, 1969.
434. Wurtman, R. J., Axelrod, J., and Fischer, J. E.: Melatonin synthesis in the pineal gland effect of light mediated by the sympathetic nervous system. *Science 143*:1328–1330, 1964.
435. Wurtman, R. J. and Fernstrom, J. D.: L-Tryptophan, L-tyrosine and the control of brain monoamine biosynthesis. In Snyder, S. (Ed.): *Perspectives in Neuropharmacology*. New York: Oxford University Press, 1972, pp. 143–193.
436. Wyatt, R. J.: The serotonin-catecholamine-dream bi-cycle: A clinical study. *Biol. Psychiatry* 5:33–63, 1972.
437. Wyatt, R. J., Engelman, K., Kupfer, D. J., et al.: Effect of parachlorophenylalanine on sleep in man. *Electroencephalogr. Clin. Neurophysiol.* 27:529–532, 1969.
438. Wyatt, R. J., Zarcone, V., Engelman, K., et al.: Effects of 5-hydroxytryptophan on the sleep of normal human subjects. *Electroencephalogr. Clin. Neurophysiol.* 30:505–509, 1971.
439. Yamaguchi, N., Marczynski, T. J., and Ling, G. M.: The effect of electrical and chemical stimulation of the preoptic region and some non-specific thalamic nuclei in unrestrained, waking animals. *Electroencephalogr. Clin. Neurophysiol.* 15:154, 1963.
440. Yamaguchi, N., Ling, G. M., and Marczynski, T. J.: The effects of chemical stimulation of the preoptic region, nucleus centralis medialis or brain stem reticular formation with regard to sleep and wakefulness. *Rec. Adv. Biol. Psychiatr.* 6:9–20. 1964.
441. York, J. L.: Role of Brain Catecholamines and Serotonin in Morphine Analgesia. Ph.D. Thesis, University of Illinois at the Medical Center, Chicago, Illinois, 1972.
442. Yuwiler, A., Geller, F., and Slater, G. C.: On the mechanism of the brain serotonin depletion in experimental phenylketonuria. *J. Biol. Chem.* 240:1170–1174, 1965.
443. Zarcone, V. P., Jr., Gulevich, G., and Pivik, T.: Partial REM sleep deprivation and schizophrenia. *Arch. Gen. Psychiatr.* 18:94, 1968.
444. Zemlan, F. P., Ward, I. L., Crowley, W. R., et al.: Activation of lordotic responding in female rats by suppression of serotonergic activity. *Science 179*:1010–1011, 1973.
445. Zuber, B. L. and Stark, L.: Saccadic suppression: Elevation of visual threshold associated with saccadic eye movements. *Exp. Neurol.* 16:65–79, 1966.

Index

Acetylaspartate, concentration during brain development, 93
Acetylcholine, 22
  bicuculline, 48
  microiontophoresis, 173–176
  release, 163–171
    from cerebral cortex, 164–165
    pharmacological studies of, 169–171
    physiological studies of, 163–169
    quantal, 12
  retinal illumination, 165
  synaptic transmission in CNS, 159–213
  synthesis, 170
Acetylcholine-dopamine balance, 282–285
Acetylcholinesterase distribution in spinal cord, 178
Acetyltransferase, 170
Adenosine 3',5'-monophosphate, cyclic, as mediator, 337–339
Adenylcyclase, dopamine receptor, 263–266
Adlumine
  as GABA antagonist, 40
  structure, 41
Affective disorders
  dopamine, 280–282
  pathoneurobiology, 399–402
Akinesia, dopamine, 267–269
Alanine, 57
L-α-Alanine in CNS, 62
β-Alanine
  in CNS, 65
  and GABA transaminase, 34
Alcohol intake, volitional
  and 5-hydroxytryptamine, 391
  and p-chlorophenylalanine, 391
Allylglycine, decreases in brain GABA due to, 34
Amacrine cells, dopamine, 218
Amanita muscaria, muscimol isolated from, 49

Amino acids
  acidic
    and aortic occlusion, 90
    axonal transport, 90
    distribution, 85–86
    presynaptic action, 136
    receptors, 118
    relative potencies, 114
    release, 104–108
    in spinal cord, 88–89
    topical application, 108–110
    uptake, 98
  excitatory, 83–158
    developmental aspects, 91–94
    distribution, 83–91, 92
    enzyme inhibitors of, 132
    inactivation, 129–136
    iontophoretic studies, 110–120
    mode of action, 120–124
    neurophysiology and pharmacology, 108–129
    stereospecificity of receptors, 120
    transport, 97–104
  inhibitory transmitters in CNS, 31–81
  pools, 94–97
γ-Aminobutyric acid: see GABA
γ-Aminobutyrylcholine: see GABACh
ε-Aminocaproic acid, 57
  and brain synapses, 118
  on cortical neurons, 119
ω-Aminocaprylic acid
  and brain synapses, 118
  on cortical neurons, 119
γ-Amino-β-hydroxybutyric acid: see GABOB
Aminooxyacetic acid, increase of GABA levels by, 36
3-Aminopropanesulfonic acid, 57
4-Aminotetrolic acid, 40
  inhibition of GABA uptake by, 35
δ-Aminovaleric acid, 57
  and GABA transaminase, 34

431

Amphetamine
    acetylcholine release, 169
    stereotypies, 277
    striatal neurons, 261–262
Amygdala
    acetylcholine, 175
    choline acetyltransferase, 187
    5-hydroxytryptamine, 356
Anamirta cocculus, 38
Apomorphine, stereotypies, 277–279
Ascending cholinergic pathways, 177
    Acetylcholine, 165
Aspartate, spinal levels, 52
L-Aspartate
    distribution, 84
    exciting of neurons by, 118
    and HA-966, 51
    metabolism, 94
    microiontophoresis, 111–112
    release, 104
    uptake, 98
Aspartic acid, 23
Atropine
    acetylcholine, 179
    acetylcholine release, 169, 170
    Renshaw cells, 177–178
Auditory cortex
    acetylcholine, 176
    acetylcholine release, 166
Augmenting and recruiting pathways, acetylcholine, 166
Autoradiography, acetylcholine synthesis, 163
Axoaxodendritic complexes, 21
Axoplasmic flow, 7

Basal ganglia
    acetylcholine, 186
    pathophysiology, 285–288
    serotonin, 350
Basket cells
    acetylcholine, 174
    DL-homocysteic acid, 181
Behavior
    aggressive
        drug-induced, 389–390
        stress-induced, 389
    consummatory
        5-hydroxytryptamine, 390–391
        p-chlorophenylalanine, 390–391
    sexual, 5-hydroxytryptamine 387–389
Benzoquinonium, acetylcholine receptors, 180
Benzyl penicillin
    epileptogenic action, 42
    structure, 39
Betz cell, acetylcholine, 176
Bicucine
    as GABA antagonist, 40
    structure, 41
Bicucine methyl ester, structure, 41
Bicuculline, 16
    as antagonist of GABOB, 47
    blocking of GABA action by, 45
    cerebellar cortex, 42
    cerebral cortex, 42
    cuneate nucleus, 42
    Dieter's neurons, 42
    as GABA antagonist, 40–42
    hippocampal cortex, 42
    lateral geniculate nucleus, 42
    and muscimol, 49
    olfactory bulb, 42
    Purkinje cells, 44
    pyramidal cells, 42
    pyramidal tract neurons, 45
    structure, 39, 41
    thalamus, 42
    vestibular nucleus, 42
(-) Bicuculline, structure, 41
Bicullinediol
    as GABA antagonist, 40
    structure, 41
Bicuculline methochloride, structure, 41
Brain
    human
        2-aminoethanephosphonic acid, 65
        cystathionine levels, 61
        single neurons of, 171–191

2-Brono-lysergic acid diethylamide
as L-glutamate antagonist,
127
Brucine, 58
olivocochlear bundle, 180
1,4-Butanediol, CNS depressant
effects, 49
γ-Butyrobetaine
in brain, 48
conversion of GABA to, 34
γ-Butyrolactone, CNS depressant
effects, 49
Butyrylcholine in brain, 160

Cajal, 2
Calcium, acetylcholine release, 171
Calcium chelating agents, neurons,
123
Capnoidine
as GABA antagonist, 40
structure, 41
Carbachol, cholinesterase, 181
Carnitine in brain, 47
Catalepsy, dopamine, 267–268
Caudate nucleus
acetylcholine, 175
acetylcholine release, 167–168
dopamine, 256–260
Central nervous system
acetylcholine and synaptic transmission in, 159–213
amino acid inhibitory transmitters in, 31–81
GABA levels in, 32–33
serotonin, 349–429
Central nervous system transmitters,
identification of, 136–141
Cerebellar cortex
acetylcholine release, 167
β-alanine, 65
bicuculline, 42
cholinesterase, 182
GABA synapses, 43
Cerebellar peduncles, choline acetyltransferase, 181
Cerebellum
acetylcholine, 180
GABA as inhibitory transmitter, 44

Cerebral cortex
acetylcholine, 187–191
acetylcholine release, 164, 167
β-alanine, 65
bicuculline, 42
GABA, 45
GABA synapses, 43
glycine, 56
5-hydroxytryptamine and glutamate, 127
inhibition of pyramidal tract
neurons, 45
specific pathways to, 165
Cerebral ventricles, acetylcholine release, 167
Cerebrum, physiology and pharmacology of dopamine
neurons, 215–325
Chih-shih-hu, 59
p-Chloroamphetamine, preference
for ethanol, 391
2-Chloro-GABA, inhibition of GABA
uptake by, 35
p-Chloromercuriphenylsulfonate
action of GABA potentiated by,
46
and glycine, 55
inhibition of GABA uptake by,
36
Renshaw cells, 135
p-Chlorophenylalanine, 5-hydroxytryptamine, 373
β-(p-Chlorophenyl)-γ-aminobutyric
acid in treatment of spasticity, 49
Chlorpromazine, inhibition of GABA
uptake by, 36
Choline acetyltransferase
activity, 183
in cerebellum, 182
distribution, 163
Cholinesterase
activity in cerebellum, 183
ascending pathways, 162
in cerebellar cortex, 184
in CNS, 161
distribution of activity, 184
inhibition, 170
pseudocholinesterase, 162

Cholinesterase inhibitors, acetylcholine release, 170
Cochlear nucleus
 acetylcholine, 173, 179
 cholinergic mossy afferents, 182
Colliculi, superior and inferior, acetylcholine release, 168
Conductance, resting, 13
Convulsants, strychnine-like, 59
Coriamyrtin, 39
Corlumine, structure, 41
Cortical slices, acetylcholine release, 170
Corydalis species, 40
Cuneate nucleus
 acetylcholine, 174
 bicuculline, 42
 cholinergic mossy afferents, 182
 glutamate, 117
 and HA-966, 129
 presynaptic action of acidic amino acids, 136
Cystathionine, cystathioninase, 61
L-Cystathionine
 blood-brain barrier, 61
 isomers, 61
 regional distribution, 61
Cystathioninuria, and mental retardation, 61
L-Cysteate
 distribution, 84
 microiontophoresis, 111–112

Deiters' neurons, bicuculline, 42
Deiters' nucleus
 acetylcholine, 179
 bicuculline, 44
 GABA, 38
 GABA synapses, 43
 glutamate decarboxylase, 44
 glycine, 56
 picrotoxin, 44
Dendrobine, 40
 glycine, 59
Dentate gyrus, acetylcholine, 175
Dentate nucleus, acetylcholine, 180
Depolarization
 primary afferent, 18
 primary afferent terminals, 16

Depolarization block of spinal neurons, 119
Depression
 5-hydroxytryptamine system, 399
 spreading, acidic amino acids, 108–109
Descending pathways, cholinergic, 177
Diaboline, 58
L-2,4-Diaminobutyric acid in brain, 48
Dibutyryl cyclic adenosine monophosphate, inhibition of GABA uptake by, 36
Dicentra cucullaria, 40
Dihydro-β-erythroidine
 acetylcholine, 179
 dorsal root potential, 178
 Renshaw cells, 177
Dopamine, 23
 conditioned avoidance, 281
 drugs depleting, 249–250
 feedback regulation, 229–231
 indirect mimetic substances, 244–246
 metabolism, 222–228
 pathways and cellular localization, 218
 receptor, 263–266
 receptor agonists, 241–244
 receptor antagonists, 246–249
 regional distribution in the brain, 216
 regulation of synthesis, 228
 release and uptake, 233–241
 sensorimotor functions, 267–277
 subcellular localization, 222
 synthesis inhibitors, 250–251
 transmitter action, 288–289
 turnover, 231–233
Dorsal root potential, 178
Dorsal tegmental pathway of Shute and Lewis, 186

Electroencephalogram
 acetylcholine, 164, 185
 morphine, 169
Electrotonic coupling, 23

Epilepsy
    acetylcholine, 193–194
    5-hydroxytryptamine, 382
Equilibrium potentials, 13, 15
Ethyl 2-allyl-2-phenyl-4-diethyl-
        aminobutyrate and GABA,
        51
Excitation, spherical, 19

Fastigial nucleus, acetylcholine, 180
Fenfluramine, 5-hydroxytryptamine,
        374
Formatio reticularis, acetylcholine,
        174

GABA, 16, 18, 19, 22, 23, 31
    action on spinal interneurons, 39
    antagonists, 38–43
    anticonvulsant action, 33
    compounds structurally related
        to, 46–51
    conformation, 32, 33
    degradation, 34
    in globus pallidus, 32
    levels in CNS, 32–33
    neurochemistry, 32–38
    neuropharmacology, 38–43
    postsynaptic action, 38
    postsynaptic inhibition, 43
    precursors, 33
    presynaptic inhibition, 44
    pyramidal tract neurons, 46
    pyridoxal kinase, 37
    release, 37
    reversal potentials, 38
    subcellular localization, 36–37
    in substantia nigra, 32
    synapses, 43–46, 50
    synaptic plasticity, 37
    synthesis, 33
    uptake, 35–36
GABACh
    in brain, 47
    conversion of GABA to, 34
GABA aminotransferase in spinal
        cord tissue, 96
GABA shunt, 35

GABA transaminase, 34
    inhibitors of, 34
    and muscimol, 49
    regional distribution, 45
GABOB
    in CNS, 42, 47
    conversion of GABA to, 35
    and GABA transaminase, 34
    glycine, 57
GAD
    adrenalectomy, 33
    ATP, 33
    carbonyl trapping reagents, 33
    electroconvulsive shock treat-
        ment, 33
    forms I and II, 33
    inhibition of, by ATP, 33
    Parkinson's disease, 33
    pyridoxal phosphate as cofactor,
        33
    regional distribution, 33
    in spinal cord tissue, 96
    subcellular distribution, 33
GAD I
    activity, 37
    levels in hippocampus, 45
    regional distribution, 34
Gaddum push-pull cannula, 164, 233
Ganglia, autonomic, GABA, 46
Ganglion cells, sensory, GABA, 46
Gelsemine
    glycine, 58
    structure, 59
Geniculate nuclei, acetylcholine,
        184–186
Geniculocortical cholinergic fibers,
        acetylcholine, 166
Glial cells, pseudocholinesterase, 162
Globus pallidus/putamen, acetylcho-
        line, 175
Glutamate, spinal levels, 52
D-Glutamate
    and crustacean muscle, 118
    and insect muscle, 118
    uptake, 98, 101
L-Glutamate
    antagonists, 124–129
    diethylester of, 124
    distribution, 84

equilibrium potential, 121
and HA-966, 51
metabolism, 94
microiontophoresis, 111–112
primary afferent fibers, 87
release, 104
  calcium-dependent, 105
reticulocortical system, 107
retinal transmitter, 107
reversal potentials, 121, 123
small pool, 95
spreading depression, 107
stereoisomers, 117
subcellular distribution, 96
synaptosomes, 102
transmitter pool, 96
transport, 102
uptake, 98
uptake inhibitors, 99–100
Glutamate decarboxylase: *see* GAD
Glutamate dehydrogenase distribution, 95, 96
Glutamate receptors, stereospecificity, 117
Glutamic acid, 23
L-Glutamic acid, 15
Glutamic decarboxylase, -glutamate, 132
Glutamine, spinal levels, 52
L-Glutamine
  distribution, 91
  uptake, 98
Glutamine synthetase
  distribution, 95
  subcellular distribution, 96
Glycine, 16, 23, 31
  action on spinal interneurons, 39
  antagonists, 40, 57
  aortic occlusion, 52
  autoradiographic studies, 55
  bulbar reticular neurons, 56
  cerebral cortex, 56
  in CNS, 51–63
  cuneate neurons, 56
  Deiters' nucleus, 56
  electroshock threshold, 52
  ethanol-induced sleeping time, 52
  hind limb rigidity, 52
  hippocampal pyramidal neurons, 45
  hyperglycinemia, 52

medial geniculate nucleus, 56
mediated transport, 52
in medulla, 52
membrane conductance, 57
metabolism, 52–54
motoneurons, 56
neurochemistry, 52–56
neurons in red nucleus, 56
neuropharmacology, 56–59
postsynaptic action, 56
release, 56
  evoked, 56
Renshaw cells, 56
reversal potentials, 57
in spinal cord, 52
spinal interneurons, 56
spinal postsynaptic inhibition, 51
structure, 51
synapse, 60
uptake, 54–56
uptake inhibitors, 55
Glyoxylate, thiamine deficiency, 62
Gonadotropin release, 5-hydroxytryptamine nerve terminals, 385
Gracile nucleus
  acetylcholine, 174
  cholinergic mossy afferents, 182
Granule cells
  acetylcholine, 174
  DL-homocysteic acid, 181
Granule layer, acetylcholine, 181
γ-Guanidinobutyric acid
  in brain and other organs, 48–49
  conversion of GABA to, 34

HA-966, depressant effects, 129
Haloperidol
  and GABA, 49
  inhibition of GABA uptake by, 36
Hemicholinium
  acetylcholine synthesis, 163
  acetylcholine release, 171
  acetylcholine synthesis, 172
Hemicholinium-3
  acetylcholine, 179
  action of, 177
  olivocochlear bundle, 180

Hippocampal cortex, 65
  bicuculline, 42
  GABA synapses, 43
Hippocampus
  acetylcholine, 175
  acetylcholine release, 168
  cholinesterase, 186
  GABA, 45
  5-hydroxytryptamine, 356
  5-hydroxytryptamine and glutamate, 127
Histochemical studies, cerebellum, 182
Histochemistry, cholinesterase, 161
Homoanserine in brain, 48
Homocarnosine in brain, 48
DL-Homocysteate, 117
  cortical neurons, 119
  N-methyl-D-aspartate, 122
  pyramidal cell firing, 118
DL-Homocysteic acid, granule and basket cells, 181
Huntington's chorea
  acetylcholine, 192−193
  dopamine, 277
Hydrastine
  as GABA antagonist, 40
  structure, 41
3-Hydrazinopropionic acid and GABA, 51
1-Hydroxy-3-amino-pyrrolidone-2, 50−51
γ-Hydroxybutyric acid
  conversion of GABA to, 35
  dopamine release, 49
4-Hydroxybutyric acid as precursor of GABA, 33
2-Hydroxy-GABA, inhibition of GABA uptake by, 35
5-Hydroxytryptamine: *see also* Serotonin
  metabolism, 352−355
  precursors, 368−373
  release and uptake, 355−356
  synaptic action, 356−357
  synthesis inhibitors, 373−374
Hyoscine, acetylcholine release, 169
Hypercapnia, acetylcholine release, 168
Hyperkinesias, dopamine, 271−277

Hypothalamus
  acetylcholine, 174
  acetylcholine and choline acetyltransferase, 184
  acetylcholine release, 168
  serotonin, 350
Hypoxia, acetylcholine release, 168

DL-Ibotenate, glutamate, 123
DL-Ibotenic acid, excitant action, 120
Imidazole-4-acetic acid, 48
  in CNS, 42
Imipramine, inhibition of GABA uptake by, 36
Inferior colliculus, acetylcholine, 174
Inferior olive, climbing fibers from, 182
Inhibition
  basket cell, 45
  flattened vesicles, 19
  GABA, 44
  granule cells, 45
  of hippocampal pyramidal neurons, 45
  postsynaptic, 18, 31
  presynaptic, 16, 18, 19, 31, 46
    in cuneate nucleus, 46
    flattened vesicles, 21, 22
    round vesicles, 21, 22
  of pyramidal tract neurons, 45
Inhibitory feedback circuit, cholinergic neurons, 170
Injection, intracellular, of anions and cations, 38
Interpositus nucleus, acetylcholine, 180
Ionophores, acidic amino acid receptors, 117−118

Juglone, inhibition of GABA uptake by, 36

Lambda waves, REM sleep, 375
Lateral geniculate
  acetylcholine, 175
  5-hydroxytryptamine, 356

Lateral geniculate neurons, glutamate, 117
Lateral geniculate nucleus
  acetylcholine, 166
  bicuculline, 42
Lateral reticular nucleus, cholinergic mossy afferents, 182
Lateral vestibular nucleus, acetylcholine, 173
Laudanosine
  glycine, 58
  as glycine antagonist, 40
  structure, 59
Leptazol, acetylcholine release, 169
Levallorphan, acetylcholine release, 169
Limbic system, serotonin, 350
Locomotion, dopamine, 271–277
Loewi, 23
Lysergic acid diethylamide
  as L-glutamate antagonist, 127
  5-hydroxytryptamine, 356
L-Lysine uptake, 98

Magnesium
  acetylcholine release, 171
  L-Glutamate, 136
Medial geniculate, acetylcholine, 175
Medial geniculate nucleus
  acetylcholine release, 166
  glycine, 56
Medulla, acetylcholine, 179
Membrane conductance, 12
  glycine, 57
3-Mercaptopropionic acid, decreases in brain GABA due to, 34
Mesencephalic reticular formation, L-glutamate, 107
Mesocortex, serotonin, 350
1-Methionine-DL-sulfoximine
  convulsant action, 134
  as L-glutamate antagonist, 127
2-Methoxyaporphine as L-glutamate antagonist, 127
N-Methyl-D-aspartate
  depolarizaton during action of, 121
  L-glutamate and L-aspartate, 135
N-Methyl-DL-aspartate, glutamate, 123
N-Methylbicuculleine
  as glycine antagonist, 40
  structure, 59

2-Methyl-GABA, inhibition of GABA uptake by, 35
4-Methyl-GABA, inhibiton of GABA uptake by, 35
α-Methyl-DL-glutamate as L-glutamate antagonist, 124
Methysergide as L-glutamate antagonist, 127
Microiontophoresis
  acetylcholine, 160
  L-α-alanine, 62
  β-alanine, 65
  α-amino acids, 61
  bicuculline, 46
  bicuculline methochloride, 40
  GABA, 38, 44
  glycine, 57
  glyoxylate, 63
  γ-guanidiobutyric acid, 48–49
  imidazole-4-acetic acid, 48
  muscimol, 50
  picrotoxin, 39
  strychnine, 58
  taurine, 64
  taurine antagonists, 64
Midpontine transections, acetylcholine release, 167
Mangolism, taurine, 63
Monoamine oxidase, 10
Monoamine oxidase inhibitors, raphe nuclei, 357
Morphine
  acetylcholine release, 169
  physical dependence on, 395–396
Morphine analgesia, 5-hydroxytryptamine system, 392–395
Mossy fibers, acetylcholine, 182
Motoneurons
  acetylcholine, 179
  glycine, 56
  spinal
    L-glutamate, 119
    strychnine, 58
Muscimol, 40
  and GABA, 49
Mytelase, acetylcholine receptors, 180

Nalorphine, acetylcholine release, 169
Naloxone, acetylcholine release, 169
Narcolepsy, idiopathic, 370

Narcotine
  as GABA antagonist, 40
  structure, 41
Neocortex, serotonin, 350
Neurochemistry of GABA, 32–38
Neuroendocrine function, 5-hydroxy-
    tryptamine, 385–387
Neuroleptics
  dopamine receptor antagonists, 246
  operant conditioning and intra-
    cranial self-stimulation, 281
  striatal neurons, 262–263
Neurons
  bulbar reticular, glycine, 56
  central, norepinephrine, 327–348
  cerebral dopamine, physiology and
    pharmacology of, 215–325
  cuneate, glycine, 56
  defined postsynaptic, 333–337
  norepinephrine, 327–328
  in red nucleus, glycine, 56
  spinal, receptors for L-glutamate,
    119
  striatal, drug effects, 261–263
Neuropharmacology of GABA, 38–43
Nialamide and GABA, 51
Nicotine, cerebral cortex, 189
Noradrenaline, 22
Norepinephrine
  central neurons, 327–348
  microiontophoresis, 330
Nucleus cuneatus, sensitivity to gulta-
    mate, 114
Nucleus gracilis, sensitivity to gluta-
    mate, 114
Nucleus reticularis pontis oralis,
    cholinergic mossy afferents,
    182
Nystagmus, central, 51

Olfactory bulb
  acetylcholine, 175
  acetylcholine and choline acetyl-
    transferase, 187
  bicuculline, 42
  glutamate receptors, 114
Olivocochlear bundle, cholinesterase,
    179–180
Optic nerve, cholinergic fibers, 185
Ouabain, release of L-glutamate, 104

Pallidum, dopamine, 261
Paramedian nucleus, acetylcholine,
    173
Paramedian reticular nucleus, acetyl-
    choline, 179
Paraventricular nucleus
  acetylcholine, 174
  neurosecretory cells, 184
  staining intensity, 184
Parkinson's disease
  acetylcholine, 191–192
  atropine, 186
  GAD, 33
Pathoneurobiology, 397–402
Pentobarbital, acetylcholine release,
    169
Pentylenetetrazol, topical application
    of amino acids, 108
Pericruciate cortex, acetylcholine, 176
Phenylhydrazine, nerve-muscle syn-
    apses, 134
Phenylketonuria, 5-HIAA and
    5-hydroxytryptamine, 369
Photic stimulation, acetylcholine re-
    lease, 167
Picrotin, 38–39
Picrotoxin, 16
  acetylcholine release, 169
  bicuculline, 22
  blocking of GABA action by, 46
Picrotoxinin
  as GABA antagonist, 38–40
  glycine, 59
  structure, 39
Piriform cortex
  acetylcholine, 175
  choline acetyltransferase, 187
Pons, and acetylcholine, 179
Postsynaptic inhibition, 12, 15
Postsynaptic potentials
  excitatory, 13, 14, 15, 16
  inhibitory, 13, 15
  slow, 12
Preoptic area, and acetylcholine, 174
Preoptic cells, acetylcholine and warm-
    and cold-sensitive neurons,
    184
Prepontine transection, acetylcholine
    release, 167
Presynaptic facilitation, glutamate,
    110
Presynaptic inhibition in dorsal roots,
    178

Primary afferent fibers, HA-966, 129
Primary receiving areas, cholinergic arousal system, 166
Probenecid, depression, 399
Procion yellow, Renshaw cells, 178
Prolactin, 5-hydroxytryptamine or melatonin, 386
Propionylcholine in brain, 160
Protoveratrines, inhibition of GABA uptake by, 36
Pseudocholinesterase in neurons, 162
Purkinje cells
　acetylcholine, 174, 181
　carbachol, 181
　GABA, 44
　GABA synapses, 43
　noncholinergic transmission, 182
　noradrenergic projection, 334
Push-pull cannulas, acetylcholine release, 168
Putamen, dopamine, 260–261
Putrescine as precursor of GABA, 33
Pyramidal cells, bicuculline, 42
Pyramidal tract neuron, firing of, 128
Pyridoxal phosphate as cofactor for GAD, 33
Pyridoxal-5-phosphate, L-glutamate, 132

Raphe nuclei
　acetylcholine, 174
　afferent connections, 352
　ascending pathways, 350–351
　bulbospinal pathways, 351–352
　lesions, 360
Receptors
　muscarinic
　　and acetylcholine, 172
　　activation of, 178
　nicotinic
　　and acetylcholine, 172
　　desensitization of, 178
Red nucleus, and acetylcholine, 174
Renshaw cells
　acetylcholine release, 168
　acetylcholinesterase, 177
　and atropine, 177–178
　p-chloromercuriphenylsulfonate, 135
　excitant amino acid sensitivity, 120
　GABA action on, 46
　GABACh, 47
　L-glutamate, 119
　glycine, 56
　hemicholinium and acetylcholine synthesis, 172
　identification of, 178
　strychnine, 58
Reserpine, dopamine depletion, 249
Reticular formation, acetylcholine, 165
Retina
　aspartate, 114
　dopamine, 218
　glutamate, 114
Reversal potentials, 15
　glycine, 57
Rigidity, dopamine, 269–270

Schizophrenia
　dopamine, 279–280
　5-hydroxytryptamine deficit, 374
　pathoneurobiology, 397–399
Self-stimulation, norepinephrine, 340–341
Semicarbazide
　nerve-muscle synapses, 134
　presynaptic inhibition, 46
Septum, acetylcholine, 164, 175
L-Serine
　in CNS, 62
　uptake, 98
Serotonin
　central nervous system, 349–429
　topography and neurochemistry of brain neurons, 350–357
Sherrington, 3
Sleep
　norepinephrine-containing neurons, 341–342
　paradoxical, acetylcholine, 164
　raphe neuron system, 360–380
　slow wave, acetylcholine, 164
Sodium azide, L-glutamate release, 104
Soma-axonal junction, 15
Spasticity
　tetanus, relief of, 59

treatment of, 49
Spinal cord
  and acetylcholine, 177
  acetylcholine release, 168
  choline acetyltransferase, 177
  GABA, 46
  L-glutamate, 87
  presynaptic inhibition, 46
  single neurons of brain and, 171–191
Spinal interneurons, glycine, 56
Spinal motoneurons, GABA, 46
Spinal neurons, action of glycine on, 39
Stain, cholinesterase, 161
Stereotypies, dopamine receptor stimulation, 277
Strychnine, 16
  acetylcholine and olivocochlear bundle, 180
  acetylcholine release, 169
  antagonism of postsynaptic inhibitions, 51
  as antagonist of GABOB, 47
  and glycine, 54–55
  as glycine antagonist, 57–58
  glycine-like amino acids, 57
  glycine receptors, 58
  medullary reticular formation, 58
  and muscimol, 49
  pyramidal tract neurons, 45
  Purkinje cells, 45
  recurrent inhibition, 58
  Renshaw cells, 58
  spinal cord, 58
  structure, 59
Substantia nigra
  acetylcholine release, 167
  surgical and drug-induced lesions, 252
Superior colliculus, and acetylcholine, 174
Superior olive, and acetylcholine, 179
Supraoptic nucleus
  and acetylcholine, 174, 184
Sympathetic ganglia, acetylcholine synthesis and release, 171
Synapses
  axoaxonic, 18, 19, 20, 22
    hyperpolarization by GABA at, 46
  axodendritic, 15
  axosomatic, 15
  cholinergic, 170
    and acetylcholine, 172, 180
  GABA, 50
  glycine, 60
  norepinephrine, 329–330
  quantum, 9
  relationships between structure and function in the system, 1–29
  Renshaw cells, 177
  type 1, 4
  type 2, 4
  types, 4
Synaptic noise, 7
Synaptic transmission in CNS, acetylcholine, 159–213
Synaptic vesicle
  complex, 7
  dense-cored granules, 10
  F-type, 4
  reserpine, 10
  S-type, 4
Synaptosomes
  L-aspartate, 102
  calcium-dependent release, 107

Taurine, 31
  in CNS, 42
  conformation, 63
  and GABA transaminase, 34
  glycine, 57
  levels in CNS, 63
  metabolism, 63–64
  neurochemistry, 63–64
  neuropharmacology, 64
  postsynaptic action, 64
  release, 64
  structurally related compounds, 65
  synaptic inhibitions, 64
  uptake, 64
Tegmental reticular nucleus, cholinergic mossy afferents, 182

Teratology, biochemical, 5-hydroxytryptamine neuronal system, 396–397
Tetanus toxin
  and GABA synapses, 43
  Renshaw cells, 59
  spinal motoneurons, 59
Tetrahydroisoquinoline alkaloids, glycine, 58
Tetrodotoxin
  acetylcholine release, 170
  influx of sodium ions into cerebrum, 123
Thalamus
  and acetylcholine, 175, 184–186
  acetylcholine release, 168
  β-alanine, 65
  bicuculline, 42
  glutamate, 117
  responsiveness to L-glutamate, 114
  serotonin, 350
Thebaine
  glycine, 58
  structure, 59
Thermoregulation, and 5-hydroxytryptamine, 380–382
Thioproperazine, dopamine, 239
Thiosemicarbazide
  as carbonyl trapping reagent for GAD, 33
  decreases in brain GABA due to, 34
  effects of L-glutamate on neurons after, 132
Trans-4-aminocrotonic acid, 40
  inhibition of GABA uptake by, 35
Tremor, dopamine, 269–270
Tremorine and GABA, 51
Triethylcholine, acetylcholine release, 171
Tuborcurarine, cerebral cortex, 189
Tutin, 39
Tyramine, 239

Veneridae, acetylcholine antagonists, 180
Vesicles, flat, 55
Vestibular nucleus
  and acetylcholine, 173
  bicuculline, 42
  cholinergic mossy afferents, 182
Visual cortex
  and acetylcholine, 176
  acetylcholine release, 167
Visual system, acetylcholine, 166

Wieland-Gumlich aldehyde, glycine, 58
Wilson's disease, acetylcholine, 193